# Smart Innovation, Systems and Technologies

## Volume 59

**Series editors**

Robert James Howlett, KES International, Shoreham-by-sea, UK
e-mail: rjhowlett@kesinternational.org

Lakhmi C. Jain, University of Canberra, Canberra, Australia;
Bournemouth University, UK;
KES International, UK
e-mails: jainlc2002@yahoo.co.uk; Lakhmi.Jain@canberra.edu.au

*About this Series*

The Smart Innovation, Systems and Technologies book series encompasses the topics of knowledge, intelligence, innovation and sustainability. The aim of the series is to make available a platform for the publication of books on all aspects of single and multi-disciplinary research on these themes in order to make the latest results available in a readily-accessible form. Volumes on interdisciplinary research combining two or more of these areas is particularly sought.

The series covers systems and paradigms that employ knowledge and intelligence in a broad sense. Its scope is systems having embedded knowledge and intelligence, which may be applied to the solution of world problems in industry, the environment and the community. It also focusses on the knowledge-transfer methodologies and innovation strategies employed to make this happen effectively. The combination of intelligent systems tools and a broad range of applications introduces a need for a synergy of disciplines from science, technology, business and the humanities. The series will include conference proceedings, edited collections, monographs, handbooks, reference books, and other relevant types of book in areas of science and technology where smart systems and technologies can offer innovative solutions.

High quality content is an essential feature for all book proposals accepted for the series. It is expected that editors of all accepted volumes will ensure that contributions are subjected to an appropriate level of reviewing process and adhere to KES quality principles.

More information about this series at http://www.springer.com/series/8767

Vladimir L. Uskov · Robert J. Howlett
Lakhmi C. Jain

Editors

# Smart Education and e-Learning 2016

 Springer

*Editors*
Vladimir L. Uskov
Computer Science and Information Systems,
  InterLabs Research Institute
Bradley University
Peoria, IL
USA

Robert J. Howlett
KES International
Shoreham-by-sea
UK

Lakhmi C. Jain
University of Canberra
Canberra
Australia

and

Bournemouth University
Poole
UK

and

KES International
Shoreham-by-sea
UK

ISSN 2190-3018          ISSN 2190-3026   (electronic)
Smart Innovation, Systems and Technologies
ISBN 978-3-319-81942-6          ISBN 978-3-319-39690-3   (eBook)
DOI 10.1007/978-3-319-39690-3

Printed on acid-free paper

This Springer imprint is published by Springer Nature
The registered company is Springer International Publishing AG Switzerland

# Preface

Smart Education and e-Learning (SEEL) are emerging and rapidly growing areas that represent an integration of smart objects and systems, smart technologies, smart environments, smart features or smartness levels, smart pedagogy, smart learning and teaching analytics, various branches of computer science and computer engineering, state-of-the-art smart educational software and/or hardware systems. This is the main reason that in June 2013, a group of enthusiastic and visionary scholars from all over the world arrived with the idea to organize a new professional event that would provide an excellent opportunity for faculty, scholars, Ph.D. students, administrators, and practitioners to meet well-known experts and discuss innovative ideas, findings and outcomes of research projects, and best practices in smart education and e-learning.

The main research, design and development topics in SEEL area include, but are not limited to, (1) conceptual frameworks for Smart Education (SmE) and Smart e-Learning (SeL), (2) infrastructure, main characteristics and features of Smart Universities (SmU) and Smart Classrooms (SmC), (3) SmU-wide software, hardware, security, safety, communication, collaboration and management systems, (4) SmE analytics, (5) innovative learning and teaching strategies as components of smart pedagogy, (6) SeL strategies, approaches and environments, (7) smart learner modelling, (8) assessment and quality assurance in SmE and SeL, (9) social, cultural and ethical dimensions and challenges of SmE and SeL, (10) applications of various innovative technologies–Internet of Things, cloud computing, Ambient Intelligence (AmI), smart agents, sensors, wireless sensor networks, context-awareness technology, etc.—and smart software/hardware systems in universities and classrooms, and numerous other topics. We hope that active and open discussion of those topics within SEEL professional research and academic communities will help us to (a) organize mutually beneficial partnerships, stimulate national and international research, design and development projects in SEEL area, (b) propose innovative pedagogy, teaching and learning strategies, standards and policies in SEEL, (c) identify tangible and intangible benefits of SEEL.

The inaugural international KES conference on Smart Technology-based Education and Training (STET) has been held at Chania, Crete, Greece, during 18–20 June 2014. The 2nd international KES conference on Smart Education and e-Learning took place in Sorrento, Italy, during 17–19 June 2015. This book contains the contributions presented at the 3rd international KES conference on Smart Education and e-Learning, which took place in Puerto de la Cruz, Tenerife, Spain, during 15–17 June 2016. Book chapters, a total of 56 peer-reviewed chapters, are grouped into several parts, as follows: Part I—Smart University: Conceptual Modelling, Part II—Smart Education: Research and Case Studies, Part III—Smart e-Learning, Part IV—Smart Education: Software and Hardware Systems, and Part V—Smart Technology as a Resource to Improve Education and Professional Training.

We would like to thank scholars who dedicated a lot of efforts and time to make SEEL international conference a great success: Dr. Luis Anido (Spain), Dr. Elena Barbera (Spain), Dr. Claudio da Rocha Brito (Brazil), Dr. Dumitru Burdescu (Romania), Dr. Nunzio Casalino (Italy), Prof. Melany Ciampi (Brazil), Mr. Marc Fleetham (UK), Dr. Ekaterina Prasolova-Førland (Norway), Dr. Mikhail Fominykh (Norway), Dr. Brian Garner (Australia), Prof. Natalya Gerova (Russia), Dr. Jean-Pierre Gerval (France), Dr. Karsten Henke (Germany), Dr. Alexander Ivannikov (Russia), Dr. Marina Lapenok (Russia), Dr. Aleksandra Klasnja-Milicevic (Serbia), Prof. Andrew Nafalski (Australia), Dr. Enn Õunapuu (Estonia), Dr. Elvira Popescu (Romania), Dr. Valeri Pougatchev (Jamaica), Prof. Jerzy Rutkowski (Poland), Dr. Danguole Rutkauskiene (Lithuania), Prof. Adriana Burlea Schiopoiu (Romania), Prof. Masanori Takagi (Japan), Dr. Gara Miranda Valladares (Spain), Dr. Heinz-Dietrich Wuttke (Germany), and Dr. Larisa Zaiceva (Latvia).

We are indebted to many international collaborating organizations that made SEEL international conference possible, specifically the following: KES International (UK), InterLabs Research Institute, Bradley University (USA), Institut Superieur de l'Electronique et du Numerique ISEN-Brest (France), Silesian University of Technology (Poland), and Multimedia Apps D&R Center, University of Craiova (Romania).

It is our sincere hope that this book will serve as a useful source of valuable data and information, and provide a baseline of further progress and inspiration for research projects and advanced developments in SEEL area.

June 2016                                                                        Vladimir L. Uskov
                                                                                      Robert J. Howlett
                                                                                      Lakhmi C. Jain

# Contents

# Part I
# Smart University: Conceptual Modeling

# Smart University Taxonomy: Features, Components, Systems

Vladimir L. Uskov, Jeffrey P. Bakken, Akshay Pandey,
Urvashi Singh, Mounica Yalamanchili and Archana Penumatsa

**Abstract** Smart education creates unique and unprecedented opportunities for academic and training organizations in terms of higher standards and innovative approaches to (1) learning and teaching strategies—smart pedagogy, (2) unique highly technological services to local on-campus and remote/online students, (3) set-ups of innovative smart classrooms with easy local/remote student-to-faculty interaction and local/remote student-to-student collaboration, (4) design and development of Web-based rich multimedia learning content with interactive presentations, video lectures, Web-based interactive quizzes and tests, and instant knowledge assessment. This paper presents the outcomes of an ongoing research project aimed to create smart university taxonomy and identify main features, components, technologies and systems of smart universities that go well beyond those in a traditional university with predominantly face-to-face classes and learning activities.

**Keywords** Smart university · Smartness features · Smart university components · Systems · Smart pedagogy

## 1 Introduction

The "smart university" (SmU) concept and several related concepts, such as smart learning environment, smart campus, smart education, smart e-learning, smart training, and smart classrooms were introduced just several years ago; they are in permanent evolution and improvement since that time [1, 2].

V.L. Uskov (✉) · A. Pandey · U. Singh · M. Yalamanchili · A. Penumatsa
Department of Computer Science and Information Systems, InterLabs Research Institute,
Bradley University, Peoria, Illinois, USA
e-mail: uskov@fsmail.bradley.edu; uskov@bradley.edu

J.P. Bakken
The Graduate School, Bradley University, Peoria, Illinois, USA
e-mail: jbakken@fsmail.bradley.edu

© Springer International Publishing Switzerland 2016
V.L. Uskov et al. (eds.), *Smart Education and e-Learning 2016*,
Smart Innovation, Systems and Technologies 59,
DOI 10.1007/978-3-319-39690-3_1

Smart education is rapidly gaining popularity among the world's best universities because modern, sophisticated smart technologies, smart systems and smart devices create unique and unprecedented opportunities for academic and training organizations in terms of higher standards and innovative approaches to (1) education, learning and teaching strategies, (2) unique services to local on-campus and remote/online students, (3) set-ups of highly technological smart classrooms with easy local/remote student-to-faculty interaction and local/remote student-to-student collaboration, (4) design and development of Web-based rich multimedia learning content with interactive presentations, video lectures, Web-based interactive quizzes and tests, instant knowledge assessment, etc. Additionally, "the analysts forecast the global smart education market to grow at a CAGR of 15.45 % during the period 2016–2020" [3]. "Markets and Markets forecasts the global smart education & learning market to grow from \$105.23 Billion in 2015 to \$446.85 Billion in 2020, at a Compound Annual Growth Rate (CAGR) of 24.4 %" [4].

Therefore, it is necessary to perform active research and obtain a clear understanding of what main features, components, technologies, software, hardware, pedagogy, faculty, etc. will be required by SmUs in the near future.

## 2   Smart University: Literature Review

Recently, various creative researchers and developers began presenting their vision of SmU concepts and principles; a brief summary of several remarkable publications on such concepts is given below.

**Smart University**. Tikhomirov's [5] vision is that "*Smart University* is a concept that involves a comprehensive modernization of all educational processes. ... The *smart education* is able to provide a new university, where a set of ICT and faculty leads to an entirely new quality of the processes and outcomes of the educational, research, commercial and other university activities. ... The concept of *Smart* in education area entails the emergence of technologies such as smart boards, smart screens and wireless Internet access from everywhere".

**Smart Learning Environment**. Hwang [6] presented a concept of *smart learning environments* "... that can be regarded as the technology-supported learning environments that make adaptations and provide appropriate support (e.g., guidance, feedback, hints or tools) in the right places and at the right time based on individual learners' needs, which might be determined via analyzing their learning behaviors, performance and the online and real-world contexts in which they are situated. ... (1) A smart learning environment is context-aware; that is, the learner's situation or the contexts of the real-world environment in which the learner is located are sensed... (2) A smart learning environment is able to offer instant and adaptive support to learners by immediate analyses of the needs of individual learners from different perspectives... (3) A smart learning environment is able to adapt the user interface (i.e., the ways of presenting information) and the subject contents to meet the personal factors (e.g., learning styles and preferences) and learning status (e.g., learning performance) of individual learners".

**Smart Education**. IBM [7] defines *smart education* as follows: "A smart, multi-disciplinary student-centric education system—linked across schools, tertiary institutions and workforce training, using: (1) adaptive learning programs and learning portfolios for students, (2) collaborative technologies and digital learning resources for teachers and students, (3) computerized administration, monitoring and reporting to keep teachers in the classroom, (4) better information on our learners, (5) online learning resources for students everywhere".

Cocoli et al. [8] described *smart education* as follows: "Education in a smart environment supported by smart technologies, making use of smart tools and smart devices, can be considered smart education... . In this respect, we observe that novel technologies have been widely adopted in schools and especially in universities, which, in many cases, exploit cloud and grid computing, Next Generation Network (NGN) services and portable devices, with advanced applications in highly interactive frameworks ... smart education is just the upper layer, though the most visible one, and other aspects must be considered such as: (1) communication; (2) social interaction; (3) transport; (4) management (administration and courses); (5) wellness (safety and health); (6) governance; (7) energy management; (8) data storage and delivery; (9) knowledge sharing; (10) IT infrastructure".

**Smart Campus**. Kwok [9] defines *intelligent campus* (*i-campus*) "... a new paradigm of thinking pertaining to a holistic intelligent campus environment which encompasses at least, but not limited to, several themes of campus intelligence, such as holistic e-learning, social networking and communications for work collaboration, green and ICT sustainability with intelligent sensor management systems, protective and preventative health care, smart building management with automated security control and surveillance, and visible campus governance and reporting".

Xiao [10] envisions smart campus as follows: "*Smart campus* is the outcome of the application of integrating the cloud computing and the internet of things. ...The application framework of smart campus is a combination of IoT and cloud computing based on the high performance computing and internet".

**Smart Teachers**. Abueyalaman [11] argues "A smart campus depends on an overarching strategy involving people, facilities, and ongoing faculty support as well as effective use of technology.... A smart campus deploys *smart teachers* and gives them smart tools and ongoing support to do their jobs while assessing their pedagogical effectiveness using smart evaluation forms".

**Smart Learning Communities**. Adamko et al. [12] describe features of smart learning community applications as follows: "... the requirements of the smart community applications are the following: (1) sensible—the environment is sensed by sensors; (2) connectable—networking devices bring the sensing information to the web; (3) accessible—the information is published on the web, and accessible to the users; (4) ubiquitous—the users can get access to the information through the web, but more importantly in mobile any time and any place; (5) sociable—a user can publish the information through his social network; (6) sharable—not just the data, but the object itself must be accessible and addressable; (7) visible/augmented—make the hidden information seen by retrofitting the physical environment".

**Smart Classrooms**. An overview of smart classrooms of the first generations and requirements for second generation smart classrooms is available [13].

## 3 Research Project Goal and Objectives

The performed analysis of these and multiple additional existing publications and reports relevant to (1) smart systems, (2) smart technologies, (3) smart devices, (4) smart universities, (5) smart campuses, (6) smart classrooms, and (7) smart learning environments undoubtedly shows that "smart university" as a topic should be in the center of multiple research, design and development projects in upcoming years. It is expected that, in the near future, SmU concepts, features, hardware/software solutions and technologies will have a significant role and be actively deployed by leading academic intuitions—smart universities in the world.

**Project Goal**. The overall goal of the ongoing multi-aspect research project is to create a taxonomy of a smart university, i.e. to identify and classify a SmU's main (1) features, (2) components (smart classrooms, technological resources—systems and technologies, human resources, financial resources, services, etc.), (3) relations (links) between components, (4) interfaces, (5) inputs, (6) outputs, and (7) limits/constraints. The premise it that to-be-developed SmU taxonomy will (1) enable us to identify and predict most effective software, hardware, pedagogy, teaching/learning activities, services, etc. for the next generation of a university—smart university, and (2) help traditional universities to understand, identify and evaluate paths for a transformation into a smart university.

**Project Objectives**. The objectives of this project were to identify an SmU's main (1) features, (2) components, and (3) systems that go well beyond those in a traditional university with predominantly face-to-face classes and learning activities. Due to limited space, we present a summary of up-to-date research outcomes below.

## 4 Research Project Outcomes

### 4.1 Smart University: Distinctive Features

Our vision of SmUs is based on the idea that SmUs—as a smart system—should implement and demonstrate significant maturity at various "smartness" levels or smart features, including (1) adaptation, (2) sensing (awareness), (3) inferring (logical reasoning), (4) self-learning, (5) anticipation, and (6) self-organization and re-structuring (Table 1).

**Table 1** SmU distinctive features (that go well beyond features of a traditional university)

| SmU smartness levels | Details | Possible examples (limited to 3) |
|---|---|---|
| Adaptation | SmU ability to automatically modify its business functions, teaching/learning strategies, administrative, safety, physical, behavioral and other characteristics, etc. to better operate and perform its main business functions (teaching, learning, safety, management, maintenance, control, etc.) | • SmU easy adaptation to new style of learning and/or teaching (learning-by-doing, flipped classrooms, etc.) and/or courses (MOOCs, SPOCs, open education and/or life-long learning for retirees, etc.) |
| | | • SmU easy adaptation to needs of students with disabilities (text-to-voice or voice-to-text systems, etc.) |
| | | • SmU easy network adaptation to new technical platforms (mobile networking, tablets, mobile devices with iOS and Android operating systems, etc.) |
| Sensing (awareness) | SmU ability to automatically use various sensors and identify, recognize, understand and/or become aware of various events, processes, objects, phenomenon, etc. that may have impact (positive or negative) on SmU's operation, infrastructure, or well-being of its components—students, faculty, staff, resources, properties, etc. | • Various sensors of a Local Action Services (LAS) system to get data regarding power use, lights, temperature, humidity, safety, security, etc. |
| | | • Smart card (or biometrics) readers to open doors to mediated lecture halls, computer labs, smart classrooms and activate features/software/hardware that are listed in user's profile |
| | | • Face, voice, gesture recognition systems and corresponding devices to retrieve and process data about students' class attendance, class activities, etc. |
| Inferring (logical reasoning) | SmU ability to automatically make logical conclusion(s) on the basis of raw data, processed information, observations, evidence, assumptions, rules, and logic reasoning | • Student Analytics System (SAS) to create (update) a profile of each local or remote student based on his/her interaction, activities, technical skills, etc. |
| | | • Local Action Services (LAS) campus-wide system to analyze data from multiple sensors and make conclusions (for ex: activate actuators and close/lock doors in all campus buildings and/or labs, turn off lights, etc.) |
| | | • SAS can recommend administrators take certain pro-active measures regarding a student |

(continued)

**Table 1** (continued)

| SmU smartness levels | Details | Possible examples (limited to 3) |
|---|---|---|
| Self-learning | SmU ability to automatically obtain, acquire or formulate new or modify existing knowledge, experience, or behavior to improve its operation, business functions, performance, effectiveness, etc. (A note: Self-description, self-discovery and self-optimization features are a part of self-learning) | • Learning from active use of innovative software/hardware systems—Web-lecturing systems, class recording systems, flipped class systems, etc. |
| | | • Learning from anonymous Opinion Mining System (OMS) |
| | | • Learning from different types of classes—MOOCs, blended, online, SPOCs, etc. |
| Anticipation | SmU ability to automatically think or reason to predict what is going to happen, how to address that event, or what to do next | • Campus-wide Safety System (CSS) to anticipate, recognize and act accordingly in case of various events on campus |
| | | • Enrollment Management System to predict, anticipate, and control variations on student enrollment |
| | | • University-wide Risk Management System (snow days, tornado, electricity outage, etc.) |
| Self-organization and configuration, re-structuring, and recovery | SmU ability automatically to change its internal structure (components), self-regenerate and self-sustain in purposeful (non-random) manner under appropriate conditions but without an external agent/entity. (A note: Self-protection, self-matchmaking, and self-healing are a part of self-organization) | • Automatic configuration of systems, performance parameters, sensors, actuators and features in a smart classroom in accordance with instructor's profile |
| | | • Streaming server automatic closedown and recovery in case of temp electrical outage |
| | | • Automatic re-configuration of wireless sensor network (WSN) because nodes may join or leave spontaneously (i.e. evolving network typology), university-wide cloud computing (with multiple clients and services), etc. |

## 4.2 Smart University: Distinctive Main Components

SmUs may have numerous components of a traditional university; however, it must have multiple additional components to implement and maintain SmU distinctive features that are described in Table 1. Based on our vision of SmUs and outcomes of our research, the SmU main distinctive components should include at least those that are described in Table 2 below.

**Table 2** SmU main components and main distinctive sub-components (that go well beyond components of a traditional university)

| SmU components | SmU distinctive sub-components (that go well beyond those in a traditional university) |
| --- | --- |
| Software systems | • Web-lecturing systems (with video capturing and computer screen capturing functions) for learning content development pre-class activities<br>• Smart classroom in-class activities recording systems<br>• Smart cameraman software systems<br>• Systems for seamless collaborative learning (of both local and remote students) in smart classroom and sharing learning content/documents<br>• Collaborative Web-based audio/video one-to-one and many-to-many communication systems<br>• Systems to host, join, form and evaluate group discussions (including both local and remote students)<br>• Systems to replay automatically recorded class activities and lectures for post-class review and activities (by both local and remote students)<br>• Repositories of digital learning content and online (Web) resources, learning portals<br>• Smart learning analytics and smart teaching analytics systems<br>• Speaker/instructor motion tracking systems<br>• Speech/voice recognition systems<br>• Speech-to-text systems<br>• Text-to-voice synthesis systems<br>• Face recognition systems<br>• Emotion recognition systems<br>• Gesture (activity) recognition systems<br>• Context (situation) awareness systems<br>• Automatic translation systems (from/to English language)<br>• Intelligent cyber-physical systems (for safety and security)<br>• Various smart software agents<br>• Power/light/HVAC consumption monitoring system(s) |
| Technology | • Internet-of-Things technology<br>• Cloud computing technology<br>• Web-lecturing technology<br>• Collaborative and communication technologies<br>• Ambient intelligence technology<br>• Smart agents technology<br>• Smart data visualization technology<br>• Augmented and virtual reality technology<br>• Computer gaming (serious gaming) technology<br>• Remote (virtual) labs<br>• 3D visualization technology<br>• Wireless sensor networking technology<br>• RFID (radio frequency identification) technology<br>• Location awareness technologies (indoor and outdoor)<br>• Sensor technology (motion, temperature, light, humidity, etc.) |
| Hardware/equipment | • Panoramic video cameras<br>• Ceiling-mounted projectors (in some cases, 3D projectors)<br>• SMART boards and/or interactive white boards<br>• Smart pointing devices<br>• Controlled and self-activated microphones and speakers<br>• Interconnected big screen monitors or TVs ("smart learning cave")<br>• Interconnected laptops or desktop computers<br>• Smart card readers<br>• Biometric-based access control devices<br>• Robotic controllers and actuators |

(continued)

**Table 2** (continued)

| SmU components | SmU distinctive sub-components (that go well beyond those in a traditional university) |
|---|---|
| Smart curricula | • Adaptive programs of study—major and minor programs, concentration and certificate programs with variable structures adaptable to types of students/learners, smart pedagogy, etc.<br>• Adaptive courses, lessons and learning modules with variable components and structure suitable for various types of teaching—face-to-face, blended, online, types of students/learners, smart pedagogy, etc. |
| Students, learners, faculty | • Students and/or learners with blended or flexible learning<br>• Fully remote (or fully online) students and/or learners<br>• Life-long learners (retirees) in open education<br>• Students with disabilities<br>• Smart faculty (smart instructors) |
| Smart pedagogy | Active utilization and, if needed, adaptable combination of the following innovative types of pedagogy (teaching strategies):<br>• Learning-by-doing (including active use of virtual labs)<br>• Collaborative learning<br>• e-Books<br>• Learning analytics<br>• Adaptive teaching<br>• Student-generated learning content<br>• Serious games- and gamification-based learning<br>• Flipped classroom<br>• Project-based learning<br>• Bring-Your-Own-Device<br>• Smart robots (robotics) based learning |
| Smart classrooms | Smart classrooms with corresponding technologies, software hardware systems, and smart pedagogy for smart education |

## 4.3   Smart University: Distinctive Software Systems

As a part of this research project, for several classes of selected software systems, in Table 2 we

(1) analyzed about 10–15 existing systems usually—including both open source and commercial systems—by means of (a) review of system's functions and features, (b) review of system's demo version, (c) installation and testing of the systems, and (d) review of users and analysts' feedback,

(2) identified a list of main functions of those systems—functions to be required by SmUs, and (3) evaluated and ranked those systems. A brief summary of our research outcomes for selected classes of software systems for SmUs is presented in Table 3 below. A detailed list of references to all analyzed and mentioned below systems is available at Towards Smart University project web site at Bradley University at [14].

**Table 3** Selected classes of software systems to be used by SmUs [14]

| Class of systems | Open-source systems | Commercial systems | Our choice (1-best) |
|---|---|---|---|
| In-class activities recording systems | • Opencast | • Panopto | 1—Opencast |
| | • ClassX | • Echo360 Lect. Cap. | 1—Panopto |
| | • Kaltura | • Camtasia Studio | 2—Kaltura |
| | • openEyA | • Mediasite Lecture C. | 2—Mediasite |
| | • Lecture Record.x2 | • Tegrity | 3—ClassX |
| | • VSDC Video Ed. | • Valt | 3—Echo360 L.C. |
| | • CamStudio | • Adobe Presenter 11 | |
| | • SameView | • YuLa Lecture/Room C. | |
| Instructor-to-remote students audio/video conferencing systems (one-to-many, many-to-many) | • Skype | • WebEx Meeting Center | 1—Hangouts |
| | • BigBlueButton | • TurboMeeting | 1—BlackBoard C. |
| | • Open meetings | • Adobe Connect | 2—BigBlueButton |
| | • DimDim | • Citrix | 2—Adobe connect |
| | • Mconf | • Netop Vision ME | 3—Skype |
| | • BlueJeans | • AB Tutor | 3—GlobalMeet |
| | • Jitsi | • SoftLink | |
| | • Hangouts | • LAN School | |
| | • JoinMe | • GoToMeeting | |
| | • MeetingBurner | • GlobalMeet | |
| | • WebHuddle | • AnyMeeting | |
| | • Zoom | • BlackBoard Collabor. | |
| Web lecturing systems for pre-class learning content development activities | • InterLabs | • Camtasia Studio | 1—CamStudio |
| | • ActivePresenter | • Adobe Presenter 11 | 1—Camtasia Stud. |
| | • Jing | • Movavi Studio V7 | 2—Ezvid Scr.Rec. |
| | • Webinaria | • CamVerse 1.95 | 2—Adobe Pres 11 |
| | • Rylstim | • WM Recorder Bundle | 3—Screen-O-Mat. |
| | • IceCream screen rec. | • Debut Video Capture | 3—Movavi Stud. |
| | • CamStudio | • Fraps3.5.99 | |
| | • Screen-O-Matic | • Snagit 12 | |
| | • Flash Back Exp. Rec. | • 1AVCapture | |
| | • Ezvid Screen Rec. | • ScreenPresso | |
| Instructor motion tracking systems | • Motion | • Qualisys | 1—Motion |
| | • Genious Vis. NVR | • Bosh Security | 1—Bosh Security |
| | • iSpy | • Honeywell Mot.Sens. | 2—Voodoo C.T. |
| | • OptiTrack | • Camera Viewer Pro | 2—Qualisys |
| | • Zoneminder | • Netcam Studio | 3—OptiTrack |
| | • Voodoo Camera Tr. | | 3—Netcam Studio |

(continued)

**Table 3** (continued)

| Class of systems | Open-source systems | Commercial systems | Our choice (1-best) |
|---|---|---|---|
| Speech/voice recognition systems | • HDecode | • Dragon Natur.Sp. | 1—Jasper |
| | • JULIUS | • IBM ViaVoice | 1—Dragon N.S. |
| | • KALDI | • LH Voice Express | 2—CSLU TK |
| | • CMU Sphinx | • Briana | 2—Naunce Rec. |
| | • SHoUT Toolkit | • Kurzweil 3000 | 3—CMU Sphinx |
| | • SIMON | • IVR with SR | 3—ViaTalk |
| | • eSpeak | • Tazti | |
| | • Jasper | • Speechlogger | |
| | • EmacSpeak | • iSpeech Translator | |
| | • MARF | • Rubidium | |
| | • IVONA | • ViaTalk | |
| | • CSLU Toolkit | • ClapCommander | |
| | • iListen | • Naunce Recognizer | |
| Gesture recognition systems | • OpenGesture | • GestureTek | 1—GRT |
| | • GRT | • Cognitec | 1—Myo |
| | • GR Engine | • Omek | 2—HandVu |
| | • iGesture | • PointGrab | 2—GestureTek |
| | • HandVu | • SoftKinetic | 3—iGesture |
| | • LinHand | • Myo | 3—Rithmio |
| | • GestureWorks | • Rithmio | |
| Face recognition systems | • OpenBR | • Cognitec FaceVACS | 1- OpenBR |
| | • OpenCV | • EmoVu | 1—FaceVACS |
| | • Skybiometry | • Kairos | 2—FaceMark |
| | • FaceMark | • Eyeface | 2—EmoVu |
| | • Libface | • Rekognition | 3—Liccv |
| | • Libccv | • Face++ | 3—Kairos |
| Collaborative learning systems | • Cynapse | • Mikogo | 1—Cynapse |
| | • Voki | • Socrative | 1—Socrative |
| | • Storybirds | • Weebly | 2—Sakai |
| | • Moodle | • Edmodo | 2—ClassDojo |
| | • Sakai | • ClassDojo | 3—Moodle |
| Context/situation awareness systems | | • SARA | 1—Qognify |
| | | • Magitti | 2—Magitti |
| | | • Qognify | 3—SARA |

# 5 Conclusions

The performed research, and obtained research findings and outcomes enabled us to make the following conclusions:

(1) Leading academic intuitions all over the world are investigating ways to transform the traditional university into a smart university and benefit from the

advantages of a smart university. Smart University concepts, principles, technologies, systems, and pedagogy will be essential parts of multiple research, design and development projects in upcoming years.

(2) It is necessary to create a taxonomy of a smart university, i.e. to identify and classify SmU main (1) features, (2) components (smart classrooms, technological resources—systems and technologies, human resources, financial resources, services, etc.), (3) relations (links) between components, (4) interfaces, (5) inputs, (6) outputs, and (7) limits/constraints. The premise it that to-be-developed SmU taxonomy will (1) enable us to identify and predict most effective software, hardware, pedagogy, teaching/learning activities, services, etc. for the next generation of a university—smart university, and (2) help traditional universities to understand, identify and evaluate paths for a transformation into a smart university.

(3) Our vision of SmUs is based on the idea that SmUs—as a smart system—should implement and demonstrate significant maturity at various "smartness" levels or distinctive smart features, including (1) adaptation, (2) sensing (awareness), (3) inferring (logical reasoning), (4) self-learning, (5) anticipation, and (6) self-organization and re-structuring—the corresponding research outcomes are presented in Table 1.

(4) Based on our vision of SmUs, the identified SmU main components are presented in Table 2, and multiple analyzed and ranked software systems of selected classes to be used by SmU in Table 3.

Based on obtained research findings and outcomes, and developed SmU features, components and systems, the future steps in this research project are to (a) implement, test, validate, and analyze various identified software and hardware systems, technologies and smart pedagogy in smart classroom environment, (b) perform summative and formative evaluations of local and remote students and gather sufficient data on the quality of SmU main components—hardware, software, technologies, services, etc.).

**Acknowledgments** The authors would like to thank Ms. Colleen Heinemann, Mr. Rajat Palod, Mr. Srinivas Karri, Ms. Supraja Talasila, Mr. Siva Margapuri, Ms. Aishwarya Doddapaneni, Mr. Harsh Mehta, Mr. Priynk Bondili, Ms. Divya Doddi, and Ms. Rekha Kondamudi—the research associates of the InterLabs Research Institute and/or graduate students of the Department of Computer Science and Information Systems at Bradley University—for their valuable contributions into this research project.

This research is partially supported by grant REC # 1326809 at Bradley University [14].

# References

1. Neves-Silva, R., Tshirintzis, G., Uskov, V., Howlett, R., Lakhmi, J.: Smart Digital Futures. In: Proceedings of the 2014 International Conference on Smart Digital Futures. IOS Press, Amsterdam, The Netherlands (2014)

2. Uskov, V.L., Howlet, R. Jain, L. (eds.): Smart Education and Smart e-Learning. In: Proceedings of the 2nd International Conference on Smart Education and e-Learning SEEL-2016, 17–19 June 2015, Sorrento, Italy. Springer, Berlin (2015)
3. Global Smart Education Market 2016–2020. Research and Markets (2016). http://www.researchandmarkets.com/research/x5bjhp/global_smart
4. Smart Education and Learning Market—Global Forecast to 2020 (2015). http://www.marketsandmarkets.com/Market-Reports/smart-digital-education-market-571.html
5. Tikhomirov, V., Dneprovskaya, N.: Development of strategy for smart University. Open Education Global International Conference, Banff, Canada, 22–24 April 2015 (2015)
6. Hwang, G.J.: (2014).: Definition, framework and research issues of smart learning environments—a context-aware ubiquitous learning perspective. Smart Learn. Environ. Springer Open J. **1**, 4 (2014)
7. IBM: Smart Education. https://www.ibm.com/smarterplanet/global/files/au__en_uk__cities__ibm_smarter_education_now.pdf
8. Coccoli, M., Guercio, A., Maresca, P., Stanganelli, L.: Smarter Universities: a vision for the fast changing digital era, J. Vis. Lang. Comput. **25**, 1003–1011 (2014)
9. Kwok, L.: A vision for the development of i-campus. Smart Learn. Environ. Springer Open J. **2**, 2 (2015)
10. Xiao, N.: Constructing smart campus based on the cloud computing platform and the internet of things. In: Proceedings of 2nd International Conference on Computer Science and Electronics Engineering (ICCSEE 2013), Atlantis Press, Paris, France, pp. 1576–1578 (2013)
11. Abueyalaman, E.S., et al.: Making a smart campus in Saudi Arabia. EDUCAUSE Q. **2**, 1012 (2008)
12. Adamko, A., Kadek, T., Kosa, M.: Intelligent and adaptive services for a smart campus visions, concepts and applications. In: Proceedings of 5th IEEE International Conference on Cognitive Infocommunications, 5–7 Nov 2014. IEEE, Vietri sul Mare, Italy (2014)
13. Uskov, V.L., Bakken, J.P. & Pandey, A. The Ontology of Next Generation Smart Classrooms. In: Proceedings of the 2nd International Conference on Smart Education and e-Learning SEEL-2016, June 17–19, 2015, Sorrento, Italy, Springer, pp. 1–11 (2015)
14. Smart University—Systems: a list of references, Research, Analysis and Design of Innovative Smart Education and Smart e-Learning at Bradley University. http://www.interlabs.bradley.edu/Smart_Education_Project/Systems

# Smart Universities, Smart Classrooms and Students with Disabilities

Jeffrey P. Bakken, Vladimir L. Uskov, Archana Penumatsa
and Aishwarya Doddapaneni

**Abstract** To better educate in-classroom and remote students we will need to approach education and how we teach various types of students differently. In addition, students these days are more technological than ever and are demanding new and innovative ways to learn. One of the most promising approaches is based on design and development of smart universities and smart classrooms. This paper presents the up-to-date outcomes of research project that is aimed on analysis of students with disabilities and how they might benefit from smart software and hardware systems, and smart technology.

**Keywords** Smart university · Smart classroom · Learning disabilities · Visual impairments · Hearing impairments · Speech and language disabilities · Smart system

## 1 Introduction

Smart universities (SmU) and smart classrooms (SmC) can create multiple opportunities for students to learn material in a variety of ways. In addition, they can give access to materials in a variety of ways. Although not designed or even

J.P. Bakken (✉)
The Graduate School, Bradley University, Peoria, Illinois, USA
e-mail: jbakken@fsmail.bradley.edu

V.L. Uskov · A. Penumatsa · A. Doddapaneni
Department of Computer Science and Information Systems,
InterLabs Research Institute, Bradley University, Peoria, Illinois, USA
e-mail: uskov@fsmail.bradley.edu

A. Penumatsa
e-mail: apenumatsa@fsmail.bradley.edu

A. Doddapaneni
e-mail: adoddapaneni@fsmail.bradley.edu

© Springer International Publishing Switzerland 2016
V.L. Uskov et al. (eds.), *Smart Education and e-Learning 2016*,
Smart Innovation, Systems and Technologies 59,
DOI 10.1007/978-3-319-39690-3_2

conceptualized to benefit students with disabilities, this concept would definitely have an impact on the learning and access to material for students with disabilities.

## 1.1  Smart Classrooms: Literature Review

Pishva and Nishantha define a SmC as an intelligent classroom for teachers involved in distant education that enables teachers to use a real classroom type teaching approach to teach distant students. "Smart classrooms integrate voice-recognition, computer-vision, and other technologies, collectively referred to as intelligent agents, to provide a tele-education experience similar to a traditional classroom experience" [1].

Glogoric, Uzelac and Krco addressed the potential of using Internet-of-Things (IoT) technology to build a SmC. "Combining the IoT technology with social and behavioral analysis, an ordinary classroom can be transformed into a smart classroom that actively listens and analyzes voices, conversations, movements, behavior, etc., in order to reach a conclusion about the lecturers' presentation and listeners' satisfaction" [2].

Slotta, Tissenbaum and Lui described an infrastructure for SmC called the Scalable Architecture for Interactive Learning (SAIL) that "employs learning analytic techniques to allow students' physical interactions and spatial positioning within the room to play a strong role in scripting and orchestration" [3].

Koutraki, Efthymiou, and Grigoris developed a real-time, context-aware system, applied in a SmC domain, which aims to assist its users after recognizing any occurring activity. The developed system "…assists instructors and students in a smart classroom, in order to avoid spending time in such minor issues and stay focused on the teaching process" [4].

Given all the research available that focus on SmC, no literature was located that dealt with analysis of possible impact of SmCs concepts, features and functionality on students with disabilities.

## 1.2  Smart Universities: Literature Review

Primary focus of smart universities is in the education area, but they also drive the change in other aspects such as management, safety, and environmental protection. The availability of newer and newer technology reflects on how the relevant processes should be performed in the current fast changing digital era. This leads to the adoption of a variety of smart solutions in university environments to enhance the quality of life and to improve the performances of both teachers and students. Nevertheless, we argue that being smart is not enough for a modern university. In fact, all universities should become smarter in order to optimize learning. By "smarter university" we mean a place where knowledge is shared between employees, teachers, students, and all stakeholders in a seamless way [5].

Aqeel-ur-Rehman et al. in [6] present the outcomes of their research on one feature of future SmU—sensing with RFID (Radio frequency identification) technology; it should benefit students and faculty with identification, tracking, smart lecture room, smart lab, room security, smart attendance taking, etc.

Lane and Finsel emphasize an importance of big data movement and how it could help to build smarter universities. "Now is the time to examine how the Big Data movement could help build smarter universities—in situations that can use the huge amounts of data they generate to improve the student learning experience, enhance the research enterprise, support effective community outreach, and advance the campus's infrastructure. While much of the cutting-edge research being done with Big Data is happening at colleges and universities, higher education has yet to turn the digital mirror on itself to innovate the academic enterprise" [7]. Big data analytics systems will strongly support inferring feature of a SmU.

Al Shimmary et al. analyzed advantages of using RFID and WSN technology in development of SmU. "The developed prototype shows how evolving technologies of RFID and WSN can add in improving student's attendance method and power conservation" [8]. RFID, WSN as well as Internet-of-Things technology are expected to be significant parts of a SmU and strongly support sending characteristics of smart universities.

Doulai in [9] presents a developed system for a smart campus. This system "... offers an integrated series of educational tools that facilitate students' communication and collaboration along with a number of facilities for students' study aids and classroom management. The application of two technologies, namely dynamic web-based instruction and real-time streaming, in providing support for "smart and flexible campus" education is demonstrated. It is shown that the usage of technology-enabled methods in university campuses results in a model that works equally well for distance students and learners in virtual campuses".

Yu et al. argue that "... with the development of wireless communication and pervasive computing technology, smart campuses are built to benefit the faculty and students, manage the available resources and enhance user experience with proactive services. A smart campus ranges from a smart classroom, which benefits the teaching process within a classroom, to an intelligent campus that provides lots of proactive services in a campus-wide environment" [10]. The authors described 3 particular systems–Wher2Study, I-Sensing, and BlueShare—that provide sensing, adaptation, and inferring smart features of a SmU.

## 1.3 Research Project Goal and Objectives

The performed analysis of these and multiple additional publications and reports relevant to (1) SmU, (2) SmC, (3) smart learning environments (SmLE), (4) smart technologies, and (5) smart systems undoubtedly shows that (a) SmU, (b) SmC,

(c) smart pedagogy, and (d) smart faculty topics will be essential themes of multiple research, design and development projects in the upcoming 5–10 years. It is expected that in the near future SmC concepts and hardware/software solutions will have a significant role and be actively deployed by leading academic intuitions—smart universities—in the world.

Unfortunately, all analyzed publications are lacking a systematic approach to "smartness levels" of a smart educational system (i.e., school, college, university). Additionally, all analyzed publications are focused on traditional students/learners; however, we could not find publications on detailed analysis of "SmU, SmC and students with disabilities".

The goal of ongoing research project at the InterLabs Research Institute at Bradley University (Peoria, IL, U.S.A.) is to perform a detailed analysis and identify potential benefits of SmU and SmC components, features, systems, and technology for special type of students—students with various types of disabilities.

The objectives of this particular research project include but are not limited to:

(1) identification of smartness levels in a smart educational system;
(2) identification of characteristics of students with various types of disabilities;
(3) identification of software and hardware systems and technology to aid students with disabilities in highly technological SmCs.

The up-to-date outcomes of this research project are presented below.

## 2  Smart University and Students with Disabilities: Analysis Phase

SmU and SmC can create multiple opportunities for students to learn material in a variety of ways. In addition, they can give access to materials in a variety of ways. Although not designed or even conceptualized to benefit students with disabilities, this concept would definitely have an impact on the learning and access to material for students with disabilities.

### 2.1  Smart Educational System: Smartness Levels

Based on our vision of a SmU and up-to-date obtained research outcomes, we believe that a SmU should significantly emphasize not only software/hardware/technology features but also "smart" features and functionality of smart systems (Table 1) [11].

In order for SmU and SMC to be effective and efficient for various types of students and learners there are certain smartness levels (Table 1) that should be addressed. These levels or features should guide designers and developers of SmC,

**Table 1** Classification of levels of "smartness" of a smart system [11]

| Smartness levels (i.e. ability to ...) | Details |
|---|---|
| Adapt | Ability to modify physical or behavioral characteristics to fit the environment or better survive in it |
| Sense | Ability to identify, recognize, understand and/or become aware of phenomenon, event, object, impact, etc. |
| Infer | Ability to make logical conclusion(s) on the basis of raw data, processed information, observations, evidence, assumptions, rules, and logic reasoning |
| Learn | Ability to acquire new or modify existing knowledge, experience, behavior to improve performance, effectiveness, skills, etc. |
| Anticipate | Ability of thinking or reasoning to predict what is going to happen or what to do next |
| Self-organize | Ability of a system to change its internal structure (components), self-regenerate and self-sustain in purposeful (non-random) manner under appropriate conditions but without an external agent/entity |

smart labs, smart libraries, smart offices, etc. In doing so, we can then identify the most effective hardware, software, pedagogy and learning activities for all students, including students with disabilities...

## 2.2  Characteristics of Students with Disabilities

Types of students with disabilities that SmU and SmC can impact include students with (1) learning disabilities, (2) speech or language impairments, (3) visual impairments and (4) hearing impairments. Brief characteristics of each designated type of disability are given below.

**Learning Disabilities** [12, 13]. Learning disabilities are associated with many different problems that include difficulties in listening, reasoning, memory, attention, selecting and focusing on relevant stimuli, and the perception and processing of visual and/or auditory information. These perceptual and cognitive processing difficulties are assumed to be the underlying reason why students with learning disabilities experience one or more of the following characteristics: reading problems, deficits in written language, and underachievement in math. Not all students with learning disabilities will exhibit these characteristics, and many students who demonstrate these same behaviors are quite successful in the classroom. These students are a diverse group of individuals, exhibiting potential difficulties in many different areas. For example, one child with a learning disability may experience significant reading problems, while another may experience no reading problems whatsoever, but has significant difficulties with written expression. Learning disabilities may also be mild, moderate, or severe which complicates instruction for these students in the classroom even further.

**Speech or Language Impairments** [14, 15]. The characteristics of speech or language impairments will vary depending upon the type of impairment involved. There may also be a combination of several problems. Students could have difficulties with articulation (difficulty making certain sounds), fluency (something is disrupting the rhythmic and forward flow of speech), or voice (problems with the pitch, loudness, resonance, or quality of the voice). Students may also have difficulties with language. Language has to do with meanings, rather than sounds. A language disorder refers to an impaired ability to understand and/or use words in context [14]. A child may have an expressive language disorder (difficulty in expressing ideas or needs), a receptive language disorder (difficulty in understanding what others are saying), or a mixed language disorder (which involves both). Some characteristics of language disorders include: (1) improper use of words and their meanings, (2) inability to express ideas, (3) inappropriate grammatical patterns, (4) reduced vocabulary, and (5) inability to follow directions. Children may hear or see a word but not be able to understand its meaning. They may also have trouble getting others to understand what they are trying to communicate.

**Visual Impairments** [14, 16, 17]. Total blindness is the inability to tell light from dark, or the total inability to see. Visual impairment or low vision is a severe reduction in vision that cannot be corrected with standard glasses or contact lenses and reduces a person's ability to function at certain or all tasks. Legal blindness (which is actually a severe visual impairment) refers to a best-corrected central vision of 20/200 or worse in the better eye or a visual acuity of better than 20/200 but with a visual field no greater than 20° (e.g., side vision that is so reduced that it appears as if the person is looking through a tunnel) [16]. Being able to see gives us tremendous access to learning about the world around us. That's because so much learning typically occurs visually. When vision loss goes undetected, children are delayed in developing a wide range of skills. While they can do virtually all the activities and tasks that sighted children take for granted, children who are visually impaired often need to learn to do them in a different way or using different tools or materials. Central to their learning will be touching, listening, smelling, tasting, moving, and using whatever vision they have [17].

**Hearing Impairments** [18, 19]. The term "hearing impaired" refers to any person with any type or degree of hearing loss. The term may be used with qualifying adjectives such as "mild," "moderate," "severe," and "profound" to denote the degree of impairment. "Deaf" refers to a hearing-impaired person in whom the auditory sense is sufficiently damaged to preclude the auditory development and comprehension of speech and language with or without sound amplification. "Hard of hearing" is used to define a hearing-impaired person in whom the sense of hearing, although defective, is functional with or without a hearing aid and whose speech and language, although deviant, will be developed through an auditory base. The major challenge faced by students with hearing impairments is communication. Hearing-impaired students vary widely in their communication skills. Age of onset plays a crucial role in the development of language. Persons with prelingual hearing loss (present at birth or occurring before the acquisition of language and the development of speech patterns) are more functionally disabled than those who lose some degree of hearing after the development of language and speech. Many students with hearing impairments can and do speak. Most deaf students have normal speech organs and have learned to use them through speech therapy. Some deaf students cannot monitor or automatically control the tone and volume of their speech, so their speech may be initially difficult to understand. Understanding improves as one becomes more familiar with the deaf student's speech pattern.

## 3   Smart University and Students with Disabilities: Design Phase

### 3.1   Considerations for Students with Disabilities

The implementation of a SmC model could potentially have a huge impact on the learning of students with disabilities in general and more specifically students with learning disabilities, speech and language impairments, visual impairments, and hearing impairments. Many of the smart features of SmC are the exact areas where students with these disabilities have documented weaknesses. Most noted are deficiencies with learning, inferring, and self-organizing. Thus, the SmC should be considered when working with students with all of these disabilities [20].

Although we cannot create an exhaustive list of software and hardware technologies that should be incorporated into a SmC, we can suggest some things to consider. One must realize that one technology will not necessarily work or be effective with all students with disabilities, but when choosing software one must choose the software that will benefit the most students. As students enter your classrooms with more specific needs then those can be dealt with at that time. For example, some examples of objectives, hardware, and software of a SmC [21] that could be beneficial to students with disabilities are presented in Table 2 below.

### 3.2   Students with Disabilities and SMART Boards

Given the difficulties that students with disabilities encounter during their lives and in school SmC would benefit them and help them learn more efficiently and effectively. Where traditional classrooms do not specifically address the levels of smartness unless specific lessons focus on them, the implementation of SmC would be suggested to meet the difficulties students with learning disabilities encounter. This way, the exact areas that are of difficulty for students with learning disabilities would be addressed often and continuously in the classroom.

**Table 2** SmC Objectives, hardware and software systems for students with disabilities

| Scope | Main functions or features [21] |
|---|---|
| Objectives | • Seamlessly connect several remote SmCs to share lectures and information via networking<br>• Seamless connect various types of users' mobile smart devices and technical platforms; provide scalability and timely update of software systems and applications used by various users<br>• Automatically record all class activities and provide students with post-class review activities, for example, to review/learn content at student's own pace and comfort level<br>• Accommodate, adapt and implement newest and emerging technologies and innovative trends, for example, computer vision, face recognition, speech recognition, noise cancellation, gesture recognition, etc.<br>• Provide voice recognition, quality and fats automatic translation from English language to other languages, and visa versa<br>• Empower instructor with voice recognition, face recognition, gestures and smart pointing devices and boards to navigate, edit and display information on smart boards<br>• Provide remote students a regular face-to-face learning like experience to online/remote students logging into a session in a SmC/lab |
| Hardware | • Array of video cameras installed to capture main classroom activities, movements, discussions, expressions, gestures, etc.<br>• Ceiling-mounted projector(s) with 1 or 2 big size screen to display main activities in actual classroom; in some cases—3D projectors<br>• Student boards (big screen displays or TV) to display images of remote/online students from different locations<br>• Bluetooth and Internet enabled devices like cell phones, smart phones, PDAs and laptops to facilitate communication and information exchange<br>• Sensors (location detection, voice detection, motion sensors, thermal sensors, humidity, sensors for facial and voice recognition, etc. |
| Software | • Learning management system (LMS) or access to university wide LMS<br>• Advanced software for rich multimedia streaming, control and processing<br>• Software systems to address needs of special students, for example, visually impaired students (speech and gesture based writing/editing/navigation and accessibility tools to facilitate reading and understanding)<br>• Smart cameraman software (for panoramic cameras)<br>• Recognition software: face, voice, gesture<br>• Motion or hand motion stabilizing software<br>• Noise cancellation software |

We know the more opportunities provided to students will give them a better chance to learn so having this type of system implemented and part of daily instruction would give the more practice and learning situations to improve the exact areas that they need to improve and work on. For example, a list of possible impacts of SMART boards on students with academic difficulties is presented in Table 3.

**Table 3** Impact of SMART boards on academic difficulties

| Scope | Main functions or features [22] |
|---|---|
| Reading comprehension | • Enlarging text on the SMART board to make it more legible<br>• Highlighting parts of the text with digital highlighter ink or the SMART pen tools<br>• Using the "Spotlight" feature to only reveal certain, relevant areas of text<br>• Integrating a SMART document cameras to display text book pages and other hard copy literature on the SMART board<br>• Change text colors and backgrounds to make it more readable on the SMART board |
| Writing compre-hension | • Students using the SMART board can write on it using a finger, a pen from the SMART pen tray or a soft object like a tennis ball or hacky sack to practice their handwriting. The SMART notebook software can convert handwritten letters to text using its handwriting recognition capabilities<br>• Teachers can show pre-lined paper templates on the SMART board to make it easy for students to keep writing on the SMART board straight<br>• Teachers can help students with constructing letter forms by asking them to trace over built in alphabet letters and numerical symbols that are included with SMART notebook software<br>• Students can annotate over web pages, images and electronic documents on the SMART board to practice their writing skills |

## 3.3 Students with Disabilities and Systems/Technologies to Be Used in SmC

When looking at SmU/SmC and the possible impact on students with disabilities the outlook is very good. Not all students with disabilities will probably attend a university, but it is very likely that (1) students with learning disabilities, (2) students with speech or language impairments, (3) students who are blind or visually impaired, and (4) students who are deaf or hearing impaired will potentially attend a university. These students combined make up 58 % of the total population of students with disabilities and about 8 % of the total school population [23]. As a result, those students are in the primary focus of our current research. Based on performed analysis, below there are examples of software/hardware systems or technologies available for designated types of students with disabilities (Tables 4, 5, 6 and 7). Active utilization of those systems can serve as a starting point for colleges and universities to aid students with disabilities to learn in highly technological SmC.

(A note: if accepted, conference full paper will include a comparison of main functions of existing SW systems for various groups of students with disabilities in terms of functions' relevance to features and characteristics of SmC).

**Table 4** Technologies to aid students with learning disabilities in SmC

| Scope | Main functions or examples of existing systems |
|---|---|
| Reading-assistive technology | • Text-to-Speech<br>• Wynn Reader (from Scientific Freedom)<br>• TextAloud (from NextUp Technologies)<br>• NaturalReader (from AT&T)<br>• ReadPlease (ReadPlease Corporation)<br>• Kurzweil 3000 (Kurzweil Educational Systems, Inc.)<br>• Dragon Naturally Speaking from Nuance |
| Writing-assistive technology | • Neo2<br>• Writers Plus<br>• Inspiration®<br>• Kidspiaration<br>• Webspiaration |
| Math assistive technology | • Various calculators (4-function, graphing and scientific calculators)<br>• Manipulatives (geoboards, pictorial representations, symbol and virtual manipulatives, Cuisenairerods)<br>• Computer-assisted instruction<br>• Arithmetic-focused software<br>• Assistive technology for students with learning disabilities |

**Table 5** Technologies to aid students with speech and language disabilities in SmC

| System name | Main functions or features |
|---|---|
| DynaVox 3100 [14] | The DynaVox 3100 is a hardware/software application which assists the user in carrying a conversation by speaking for them. Words, pictures, sentences and ideas can be selected via a touch screen, mouse, joystick and multiple switches. The DynaVox can also be used in multiple languages. This communication device will greatly reduce the frustration for children who cannot speak or whose speech is unrecognizable [24] |
| CH-7KIVORY [25] | A handset designed to assist someone weak speech by amplifying their outbound speech. An example of how this type of technology would assist a student is if they were using a telephone for research, interviewing a professional and many other educational purposes |
| Chattervox [25] | A portable voice amplification system designed to raise the vocal output of people with temporary or permanent voice impairments. This device can assist students who cannot produce enough decibels naturally so that they can be heard |
| Servox [25] | An artificial larynx that can assist anyone who has lost their voices due to injury and illness as well as those who have to rest their vocal cords or are attached to a respiration device |

**Table 6** Technologies to aid students who are visually impaired or blind in SmC

| System name or scope | Main functions or features [26] |
| --- | --- |
| Lunar | Lunar is a screen magnification software system for computer users. It has a number of advanced features to help you manage the enlarged screen more efficiently. Magnification from 2x to 32x with five different viewing modes |
| Screen readers | JAWS (Job Access With Speech) is a screen reader, developed for computer users whose vision loss prevents them from seeing screen content or navigating with a mouse. JAWS provides speech and Braille output for the most popular computer applications on your PC |
| Duxbury | Grade 2 braille editing and translation software. It is available in versions for DOS, Windows and Macintosh computers. Duxbury is easy to use and is compatible with speech and braille output. It supports dozens of word processors through highly accurate ASCII and WordPerfect import bridges |
| Kurzweil 1000 | Software that works on your personal computer and a scanner to convert the printed word into speech. It has the ability to find key words or phrases within a document, editing of scanned text, magnification of scanned documents to accommodate users with visual impairments, and the ability to specify unlimited bookmarks within a document |

**Table 7** Technologies to aid students who are hearing impaired or deaf in SmC

| Scope | Main functions or features [27] |
| --- | --- |
| FM systems | Use radio signals to transmit amplified sounds. They are often used in classrooms, where the instructor wears a small microphone connected to a transmitter and the student wears the receiver, which is tuned to a specific frequency, or channel. FM systems can transmit signals up to 300 feet and are able to be used in many public places |
| Infrared systems | Use infrared light to transmit sound. A transmitter converts sound into a light signal and beams it to a receiver that is worn by a listener. The receiver decodes the infrared signal back to sound |
| Handwriting recognition | There are commercially available products that convert hand written materials into computer-generated text. Depending on the device, the information can be saved and printed as written or can convert the hand written materials into printed text for easier reading similar to a voice recognition system |
| Dragon Naturally SpeakingTM | A voice recognition software package that was developed for general public use that can be beneficial for deaf and hard of hearing individuals by creating text documents out of voice files |

# 4  Conclusions. Future Steps

**Conclusions**. The performed research, helped us identify new ways of thinking and our research findings enabled us to make the following conclusions:

1. SMU and SmC can significantly benefit students with disabilities even though they are not the focus.
2. Many technologies geared towards students without disabilities in SmC will actually impact the learning of students with disabilities.
3. Some students with disabilities may need specialized technology to be successful.
4. Some technologies focusing on the success of students with disabilities may help students without disabilities to be successful.
5. More research needs to be completed addressing SmC and access for students with disabilities.

**Future steps**. Based on obtained research findings and outcomes, the future steps of this research are to (1) test, evaluate and analyze different software (commercial and open source) and hardware applications for students with disabilities, (2) conduct assessments on the effectiveness of different technological applications, (3) design and develop components of a SmC for students with and without disabilities in local or in a distance learning environments, and (4) create a list of what technologies should be in an optimal SmC.

# References

1. Pishva, D., Nishantha, G.G.D.: Smart classrooms for distance education and their adoption to multiple classroom architecture. J. Netw. **3**(5) (2008)
2. Gligorić, N., Uzelac, A., Krco, S. Smart classroom: real-time feedback on lecture quality In: Proceedings of 2012 IEEE International Conference on Pervasive Computing and Communications Workshops (PERCOM Workshops), 19–23 Mar 2012, pp. 391–394. IEEE, Lugano, Switzerland (2012). doi:10.1109/PerComW.2012.6197517
3. Slotta, J., Tissenbaum, M., Lui, M.: Orchestrating of complex inquiry: three roles for learning analytics in a smart classroom infrastructure. In: Proceedings of the Third International Conference on Learning Analytics and Knowledge LAK'13, pp. 270–274. ACM, New York, NY, USA (2013). doi:10.1145/2460296.2460352
4. Maria, K., Vasilis, E., Grigoris, A.: S-CRETA: smart classroom real-time assistance. In: Ambient Intelligence—Software and Applications. Advances in Intelligent and Soft Computing, vol. 153, pp. 67–74, Springer (2012)
5. Coccoli, M., et al.: Smarter universities: a vision for the fast changing digital era. J. Vis. Lang. Comput. **25**, 103–1011 (2014)
6. Rehman, A.U., Abbasi, A.Z., Shaikh, Z.A.: Building a smart university using RFID technology. In: 2008 International Conference on Computer Science and Software Engineering (2008)
7. Lane, J., Finsel, A.: Fostering smarter colleges and universities data, big data, and analytics. State University of New York Press (2014). http://www.sunypress.edu/pdf/63130.pdf

8. Al Shimmary, M.K., Al Nayar, M.M., Kubba, A.R.: Designing Smart University using RFID and WSN (2008). https://www.researchgate.net/publication/221195787_Building_a_Smart_University_Using_RFID_Technology
9. Doulai, P.: Smart and flexible campus: technology enabled university education. In: Proceedings of the World Internet and Electronic Cities Conference (WIECC), Kish Island, Iran, 1–3 May 2001, pp. 94–101 (2001)
10. Yu, Z., et al.: Towards a smart campus with mobile social networking. In: Proceedings on the 2011 International Conference on Cyber, Physical and Social Computing, 19–21 Oct 2011, pp. 162–169. IEEE, Dalian, China (2011)
11. Derzko, W.: Smart Technologies (2007). http://archives.ocediscovery.com/discovery2007/presentations/Session3WalterDrezkoFINAL.pdf
12. Fletcher, M.: Learning Disabilities: From Identification to Intervention, 324 p. The Guilford Press (2006)
13. Smith, C., Strick, L.: Learning Disabilities: A to Z: A Complete Guide to Learning Disabilities from Preschool to Adulthood, 528 p. Free Press (2010)
14. Center for Parent Information and Resources. http://www.parentcenterhub.org
15. Bishop, D., Leonard, L.: Speech and Language Impairments in Children: Causes, Characteristics, Intervention and Outcome, 320 p. Psychology Press (2001)
16. Mason, H., Stone, J.: Visual Impairment: Access to Education for Children and Young People, 495 p. David Fulton Publishers (1997)
17. The Free medical Dictionary. http://medical-dictionary.thefreedictionary.com/Visual+Impairment
18. Ling, D.: Speech and the Hearing-Impaired Child: Theory and Practice, 2nd edn., 402 p. Deaf and Hard of Hearing (2002)
19. Allegheny College. http://sites.allegheny.edu/disabilityservices/students-who-are-deaf-or-hard-of-hearing/
20. National Center for Educational Statistics. http://nces.ed.gov/programs/coe/indicator_cgg.asp
21. Uskov, V.L., Bakken, J.P., Pandey, A.: The ontology of next generation smart classrooms. In: Proceedings of the 2nd International Conference on Smart Education and e-Learning SEEL-2016, 17–19 June 2015, pp. 1–11. Springer, Sorrento, Italy (2015)
22. Information about the features of SMART Board interactive whiteboards that can help students who have special learning needs. https://www.blossomlearning.com
23. National Center for Education Statistics. Fast facts: Students with disabilities (2015). http://nces.ed.gov/fastfacts/display.asp?id=64
24. Courtad, C.A., Bouck, E.M.: Assistive technology for students with learning disabilities. In: Bakken, J.P., Obiakor, F.E., Rotatori, A.F. (eds.) Advances in Special Education: Learning Disabilities: Practice Concerns and Students with LD, vol. 25, pp. 153–174. Emerald Group Publishing Limited, Bingley, United Kingdom (2013)
25. HITEC company. http://www.hitec.com
26. Texas School for the Blind and Visually Impaired. http://www.tsbvi.edu/math/67-early-childhood/1074-overview-of-technology-for-visually-impaired-and-blind-students
27. Assistive Technology for Students who are Deaf or Hard of Hearing, http://www.wati.org/content/supports/free/pdf/Ch13-Hearing.pdf

# Innovative Approaches Toward Smart Education at National Institute of Technology, Gifu College

Nobuyuki Ogawa and Akira Shimizu

**Abstract** The current educational policy of Japan toward higher education institutions is to promote education using ICT-driven equipment. Japanese universities/colleges are promoting systematic curriculum that ensures their educational quality due to the Deregulation of University Act introduced by the former Education Ministry in 1991. As for National Institutes of Technology, our college, as a leading position among higher education institutions such as the other National Institutes of Technology, universities and colleges, is promoting the improvement of the educational environment using ICT-driven equipment as well as education based on unique curriculum called "Model Core Curriculum" that ensures our educational quality. We describe the present actual practice conducted at our college as a model for the other higher education institutions that will follow the same approach in future.

**Keywords** Smart education · Model core curriculum · Acceleration program for university education rebuilding · ICT-driven education · Teaching materials

## 1 Introduction

Not only hardware and software but also smart functions of a smart system are considered important for the next generation of smart education and smart university. The next generation of a smart classroom system is supposed to consider (1) adaptation, (2) sensing (awareness), (3) inferring (logical reasoning), (4) self-learning, (5) anticipation, and (6) self-organizations and restructuring [1].

N. Ogawa (✉)
Department of Architecture, National Institute of Technology, Gifu College,
Motosu, Japan
e-mail: ogawa@gifu-nct.ac.jp

A. Shimizu
General Education, National Institute of Technology, Gifu College, Motosu, Japan
e-mail: ashimizu@gifu-nct.ac.jp

© Springer International Publishing Switzerland 2016　　　　　　　　　　　　29
V.L. Uskov et al. (eds.), *Smart Education and e-Learning 2016*,
Smart Innovation, Systems and Technologies 59,
DOI 10.1007/978-3-319-39690-3_3

The introduction of smart education can make the classroom environment beneficial and realize real time response using ICT-driven equipment [2–4]. Our institute has been improving the classroom environment for the past fifteen years to realize the beneficial classroom environment and real time response by means of e-Learning, blended learning and ICT-driven education. Our ICT-driven education practiced for the past fifteen years was highly evaluated, and our college was picked up as a leading position among the other National Institutes of Technology, universities and colleges by the Ministry of Education, Culture, Sports, Science and Technology, Japan (MEXT) in 2014. Thus, funded by MEXT for five years, we started college-wide "smart education". We have already improved hardware, software, curriculum and teaching materials that matches educational curriculum respectively, covering (1) adaptation, (2) sensing (awareness), (3) inferring (logical reasoning), (4) self-learning, (5) anticipation, and (6) self-organizations. At this time (the second year of the program), we are now re-examining (1) to (6) for further improvement. The current state of our efforts is described below.

## 2 The Improvement of the Environment for Smart Education in Terms of Hardware at National Institute of Technology, Gifu College

At National Institute of Technology (NIT), Gifu College, we have examples of various efforts concerning ICT-driven education and have been practicing various smart education, e.g., [5–7]. And, in the academic year 2014 and 2015, with an intention of spreading these methods in the whole school, we introduced the following six kinds of equipment for ICT-driven education through bids at the expense of the "Acceleration Program for Rebuilding of University Education (AP)" budget. (1) Projectors with an electronic blackboard system: introduction into about three fifth of all the classrooms within college (2) Software for making teaching materials, STORM Maker (3) Lease and maintenance operations of wireless LAN switch (4) Tablet computers (Toshiba): introduction of more than 160 units (5) notebook PC (Fujitsu): introduction of 50 units (6) LMS server (Moodle) and DB server +FileMaker.

In the academic year 2014, we replaced a chalkboard placed at the back of the first-grade classrooms of the five departments with a whiteboard and introduced projectors with an electronic blackboard system of short focal length (Epson). In the academic year 2015, we introduced the same ICT environment into the second and third classrooms of the five departments. The introduced electronic blackboard makes it possible to draw and write on its whiteboard with a dedicated electronic pen without connecting a personal computer, and digital data of drawing and writing can be recorded and stored in a file server connected to the network. (Figure 1) Moreover, the linkage function of a tablet makes it possible for teachers to arbitrarily select students' tablet up to 50 units from teachers' tablet and display

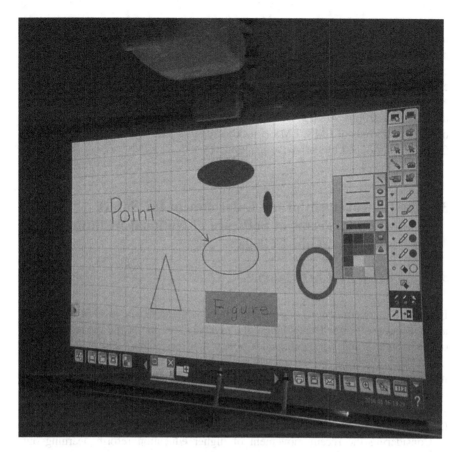

**Fig. 1** Drawing on an introduced electronic blackboard with an electronic pen

them by up to four screens. Using the projector control toolbar displayed on the projection screen of the electronic blackboard, teachers can easily select and control students' tablet screen by operating on the screen.

The assignment of Software for making teaching materials, STORM Maker is as follows: Two licenses for the departments of liberal arts and natural science respectively. Three licenses for the specialized five departments respectively. The special characteristic of STORM Maker, which has an automatic voice synthesis function, is that we can easily make content based on materials. The software is intended to be used not only by teachers for making content, but also by students for making content for future use. In addition to that students' creative activities themselves can also be considered active learning, while the teaching materials they produce are implemented in classes of active learning. The wireless LAN device was set up for use in all the 25 classrooms of all the grades (from the first to the fifth grade) of all the five departments. The system was developed by providing two access points for the wireless LAN in each classroom and by controlling using MAC address to prevent injustice access. More than 160 tablet computers (Toshiba) were introduced so that we can use them in four classrooms at the same time, and about 50 notebook computers (Fujitsu) were introduced. Moreover, all of them were set up for connecting to all the access points of the wireless LAN of the above-mentioned 25 classrooms. Though tablet computers are stored near the first grade classrooms, they can be used in the classrooms of the second, third, fourth and fifth grades by carrying them to a classroom. At NIT, Gifu College, we are promoting active learning while utilizing two kinds of LMS; LMS (Blackboard) of started by NIT, Japan and LMS (Moodle) introduced by the above-mentioned "AP".

## 3 The Improvement of the Environment for Smart Education in Terms of Software at National Institute of Technology, Gifu College

### 3.1 Model Core Curriculum

In Japan, the government is, in principle, supervising academic level of universities and colleges [8, 9]. The Deregulation of University Act liberalized the requirements of the standard, and also presented a policy that universities and colleges would assure their quality of educational and research performance, which obliged them to make efforts to do self-evaluation regarding their quality [10, 11]. NIT, colleges have been positive toward the system of assuring the educational quality. They started to get certified by the Japan Accreditation Board for Engineering Education (JABEE) [12].

Considering the recent movement of higher education reform, learning outcomes, and systematization and structuring of curriculum towards the achievement of education and research are also reflected in the standards of accreditation related to the existence as a National College. Therefore it is important to specify the minimum standards regarding the skills to be acquired and the attainment level of the engineers to be trained and administered by the Institute of National Colleges of Technology, Japan, while also responding to a social request to cultivate higher-level, practical and creative engineers. "Model Core Curriculum (MCC)" includes concrete, acquired knowledge and skills that become the "Core" as a minimum standard as well as "Model" of concrete approaches that contributes to the sophistication of education. As a whole, MCC is a combination of both concepts. With respect to "Quality Assurance", more emphasis is put on responding to a social request to cultivate "diverse, practical, creative engineers who work extensively and more sophisticatedly" rather than on meeting minimum standards (the accreditation level) to exist as a National College. With respect to the outcomes of engineer education, MCC consist of the minimum standards according to expected attained level, which were determined by referring to the accreditation standards of JABEE, Accreditation Board for Engineering and Technology

(ABET), The Engineering Council (formerly Engineering Council UK), and the CDIO Syllabus of Massachusetts Institute of Technology. Therefore, MCC was established with multiple concepts; concrete content to be studied that become "Core" and "Model" as concrete approaches that contribute to the sophistication of education.

It is sometimes difficult to clearly distinguish between "Core" positioned as the minimum standards and "Model" to respond to a social request to advance education, for the border between the two may change according to the economic and social situation and differ depending on point of view. Moreover, just take as an example "construction ability" that is considered as a vital ability for future engineers: it is difficult and inappropriate to indicate stylized study contents as an instructional method. Therefore, in MCC, the items whose attainment of skills to acquire can be indicated by concrete study contents are basically positioned as "Core" (the minimum standards), and with regard to the items that have difficulty in indicating stylized study contents, leading successful cases are shown as "Model" along with the level to be attained. MCC organizes the targets (outcomes) for students to attain that serve as a guide for curriculum design. It doesn't mean "curriculum as an input".

## 3.2 Making and Collecting Teaching Materials Consistent with Model Core Curriculum

National Institute of Technology (NIT), Japan is a large-scaled institution of higher education with 55 campuses throughout Japan, about one campus in nearly every prefecture [12]. Designated by the head office of NIT, Japan as a leading position to pro-mote active learning conducted at the other colleges, NIT, Gifu College has been making and collecting the teaching contents in regard to software mentioned above, together with NIT, Sendai and Akashi Colleges.

There are various kinds of methods for doing active learning [13–16]. Aimed for use in the practice of active learning, the educational content provides its key element. Thus, we made and collected teaching materials according to the classification of MCC and promoted our approach at National Institutes of Technology, Sendai, Akashi and Gifu Colleges, while discussing the program with the staff of NIT, Japan. In our college, performed as a schoolwide program, a total of 519 teaching materials were made and collected related to the fields of general education (liberal arts and natural science), Mechanical Engineering, Electrical and Computer Engineering, Electronic Control Engineering, Civil Engineering and Architecture. In this program, with the idea of promoting the collection of teaching materials among the other National Institutes of Technology, we stored know-how to collect them.

A variety of teaching materials are useful for promoting active learning. For example, "prior learning materials for flipped learning" and "tasks for group work".

If the teaching materials of each teacher will be shared among National Institutes of Technology, vast amounts of teaching materials will be collected, taking advantage of scale.

The collected teaching materials, if they are classified according to the achievement evaluation of MCC, can easily be developed and utilized at colleges other than the colleges the providers belong to. In addition, if they are allowed to revise shared teaching materials provided by a teacher, other teachers can improve them. In this program, from the viewpoint of using the collected teaching materials in active learning, the providers presented basic information and MCC classification information of each material in a file called a check sheet. Thus, a mechanism to use the collected teaching materials was prepared.

The providers of National Institutes of Technology, Sendai, Akashi and Gifu Colleges uploaded their teaching materials to a specific file server, and they were reviewed by the technical assistants of National Institutes of Technology, Sendai and Akashi Colleges. At our college, the entire teaching staff provided their teaching materials. On this occasion, the teachers of all the departments who are in charge of the academic affairs collected the teaching materials from each department with memory sticks and uploaded them to a file server for collecting teaching materials.

Considering different situations in class, different types of teaching materials were made and collected, such as lecture notes, materials with PowerPoint, examination questions, quiz, texts for experiments, teaching aids, videos, pictures, figures, charts, graphs, program source and e-Learning contents. The information on how to conduct active learning is also useful, so it is important to share practices of active learning with teaching materials.

In this program, a check sheet file was made for a teaching material. The number of the check sheet is as follows: 519 (NIT, Gifu College), 504 (NIT, Sendai College), 484 (NIT, Akashi College), 239 (other than the three National Institutes of Technology). The number of the total count value of MCC is more than that of the check sheet, for some teaching materials correspond to more than one item. The total count value is as follows: 532 (NIT, Gifu College), 631 (NIT, Sendai College), 489 (NIT, Akashi College), 255 (other than the three National Institutes of Technology).

The following shows the classification of all teaching materials collected in the program.

According to Table 1, the items that have 10 teaching materials or less are 31 items (57.4 %) out of all the items of middle classification of MCC, which shows that many teaching materials are biased toward part of the classifying items of MCC, though the total number of the classifying items is 1907. On the other hand, the items that have no teaching materials are about 10 items out of all the items of middle classification of MCC, which shows that teaching materials are likely to be provided for them. Therefore, it seems to be good to collect teaching materials while showing the items of MCC. In this project, teachers have already provided their lecture notes and examination papers with sensitivity to copyright laws. Though Article 35 of the Copyright Act permits us to provide teaching materials to

**Table 1** MCC classification of all teaching materials collected in the program

| Large classification | Middle classification | Total count value |
|---|---|---|
| I Mathematics | I_Mathematics | 78 |
| II Natural science | II_A_ Physics | 50 |
| | II_B_ Physical experiment | 19 |
| | II_C_Chemistry | 10 |
| | II_D_ Chemical experiment | |
| | II_E_Life Sciences·Earth sciences | 4 |
| | II_outside MCC | 3 |
| III Art and aocial science | III_A_Japanese | 33 |
| | III_B_English | 16 |
| | III_C_ Social studies | 18 |
| | III_outside MCC | |
| IV Basis of engineering | IV_A_Engineering literacy | 76 |
| | IV_B_ Engineers' ethics | 3 |
| | IV_C_ Information literacy | 30 |
| | IV_D_ History of technology | |
| | IV_E_Globalization·Cross-cultural understanding | |
| | IV_outside MCC | 33 |
| V Specialized engineering categorized by field | V_A_ Mechanical engineering-related fields | 203 |
| | V_B_ Materials-related fields | 4 |
| | V_C_ Electronics-related fields | 405 |
| | V_D_ Information-related fields | 251 |
| | V_E_Chemistry and biology-related fields | 15 |
| | V_F_Construction-related fields | 98 |
| | V_G_Architecture-related fields | 145 |
| | V_outside MCC | 103 |
| VI Engineering experiments and practical skills categorized by field | VI_A_ Mechanical engineering-related fields | 64 |
| | VI_B_ Materials-related fields | 2 |
| | VI_C_ Electronics-related fields | 63 |
| | VI_D_ Information-related fields | 26 |
| | VI_E_Chemistry and biology-related fields | 15 |
| | VI_F_ Construction-related fields | 5 |
| | VI_G_Architecture-related fields | 4 |
| | V_outside MCC | 1 |

(continued)

**Table 1**  (continued)

| Large classification | Middle classification | Total count value |
|---|---|---|
| VII Substantiation of specialized skills | VII_A_Internship | |
| | VII_B_PBL education | 27 |
| | VII_C_ Collaborative education | |
| | VII_outside MCC | 29 |
| VIII Versatile skills | VIII_A_Communication skills | 10 |
| | VIII_B_ Consensus formation | 3 |
| | VIII_C_ Information collection, use and transmission capabilities | 5 |
| | VIII_D_ Discovering problems | 9 |
| | VIII_E_ Ability of thinking logically | 4 |
| | VIII_outside MCC | 1 |
| IX Attitude·Orientation | IX_A_ Independence | 3 |
| | IX_B_ Self-management skills | 4 |
| | IX_C_ Sense of responsibility | |
| | IX_D_Group skills | 3 |
| | IX_E_Leadership | |
| | IX_F_ Morality(Respect for Originality, Public Spirit) | |
| | IX_G_Future orientation, career design skills | 5 |
| | IX_outside MCC | |
| X Comprehensive learning experience and creative thinking power | X_A_Construction skill | 1 |
| | X_B_ Engineering design skills | 3 |
| | X_outside MCC | 23 |
| Number of checksheet | | |
| 1745 | Total of classification | 1907 |

students without regard to copyright laws within the necessary scope in the case of face-to-face classes, it doesn't permit us to do so when we use teaching materials on the Net. For this reason, teachers have stated that they couldn't use the teaching materials that they had used before. However, it is possible to provide teaching materials that don't infringe copyright laws by making some alterations on the present teaching materials. Therefore, in order to continue with the program, it is considered effective to prepare a manual showing how to modify teaching materials so that they will be suitable for distribution. All National Institutes of Technology introduced Blackboard (LMS), and the use of Blackboard will promote the making of new teaching materials, which is effective for ensuring the quality and the quantity of the teaching materials provided.

Additionally, examples of practices of active learning can be useful content. By taking advantage of the vast scale of teaching resources of National Institutes of

Technology, in addition to the collection, classification and preparation of case studies as a reference for practicing active learning, our objective of enhancing effectiveness will be met.

## 4 Conclusions. Future Steps

With the advancement of society, when it comes to job hunting, students are required to show more knowledge and skills than before. It is necessary for them to acquire basic knowledge, theory, high level applied skills, creativity, cooperativeness, communication skills, etc. especially in the field of modern industry. Our college is considering smart education as next-generation education, and corresponds to the increasing needs of society. And, as mentioned earlier, our college is also promoting approaches from the viewpoint of hardware and software: the former is the improvement of educational environment and the introduction of ICT equipment; the latter is a systematic structure analysis of educational content and the establishment of an educational system that corresponds with that, and the creation of educational content based on an educational curriculum. We have already improved hardware, software, curriculum and teaching materials that matches educational curriculum respectively, covering (1) adaptation, (2) sensing (awareness), (3) inferring (logical reasoning), (4) self-learning, (5) anticipation, and (6) self-organizations. In the second year of the program, we are now re-examining them for further improvement. The next academic year falls on the intermediate period (the third year). Therefore, we regard the period as a critical stage when interim analysis/evaluation of smart education will be evaluated. We are committed to continued improvement, development and accelerative promotion performed for the remaining term. We are going to improve smart education and spread it among faculty members by spiraling up our practice over several years, and take a proactive role as an actual smart college in Japan.

**Acknowledgments** This research project is being financially supported through the AP by MEXT. We would like to express our gratitude for the support. Furthermore, the creation/collection of teaching materials based on MCC is being financially supported by the fund that NIT, Japan received from MEXT for its project. We would also like to express our gratitude for the support.

## References

1. Uskov, V.L., Bakken, J.P., Pandey, A.: The ontology of next generation smart classrooms. In: Smart Education and Smart e-Learning. Smart Innovation, Systems and Technologies, vol. 41, pp 3–14. Springer (2015). ISBN 978-3-319-19875-0

2. Gligorić, N., Uzelac, A., Krco, S.: Smart classroom: real-time feedback on lecture quality. In: Proceedings of 2012 IEEE International Conference on Pervasive Computing and Communications Workshops (PERCOM Workshops), pp. 391–394 (2012)
3. Maria, K., Vasilis, E., Grigoris, A.: S-CRETA: smart classroom real-time assistance. In: Ambient Intelligence—Software and Applications, Advances in Intelligent and Soft Computing, vol. 153, pp 67–74. Springer (2012)
4. Mercier, E., McNaughton, J., Higgins, S., Burd, E.: Orchestrating learning in the multi-touch classroom: developing appropriate tools. In: van Aalst, J., et al. (eds.) Short Papers, Symposia and Abstracts 2: the Future of Learning: 10th International Conference of the Learning Sciences (ICLS 2012) (2012)
5. Ogawa, N., Kanematsu, H., Fukumura, Y., Shimizu, Y.: Checklist system based on a web for qualities of distance learning and the operation. In: Watanabe, T. (ed.) Intelligent Interactive Multimedia: Systems & Services, vol. 14, pp. 129–141. Springer (2012)
6. Nomura, S., Yamagishi, T., Kurosawa, Y., Yajima, K., Nakahira, K.T., Ogawa, N., Irfan, C. M.A., Handri, S., Fukumura, Y.: Anticipation of the attitude of students: passive or active coping with e-learning materials result in different hemodynamic responses. In: Proceedings of the ED-MEDIA 2010, pp. 810–817 (2010)
7. Kanematsu, H., Kobayashi, T., Ogawa, N., Fukumura, Y., Barry, D.M., Nagai, H.: Nuclear energy safety project in metaverse. In: Watanabe, T. (ed.) Intelligent Interactive Multimedia: Systems & Services, pp. 411–418. Springer, Berlin (2012)
8. Yonezawa, A., "Japan" in Forest, J.F., Altbach, P.G. (eds.) International Handbook of Higher Education PartII, pp. 829–837. Springer (2006)
9. Newby, H., Weko, T., Breneman, D., Johanneson, T., Maassen, P.: OECD Reviews of Tertiary Education: Japan (2009). http://www.oecd.org/japan/42280329.pdf
10. Eades, R.G., Hada, Y.: The 'Big Bang' in Japanese Higher Education: the 2004 Reforms and the Dynamics of Change. Trans Pacific Press (2005)
11. Ministry of Education, Culture, Sports, Science and Technology (MEXT). Quality Assurance Framework in Higher Education in Japan (2009). http://www.mext.go.jp/component/english/__icsFiles/afieldfile/2011/06/20/1307397_1.pdf
12. National Institute of Technology, What is KOSEN? (2009). http://www.kosen-k.go.jp/english/what-idx.html
13. Lage, M., Platt, G., Treglia, M.: Inverting the classroom: a gateway to creating an inclusive learning environment. J. Econ. Educ. **31**(1), 30–43 (2000)
14. Bonwell, C.C., Eison, J.A.: Active learning: creating excitement in the classroom. School of Education and Human Development, George Washington University, Washington, DC (1991)
15. Khan Academy (2006). https://www.khanacademy.org/
16. Bergmann, J.; Sams, A.: Flip your classroom: reach every student in every class every day. Int. Soc. Technol. Educ (2012). ISBN: 1564843157

# University Smart Guidance Counselling

Elena Belskaya, Evgeniia Moldovanova, Svetlana Rozhkova,
Olga Tsvetkova and Mariya Chervach

**Abstract** The paper deals with the activities of university guidance counsellors.
The main aspects of their performance as well as the ways for creating a favorable
environment for successful adaptation of first-year students are considered. Particular attention is devoted to the application of innovative and smart technologies
in university guidance counselling such as training, group teaching methods,
role-playing games, using of electronic educational resources, etc.

**Keywords** Higher education · Guidance counsellor · Smart technologies

## 1 Introduction

Smart-education is gradually replacing the usual combination of traditional education and e-learning as a result of intensive development of information technology. Smart education is believed to be able to provide the highest level of education
that corresponds to the challenges and opportunities of today's world [1, 2].

E. Belskaya · E. Moldovanova (✉) · S. Rozhkova · O. Tsvetkova · M. Chervach
National Research Tomsk Polytechnic University, Tomsk, Russia
e-mail: eam@tpu.ru

E. Belskaya
e-mail: belpen@tpu.ru

S. Rozhkova
e-mail: rozhkova@tpu.ru

O. Tsvetkova
e-mail: olga_tsvetkova95@mail.ru

M. Chervach
e-mail: chervachm@tpu.ru

© Springer International Publishing Switzerland 2016                                      39
V.L. Uskov et al. (eds.), *Smart Education and e-Learning 2016*,
Smart Innovation, Systems and Technologies 59,
DOI 10.1007/978-3-319-39690-3_4

Smart is a feature of a system (or a process) that is manifested in the interaction with the environment and gives the system (or the process) the ability to:

- immediately respond to the changes of external surroundings;
- adapt to the changes of environmental conditions;
- improve self-development and self-control;
- effectively achieve goals.

Nowadays, the feature that allows quick adaptation of an object or a process to the changes in the environment is increasingly required in education. Smart education is a new educational paradigm, which involves the implementation of an adaptive educational process using informational smart technologies. Smart education should provide an opportunity to benefit from the global information society, and to meet the educational needs and interests.

The university years coincide with the period of intensive development of a personality in physical, intellectual, social, and moral sense. It could be challenging for new students to adapt to the university life, education system, and university relationships. A successful start in the learning process requires students to have such qualities as self-discipline and self-organization. They are accustomed to parents' and school teachers' assistance so, getting into new conditions, some first-year students become diffident, burnt-out, and passive and are gradually losing the prospects for further studying. However, not only students but also university staff are interested in efficient student adaptation to the university life. In many universities around the world guidance counselling services are established for effective student adaptation to university life. The guidance and counselling system plays a crucial role in the process of adaptation to new environment. The major goals of student guidance counselling are to enable facilities for successful study and social activities, to promote their personal growth and professional development [3].

The challenges facing a guidance counsellor are as follows:

1. to help students during their social, psychological and academic adaptation to university life;
2. to form soft skills, necessary professional and general cultural competencies as a part of student academic activities;
3. to help students develop their personal academic pathway;
4. to ensure students' healthy lifestyle throughout their education;
5. to assist students in their self-realization, improving their intellectual, spiritual and moral development;
6. to foster university loyalty;
7. to involve students in social activities.

A guidance counsellor serves as a link between a student and internal and external social university medium. Counsellors help students obtain what is best from the studies and quickly adapt to a new environment [4].

Academic staff is ambiguous about the educative process in university, but as practice shows this component is essential for the system of higher education [5].

Each university or institute has a unique system of activities to develop student's personality. A guidance counsellor of an academic group uses active and interactive activities in the classroom and outside, such as role-playing games, psychological training, group discussions, museum tours, excursions to the companies, conducting master classes by invited experts, meetings with representatives of companies, government, and public organizations, etc.

## 2 Academic and Social Adaptation Project

All first-year students in the Institute of Power Engineering, Tomsk Polytechnic University (TPU), must take the Adaptation Course, which runs over two terms (Autumn and Spring term). Classes meet 2 h per week in the Autumn term and 1 h per week in the Spring term, over 36 weeks giving 48 h of tuition. After taking the course students are able to:

- prepare themselves for facing new academic challenges;
- improve their time management;
- freely navigate through the information environment of the university and quickly receive all the necessary information for study and leisure;
- present information clearly and efficiently;
- make a speech confidently;
- correct their personalities after learning their traits and characteristics;
- set and achieve goals.

The course is provided by proficient guidance counsellors. At present, 17 guidance counselors work with first- and second-year students in the Institute of Power Engineering, TPU. All of them combine both positions of a subject teacher (Assistant professor, Associate professor) and a guidance counsellor. Each academic student group is supported by one guidance counsellor.

The "Adaptation Course" contains 3 modules: "Informational", "Academic", and "Social and Psychological". The "Informational Module" includes introductory sessions on the university life and the students learn to use the "New Student Guide"—guidelines for new students how to start their learning process successfully and the Website—to organize their student life including schedule service, teachers' contact hours, campus map, etc. The "Academic Module" consists of workshops such as "E-services and the Educational Process in TPU", "Techniques of Work with Information", "Public Speech", "Cloud Technologies", etc. The "Social and Psychological Module" consists of psychological trainings such as "How to Set and Achieve Goals", "Time Management", "Know Thyself", "Stress Management", "Conflict and Reconciliation Behavior", "How to Pass Exams Perfectly Well", etc.

The aim of the academic and social adaptation project is to shorten the period of student adaptation, increase stress resistance, and maintain the number of students.

## 3 Smart Technologies in Guidance Counselling

Nowadays one of the most efficient methods in guidance counselling is the use of Smart-technologies, such as e-learning, blend-learning, personalization, interactive tutorials, learning through video games, etc. [6]. This allows young people to adapt to rapidly changing conditions and ensure the transition from books to active content.

In this section, we describe the application of innovative smart technologies in the daily activities of a guidance counsellor.

### 3.1 Smart University Environment

Over the past decades, the Internet and web net have proved to be the most successful global projects, which have changed the economy and society due to new forms of communication and collaboration, implementation of innovations, new modes of work with information and knowledge. Currently electronic educational resources are an essential part of life for any person, who is connected with education [7].

Firstly, all students receive access to all information resources of the University, which will help them in solving various troubles, problems, issues, and challenges. After registration (username and password) as users of the corporate network TPU every student accepts the possibility of working with programs and services in a personalized enclosed space—an individual student online service. In the individual student online service, a student and a guidance counsellor have equal opportunities for communication together. Additionally there is the possibility that students are given online counseling.

Moreover, there is a series of tutorials, which helps students to adapt to the university life, for instance:

- "Virtual TPU". Students familiarize themselves with TPU information and education environment, namely, the resources and the services available for students;
- "The basics of work with information and library resources". Students make themselves familiar with library resources and rules and techniques for information processing. They obtain information about rare and unique resources, electronic catalogues, and options for more efficient work in the reading rooms.

Use of internet technologies allows not only to help students in their adaptation to the university life, but also to enhance efficiency in subject mastering, expanding the course content, as well as teaching and learning methods. It also increases students' motivation to learn and provides them with possibility to study independently and acquire radically new knowledge that would be of practical importance at their future workplace. Guidance counsellors are given unlimited opportunities for self-development, which, in its turn, shifts teaching process to a much higher level.

## 3.2 Group Methods

It is well known [8] that group methods, which include trainings, discussions, etc., are efficiently applied to effective group work. This activity aims to develop such student's skills as teamwork, readiness as the leader of the group to formulate team goals, to demonstrate the importance of their future profession, etc.

A cognitive process in the trainings is based on an active position of the participants and their own experience [9]. The trainer can assist in acquainting participants with each other, create an atmosphere of cooperative work, partnership, and mutual understanding. On the other hand, performing creative and challenging tasks increases communication students' skills both with each other and with the academic staff. The interactive training technologies allow students to assimilate new knowledge more effectively and to obtain more information due to an opportunity of asking questions, expressing their opinions, practicing new skills, etc.

Let us consider some trainings and their purposes [10]:

- "Student Team Building"—a communicative skills training, which aims to create a cohesive team in a student group and includes three sessions: "Make the Introductions", "Group logo", and "Development of group rules and regulations";
- "Time Management"—students learn technologies enabling them to use time according to their personal goals and values;
- "Stress Management"—students learn the causes of stress, physiological, emotional, cognitive and behavioral symptoms of stress and methods of coping with stress, examine how to avoid negative impacts of stressors;
- "Conflict and Reconciliation Behavior"—this training aims to teach students conflict management techniques and to develop negotiation skills necessary for effective conflict management;
- "SMART goals"—students examine the principles of goal setting and forming a personal life programme.

According to the opinion poll among students, 32 % of them are frequently using skills acquired during training and 46 % use the skills sometimes. In addition, 8 % of students plan for the future to use knowledge and experience gained in the training.

## 3.3 Webinars (Web Based Seminars)

This new, modern and accessible learning method through the Internet is now an integral part of the educational process at various levels. Without leaving the office or home, you are able to attend a lecture, a workshop or a seminar as well as to participate in the discussion of topical issues.

As mentioned above, the "Adaptation course" runs only over two terms. However, continuity and regularity are vital in student guidance counselling as the second-year students are also still in need of support [11]. Therefore, seminars, lectures, and trainings can be given using webinar technology. Webinar platforms suggest both web-streaming audio and phone-bridge options, yet the poll conducted in 2015 among the students of the Institute of Power Engineering found that 76 % of students prefer personal participation in trainings. And 23 % noted that they were interested in online trainings if it is impossible to attend the classroom.

## 4  Electronic Educational Resources

In TPU, modern electronic educational resources are designed and implemented, for example, software, interactive electronic documents, media resources, and educational complexes. All these resources can be put on line on a Learning Management System platform to provide the centralized management of the educational services.

In the Institute of Power Engineering, TPU, an online educational resource (OER) for guidance counsellors has been designed, which includes both interactive electronic documents and an educational complex. The resource comprises the weblog "Guidance Counselling" and the electronic course book "Communication trainings for the Adaptation Course classes".

## 4.1  Weblog "Guidance Counselling"

The personal weblog "Guidance Counselling" was designed by the senior guidance counsellor and consists of brief records, which are regularly added, represented in chronologically reversed order and include educational materials, media resources, and presentations. The resources in the blog are available for all TPU guidance counsellors.

The structure of the blog records reminds a familiar sequential structure of the log and comprises the following pages:

- "Main page"—information on up-coming events for the concerned guidance counsellors;
- "Guidance and Counselling History"—history of guidance counselling in TPU;
- "Documents"—resources for performance: the TPU Statement on the Guidance and Counselling, schedule, guidelines and work plan templates for "Adaptation Programme" courses;
- "Presentations"—large list of presentations, from which a guidance counsellor can choose the relevant one for his/her class;
- "Video"—videos for in-class activities and different video trainings for teachers;

- "Tests and Questionnaires"—tests and questionnaires are designed for the counsellors to consolidate their expertise or to estimate their level of knowledge for further professional development;
- "Wall"—a feature, which makes it possible to exchange counsellors' opinions on different topics;
- "Electronic manual"—where the electronic course book "Communication trainings: in-class application" is posted.

Blog as a means of net communication has a number of advantages in comparison with web forums, chats, and email. On the one hand, a blog reader gets information necessary for efficient performance and feels that he or she is a part of counselling community within the Institute of Power Engineering, TPU. The latter is a result of the guidance counsellor being informed about the events held at the university and knowing colleagues' attitudes to the activities performed by other guidance counsellors. On the other hand, a blog guest can make comments or dispute with the blogger, in case the blogger agrees to such mode of communication.

## 4.2 Electronic Course Book

The poll conducted in 2014 among the guidance counsellors of the Institute of Power Engineering disclosed that approximately 30 % of them considered the "Adaptation course" classes difficult to teach. This was the reason for the development of training materials, and then an electronic course book. The electronic course book aims to develop skills and knowledge for guidance counselling. After having studied the e-course book, a teacher is supposed to: know what a training session is; be capable of using training plan templates and scenarios; get experience in designing scenarios of training sessions; be able to hold a training session within a class. In terms of the structure, the electronic course book is based on three modules: introduction, basic information, and workshop and testing.

Advantages of using an electronic course book are an advanced, user-friendly and sufficiently simple navigation mechanism within the tutorial; optimization of user interface adaptation of educational material to the level of student's knowledge; inclusion of multimedia fragments (graphics, audio, and video); adaptive user interaction with the elements of the course book.

## 5  Student as a Guidance Counsellor

Within many Russian higher education institutions there is a youth social movement aimed at successful adaptation of freshmen to university life, so called "Peer Guidance". This project was launched in the Institute of Power Engineering, TPU in

2013. Currently, the project is characterized by a clear structure, strategic objectives, and feasible techniques and is successfully developing thanks to active students' support [12].

The main objective of peer guidance is to ensure educational and social adaptation of first-year students to university environment, as well as to student life, by contributing to personality development of the freshmen. To ensure successful implementation of the "Peer Guidance" project, students have designed an information and communication website on the social network "Vkontakte". The website is maintained by the student-guide under the supervision of the Senior Guidance Counsellor of the Institute and a psychologist in charge of the University Centre of Personality Development. They post records, photos, and video films on the varied events organized and hold by students. It allows students and guidance counsellors to exchange their experience, compete in achievements, and prepare the annual report, which is presented at the Guidance and Counselling Service [13].

The main advantage of peer guidance is that it was created for voluntary participation, based on the personal motivation of each student-counsellor. Thus, the student-counsellor becomes a mentor for first-year students, whom they trust, who guide them and whose opinion is significant for them. A student-counsellor not only plays an important role in first-year students' life, but helps and contributes to the development of their competencies.

## 6   Discussion

In 2015, the Institute of Power Engineering carried out a survey aimed at revealing the opinions of students about the guidance and counselling service in TPU.

The survey involved 243 students (174 first-year and 69 s-year students). Students were asked to evaluate various aspects of the guidance counsellors' performance. Overall, about 90 % of respondents assess the work as good and excellent. Figures 1, 2, and 3 show assessments in 5 point scale: 5—best possible grade, excellent; 4—above average grade, very good; 3—below average grade, improvement needed; 2—bad; 1—very bad.

The statistics shows that not all students are satisfied with the performance of the guidance counsellors. This can be explained by the fact that the academic group

**Fig. 1** Assessment of guidance counsellor support in the adaptation to university life

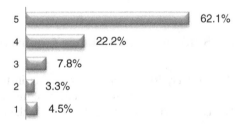

**Fig. 2** Assessment of
guidance counsellor support
in an individual educational
path choice

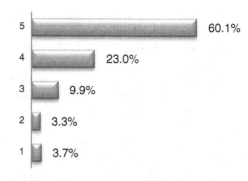

**Fig. 3** Assessment of
guidance counsellor support
in self-realization (academic
activity, social life, etc.)

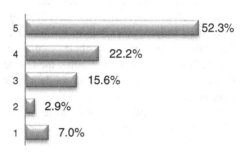

usually consists of 20–25 students and a guidance counsellor is not able to focus on such a number of wards. We suggest that one guidance counsellor support only 10 –15 students. However, the experience of involving senior students in guidance counselling showed good results.

In the contemporary world of computers and information technology, students have a lack of communication skills. This, in our opinion, is the main problem in the adaptation of first-year students to university life. Modern innovative approaches, of course, greatly expand opportunities for students to successfully adapt to the new learning environment. However, the key point was and remains a live communication in teacher-student and student-students relationships.

As noted above, the subject teachers work as guidance counsellors at TPU. This job requires special skills and experience. At the beginning teachers may meet challenges conducting classes in active and interactive ways. So it makes sense to improve training and development of specialists for guidance counselling.

# 7 Conclusions

In TPU, the guidance counselling plays the key role in providing the basis for the development of students' general and professional competences [14]. Being a connector between the freshman and a new society, a guidance counsellor intensifies the process of adaptation; creates the conditions necessary to adapt to new people, as well as to university structure, content, and requirements; contributes to socializing between the students within a group and within the university.

The work of TPU guidance counsellors proves the effectiveness of active learning methods and electronic educational resources. The undisputable advantage of the electronic educational resource is its implementation into the guidance counsellor work.

Finally, collaboration of teachers, psychologists, and students gives an effective model of university guidance counselling.

Thus, the content of the teacher's activity shows the conjugation process of self-transformation of the student and the teacher being the coordinator of the educational process. Both subjects having an active « self-concept » implement innovative training activities in multimedia learning environment. The activity of the student in this case contributes to the need for self-presentation. However, their role in this society is not the same: the teacher is organizer, coordinator and head of the educational process, and the student is an active and independent partner in the joint creative innovative teaching.

To sum up, application of smart technologies in university guidance counselling allows creating flexible system of student-centered learning.

# References

1. Tikhomirov, V., Dneprovskaya, N., Yankovskaya, E.: Three dimensions of smart education. In: Smart Innovation, Systems and Technologies, vol. 41, pp. 47–56 (2015)
2. Avdeeva, Z.K., Taratuhina, Y.V., Omarova, N.O.: Smart educational environment as a platform for individualized learning adjusted to student's cultural-cognitive profile. In: Smart Innovation, Systems and Technologies, vol. 41, pp. 69–79 (2015)
3. Nurzhanova, S.A.: Features of psychological and pedagogical support for first-year students in a high school. Way Sci. 5(15), 113–115 (2015)
4. Belskaya, E.Y., Tsvetkova, O.S.: Peer tutoring in the Institute of Power Engineering and its role in first year student adaptation. In: Linguistic and Cultural Traditions and Innovations: XIV International Science and Practice Conference, pp. 26–30, Tomsk (2014). (in Russian)
5. Andreev, G.: Education and personality development are inseparable at higher education institutions. High. Educ. Russ. 3, 61 (1996). (in Russian)
6. Teri, S., Acai, A., Griffith, D., Mahmoud, Q., Ma, D.W.L., Newton, G.: Student use and pedagogical impact of a mobile learning application. Biochem. Mol. Biol. Educ. 42(2), 121–135 (2014)
7. Osin, A.V.: Open educational module multimedia systems, vol. 328. Agency « Publishing Servica », Moscow (2010). (in Russian)
8. Burke, A.: Group work: how to use groups effectively. J. Eff. Teach. 11(2), 87–95 (2011)
9. Pokholkov, Y.P., Rozhkova, S.V., Tolkacheva, K.K.: Practice-oriented educational technologies for training engineers. In: International Conference on Interactive Collaborative Learning, ICL 2013, pp.619–620 (2013)
10. Ratochka, L.A., Petukhova, O.V., Peshkovskaya, A., Sindeyeva, N.B., Solodovnikova, O.M.: Kommunikativnyye treningi—kak forma kuratorskikh chasov. Metodicheskoye posobiye, 27, Izd-vo TPU, Tomsk (2012). (in Russian)

11. Belskaya, E.Y., Startseva, E.V.: Experience of curatorial activities in improving of Tomsk Polytechnik University first-year student adaptation. Sovremennyye problemy nauki i obrazovaniya. 5 (2015). http://www.science-education.ru/ru/article/view?id=22269 (in Russian)

12. Belskaya, E.Ya.: Electronic courseware for guidance counsellors at Tomsk Polytechnic University. Mod. Probl. Sci. Educ. **3**, 9 (2015). (in Russian)

13. Belskaya, E.Y.: The role of the guidance counselling in first-year student adaptation to university life in the institute of power engineering. In: Linguistic and Cultural Traditions and Innovations: XIV International Science and Practice Conference, pp. 20–26, Tomsk (2014). (in Russian)

14. Chung, P., Yeh, R.C., Chen, Y.-C.: Influence of problem-based learning strategy on enhancing student's industrial oriented competences learned: an action research on learning weblog analysis. Int. J. Technol. Des. Educ. doi:10.1007/s10798-015-9306-3

# Teacher's Role in a Smart Learning Environment—A Review Study

Blanka Klimova

**Abstract** The present boom of emerging information and communication technologies (ICT) and their effect on all fields of human activities undoubtedly bring significant changes also into the educational system which puts an emphasis on an independent creative potential of a student, integration of different teaching plans, or diversity of teaching methods and strategies. Thus, the traditional role of the teacher is changing. Nevertheless, the teacher still plays a decisive role in setting the learning aims and objectives, exploiting different teaching approaches and strategies, and determining whether and how to apply technologies in his/her in a new, the so-called smart learning environment The purpose of this article is to explore the role of the teacher of English as a second language in the smart learning environment and discuss benefits and limitations of smart learning environment for the EFL teacher.

**Keywords** Teacher · Smart learning environment · Benefits · Limitations

## 1 Introduction

The present boom of emerging information and communication technologies (ICT) and their effect on all fields of human activities undoubtedly bring significant changes also into the educational system which puts an emphasis on an independent creative potential of a student, integration of different teaching plans, or diversity of teaching methods and strategies. Thus, the traditional learning settings are changing. Table 1 below illustrates how ICT have generally changed the traditional approaches to teaching and learning.

As it can be seen from the table, the teacher is not an authority any more, but it is the student. Nevertheless, the teacher still plays a decisive role in setting the learning aims and objectives, exploiting different teaching approaches and

B. Klimova (✉)
University of Hradec Kralove, Rokitanskeho 62, Hradec Kralove, Czech Republic
e-mail: blanka.klimova@uhk.cz

© Springer International Publishing Switzerland 2016                        51
V.L. Uskov et al. (eds.), *Smart Education and e-Learning 2016*,
Smart Innovation, Systems and Technologies 59,
DOI 10.1007/978-3-319-39690-3_5

**Table 1** Transformation of traditional educational settings (author's own processing) [1]

| Traditional approach consists of | Technology allows more of |
|---|---|
| Teacher-centered learning | Student-centered learning |
| Mass instruction (one size fits all) | Mass customization with instruction to fit individual student needs |
| One pace applies to all students | Flexible pacing based on student abilities |
| Classroom and school building | Distributed learning possible from any place |
| Learning during school hours | Learning at anytime |
| Facts and recitation | Critical thinking in real-world contexts |
| Individual student performance | Collaboration and dialogue among students and between students and teachers |
| Textbooks | Up-to-date primary information resources |
| Parent-teacher meeting each semester | Parent-teacher communication available daily |

strategies, and determining whether and how to apply technologies in his/her in a new, the so-called smart learning environment (SLE) [2], which is perceived as technology-supported learning environment that can make adaptations and provide appropriate support (e.g., guidance, feedback, hints or tools) in the right places and at the right time based on individual learners' needs, which might be determined via analysing their learning behaviours, performance and the online and real-world contexts in which they are situated. There are three essential criteria of SLE which are characterized as minimally *context-aware* (taking into account the learner's situation or the context of the real-world environment in which the learner is located); minimally *supportive* of the learner and his online and real-world contexts; minimally *adaptive* to the user interface (e.g., ways of presenting information), subject contexts to meet the personal factors (e.g., learning preferences), and learning status (e.g., learning performance) [3–5].

Generally, SLE is an environment that is effective, efficient and engaging. In practice it means that it is the environment which can be adapted to learner's needs and personalized instruction and learning support [6].

In addition, teachers in SLE should promote collaboration, identify and help struggling students, stimulate students to learn, and provide feedback to develop students' confidence and satisfaction [7, 8].

This is also true for the teacher of English as a foreign language (EFL). The purpose of this article is to explore the role of EFL teacher in the smart learning environment and discuss benefits and limitations of SLE for the EFL teacher.

## 2 Search of Available Literary Sources

Research has shown that the traditional role of the teacher is changing under the influenced of new ICT. A number of studies exploring the role of the teacher in the new smart learning environment facilitated by ICT in the world's acknowledged databases such as the Web of Science have evidenced this. After entering the words *Teacher* AND *smart learning environment* 129 studies were found. The oldest relevant study dates back to the year of 2004 (Fig. 1).

As Fig. 1 indicates, the year of 2013 was a kind of milestone in the research on smart learning environment. At present SLE seems to be part and parcel of any learning environment. Therefore there has been a decrease in the number of publications in recent years.

The main topics of the studies found in the Web of Science with respect to the researched issue include: smart learning pedagogy [10–14], implementation of smart technologies into teaching and their effectiveness for learning process [15, 16], and smart classrooms/universities [17, 18].

## 3 The Role of EFL Teacher in SLE

Overall, teachers have to take on a number of roles when they teach. Berge [19] identified four main teachers' roles:

- *Pedagogical or intellectual roles.* These are the most important in the BL/eLearning process. The BL/e-tutor uses questions and probes for student responses that focus discussions on critical concepts, principles and skills.

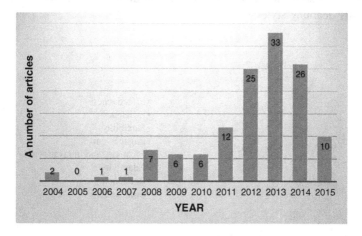

**Fig. 1** An overview of the number of studies comprising the key words *Teacher and smart learning environment* from the Web of Science from 2004 to 2015 (Author's own processing based on the data from the Web of Science) [9]

- *Social roles*. These involve the creation of friendly and comfortable social environments in which students feel that learning is possible.
- *Managerial or organizational roles*. These involve setting learning objectives; negotiating an agenda for learning activities; timetabling learning activities and tasks; clarifying procedural rules and decision-making norms.
- *Technical roles*. These are possibly the most daunting for academics. They involve becoming familiar, comfortable and competent with the ICT systems and software that comprise the eLearning environment. The ICT competences include [20]: identification of learning difficulties at an individual student; a relevant choice of technologies for teaching; verification whether the teaching materials are correct as far as the content and language are concerned; ability to competently use common software applications; and ability to conduct research with the help of ICT.

Based on the definition of SLE, described above, the teacher should consider following several attributes for a successful management of his/her teaching in SLE (Fig. 2).

- Effectivenesss—he should be able to produce acceptable and desirable outcomes for students to satisfy their needs;
- Efficiency—s/he should be cost aware in designing his/her courses;
- Flexibility—s/he should be able to flexibly adjust the goals, methods and materials to students' immediate needs and relevantly react to unpredicted situations;
- Engagement—s/he should be able to motivate his/her students to learn and successfully fulfil their learning aims and objectives;
- Creativity and Innovativeness—s/he should be able to make use of new resources, methods and technologies;
- Adaptivity—s/he should be able to adapt to different situations and different students' learning styles;

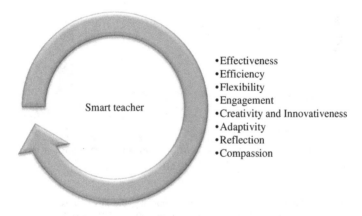

Smart teacher

- Effectiveness
- Efficiency
- Flexibility
- Engagement
- Creativity and Innovativeness
- Adaptivity
- Reflection
- Compassion

**Fig. 2** Teacher's attributes for a successful management of teaching in SLE

- Reflection—s/he should be able to reflect on his/her own teaching in order to remove insufficiencies and improve his/her teaching style;
- Compassion—s/he should be able to understand students' learning difficulties and support them in their further learning.

In addition, the EFL teacher has to be ready to integrate curricula subjects. That means that students use not only their acquired knowledge but also their skills across curricula. In practice students use English medium to study subjects such as geography, history or mathematics that were originally taught in the native language, in this case in Czech. This type of learning is exploited in the so-called Content and Language Integrated Learning (CLIL), which involves teaching a curricular subject through the medium of a foreign language. Diversity is another aspect in the current foreign language teaching, which is particularly seen in the use of different methods which focus on meeting students´ learning needs and styles. The concept of multiculturalism is being promoted. Its aim is to make students more sensitive and open to other nationalities and their culture and consequently to make them aware of their own culture, values and beliefs. The concept of multilingualism is being implemented. This means that besides the native language, each EU citizen should master other two foreign languages [21].

Thus, in this SLE the EFL teachers attempt to make students' learning effective and motivating [22, 23]. This is particularly true for the teachers who are involved in teaching young learners that still need a higher support for their learning and learn best from doing things through play and action in a warm and engaging environment where they feel safe [24]. These teachers exploit multimedia and the Internet as the main technologies for their teaching [25]. Among the most common means of technologies used in the English language classrooms are the following ones:

- interactive whiteboards (IWB);
- CD ROMs, DVDs, videos, or websites sites;
- online reference tools, e.g. online dictionaries or wikis;
- e-mail, and Skype.

Furthermore, in class the teacher can use the Internet to download model lessons, exchange knowledge with other teachers all over the world. SLE also enable easier and faster communication between the teacher and his/her students, or parents [26]. The research [27] also indicates that the use of the traditional paper textbooks is decreasing and the experienced teachers exploit more the Internet sources which are free of charge and the teachers can adapt them to the needs of their students. The most common web pages exploited by the EFL teachers in their classes in the Czech Republic are as follows:

- Youtube.com—it is a website which is widely used by English teachers because it affects most of student's senses and develops all four language skills at a time: listening, reading, writing and speaking.

- TeachingEnglish.org.uk—it is a website which was developed by the British Council and BBC. Besides teacher training, teacher development, exams in English, and various events, this site also serves as a valuable resource for L2 English teachers. It offers plans and activities, completed with worksheets to download, for primary, secondary and adult teachers.
- HelpForEnglish.cz—it is a website developed by a Czech teacher of English. It again focuses on all age levels and offers a great number of teaching resources, such as tests, grammar and vocabulary exercises, pronunciation, reading and listening activities, quizzes, and many more tips.
- BusyTeacher.org—it is another website which supplies ready-made worksheets on different everyday and seasonal topics for English teachers. In addition, it provides ESL (English as a second language) articles, classroom management worksheets, flashcards, classroom posters and other materials. Once again this website covers all age groups.
- ListentoEnglish.com—it is a podcast website for the intermediate and advanced learners of English, mostly aimed at adult learners. The podcasts on this site help to improve English vocabulary, pronunciation and listening skills. They are quite short (5 or 6 min) and delivered in clearly spoken English. Many are linked to grammar and vocabulary notes, exercises or quizzes.

Some publishing houses offer an on-line support for coursebooks. See, for example, FRAUS publishing house and the online support for the courseboook Start with Click New [28]: http://ucebnice.fraus.cz/rozsireni/on-line-podpora-start-with-click-new-2/.

**Fig. 3** An example of teacher's web page [29]

Teachers also try to make some additional materials for their students and make their own web pages in order to provide their students with more practice. See, for example, the following web page for inspiration (Fig. 3).

At university teachers run different eLearning courses which fully replace the traditional contact classes, or they are used in the so-called blended form with the traditional classes, or they serve as supporting courses for further training of the acquired knowledge and skills. The research confirms that these eLearning courses enjoy popularity among students because students can work according to their own pace, test their knowledge and they have a bigger overview of their progress and learning results [30–32].

## 4 Benefits and Limitations of SLE for the EFL Teachers

The research studies show that if SLE is used in foreign language teaching purposefully, it has significant benefits both for the EFL teacher and his/her student. Table 2 below summarizes not only the benefits of SLE for the EFL teacher, but also their limitations.

In this new environment the model of the so-called minimal competences in which teacher's basic activity is to transmit knowledge of his/her own subject is being left. And the model of the so-called broad competences is being established. In this model the teacher's profession is perceived as an expert activity. The teacher is an expert who facilitates learning processes and creates favourable conditions and stimulating opportunities to unlock students' potential [33].

**Table 2** An overview of the key benefits and limitations of SLE for the EFL teachers

| Benefits | Limitations |
|---|---|
| • Direct, fast and easy access to the target language | • Problems with technologies, internet connection |
| • Easier access to current information and abundance of materials | • Short-term memory of students since all the information is within an easy reach |
| • Authenticity of materials | • Sometimes lower quality of materials |
| • Easier modification of teaching materials | • Time-consumption for preparation of technology, materials, and assessments |
| • Fast feedback | • Pressure on teacher's flexibility and adaptability |
| • More frequent contact with his/her students and their parents | • Lower quality of family life |
| • Personalized learning | |
| • Opportunity to store a large amount of materials online | |
| • Lower costs | |

## 5 Conclusion

The findings of this study show that SLE undoubtedly offers a number of benefits for the EFL teachers, however, they require new competences and mastering of new tasks from them. Moreover, they should always consider whether the employment of SLE is relevant and purposeful in order to meet students' learning needs because the success of the exploitation of smart learning environment can be assessed according to what extent it imitates what the smart teacher does [34].

**Acknowledgments** This work is supported by SPEV project titled Impact of Mobile Technologies and Social Networks on the Development and Maintenance of Cognitive Processes run at the Faculty of Informatics and Management in Hradec Kralove.

## References

1. Michigan's State Technology Plan 1998, Michigan Department of Education (1998) www.michigan.gov/documents/Ed_Tech_40666_7.pdf
2. Lin, Y.C., Liu, T.C., Kinshuk: Research on teachers' needs when using e-textbook in teaching. Smart Learn. Env. **2**, 1 (2015)
3. Mikulecky, P.: Smart environments for smart learning. In: Proceedings of the 9th International Scientific Conference on Distance Learning in Applied Informatics, pp. 213–222, Nitra, UKF (2012)
4. Hwang, G.J., Tsai, C.C., Yang, S.J.H.: Criteria, strategies and research issues of context-aware ubiquitous learning. Educ. Technol. Soc. **11**(2), 81–91 (2008)
5. Hwang, G.J.: Definition, framework and research issues of smart learning environments—a context-aware ubiquitous learning perspective. Smart Learn. Env. **1**(4), 1–14 (2014)
6. Spector, J.M.: Conceptualizing the emerging field of smart learning environments. Smart Learn. Env. **1**, 2 (2014)
7. Klimova, B.: Assessment in smart learning environment—A case study approach. In: Uskov, V., Howlett, R.J., Jai, L.C. (eds.) Smart Innovation, Systems and Technologies, vol. 41, pp. 15–24 (2015)
8. Klimova, B., Simonova, I:. Study materials in smart leasing environment—A comparative study. In: Uskov, v., Howlett, R.J., Jai, L.C. (eds.) Smart Innovation, Systems and Technologies, vol. 41, pp. 81–92 (2015)
9. Web of Science, http://apps.webofknowledge.com/Search.do?product=UA&SID=V126d51KIqAl6uirWT2&search_mode=GeneralSearch&prID=0244f1cb-0a04–44bc-972b-c9eb186a309c (2015)
10. Cervera, M.G., Johnson, L.: Education and technology: new learning environments from a transformative perspective. Rusc-Univ. Knowl. Soc. J. **12**(2), 1–13 (2015)
11. Ha, I., Kim, C.: The research trends and the effectiveness of smart learning. Int. J. Distrib. Sensor Net. 2014(537346) (2014)
12. Jun, W., Hong, S.K.: A study on development of smart literacy standards of teachers and students in smart learning environments. J. Korean Soc. Internet Inf. **14**(6), 59–70 (2013)
13. Kim, J., Maeng, J.H.: Research of the smart education effect through students camp and teacher's training. In: Proceedings of the 1st International Mega Conference on Green and Smart Technology, vol. 338, pp. 144–150 (2012)
14. Peng, Y.: How to be a quality teacher of EFL in the new era. Soc. Sci. Soc. **7**, 58–61 (2012)

15. Heslop, P., et al.: Evaluating digital tabletop collaborative writing in the classroom. In: Human-Computer Interaction—INTERACT 2015. Lecture Notes in Computer Sciences, vol. 9297, pp. 531–548 (2015)
16. Zhang, T., Lu, S., Zhang, Z., et al.: Web-based collaboration system to improve the interactivity for mobile education through smart devices. In: Proceedings of 2013 IEEE Frontiers in Education Conference (2013)
17. Coccoli, M., Guercio, A., Maresca, P., Stanganelli, L.: Smart universities: a vision for the fast changing digital era. J. Vis. Lang. Comput. 25(6), 1003–1011 (2014)
18. Lui, M., Slotta, J.D.: Immersive simulations for smart classrooms: Exploring evolutionary concepts in secondary science. Technol. Pedagogy Educ. 23(1), 57–80 (2014)
19. Berge, Z.L.: The role of the moderator in a Scholarly Discussion Group (SDG), http://www.emoderators.com/moderators/zlbmod.html (1992)
20. Fitzpatrick, A.: Information and communication technology in foreign language teaching and learning—An overview. In: Information and Communication Technologies in the Teaching and Learning of Foreign Languages: State-of-the-Art, Needs and Perspectives, UNESCO Institute for Information Technologies in Education, Moscow, pp. 10–26 (2004)
21. Klimova, B.F.: Teaching Formal Written English. UHK, Gaudeamus (2012)
22. Yang, S.C.: Integrating computer-mediated tools into the language curriculum. J. Comput. Assist. Learn. 17, 85–93 (2001)
23. Young, S.S.C.: Integrating ICT into second language education in a vocational high school. J. Comput. Assist. Learn. 19, 447–461 (2003)
24. Harmer, J.: Essential teacher knowledge, core concepts in english language teaching. Pearson Education Limited, England (2012)
25. Warchauer, M.: CALL for the 21st century. In: IATEFL and ESADE Conference, Spain, Barcelona, http://www.gse.uci.edu/person/warschauer_m/docs/future-of-CALL.pdf (2000)
26. Ahn. D.: ICT and English education in the global age. Korean J. General Educ. 7(4), 501–527 (2013)
27. Allen, C.: Marriages of convenience? Teachers and coursebooks in the digital age. ELT J. 69(3), 249–263 (2015)
28. Karaskova, M., Sadek, J.: Start with click new 1. Fraus, Plzeň (2007)
29. An example of teacher's web page, http://www.katerinapeskova.cz/ (2015)
30. Klimova, F.B.: Traditional versus online teaching and learning, In: Proceedings of the 11th International Conference: Efficiency and Responsibility in Education 2014, pp. 132–138. Czech University of Life Sciences, Prague (2014)
31. Klimova, B., Poulova, P. Blended learning as a compromise in the teaching of foreign languages. In: Proceedings of the 13th European Conference on e-Learning (ECEL-2014), pp. 181–187. Academic Conferences and Publishing International Limited, UK, Reading (2014)
32. Psaroudaki, S., Mckay, A. Enhancing English language learning through ICT. In: Proceedings of the 7th European Conference on E-Learning, pp. 322–329, University of Cyprus, Cyprus (2008)
33. Semradova, I.: Nejvyznamnejsi konstanty pojeti ucitelske profese. In: Pelcova, N., Semradova, I. (eds.) Fenomen vychovy a etika ucitelskeho povolani. Karolinum, Praha (2014)
34. Clark, R.E.: Confounding in educational computing research. J. Educ. Comput. Res. 1(2), 445–460 (1985)

# Toward Smart Value Co-education

Vincenzo Loia, Gennaro Maione, Aurelio Tommasetti, Carlo Torre,
Orlando Troisi and Antonio Botti

**Abstract** The current environmental context, highly competitive and turbulent, has shifted the focus of scholars and managers on forms of cooperation and participation able to ensure a timely and effective response to needs of who participate in value creation processes. The paper aims to open the way to new perspectives of analysis of educational context, enabling to understand how Value Co-creation is moving emergence of a new phenomenon, Smart Value Co-education, which integrates the main and distinctive towards a markedly smart education. This suggests the elements of three different but related approaches: Value Co-creation, Co-education and Smart Education. The work also offers some insights for future researches on Smart Value Co-education, suggesting to investigate, on one hand, users' role and their active involvement for a better use of educational experience and, on the other, the factors unpredictably and rapidly influencing the emergence and development of new technologies for the dissemination of education.

**Keywords** Value Co-creation · Co-education · Smart education · Smart Value Co-education

---

V. Loia · G. Maione (✉) · A. Tommasetti · C. Torre · O. Troisi · A. Botti
Department of Business Science - Management & Information System,
University of Salerno, Fisciano, Italy
e-mail: gmaione@unisa.it

V. Loia
e-mail: loia@unisa.it

A. Tommasetti
e-mail: rettore@unisa.it

C. Torre
e-mail: ctorre@unisa.it

O. Troisi
e-mail: otroisi@unisa.it

A. Botti
e-mail: abotti@unisa.it

© Springer International Publishing Switzerland 2016
V.L. Uskov et al. (eds.), *Smart Education and e-Learning 2016*,
Smart Innovation, Systems and Technologies 59,
DOI 10.1007/978-3-319-39690-3_6

# 1    Introduction and Purposes of the Paper

The current environmental context, highly competitive and turbulent, has shifted the focus of scholars and managers on forms of cooperation and participation able to ensure a timely and effective response to needs of who participate in value creation processes. With reference to this, the work aims, on one side, to highlight the elements facilitating educational phenomenon (generated by the creation of relational and collaborative networks) and, on the other side, to propose a theoretical model allowing to reread Value Co-creation in educational context. The paper is divided into two sections. In a first phase, starting from an analysis of literature related to Value Co-creation, it pays attention to the role played by technology and users' participation in smart education. In the second section, the work aims to identify the pillars of a new phenomenon in vogue nowadays: Smart Value Co-education. As specified subsequently, it considers the output of users' interaction as a value, unique and total, created thanks to all actors' resources sharing, suggesting that value is always generated by actors' involvement in the educational process.

# 2    From Value Co-creation to Smart Education

## 2.1    Value Co-creation

Until the end of last century, value was considered as something to create for users, seen as passive recipients of the products (goods/services) offered to them. Recently, however, organizations, both public and private, have gradually become aware of the benefits arising from the involvement of users in the process of creating value [1, 2]. This tendency not to treat user as an inert recipient has encouraged the emergence and subsequent spread of a new approach, known as Value Co-creation and consisting in the benefit deriving from a series of activities and interactions with the involved actors [3, 4]. According to this approach, organizations cannot independently generate value, but they can make a value proposition and only if and when it is accepted by users, it co-creates value [5–7]. The continuous environmental changes are increasingly highlighting the need of organizations to combine different types of resources in order to create value. This integration of resources fosters the dissemination of solutions that satisfy the interests of all stakeholders, increasing the chance to generate a total and unitary value [8–12]. Over time, the importance recognized to cooperation has played an increasingly central role in the value creation process. The association of the two aforementioned concepts, cooperation and value creation, provides an even greater contribution if studied from the point of view of long-term benefits. In this perspective, the value Co-creation can be also considered as the key element of the Win-Win Logic [13], which implies a gain thanks to the mutuality of the value creation process in which several actors are involved. Therefore, value is conjointly and mutually co-created by means the combined use of the resources put into the process by a plurality of actors [14–16]. This perspective, thus, emphasizes the advantage resulting from a cooperative behavior of who participate in the creation of value. Organizations, hence, should stop considering individuals as mere passive recipients of value and start considering them as active value co-creators. This allows, on one hand, to overcome the traditional distinction between service provider and user [17–20] and, on the other, to consider the relationships among the various actors as drivers able to ensure the competitiveness and survival of organizations [21–23]. On the role played by the relational component and the involvement of all stakeholders in value creation process, several scholars have focused their attention [24, 25], fostering the birth of a recently but well established stream of research. It is known as Service-Dominant Logic and is based on the concept of Value Co-creation [26]. Table 1 shows the "Foundational Premises" of S-D Logic.

## 2.2 Co-creating Education

Although Value Co-creation was born in marketing studies, it seems to be at the foundation of both economic and social theories [23]. In recent years, for example, the great importance of the interaction among the actors involved in the same value creation process has been recognized in the educational context [27, 28]. Several authors, indeed, have dedicated their studies to the analysis of the value generated by the sharing of a plurality of actors in educational phenomenon [29], aimed at fostering a rejuvenation of institutions and a more rapid and extensive dissemination of culture. This "new" trend [30, 31], based on Value Co-creation in educational context (Co-education), has been facilitated by ICTs, which, encouraging participation, interaction and a kind of collective creativity, have addressed cultural offer toward new cooperation mechanisms. The availability of increasingly powerful tools supporting users' participation, in fact, has contributed to eliminate space and time limits that, until a few years ago, have characterized the educational phenomenon [32]. In this regard, Simon [33] argues that the generation and distribution of value in the educational process are strongly encouraged by the existence of interactive platforms, able to ensure a virtuous dialogue among service users and providers. Co-education, therefore, can be considered as an interactive process, which, thanks to users' participation (distinctive element of Value Co-creation) and the use of appropriate ICTs, fosters an easier information and knowledge sharing among suppliers and users education services.

## 2.3 Making Education "Smart"

In recent years, the rapid technological evolution has given a further impetus to the educational phenomenon, fostering the birth of *smart education* [34]. In particular, the crucial role played by customizable and ubiquitous technology has been widely

**Table 1** Service-dominant logic foundational premises (*Source* [1])

| No. | Original foundational premise | Modified/new foundational premise | Comment/explanation |
|---|---|---|---|
| (1) | The application of specialized skills and knowledge is the fundamental unit of exchange | Service is the fundamental basis of exchange | Service, considered as the application of knowledge and skills (operant resources), is the basis for all exchange. Service is exchanged for service |
| (2) | Indirect exchange masks the fundamental unit of exchange | Indirect exchange masks the fundamental basis of exchange | The service basis of exchange is not always apparent because service is provided through complex combinations of goods, money and institutions |
| (3) | Goods are a distribution mechanism for service provision | Goods are a distribution mechanism for service provision | Durable and non-durable goods derive their value through use—service they provide |
| (4) | Knowledge is the fundamental source of competitive advantage | Operant resources are the fundamental source of competitive advantage | Competition is driven by the comparative ability to cause desired change |
| (5) | All economies are services economies | All economies are service economies | Service is only now becoming more apparent with increased specialization and outsourcing |
| (6) | The customer is always a co-producer | The customer is always a co-creator of value | Value creation is interactional |
| (7) | The enterprise can only make value propositions | The enterprise cannot deliver value, but only offer value propositions | Enterprises can offer their applied resources for value creation and collaboratively create value following acceptance of value propositions, but they cannot create and/or deliver value independently |
| (8) | A service-centered view is customer oriented and relational | A service-centered view is inherently customer oriented and relational | Service is inherently customer oriented and relational because it is defined in terms of customer-determined benefit and co-created |
| (9) | Organizations exist to integrate and transform micro-specialized competences into complex services that are demanded in the marketplace | All social and economic actors are resource integrators | The context of value creation is networks of networks (resource integrators) |
| (10) | | Value is always uniquely and phenomenologically determined by the beneficiary | Value is idiosyncratic, experiential and contextual |

recognized [35]. Customizability is an important feature because it allows the user to choose the desired or needed topics [36]. This attribute ensures a great efficiency in the research and use of contents, allowing to save cognitive, timing and economic resources.

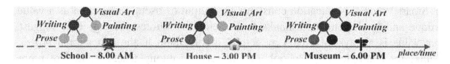

**Fig. 1** Smart education (*Source* [41])

Ubiquity, instead, allows people to enjoy the desired or needed contents anytime and anywhere [37], potentially overcoming classical education limits. So, ubiquitous environments are settings for pervasive education, where educational experience "immerse" users who are involved in formal as well as non-formal activities [38]. However, the ubiquity of the educational phenomenon, in the face of strong advantages, is an attribute requiring the design and implementation of appropriate technologies, able to foster texts multimediality, information sharing, spread of social networks, etc. [39–41]

Figure 1 shows a representation of smart education, fostered by the enjoyment of customizable and ubiquitous contents: customization is caused by user's possibility to select only the interesting content (prose, writing, visual art and/or painting); ubiquity, instead, is determined by user's opportunity to use the content at the time (morning, afternoon, evening) and in the place (school, home, museum) he or she desires.

## 3 Toward Smart Value Co-education

According to Yeo [42], a quality educational experience ensures a common growth of all the actors involved in the mutual value Co-creation and exchange process. Each person, indeed, is actively involved in each stage of educational phenomenon [43]. In light of this consideration, Bowden and D'Alessandro [44] argue that the value exchange activity is dyadic and, thus, oriented in two directions. However, Gummesson [13] considers the value Co-creation as a polyadic process, since it develops using all actors' resources [45]. All educational organizations, regardless of the sector to which they belong, consider the satisfaction-creating service experiences as important drivers to reach and/or maintain a competitive advantage. In order to achieve this goal, educational institutions should direct their activities towards the development of services capable to provide timely responses to users' needs [46]. The role of educational organizations, therefore, is not to offer a service to inert recipients, but to provide important partial inputs into the customers' experience [47], enabling them to contribute to value creation [48] and facilitating the learning control [49]. According to the aforementioned statements, the paper opens the way to new perspectives of analysis of educational context, enabling to understand how Value Co-creation is moving towards a markedly smart education [50]. This suggests the emergence of a new phenomenon, *Smart Value Co-education*, which contains and integrates the main and distinctive elements of three different but related approaches: Value Co-creation, Co-education and Smart Education (see Fig. 1).

Smart Value Co-education considers the output of users' interaction as a value, unique and total, generated thanks to all actors' resources sharing. This suggests that value is always generated by actors' involvement in the educational process (think, for example, of a school: the interaction among teachers, student representatives and students allows to fit the didactic offer to educational needs). This last statement, in turn, highlights that *every actor involved in the educational process (service provider or user) is as important* (considering the example above, it is clear that cognitive resources brought by each student, teacher or any other involved person is accorded the same weight for the definition of a more adequate didactic offer). It follows that *service provider can only make a value proposition,*

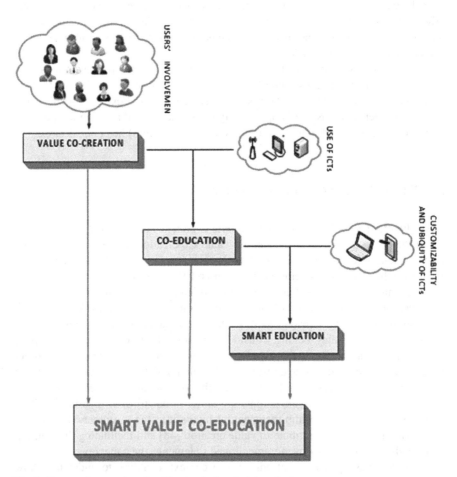

**Fig. 2** Smart value co-education (*Source* Authors' elaboration)

*which must be accepted by user so that it is possible to co-create value* (the educational offer, therefore, must be shared and accepted by students, their representatives and teachers so that it can offer a generate benefit).

From the phenomenon of Co-education, moreover, Smart Value Co-education draws on the importance of the role played by technological and human resources (users) in order to encourage the establishment and development of the educational process: it seems clear that *the interactions among the actors and the use of ICTs foster the activation and development of Co-Education processes* (think, for example, of the contribution offered by the use of slide projectors during a lesson: the projection of images in the classroom produces a visual impact helping to keep high students' attention, with a consequent benefit both for them and for teachers).

As smart education, finally, Smart Value Co-education considers contents personalization and ubiquity two crucial features, without which it would appear difficult to outline paths of development able to lead society towards a smart education. This consideration allows to believe that *Smart Value Co-Education processes presuppose the use of the only technologies ensuring educational contents personalization and ubiquity* (think, for example, of the possibility for students to choose and modify some of the contributions offered by teachers, enjoy them anytime and anywhere; it would be a great advantage both for them and for the institution as a whole) (Fig. 2).

The following figure summarizes the "five pillars" of Smart Value Co-education (Fig. 3):

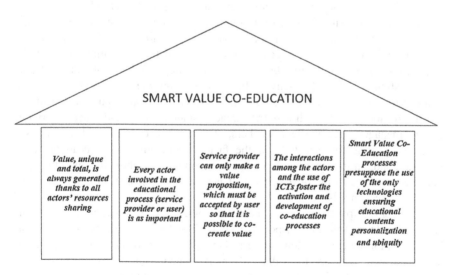

**Fig. 3** The pillars of smart value co-education (*Source* Authors' elaboration)

## 4   Conclusion and Future Works

The study shows that an appropriate use of ICTs facilitates the emergence and subsequent spread of networks and collaborative relationships in education [51–54], identifying some of the key factors that allow to reread Value Co-creation concepts in educational context. In particular, the paper highlights the main elements characterizing the conceptual change toward Smart Value Co-education: people's active participation, integrated use of all involved actors' resources, users' acceptance of suppliers' value proposition and use of technologies able to offer personalized and ubiquitous contents. Only in presence of these elements, it is possible to suppose a value Co-creation in educational context. As previously indicated (see Fig. 1), the above-mentioned elements are the characterizing attributes of three separate but related approaches: Value Co-creation, Co-education and Smart Education. This is the reason for which Smart Value Co-education is considerable as the result of their combined integration. The work also offers some insights for future researches on Smart Value Co-education, suggesting to investigate, on one hand, users' role and their active involvement for a better use of educational experience and, on the other, the factors which, unpredictably and rapidly, affect the emergence and development of new technologies for the dissemination of education. In this regard, the work highlights the great contribution that all actors involved in the co-creative processes give to the overall value creation in educational context. The understanding of this logic is considerable as the first step to head to a smart use of ICTs capable to support the provision of the educational service. In fact, in the current socio-economic scenario, strongly characterized by the rapid development and wide diffusion of artificial intelligence tools, it is neither possible nor appropriate to ignore the impact that a smart use of ICTs can produce on both individual users of educational services and the community as a whole. This consideration lays the basis for a holistic treatment of Smart Value Co-education, which makes sense only by considering the benefits arising from a value shared by all involved stakeholders. Therefore, the paper could be considered as a useful tool for both scholars (researchers, students, etc.) and practitioners (art gallery managers, headmasters, teachers, etc.). In fact, on one hand, it acts as a magnifying glass for the study of the factors stimulating the educational phenomenon and, on the other hand, it helps to become aware about the opportunity to allocate resources to the development of new ICTs capable to timely respond to changing users' needs.

However, the paper presents a main limit: it deals with Smart Value Co- Education only from a theoretical point of view. Therefore, it could be interesting to identify, in further contributions, empirical evidences able to allow to test, also from a practical viewpoint, the five pillars of Smart Value Co-education. For instance, a sample survey (through questionnaires or interviews) could be made by taking into account all typically involved actors (suppliers and users) in an educational context, such as a school (teachers, student representatives, students, etc.) or museum (visitors, director, administration manager, communications manager, librarian, etc.). Only through a careful evaluation of respondents' answers, indeed, it is possible to verify the presence of all the elements characterizing Smart Value Co-education.

# References

1. Lusch, R.F., Vargo, S.L., Malter, A.J.: Marketing as service-exchange: taking a leadership role in global marketing management. Org. Dyn. **35**, 264–278 (2006)
2. Vargo, S.L., Lusch, R.F.: Service-dominant logic: continuing the evolution. J. Acad. Mark. Sci. **36**, 1–10 (2008)
3. Lusch, R.F., Vargo, S.L.: Service-dominant logic: reactions, reflections and refinements. Market. Theory **6**, 281–288 (2006)
4. Lusch, R.F., Vargo, S.L.: The Service-Dominant Logic of Marketing: Dialog, Debate, and Directions. Routledge (2014)
5. Akaka, M.A., Vargo, S.L., Lusch, R.F.: The complexity of context: a service ecosystems approach for international marketing. J. Mark. Res. **21**, 1–20 (2013)
6. Tommasetti, A., Vesci, M., Troisi, O.: The internet of things and value Co-creation in a service-dominant logic perspective. In: Data Management in Pervasive Systems, pp. 3–18. Springer International Publishing (2015)
7. Akaka, M.A., Vargo, S.L.: Technology as an operant resource in service (eco) systems. IseB **12**, 367–384 (2014)
8. Brady, T., Davies, A., Gann, D.M.: Creating value by delivering integrated solutions. Int. J. Project Manage. **23**, 360–365 (2005)
9. Brax, S.A., Jonsson, K.: Developing integrated solution offerings for remote diagnostics: a comparative case study of two manufacturers. Int. J. Oper. Prod. Manage. **29**, 539–560 (2009)
10. Davies, A., Brady, T., Hobday, M.: Organizing for solutions: systems seller versus systems integrator. Ind. Mark. Manage. **36**, 183–193 (2007)
11. Nordin, F., Kowalkowski, C.: Solutions offerings: a critical review and reconceptualization. J. Serv. Manage. **21**, 441–459 (2010)
12. Tuli, K.R., Kohli, A.K., Bharadwaj, S.G.: Rethinking customer solutions: from product bundles to relational processes. J. Mark. **71**, 1–17 (2007)
13. Gummesson, E.: Extending the service-dominant logic: from customer centricity to balanced centricity. J. Acad. Mark. Sci. **36**, 15–17 (2008)
14. Nenonen, S., Storbacka, K.: Business model design: conceptualizing networked value co-creation. Int. J. Qual. Serv. Sci. **2**, 43–59 (2010)
15. Chandler, J.D., Lusch, R.F.: Service systems: a broadened framework and research agenda on value propositions, engagement, and service experience. J. Serv. Res. **18**, 6–22 (2015)
16. Löbler, H., Lusch, R.F.: Signs and practices as resources in it-related service innovation. Serv. Sci. **6**, 190–205 (2014)
17. Ramaswamy, V.: Leading the transformation to Co-creation of value. Strat. Leadersh. **37**, 32–37 (2009)
18. Tommasetti, A., Botti, A., Troisi, O., Vesci, M.: Customer satisfaction, commitment, loyalty ed implicazioni nella governance delle università: una ricerca esplorativa. Azienda Pubblica. **3**, 219–242 (2014)
19. Merz, M.A., He, Y., Vargo, S.L.: The evolving brand logic: a service-dominant logic perspective. J. Acad. Mark. Sci. **37**, 328–344 (2009)
20. Wieland, H., Koskela-Huotari, K., Vargo, S.L.: Extending actor participation in value creation: an institutional view. J. Strat. Mark. 1–17 (2015)
21. Barile, S., Polese, F.: Smart service systems and viable service systems: applying systems theory to service science. Serv. Sci. **2**, 21–40 (2010)

22. Bettencourt, L.A., Lusch, R.F., Vargo, S.L.: A service lens on value creation: marketing's role in achieving strategic advantage. Calif. Manage. Rev. **57** (2014)
23. Vargo, S.L., Lusch, R.F.: Evolving to a new dominant logic for marketing. J. Mark. **68**, 1–17 (2004)
24. Pellicano, M.: Il governo delle relazioni nei sistemi vitali socioeconomici: imprese, reti e territori. G. Giappichelli (2002)
25. Pellicano, M.: L'impresa relazionale. Il governo strategico dell'impresa. Giappichelli, Torino (2004)
26. Vargo, S.L., Maglio, P.P., Akaka, M.A.: On value and value Co-creation: A service systems and service logic perspective. Eur. Manage. J. **26**, 145–152 (2008)
27. Finnis, J.: Turning cultural websites inside out. changes in online user behaviour, web 2.0 and the issues for the cultural sector. In: Uzelac, A.; Cvjeticanin, B., (eds.) Digital Culture. The Changing Dynamics. Culturelink Joint Publications Series, vol. 12, pp. 151–165 (2008)
28. Tallon, W.: Digital Technologies and the Museum Experience. AltaMira Press, New York (2008)
29. Bakhshi, T.: New technologies in cultural institutions: theory, evidence and policy implications. Int. J. Cult. Policy. 1–18 (2011)
30. Vom, L.H.: Accounting for new technology in museum exhibitions. Int. J. Arts Manage. **7**, 11–21 (2005)
31. Holmberg, K., Isto, H., Kronqvist-Berg, M., Widén-Wulff, G.: What is Library 2.0? J. Doc. **65**, 668–681 (2009)
32. Medak, T.: Transformations of cultural production, free culture and the future of the internet. In: Digital Culture: The Changing Dynamics, pp. 59–69. Zagreb, Culturelink/UNESCO (2008)
33. Simon, N.: The Participatory Museum. Santa Cruz (2010)
34. Rothman, R.: City schools: How districts and communities can create smart education systems. Harvard Education Press, Cambridge (2007)
35. Graham, C.: Rethinking Curating Art after New Media. The MIT Press, Cambridge (2010)
36. Mangione, G.R., Anna, P., Salerno, S.: A model for generating personalized learning experiences. Int. J. Technol. Enhanced Learn. **1**, 314–326 (2009)
37. Kameyama, W.: Ubiquity and content/image information processing. In: J. Inst. Imag. Inf. Telev. Eng. (Kyokai Joho Imeji Zasshi) **59**, 21–26 (2005)
38. Capuano, N., Gaeta, M., Orciuoli, F., Ritrovato, P.: On-demand construction of personalized learning experiences using semantic web and web 2.0 techniques. In: Ninth IEEE International Conference on Advanced Learning Technologies. ICALT, pp. 484–488 (2009)
39. Liang, T.P., Lai, H.J., Ku, Y.C.: Personalized content recommendation and user satisfaction: theoretical synthesis and empirical findings. J. Manage. Inf. Syst. **23**, 45–70 (2006)
40. Gaeta, M., Loia, V., Mangione, G.R., Orciuoli, F., Ritrovato, P., Salerno, S.: A methodology and an authoring tool for creating complex learning objects to support interactive storytelling. Comput. Hum. Behav. **31**, 620–637 (2014)
41. D'Aniello G., Granito A., Mangione G.R., Miranda S., Orciuoli F., Ritrovato P., Rossi P.G.: a city-scale situation-aware adaptive learning system. In: IEEE Computer Society 14th International Conference on Advanced Learning Technologies (ICALT), pp. 136–137, Athens, Greece (2014)
42. Yeo, R.: Service quality ideals in a competitive tertiary environment. Int. J. Educ. Res. **48**, 62–76 (2009)
43. Binsardi, A., Ekwulugo, F.: International marketing of british education and research on students perception and the uk market penetration. Mark. Intell. Plan. **25**, 318–327 (2003)
44. Bowden, J.L., D'Alessandro, S.: Co-creating value in higher education: the role of interactive classroom response technologies. Asian Soc. Sci. **7**, 35 (2011)
45. Mele, C., Russo, S.T., Colurcio, M.: Co-creating value innovation through resource integration. Int. J. Qual. Serv. Sci. **2**, 60–78 (2010)

46. Hemsley-Brown, J., Oplatka, I.: Universities in a competitive global marketplace:a systematic review of the literature on higher education marketing. Int. J. Public Sector Manag. **19**, 316–338 (2006)
47. McColl-Kennedy, J., Vargo, S., Dagger, T., Sweeney, J.: Customers as resources integrators: styles of customer Co-creation. In: Proceedings of the Naples Forum on Services: Service Dominant Logic, Service Science and Network Theory conference, Capri (2009
48. Carvalho, S., de Oliveira Mota, M.: The role of trust in creating value and student loyalty in relational exchanges between higher education institutions and their students. J. Mark. Higher Educ. **20**, 145–165 (2010)
49. Anderson, T.: Getting the mix right again: An update and theoretical rationale for interaction. Int. Rev. Res. Open Distance Learn. **4**, 1492–3831 (2003)
50. Salah, A.M., Lela, M., Al-Zubaidy, S.: Smart education environment system. Comput. Sci. Telecommun. **44** (2014)
51. Jeong, J.S., Kim, M., Yoo, K.H.: A content oriented smart education system based on cloud computing. Int. J. Multimedia and Ubiquitous Eng. **8**, 313–328 (2013)
52. Morze, N.V., Glazunova, O.G., Grinchenko, B.: What should be e-learning course for smart education. In ICTERI, pp. 411–423 (2013)
53. De Maio, C., Fenza, G., Loia, V., Tommasetti, A., Troisi, O., Vesci, M.: Management and Computer Science Synergies: a theoretical framework for context sensitive simulation environment. In: Transactions on Computational Collective Intelligence XIX, pp. 1–16. Springer Berlin Heidelberg (2015)
54. Tommasetti, A., Troisi, O., Vesci, M.: Customer value Co-creation: a conceptual measurement model in a service dominant logic perspective. In: Naples Forum on Service (2015)

# Smart University Management Based on Process Approach and IT-Standards

**Boris Pozdneev, Filip Busina and Alexander Ivannikov**

**Abstract** The document describes state of art approaches, best practices and standards for the development of the Smart University Management System. It defines fundamental meaning of the Strategic Thinking, Risk Management and Knowledge Management in connection with the new version of the standard ISO 9001:2015. It presents a process model of the university system management that accounts for the specific features of the development of the system of electronic learning and forming of integrated Smart University environment based on fundamental IT standards and the set of standards (Information Technology for Learning, Education and Training). The implementation of the presented approach guarantees the development of the Smart University as an important component of the Smart Society.

**Keywords** Smart University · Management · Administration · Management system · Quality management system · Process approach · Standards · Information technology · e-Learning

B. Pozdneev (✉) · F. Busina
Federal State Budget Educational Institute of Higher Education,
Moscow State University of Technology «STANKIN», Moscow,
Russian Federation
e-mail: bmp@stankin.ru

F. Busina
e-mail: filipbusina@seznam.cz

A. Ivannikov
Federal State-Funded Institute of Science, Institute for Design Problems
in Microelectronics, Russian Academy of Sciences, Moscow, Russian Federation
e-mail: adi@ippm.ru

© Springer International Publishing Switzerland 2016
V.L. Uskov et al. (eds.), *Smart Education and e-Learning 2016*,
Smart Innovation, Systems and Technologies 59,
DOI 10.1007/978-3-319-39690-3_7

# 1  Introduction

The development of information and communication technologies in combination with information internet based resources became basis for the development of information society and transition to a qualitatively new level in all spheres of human activity—economy, business, industry, environment protection, science, education etc. The current development phase is associated with new possibilities for the development, modification and distribution of knowledge that are precondition for the transition to a knowledge based society—Smart Society. Knowledge now starts to be of key importance for the development of the world community, competitiveness of national economies, industries, organisations and each human being separately. This trend is particularly apparent in the field of science and education as knowledge in particular is the basis of innovative development in the Smart Society. Based on this trend and in the context of aging population and demographic projections for the next decades the civic society as well as large companies invest into life-long learning supported by ICT Dvorakova and Langhamrova [1]. Taking into account these presumptions, new education paradigm is being developed that can be characterise by new terms—Smart Education and e-Learning [2–5].

At the time being, the e-Learning industry keeps developing in many countries. It affects principles of the educational system and educational organisations (universities, colleges, schools), creation of the information and educational environment and electronic learning resources. In combination with IT support means (electronic boards, multimedia projectors, virtual labs, electronic manuals etc.), new approaches to education and process organisations, learning technologies, pedagogic design support means and principles of the interaction of knowledge holders (teachers), managers (tutors) and students are subject to development. The need to provide for the quality of new e-Learning approaches and technologies made it inevitable to standardise the important area of activity in connection with best practices [4–12].

The article deals with detailed review of current approaches to the development of the Smart University Management Systems based on standards in the field of quality management, risk management, information technologies and information technology for learning, education and training (ITLET). In terms of the development, the Smart Society—Smart University is one of key components as the core types of university activities include scientific research and learning. Knowledge resulting from research is then used in the learning activity and it gets transformed into the intellectual potential of the society—new knowledge of highly qualified human resources. In connection with new technical and technological solutions, the scientific and educational activities of leading universities play an important role in the innovative development of the Smart Society. In this context, the improvement of the university management system based on fundamental standards and best world practices will support the improvement of the efficiency of their activities, unification of information and technological interaction and more important role of universities in the Smart Society [13–16].

## 2 Smart University Management Standards

The creation of the University Management System has to include the development of electronic information and educational environment, educational process and electronic content management systems, information means and support means and systems and automation of the management process in all core types of activity for the safeguarding of the integrated management in connection with needs of next series of international standards: Quality management (ISO 9000), Environment management (ISO 14000), Information technologies for learning, education and training (ITLET); Service management (ISO/IEC 20000), Information security management (ISO/IEC 27000), Risk management (ISO/IEC 31000), Assets management (ISO 55000). At the time being, new international standards are in the process of development including such fields of activity as knowledge management, strategic and innovative management.

The management system core is the quality management system (QMS) that is basis for the sustainable development and continuous improvement of university activities as a whole. The new version of the international standard ISO 9001:2015 "Quality Management System. Requirements" focuses on integrated management and it defines complementary requirements to the role of the higher management, risk management and knowledge management. See the Fig. 1 for the development of requirements of the standard ISO 9001.

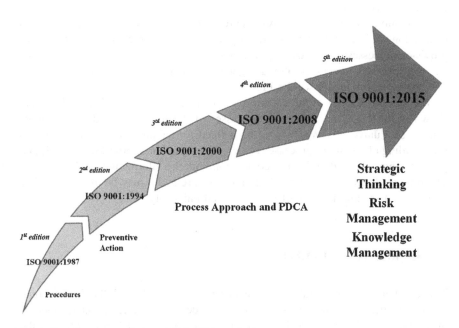

**Fig. 1** Requirements evolution of ISO 9001 standard

In 2013, ISO/IEC founded the Project Committee 288 "Educational organizations management systems. Requirements with guidance for use" that unified the area of coverage of ISO/IEC JTC1/SC36, ISO TC 176 "Quality management and quality assurance" and ISO TC 232 "Learning services outside formal education". At the time being, the above Project Committee deals with the standard ISO/IEC 21001 "Educational organization management systems. Requirements with guidance for use". The standard ISO/IEC 21001 includes basic provisions of the project of the standard ISO/IEC 36000 "Information technology. Learning, education, and training. Quality for learning, education and training. Fundamentals and reference framework" that features a version that modifies the standard ISO/IEC 19796-1.

In connection with requirements of the standard ISO/IEC 19796-1, the development of the process oriented model should be broken in four phases. The phase 1 will be implemented on the basis of the unification of current approaches to the management and it includes the strategic management, general management, quality management, risk management, knowledge management etc. The phase 2 will analyse approaches to the management for the development of the conceptual model and profile of requirements. The phase 3 will include the development of harmonised structure and general model of the description of processes that have to be implemented on the basis of the unification of approaches to the management and process models for two fundamental standards: ISO 9001 and ISO/IEC 19796-1. What will be done in the framework of the phase 4 is the localisation and adaptation of the process oriented model for the implementation of the integrated university management system.

Basic university management system processes may be divided to four groups in connection with ISO 9001: 1—Activity management processes; 2—Resources provision processes; 3—e-Learning life cycle processes; 4—Measurement, analysis and improvement processes. For the general structure of the harmonised smart university process model see the Fig. 2.

The specific features of the e-Learning in connection with ISO/IEC 19796-1 are shown in the Fig. 3 and show the processes of the life cycle of the educational activity of a university. What has to be defined for the effective management and safeguarding of quality of processes is the interconnection of processes (Fig. 4) and processes have to be described in connection with the information model (Fig. 5). The presented process model is of reference nature and it may be modified taking into account specific features of the concerned university.

## 3  Standards for ITLET

The International Organization for Standardization (ISO) and the International Electrotechnical Commission (IEC) are jointly developing international standards in the field of information technologies in the framework of the activities of the

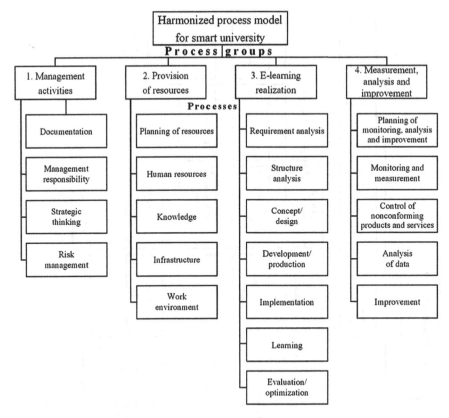

**Fig. 2** Harmonized smart university process model (ISO 9001, ISO IEC 19796-1)

Joint Technical Committee 1 (JTC1). As a part of this committee, the Subcommittee 36 (SC36) was established in 1999 for "Information Technologies for Learning, Education and Training" (ITLET). At the time being, representatives of 45 countries take part in the activities of the SC36. They are divided to seven task forces and they work on the development of international standards in the field of terminology, learning technologies, content management, e-Learning quality management etc. The functions of the permanent national working body of the Russian Federation for the ISO/IEC JTC1/SC36 are fulfilled by TC 461 "Information and Communication Technologies in the Education" (ICTE) established by GOST R in 2004 that consists of more than 100 highly qualified experts from educational and scientific research facilities, leading Russian IT companies and other organisations involved in the process who are divided to six

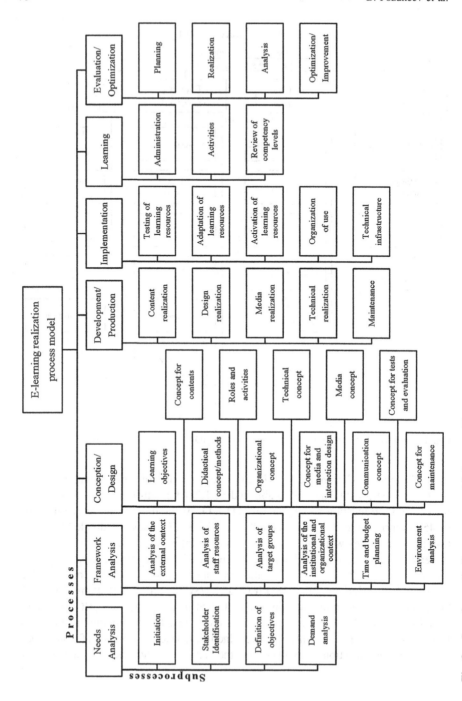

**Fig. 3** e-Learning realization process model decomposition

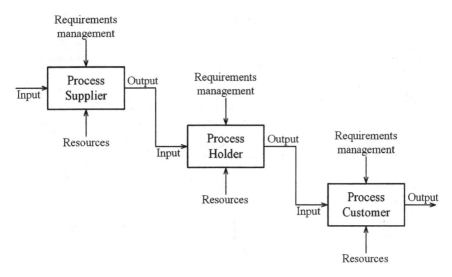

**Fig. 4** Interconnection of processes

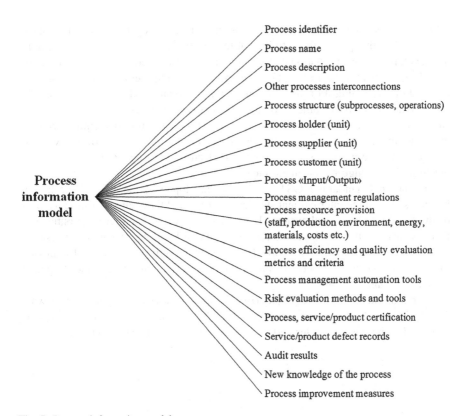

**Fig. 5** Process information model

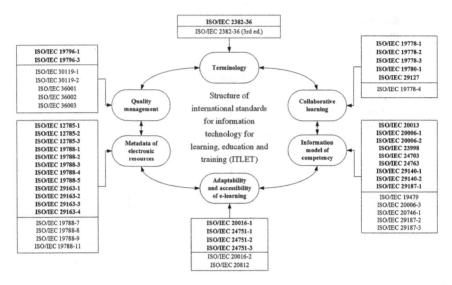

**Fig. 6** Structure of international standards for ITLET

subcommittees. Starting from 2006, Russian national delegations (TC 461) have been taking active part in the activities of the ISO/IEC JTC1/SC36, participating in the development of international standards in the field of terminology, structure of metadata, quality management and harmonisation of requirements of standards in the field of e-Learning. For the structure of international standards (ITLET) see the Fig. 6 that shows basic objects of the standardisation. Currently, SC36 actively works on the development of new standards (ISO/IEC 36000) defining requirements to the management of educational organisations, joint learning systems, competency description models, knowledge electronic testing, student's electronic portfolio, knowledge management etc.

The national standards GOST R ICTE as developed by the TC 461 are highly harmonised with the international standards ITLET and they reflect specific features of the Russian educational system (Fig. 7). A number of Russian universities use them as basis for the development and approval of e-Learning management systems and means of information support. The implementation of the unified approach made it possible for the network based cooperation of universities and corporate knowledge management.

Works on national profile project (functional standard) are currently under way. They primarily focus on the computer based Smart University management. It is supposed to be basis for later certification of university management systems.

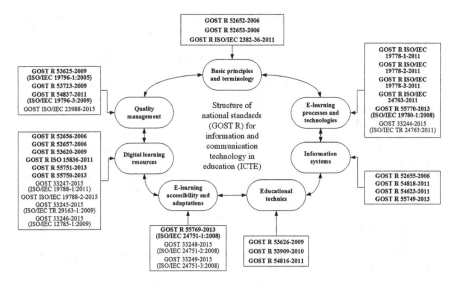

**Fig. 7** Structure of national standards (GOST R) for ICTE

# 4   Conclusion

The creation of the Smart University as one of the key components of Smart Society development is important for a qualitatively new stage in all spheres of human activity—economy, business, industry, environment protection, science, education etc.

Smart University management system should be based on the unified process model in accordance with ISO 9001 and ISO/IEC 19796-1 standards. The unified process model provides effective control of processes and resources in conformity with the specific character of the university structure and using e-Learning. Profile of requirements, based on the standards, is the basis for unification and certification of Smart University management system, which is an important factor in quality assurance in education.

# References

1. Dvorakova, Z., Langhamrova, J.: Population ageing and its human resource management consequences. In: International Days of Statistics and Economics, Prague, 19–21 Sept 2013, pp. 365–374. Melandrium, Slany. http://msed.vse.cz/files/2013/77-Dvorakova-Zuzana-paper.pdf (2013)
2. Im, D.U., Lee, J.O.: Mission-type education programs with smart device facilitating LBS. Int. J. Multimedia Ubiquitous Eng. **8**(2), 81–88 (2013)

3. European Investment Bank: JESSICA for Smart and Sustainable Cities. Horizontal Study Smart Technology based Education and Training. Smart Digital Futures. IOS Press BV, Amsterdam (2014)
4. Hwang, D.J., Yang, H., Kim, H.: e-Learning in the Republic Korea. UNESCO Institute for Information Technologies in Education, Moscow (2010)
5. Peoples, B.E.: Innovative e-Learning: information technology and standards, a current and future perspective. J. East China Normal Univ. **2**, 1–12 (2012)
6. Koole, M., McQuilkin, J., Ally, M.: Mobile learning in distance education: utility or futility? J. Dist. Educ. **24**(2), 59–82 (2010)
7. Pozdneev, B., Sosenushkin, S., Sutyagin, M.: e-Learning: quality based on standards. In: Uvaysov, S.U. (ed.) 3rd International Conference on «Innovative Information Technologies», pp. 140–149. Prague (2014)
8. Stracke, C.M.: Quality development and standards in e-Learning: benefits and guidelines for implementations. In: Proceedings of the ASEM Lifelong Learning Conference: e-Learning and Workplace Learning. ASEM, Bangkok (2009)
9. Stracke, C.M.: Quality development and standards in learning, education, and training: adaptation model and guidelines for implementations. Mag. Inf. Educ. Sci. **7**(3), 136–146 (2010)
10. Hirata, K.: e-Learning quality models with learner and competency information. IPSJ Mag. **49** (9), 1061–1067 (2008) (in Japanese)
11. Hirata, K., Laughton, S., Seta, K., Stracke, C.M.: A content analysis and information model for the European Qualifications Framework (EQF). In: Stracke, C.M. (ed.) The Future of Learning Innovations and Learning Quality. How Do They Fit Together? pp. 51–60. GITO Verlag, Berlin (2012)
12. Shimoda, E., Hirata, K.: Cognitive task model and learning sequence model for cognitive competency modelling. In: Stracke, C.M. (ed.) Competence Modelling for Human Resources Development and European Policies. Bridging Business, Education, and Training, pp. 119–130. GITO Verlag, Essen (2011)
13. Kucharčíková, A., Tokarčíková, E., Blašková, M.: Human capital management—aspect of the HC efficiency in university education. In: Global Conference on Contemporary Issues in Education, GLOBE-EDU 2014, 12–14 July 2014, Las Vegas (2014)
14. Pozdneev, B.M., Tikhonov, A.N., Ivannikov, A.D., Shatrov, A. F.: National e-Learning industry development trends, based on national and international standards. In: 4th International Conference for e-Learning Technologies «MOSCOW Education Online 2010», pp. 15–19. Moscow (2010)
15. Pozdneev, B.M., Kosulnikov, Y.A., Sutyagin, M.V.: Innovative development of the Russian education system based on standardization and certification of e-Learning. In: Stracke, C.M. (ed.) The Future of Learning Innovations and Learning Quality. How Do They Fit Together? pp. 85–96. GITO Verlag, Berlin (2012)
16. Kosulnikov, Y.A., Pozdneev, B.M., Sutyagin, M.V.: Competence modeling and standardization of terminology in the field of e-Learning. In: Stracke, C.M. (ed.) Competence Modelling for Human Resources Development and European Policies. Bridging Business, Education, and Training, pp. 97–106. GITO Verlag, Essen (2011)

# A Formal Algebraic Approach to Modeling Smart University as an Efficient and Innovative System

Natalia A. Serdyukova, Vladimir I. Serdyukov, Vladimir L. Uskov, Vladimir V. Ilyin and Vladimir A. Slepov

**Abstract** The development of Smart University ideas and concepts started just several years ago. Despite obvious progress in this area, though, the concepts and principles of this new trend are not clarified in full yet due to obvious innovativeness of this concept, as well as numerous types of smart systems, smart technologies and smart devices available to academic institutions, students and learners. This paper presents the outcomes of a research project aimed at the development of formal methodology for a description and modeling of smart universities as a system based on an algebraic formalization of general systems' theory, theory of algebraic systems, theory of groups, and generalizations of purities. The ultimate goal of this research project is to identify formal mathematical conditions for a system—smart university—to become an efficient and/or innovative system.

**Keywords** Smart University · Mathematical modeling · Systems theory · Algebraic formalization · Theory of groups · Efficient system · Innovative system

N.A. Serdyukova (✉) · V.A. Slepov
Plekhanov Russian University of Economics, Moscow, Russia
e-mail: nsns25@yandex.ru

V.A. Slepov
e-mail: vlalslepov@yandex.ru

V.I. Serdyukov
Institute of Management of Education, Russian Academy of Education,
Moscow State Technical University n.a. N.E. Bauman, Moscow, Russia
e-mail: wis24@yandex.ru

V.L. Uskov
Department of Computer Science and Information Systems, Bradley University,
Peoria, Illinois, USA
e-mail: uskov@bradley.edu

V.V. Ilyin
The State Research Institute of System Analysis, The Account Chamber of the
Russian Federation, Moscow, Russia
e-mail: vladimir_v_ilyin@hotmail.com

© Springer International Publishing Switzerland 2016
V.L. Uskov et al. (eds.), *Smart Education and e-Learning 2016*,
Smart Innovation, Systems and Technologies 59,
DOI 10.1007/978-3-319-39690-3_8

# 1 Introduction

The progress of advanced technology-based education, e-Learning, university-wide software and hardware systems of various scope (education, teaching, learning, management, communication, security, etc.), modern pedagogical approaches, innovative teaching and learning techniques (that are based on active use of advanced technology) contribute to a formation of a new type of university—Smart University (SmU).

The development of the SmU concepts started just several years ago. Despite obvious progress in this area, though, the concepts and principles of this new trend are not clarified in full yet due to obvious innovativeness and complexity of this concept.

## 1.1 Previous Research: Literature Review

Most researchers are focused on perspectives of contemporary higher education and tendencies that correspond to the concept of smart education. The authors of [1–3] discuss various aspects of contemporary universities and their future perspectives in the context of applications of smart information and communication technologies (ICT) in education.

One of the first researches on new opportunities for universities in the context of smart education is presented in [4].

The other significant part of the research is focused on the problem of educational outcomes in the contemporary educational systems—smart learning environments. The concept of outcomes-based education is proposed and this approach can be regarded as a part of the smart education paradigm. The main part of such outcomes includes the studies of cognitive abilities, needs, skills and their training through e-Learning, or, in general, to 21th century skills [5]. There are also resources available on instructional design and cognitive science, which considers the problem of learning, structuring material, communication, and forming cognitive competence in this group, for example [6].

One more group of researches is devoted to the problem of smart learning and smart e-Learning. There is an attempt to analyze different possible ways of using this concept in [7]. Several other publications in this area (especially [8]) explore the problem of implementation of principles of smart e-Learning in different educational services. Some authors propose possible models and schemes of smart educational systems.

Different attempts to make conceptual model of ICT infrastructure of SmU and smart e-Learning systems are under active research now; for example, the topics concerning smart e-Learning standards, smart gadgets and learning equipment are discussed [9, 10]. Several research projects were aimed at organizational aspects of SmU, smart education such as organizational structure, educational trajectories, learning strategies, etc. These projects usually emphasize the fact that many aspects of contemporary education need new flexible organizational structures, which can be referred to as "smart" [10, 11].

## 1.2  Research Project Goal and Objectives

The performed analysis of these and multiple additional publications and reports relevant to (1) smart universities, (2) university-wide smart systems and technologies, (3) smart classrooms, (4) smart learning environments, and (5) smart educational systems undoubtedly shows that smart university will be essential topics of multiple research, design and development projects in upcoming 5–10 years. It is expected that in near future SmU concepts and hardware/software solutions will have a significant role and be actively deployed by leading academic intuitions—Smart Universities—in the world.

**"Smartness Levels"**. Based on our vision of SmUs and up-to-date obtained research outcomes, we believe that SmUs as systems should significantly emphasize not only software/hardware features, but also "smart" features and functionality of smart systems (Table 1) [12, 13]. Therefore, the designers of the next generation of smart classrooms should pay more attention to the implementation of "smartness" maturity levels or "intelligence" levels, and the abilities of various smart technologies [23].

**Project goal**. The goal of the performed research was to develop a methodology of SmU modeling as a system based on an algebraic formalization of general systems' theory [13], theory of algebraic systems [14], theory of groups [15], and

**Table 1**  Classification of levels of "smartness" of a smart system [12, 13]

| Smartness levels (i.e. ability to ...) | Details |
| --- | --- |
| Adapt | Ability to modify physical or behavioral characteristics to fit the environment or better survive in it |
| Sense | Ability to identify, recognize, understand and/or become aware of phenomenon, event, object, impact, etc. |
| Infer | Ability to make logical conclusion(s) on the basis of raw data, processed information, observations, evidence, assumptions, rules, and logic reasoning |
| Learn | Ability to acquire new or modify existing knowledge, experience, behavior to improve performance, effectiveness, skills, etc. |
| Anticipate | Ability of thinking or reasoning to predict what is going to happen or what to do next |
| Self-organize | Ability of a system to change its internal structure (components), self-regenerate and self-sustain in purposeful (non-random) manner under appropriate conditions but without an external agent/entity |

generalizations of purities [16], and identify formal mathematical conditions for a system—in this case, SmUs—to become efficient and/or innovative.

**Project Objectives**. The objectives of this multi-aspect research project were to use authors' research outcomes [17–23], and

(1) identify SmU main components, and describe SmUs as a system;
(2) build a group of abstract system links and highlight effective links of a system,
(3) define the notion of a final state of a system and to determine on this basis the number of different ways of system development in the case when the system can be represented with the help of a finite group,
(4) get the scheme of an algebraic formalization of knowledge represented as a compound system, and
(5) construct a scheme of the model of risks' scenarios of the development of the education system comprising of six factors.

The premise is that the development of designated mathematical modeling methodology will enable researchers to identify formal conditions and algorithms for SmU to become efficient and/or innovative system.

## 2 Conceptual Modeling of Smart University

### 2.1 Theoretical Preliminaries

The conceptual model of a SmU—*CM-SmU*—can be described as follows.

**Definition 1** Smart University is described as *n*-tuple of n elements which can be chosen from the following main sets:

$$
\begin{aligned}
CM - SmU = \langle &\{SmU\_STUDENTS\}, \{Sm\_CURRICULA\}, \\
&\{Sm\_FACULTY\}, \{Sm\_PEDAGOGY\}, \\
&\{Sm\_CLASSES\}, \{Sm\_SOFTWARE\}, \\
&\{Sm\_HARDWARE\}, \{Sm\_TECHNOLOGY\}, \\
&\{Sm\_RESOURCES\} \rangle
\end{aligned}
\tag{1}
$$

where:

| | |
|---|---|
| SmU_STUDENTS | a set of types of students at SmU (local, remote, special types of students, undergraduate, graduate, etc.); |
| Sm_CURRICULA | a set of smart programs of study and smart courses at SmU —those that can change its structure in accordance with identified requirements, for example, types of students in those programs and courses; |

| | |
|---|---|
| Sm_FACULTY | a set of faculty (instructors) at SmU who are trained and actively use smart systems, smart technology, smart boards, smart classrooms, etc., |
| Sm_PEDAGOGY | a set of pedagogical styles (strategies) to be used at SmU; in general case, it may include a great variety of innovative teaching and learning styles such as collaborative (local and remote students together), online learning, learning-by-doing, project-based learning, games-based learning, etc., |
| Sm_CLASSES | a set of smart classrooms, smart labs, smart departments and offices at SmU; |
| Sm_SOFTWARE | a set of university-wide smart software systems at SmU; in general case, it may include a great variety of software systems for smart classrooms, smart labs, university-wide learning management systems, security systems, identification systems, etc., |
| Sm_HARDWARE | a set of university-wide smart hardware systems, devices, equipment and technologies used in SmU software systems at SmU; in general case, it may include a great variety of sensors, video cameras, smart interactive boards, etc., |
| Sm_TECHNOLOGY | a set of university-wide technologies to facilitate main functions of SmU; in general case it may include sensor technology, RFID technology, ambient intelligence technology, Internet-of-Things technology, wireless sensor networks, security and safety technology, etc., |
| SmU_RESOURCES | a set of various resources of SmU (financial, technological, human, etc.). |

In general, SmUs may have multiple additional minor sets; however, for the purpose of this research project we will limit a number of SmU sets as presented in (1). The designated SmU components (below-factors) may have a great variety of relations (below-connections), for example, *SmU_students-to-Sm_faculty, Sm_faculty-to-Sm_software, Sm_curricula-to-Sm_pedagogy,* etc.

Various SmU components–*factors*—may form a *group*; as a result, we present below several definitions from [20] for the convenience of readers.

**Definition 2** [20] Under the algebra of factors, a system will be understood as algebra $\bar{A} = \langle A | \{f_\alpha^{n_\alpha} | \alpha \in \Gamma\} \rangle$ with a fundamental set of factors $A$ and a set of operations $\{f_\alpha^{n_\alpha} | \alpha \in \Gamma\}$ that describe connections between factors.

**Definition 3** [20] A sub-algebra $\bar{B} = \langle |B\{f_\alpha^{n_\alpha} | \alpha \in \Gamma\} \rangle$ of algebra $\bar{A} = \langle |A\{f_\alpha^{n_\alpha} | \alpha \in \Gamma\} \rangle$ is called *P*-pure in $\bar{A}$ (or an embedding $\varphi$ of a sub-algebra $\bar{B}$ into an algebra $\bar{A}$ is *P*-pure, if (1) every homomorphism $\bar{B} \rightarrow {}^\alpha \bar{C}$ of the subalgebra $\bar{B}$ into $\bar{C}$ (where $\bar{C}$ is an algebra of the signature $\{f_\alpha^{n_\alpha} | \alpha \in \Gamma\}$ of $\bar{A}$, and (2) $P(\bar{C})$ is true, (3) $P$ is a predicate on the class of algebras of the signature $\{f_\alpha^{n_\alpha}|\alpha \in \Gamma\}$ closed under taking subalgebras and factor algebras, can be continued to a homomorphism $\beta$ of $\bar{A} = \langle A|\{f_\alpha^{n_\alpha}|\alpha \in \Gamma\}\rangle$ into $\bar{C} = \langle C|\{f_\alpha^{n_\alpha}|\alpha \in \Gamma\}\rangle$ in such a way that the following diagram is commutative:

$$0 \to \bar{B} = \langle B|\{f_\alpha^{n_\alpha}|\alpha \in \Gamma\}\rangle \xrightarrow{\varphi} \bar{A} = \langle A|\{f_\alpha^{n_\alpha}|\alpha \in \Gamma\}\rangle$$

$$\bar{C} = \langle C|\{f_\alpha^{n_\alpha}|\alpha \in \Gamma\}\rangle$$

with $\alpha$ on the left, $\beta$ on the right. (2)

that is $\beta\varphi = \alpha$. (A note: The general operations of the same type in algebraic systems will be denoted in identical manner).

In fact, purities are the fractality of links. In this case, $P$-purities are the fractality of links with the property $P$.

Now we need to provide a formalization of an axiomatic description of a system.

**Definition 4** [20] By a system we understand a two-dimensional vector

$$S = \langle\{\langle S_\alpha, Q_\alpha, U_\alpha\rangle|\alpha \in A\}, I(S) = \langle\{a_\beta|\beta \in B\}|_S = \left\{f_\gamma^{n_\gamma}|\gamma \in \Gamma\right\}\rangle\rangle \quad (3)$$

where

- $\{S_\alpha|\alpha \varepsilon A\}$ is a set of all system's statuses which are possible as a result of system $S$ operation,
- $\{Q_\alpha|\alpha \varepsilon A\}$ is a set of all system's statuses $Q$ upon which system $S$ is affected,
- $\{U_\alpha|\alpha \varepsilon A\}$ is a set of all statuses of an external environment which are possible as a result of system $S$ operation,
- $\{a_\beta|\beta \varepsilon B\}$ is a set of all internal factors that influence system $S$ behavior.

If a composition of factors $a_1^\circ a_2^\circ \ldots^\circ a_{n_\gamma} = a$, than let $f_\gamma^{n_\gamma}(a_1, a_2, \ldots, a_{n_\gamma}) = a$, where $\{f_\gamma^{n_\gamma}|\gamma \varepsilon \Gamma\}$ is a set of operations on set of factors $\{a_\beta|\beta \varepsilon B\}$, $f_\gamma^{n_\gamma}$—$n_\gamma$-argument operation, $I(S) = \langle\{a_\beta|\beta \varepsilon B\}|\Omega_S = \{f_\gamma^{n_\gamma}|\gamma \varepsilon \Gamma\}\rangle$—an algebraic system of internal factors of system $S$.

Let $\{P_i|i \in I\}$ be a set of all properties of system $S$, which it holds as a result of its operation, $\{B_j|j \in J\}$—a set of all subsystems of system $S$, $\{v_m^n|m \in M, n \in N\}$—a set of all connections in system $S$, *—a composition operation. In this case, $G$ (S) is a goal of system $S$. As a result, we get an algebra $\langle\{P_i|i \in I\}|*\rangle$ under the assumption that a set of all system $S$ properties is closed under operation of composition *. In its own turn, it means that is that we have a full description of system $S$.

## 2.2 Factors to Define a Group

Algebra $\bar{A} = \langle A | \{ f_\alpha^{n_\alpha} | \alpha \in \Gamma \} \rangle$ with main set of factors $A$ and a set of operations $\{ f_\alpha^{n_\alpha} | \alpha \in \Gamma \}$ (that actually describe connections of factors in a system), is group $\bar{A} = \langle A |^\circ, ^{-1}, e \rangle$, where $^\circ$–is a composition of factors (i.e. the consistent implementation of factors, $^{-1}$ is an operation of taking a reverse factor, and $e$ is a neutral factor. In this case, $\bar{A} = \langle A | \{ f_\alpha^{n_\alpha} | \alpha \in \Gamma \} \rangle$ is called a group of factors that represent system $S$.

Let us consider the meaning of purities and examples of $P$-purities in a class of all groups. Formula (2) has the following meaning in a class of groups: epimorphic images of $B$ and $A$ are the same in the class of all finite groups [15]. For $P$-purities the meaning is as follows: $B$ and $A$ have the same epimorphic images in class of all groups satisfying the condition $P$.

The possible examples include but are not limited to:

(1) $P$ allocates the class of all finite groups in the class of Abelian groups, get the usual purity in the class of all Abelian groups;
(2) $P$ allocates the class of all Abelian groups in the class of all groups;
(3) $P$ allocates the class of all finite groups in the class of all groups;
(4) $P$ highlights the diversity in the class of all groups i.e. the class of groups closed under subgroups, homomorphic images and Cartesian products, such as Burnside's variety of all groups of the exponent (indicator) $n$ defined by the identity $x^n = 1$, the variety of nilpotent groups of class of nilpotent is not more than $n$, soluble groups of length not exceeding the number $l$, etc.

## 2.3 A Group of Links of a System

Let's consider the decomposition of a system $S$ at elements: $S = \coprod_i S_i$, where every $S_i$ is a subsystem of a system $S$ or an atomic element of a system $S$. Usually, the flat graphs are used to represent links of a system. A flat graph is a picture on a plane that consists of a set of points $\{a\}$ (which are the vertices of a graph with one to one corresponding to a set $\{a = S_i\}$) and a set of edges of a graph (which are directed arks $v_{ab}$) if there exists a link between $a = S_i$ and $b = S_j$ in the system $S$. If a link is two-dimensional, then arc $v_{ab}$ is non-direct one. There are a lot of examples that show that it is necessary to use multigraphs (or pseudo-graphs)—graphs in which multiple edges are allowed, i.e. edges with the same vertices—during a simulation. For example, modeling a possible trajectory of an aircraft requires the utilization of a multigraph. However, a multigraph, which represents a system, does not provide information to identify quality of system's links. Various methods, that are used in mathematical statistics and econometrics, are able to evaluate quantitative characteristics of system's links quite strong constrains. However, they are useless to analyze systems' qualitative characteristics.

**Definition 5** Let $G(S) = \langle G, \cdot, ^{-1} \rangle$ be a group of factors that represents system $S$, and $G_S = \{a, b, \ldots\}$ be a main set of group $G(S)$. Let $\Gamma'_S = \{v_{ab}|a, b \in G_S\}$ be the Cayley's graph for $G(S)$, or, in other words, there exists an edge $v_{ab}$ if and only if $b = a*h$, $h \in H \cup H^{-1}$, where $H$ is a set of generators of a group $G(S)$. Let $\Gamma_S = \{v_{aa}|a \in G_S\} \cup \Gamma'_S$. Let $a, b \in G_S$ and $V_{ab} = \{v^h_{ab}|a, b \in G_S \,\&\, \exists h \in H^{-1} \cup H : b = ah\}$. In this case, $V = \cup_{a \in G_S} V_{ab}$ is a set of all links between elements, $b \in G_S$ such that $\exists h \in H^{-1} \cup H : b = ah$.

Let $M = V \cup \{v_{aa}a \in G_S\}$. Let $G_a = \langle ah|h \in H^{-1} \cup H \rangle$ be a subgroup of $G$ $(S)$ generated by the set $\{ah|h \in H^{-1} \cup H\}$. In this case, $G_a \cap G_b \neq E \Leftrightarrow a^{-1}b = hk^{-1}, h, k \in H^{-1} \cup H$, where $E$ is a unit subgroup of $G(S)$.

Let us consider a free product $F_S = {}^*_a G_a$, where a free product is taken over all $a \in G_S$ in such a way that for every free factors $G_a$ and $G_b$, where $a, b \in G_S$ we will get has $G_a \cap G_b = E$. In this case, $F_S$ is called a group of links of system $S$. (A note: a free product operation defines all possible best combinations of links).

## 3  Risk Modeling in a Smart University

### 3.1  System's Final State and Sub-systems

It is necessary to determine a notion of system's final state.

**Definition 6** Let $S(t) = S'$ be a transition of system $S$ at the moment $t$ into a new state $S'$. A set of all system $S$ states, which occur during its operation, is denoted as $\{S(t_i)|i \varepsilon I\}$, where $t_i$ is a point of time for every $i \varepsilon I$. During its operation the system $S$ converts itself into system $S_f$ at the point of time $t_f$. After that time a system stops any changes, or, in other words, a group of factors $G_S = G_S(t_f)$ that represent system $S$ stop to change. Let us call $G_S = G_S(t_f)$ as the final state of system $S$. It is possible to determine the number of final states of a system as a number of non-isomorphic groups of the order of the group of the factors representing the system; this is because the described group of factors, which represent a system, is finite. Each final state of a system, in its own turn, determines a predictive variant of development of a system.

Consider a system of education from this point of view. In general, a system of education may contain a huge number of subsystems each of which can represent one or a part or several SmU components that are described in (1). However, for the purpose of this research project and in order to simplify the explanation of developed methodology, we will take into consideration only the following sub-systems of system $S$:

(1)  $S_1$—a subsystem of learning content or knowledge (let us call it informational subsystem); it corresponds exactly to the set {Sm_CURRICULA} in (1);

(2) $S_2$—a subsystem of pedagogy or methodological and methodical subsystem (let us call it adaptive subsystem); it corresponds exactly to the (Sm_PEDA-GOGY} set in (1);

(3) $S_3$—a subsystem of students (let us call it target subsystem); it corresponds exactly to the {SmU_STUDENTS} set in (1);

(4) $S_4$—a financial subsystem (let us call it providing subsystem); it corresponds to a part of {SmU_RESOURCES}set in (1).

In this case, the connections between designated subsystems may be carried out as follows:

1. Stage 1. $S_2$ affects $S_1$; an improvement of learning content takes place; the outcome of this action is designated as $S_2(S_1,)$.

2. Stage 2. $S_2(S_1)$ affects $S_3$; as a result, there is a transformation of $S_3$ into $(S_2(S_1))(S_3)$.

The overall goal of a system of education is to find the optimal final state $(S_2(S_1,))(S_3)$ for a system in accordance with the following selected criteria:

(1) a completeness, necessity and sufficiency of an information subsystem $S_1$;

(2) a clarity, simplicity and accessibility for a perception of an adapting subsystem $S_2$;

(3) a required level of availability of a target subsystem $S_3$;

(4) a sufficient level of support by providing subsystem $S_4$.


## 3.2 A Selection of Factors to Determine Long-Term Risks

It is necessary to identify indicators $s_i$ that will help us determine the long-term risks of a system of education; in general, a system may have numerous indicators. However, for the purpose of this research project and in order to simplify an explanation of developed methodology, below we will take into a consideration only the following indicators $s_i$

(1) $S_1$—an indicator of system's financial status (for example, university's total assets that may include tuition and fees, government grants and appropriations, investments, contributions, property, equipment, etc.);

(2) $S_2$—an indicator of system's target subsystem (for example, a total number of students of various types such as undergraduate students, graduate students, life-long learners, etc.);

(3) $S_3$—an indicator of systems' adaptive subsystem (as the indicator one can chose the level of the development of methodological and methodical support of the knowledge system, it can be measured by the frequency of using the methodological support of the knowledge system by the personnel of higher qualification and by the target audience);

(4) $S_4$—an indicator of the required level of availability for target audience (test control of knowledge of the target audience and its monitoring);

(5) $S_5$—an indicator of sufficiency of the financial support for knowledge subsystem (for example, external funds for a system of education, or funds for special projects—disabled students, life-long learners, etc.);

(6) $S_6$—an indicator of accessibility of a system of education to students (for example, TOEFL level for international students, tuition fee for a credit hour, etc.)

## 3.3 An Algebraic Formalization of Factors' Selection

Let us consider two possible scenarios under the assumption of a system's closeness and associativity.

**Scenarios of system operation.** Scenario # 1 corresponds to the cyclic group of the order 6, i.e. group $Z_6$ (Fig. 1). Scenario # 2 corresponds to the symmetric group of permutations of the order 3, i.e. group $S_3$ (Fig. 2).

| $\circ$ | $S_1$ | $S_2$ | $S_3$ | $S_4$ | $S_5$ | $S_6$ |
|---|---|---|---|---|---|---|
| $S_1$ | $S_1$ | $S_2$ | $S_3$ | $S_4$ | $S_5$ | $S_6$ |
| $S_2$ | $S_2$ | $S_3$ | $S_4$ | $S_5$ | $S_6$ | $S_1$ |
| $S_3$ | $S_3$ | $S_4$ | $S_5$ | $S_6$ | $S_1$ | $S_2$ |
| $S_4$ | $S_4$ | $S_5$ | $S_6$ | $S_1$ | $S_2$ | $S_3$ |
| $S_5$ | $S_5$ | $S_6$ | $S_1$ | $S_2$ | $S_3$ | $S_4$ |
| $S_6$ | $S_6$ | $S_1$ | $S_2$ | $S_3$ | $S_4$ | $S_5$ |

**Fig. 1** Cyclic group of the order 6

| $\circ$ | $S_1$ | $S_2$ | $S_3$ | $S_4$ | $S_5$ | $S_6$ |
|---|---|---|---|---|---|---|
| $S_1$ | $S_1$ | $S_2$ | $S_3$ | $S_4$ | $S_5$ | $S_6$ |
| $S_2$ | $S_2$ | $S_1$ | $S_6$ | $S_5$ | $S_4$ | $S_3$ |
| $S_3$ | $S_3$ | $S_5$ | $S_1$ | $S_6$ | $S_2$ | $S_4$ |
| $S_4$ | $S_4$ | $S_6$ | $S_5$ | $S_1$ | $S_3$ | $S_2$ |
| $S_5$ | $S_5$ | $S_3$ | $S_4$ | $S_2$ | $S_6$ | $S_1$ |
| $S_6$ | $S_6$ | $S_4$ | $S_2$ | $S_3$ | $S_1$ | $S_5$ |

**Fig. 2** Symmetric group of the order 3

Let's show how to use the tables. The analysis of scenarios 1 and 2 gives us the following. The equation $s_i^o s_j = s_k$ shows that the changes of factors $S_i$ and $S_j$ leads to the change of $S_k$.

**The difference between scenarios.** Every table contains 36 squares. While searching identical squares, it makes sense to consider squares with coordinates $(i, j)$, where $i, j \geq 2$, $i, j = 1, \ldots, 6$. These are the following 17 squares: diagonal squares with coordinates $(2,2)$, $(3,3)$, $(5,5)$ and non-diagonal squares with coordinates $(3,2)$, $(4,2)$, $(5,2)$, $(6,2)$, $(2,3)$, $(4,3)$, $(5,3)$, $(2,5)$, $(3,5)$, $(4,5)$, $(2,6)$, $(3,6)$, $(4,6)$, $(5,6)$. The other 19 squares in the 1st and 2nd scenarios are the same. As a result, we get the following meta-scenarios: the first scenario $s_1$ changes when $s_3$ and $s_5$ are changed; the second scenario $s_2$ changes and $s_1$ does not change when $s_3$ and $s_5$ are changed, and so on.

It is possible to identify possible points of time for risk regulation if the dynamics of each indicator is known. One scenario can be changed into another one in these points.


# 4 Formalization of Innovation and Effectiveness Concepts

## 4.1 Preliminaries

Let us present several definitions from [16–21] in order to provide consistency in mathematical logic of statements described below.

**Definition 7** Let $S$ be a system with integrity property $P$. Then system $S$ is called an innovative system with deciding integrity property $P$ if system $S$ can be off line in every super system $S'$, such that $S$ is $P$-pure in $S'$, or, in other words, it contains it without a distortion of $P$-connections.

An innovative system $S$ with integrity property $P$ differs from a system with integrity property $P$ by higher (not lower than in super system containing it) implementation performance numeric innovative properties, or the fact that analogues of a system with property $P$ does not exist.

The main indicators of innovation are: newness, the degree or level of newness, consumer value, degree of implementation in practice, effectiveness, the presence of a single indicator of efficiency, the phenomenon of "flash" (a synergistic effect) characterizing the beginning of the autonomous work of the innovation system, the graph of all links of all innovations.

The possible examples of innovative systems include but are not limited to ICT technologies, expert systems, autonomous work in digital libraries, and so on. The concept of an effective system follows the *Occam's Razor* principle: "When you have two competing theories that make exactly the same predictions, the simpler one is the better".

The possible examples of effective systems are as follows: an application of scenario 1 in accordance with property $P$ which saves all numerical indicators of a system.

Our goal now is to highlight the main properties and attributes of an effective system and to formalize them. The following are the properties of an effective system:

(1) The system works effectively if it operates in accordance with external conditions.
(2) The existence of an effective system is as much as possible.
(3) The system works $P$-effectively if it operates in accordance with external conditions, preserving $P$-links.
(4) The existence of $P$-effective system is as much as possible to preserve $P$-links.

**Definition 8** The system $S$ is effective in the system $S'$ if $S$ works offline and there is no other subsystem in $S$ which works offline.

**Definition 9** The system $S$ is $P$-effective in the system $S'$ if $S$ works offline and there is no $P$-pure subsystem in $S$ which works offline.


## 4.2 Efficient and P-Efficient Links of a System

**Definition 10** Let (a) $P$ be a predicate on the class of groups closed under taking subgroups and factor groups, (b) $G(S) = \langle G, \cdot, ^{-1} \rangle$ be a group of factors that represents system $S$, and (c) $F_S$ be a group of links of the system $S$. A subgroup $H_S$ of a group $F_S$ is called a subgroup of effective links of a system $S$ if $H_S$ is $P$-effective in $F_S$. (A note: a notion of "effective links" is worth considering if these links form a structure; for example in this paper we discuss a group of links).


## 4.3 Inverse Limits and Systems with Full Implementation of P-Links

Generalizations of the theorem on the structure of algebraically compact groups and the theorem that any reduced Abelian group can be embedded as a pure subgroup into an algebraically compact group (that is into an inverse limit of cyclic (atomic) groups) make it possible to build $P$-pure embedding of a system into a system with full implementation of $P$-links. In addition, this approach allows one to prove an analogue of the theorem that every algebraically compact Abelian group is a direct summand of the group consisting of it as a pure subgroup for the general systems theory: every $P$-pure subsystem with the full implementation of $P$-links is a retract of any system containing it. As a result, we can get the following theorems from the Definitions 7 and 8 above.

**Theorem 1** Every system with full implementation of $P$-links can be off line.

**Theorem 2** Every model of closed associative system which consists of 5 factors is an effective one.

**Theorem 3** Every model of closed associative system which consists of $p$ factors where $p$ is a prime number is an effective one.

From the point of view of the dynamics in the process it can be considered as an effective strategy, from the point of view of statics—as an effective system or subsystem.

## 5 Conclusions. Future Steps

**Conclusions**. The goal of the performed research was to develop a methodology of SmU modeling as a system based on an algebraic formalization of general systems' theory [14], theory of algebraic systems [15], theory of groups [16], and generalizations of purities [17], as well as identify formal mathematical conditions for a system—in this case SmUs—to become efficient and/or innovative.

The main obtained research outcomes are presented in Sects. 2 and 3 above, specifically:

(1) The main components of SmU were identified; as a result, it enabled us to describe SmU in terms of general systems' theory—Eq. (1) above;
(2) define the notion of a final state of a system, and, on this basis, determine the number of different ways of the system development in the case when the system can be represented with the help of a finite group (Sects. 2.1–2.3) above;
(3) build a group of links of an abstract system and highlight effective links of a system (Sect. 2.2);
(5) construct a scheme of the model of risks' scenarios of the development of the education system comprising of six factors (Fig. 1);
(6) algebraically formalize the concepts of "effective system" and "innovative system" (items 4.2 and 4.3 above).

**Next step**. The next step of this research project is to develop a software system that will implement the developed formal methodology for a description and modeling of a smart university as a system.

## References

1. Coccoli, M., Guercio, A., Maresca, P., Stanganelli, L.: Smarter universities: a vision for the fast changing digital era. J. Visual Lang. Comput. **6**, 1003–1011 (2014)
2. Barnett, R.: The Future University. Routlege, New York (2012)

3. Temple, P. (ed.): Universities in the Knowledge Economy: Higher Education Organization and Global Change. Routledge, New York (2011)
4. Tikhomirov, V.P.: Mir na puti k smart education. Novye vozmozhnosti dlia razvitiia [The World on the Way to Smart Education. New opportunities for Development]. Otkytoe obrazovanie [Open Education] **3**, 22–28 (2011) (in Russian)
5. Hilton, M.: Exploring the Intersection of Science Education and 21st Century Skills: A Workshop Summary. National Research Council (2010)
6. Richey, R.C., Klein, J.D., Tracey, M.W.: The Instructional Design Knowledge Base: Theory, Research, and Practice. Routledge, New York (2010)
7. Spector, J.: Conceptualizing the emerging field of smart learning environments. Smart Learn. Environ. **1** (2014)
8. Gamalel-Din, S.A.: Smart e-Learning: a greater perspective: from the fourth to the fifth generation e-Learning. Egypt. Inf. J. **11**, 39–48 (2010)
9. Ke, C.-K., Liu, K.-P., Chen, W.-C.: Building a Smart e-Portfolio Platform for Optimal e-Learning Objects Acquisition. Mathematical Problems in Engineering
10. Tikhomorov, V., Dneprovskaya, N., Yankovskay, E.: Three dimensions of smart education. In: SEEL2015, Smart Education and Smart e-Learning, Smart Innovation, Systems and Technologies, vol. 41, pp. 47–56. Springer (2015)
11. Ruberg, L.: Transferring smart e-Learning strategies into online graduate courses. In: SEEL2015, Smart Education and Smart e-Learning, Smart Innovation, Systems and Technologies, vol. 41, pp. 243–254. Springer (2015)
12. Derzko, W.: Smart Technologies. http://archives.ocediscovery.com/discovery2007/presentations/Session3WalterDrezkoFINAL.pdf (2007)
13. Uskov, A., Sekar, B. Smart gamification and smart serious games. In: Sharma, D., Jain, L., Favorskaya, M., Howlett, R. (eds.) Fusion of Smart, Multimedia and Computer Gaming Technologies, Intelligent Systems Reference Library, vol. 84, pp. 7–36. Springer (2015). doi 10.1007/978-3-319-14645-4_2. ISBN: 978-3-319-14644-7
14. Mesarovich, M., Takahara, Y.: General System Theory: Mathematical Foundations. Mathematics in Science and Engineering, vol. 113. Academic Press, New York, San Francisco, London (1975)
15. Maltcev, A.I.: Algebraic Systems, 392 pp. Nauka, Moscow (1970) (in Russian)
16. Kurosh, A.G.: Theory of Groups, 648 pp. Moscow, Nauka (1967) (in Russian)
17. Serdyukova, N.A.: On generalizations of purities. Algebra Logic **30**(4), 432–456 (1991)
18. Serdyukova, N.A.: Optimization of Tax System of Russia, Parts I and II. Rostov State Economic University, Budget and Treasury Academy (2002). (in Russian)
19. Serdyukova, N.A., Serdyukov, V.I.: The new scheme of a formalization of an expert system in teaching. In: ICEE/ICIT 2014 Proceedings, Paper 032, Riga (2014)
20. Monitoring and Evaluation of Research in Learning Innovations: Final Report of project HPHA-CT2000-00042 funded under the Improving Human Research Potential & the Socio-economic Knowledge Base Directorate General Science, Research and Development EUROPEAN COMMISION (MERLIN)
21. Serdyukova, N.A., Serdyukov, V.I., Slepov, V.A.: Formalization of knowledge systems on the basis of system approach. In: SEEL2015, Smart Education and Smart e-Learning, Smart Innovation, Systems and Technologies, vol. 41, pp. 371–380. Springer (2015)
22. Serdyukova, N.A., Serdyukov, V.I., Modeling, simulations and optimization based on algebraic formalization of the system. In: 19th International Conference on Engineering Education, 20–24 July 2015, Zagreb, Zadar (Croatia), New Technologies and Innovation in Education for Global Business, Proceedings, pp. 576–582, ICEE2015
23. Uskov, V.L., Bakken, J.P., Pandey, A.: The ontology of next generation smart classrooms. In: SEEL2015, Smart Education and Smart e-Learning, Smart Innovation, Systems and Technologies, vol. 41, pp. 3–14. Springer (2015)

# Conceptual Framework for Feedback Automation in SLEs

Estefanía Serral and Monique Snoeck

**Abstract** Although feedback is a very time-consuming task, it is crucial to improve learning and foster students' development. For this reason, automating feedback has become one of the most pursued challenges within the field of smart learning environments -SLEs-. After performing a literature study about which dimensions of feedback are relevant to properly automate it, we present a novel conceptual framework that structures, explains, and analyses these dimensions. The framework can be key to appropriately contextualize future research about feedback, and can also be a guide for educators and developers in order to design and implement quality feedback in a more methodical way in SLEs.

**Keywords** Feedback automation · Conceptual framework · Smart learning environments

## 1 Introduction

Feedback to the learners has always been viewed as a critical factor for improving knowledge and acquiring skills; however, students in higher education report deficits in the amount and quality of feedback they receive [1]. To deal with this issue, and considering the lack of teaching resources and the growing number of students with heterogeneous profiles, personalized and automated feedback has become one of the most pursued challenges within the field of smart learning environments (SLEs) [2]—technology-supported learning environments that provide personalized support to the learners, including guidance and feedback based on their needs [3].

E. Serral (✉) · M. Snoeck
Leuven Institute for Research on Information Systems (LIRIS),
KU Leuven Naamsestraat 69, Box 3500, 3000 Leuven, Belgium
e-mail: estefania.serralasensio@kuleuven.be

M. Snoeck
e-mail: monique.snoeck@kuleuven.be

© Springer International Publishing Switzerland 2016
V.L. Uskov et al. (eds.), *Smart Education and e-Learning 2016*,
Smart Innovation, Systems and Technologies 59,
DOI 10.1007/978-3-319-39690-3_9

How to properly give feedback to learners is a very investigated topic and we can already find in the literature excellent research that studies it from different angles [4–8]. In addition, several advance learning environments have been developed featuring certain forms of feedback automation; e.g.: [9] presents JMermaid, a learning environment for teaching conceptual modelling that provides several forms of automated feedback; [10, 11] have developed learning systems for teaching SQL which provides domain-specific feedback to help students in constructing correct SQL queries; [12] studies the combination of learning analytics and formative assessment to provide students with immediate detailed feedback about their performance.

However, to the best of our knowledge, no previous work has focused on identifying the most important aspects that have influence on the quality of automated feedback in SLEs. Through a literature study, the results of which were combined with the authors' experience and reflection, we have created a novel conceptual framework that identifies and analyses the most important dimensions to consider for automating feedback in SLEs. By considering these dimensions, SLEs can apply a more methodological approach to design and implement quality feedback. In addition, the framework will also serve to clearly define the specific forms of feedback to which future research works contribute and can be compared with.

The rest of the paper is organized as follows. Section 2 summarizes the literature reviews related to feedback for learners. Section 3 presents the proposed conceptual framework describing all the important dimensions for feedback automation. Section 4 concludes the paper and explains our further work.

## 2   Related Work

Several interesting literature reviews have been performed that contribute to feedback research. For instance, in [13] the authors present a systematic qualitative review on online formative assessment in higher education, in which they conclude that effective online formative assessment focused on formative feedback can foster learning and that accurate assessment activities and interactive formative feedback are very important to address threats to validity and reliability in online formative assessment. [14, 15] review the literature on how to encourage feedback-seeking behaviour and feedback engagement. [16] studies the effects of feedback provided to teams in higher education or organizational settings. [17] formulates several rules of thumb to reflect what feedback a majority of learning theories suggested as effective for learning and shows that feedback processes are complicated and many variables influence and mediate the processes. Finally, [18] analyses the research evidence on the feedback that students receive within their coursework from multiple sources.

We have considered these works as part of our literature study since they provide remarkable findings about feedback. However, no paper has yet identified the most important dimensions for automating feedback.

# 3   Conceptual Framework for Feedback Automation

The analysis of a thorough literature review, contrasted to our experience as educators and computer scientists, has led to the construction of a conceptual framework that identifies 9 essential dimensions to be considered in order to automatically provide feedback: *purpose, level, nature, domain knowledge, presentation format, timing, recipient, classroom, and use and impact.* As shown in Fig. 1, the conceptual framework classifies these dimensions within 4 categories: *content, delivery, context, and use and impact.*

## *3.1   Content*

The content of the feedback must be carefully composed to maximize its effect. The content depends on the purpose of the feedback, at which level it should be provided, its nature, and the domain knowledge to which it refers.

### 3.1.1   Purpose

According to the purpose of the feedback, also denoted by some authors as focus or types of feedback, we have found the following different options referred in the literature:

- *Summative* [4, 7, 8, 19]: This feedback evaluates the learner with regard to a set of criteria and presents up-to-date success or failure information. Besides informing

**Fig. 1** Conceptual framework for feedback automation

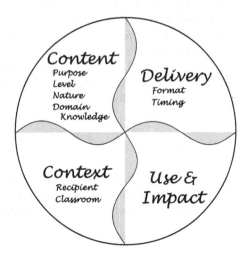

individuals about their status, such feedback can also provide information about a learner's progress.

- *Formative or instructional* [4, 6, 8, 20–23]: This feedback provides learners with information and guidelines to improve their answers to a particular problem or task while it is still being performed or completed. Information may also be provided to address particular errors or misconceptions or to give detailed metalinguistic explanations to learners regarding their performance [21]. The purpose of formative feedback is to bridge the gap between current and desired performance [22, 23].

- *Motivational* [6, 8, 24]: This feedback is explicitly given to motivate the learners. It usually provides positive or negative reinforcement to promote engagement. Examples of motivational indicators and strategies include usage statistics to summarize achievements or engagement in terms of persistence, feedback from peers or more able partners, the use of awards, etc. Personalized feedback in e-learning has already a motivational effect because it helps to increase the student's feeling that he/she is supported during the learning process [6].

- *Informative* [4, 8, 25]: This feedback goes beyond reinforcement and provides an elaborate picture of where a learner stands in reference to others, certain criteria, or his/her previous performance. A similar kind of feedback named *Ipsative* is defined in [4]. It compares the current state with previous performance and links it to long-term progress. In the ipsative feedback approach it is important to give not only feedback about the current assignment, but also some information about the student evolution in the course or the current module. Informative feedback could be subclassified as descriptive, comparative, or evaluative [25].

- *Verification/Corrective* [6, 26]: This feedback assesses whether the answer is correct or not and it can provide knowledge of the correct response (e.g., "this concept is not correctly defined"; "the correct answer is"; etc.).

- *Suggestive* [26]: This feedback includes advice on how to proceed or progress and invites the learner to explore, expand or improve an idea (e.g.,"using a structure of introduction, development and conclusion, would make your essay clearer"). A similar kind of feedback named *reflective* is used in [27]. It is given to make students reflect on specific problems, ideas, or solutions.

- *Elaborative* [28]: This feedback addresses the answer given by the learner and its errors, and provides explanation, examples or/and guidance that can help the learner to improve. Its definition is very close to formative feedback.

### 3.1.2  Level

The level at which feedback is provided was introduced by Hattie and Timperley [5] as an important dimension to classify feedback, and it is very well accepted in the education community. This classification distinguishes the following 4 feedback levels:

- *Task level*: it addresses how well tasks are understood, performed or accomplished. It focuses on faults in the interpretation of the task or in the outcome produced.
- *Process level*: it addresses the process that needs to be executed to understand or perform a task. Such feedback should be related to the student's own error-detection strategies, and has to serve as an "advance-organiser", leading to more effective information search or to better strategies.
- *Self-regulation level*: it refers to the regulatory or metacognitive process level. It consists of self-monitoring, directing, and regulation of actions, and leads to enhance self-efficacy and to further engagement with or investing more effort into a task.
- *Self level*: it addresses issues of personal evaluation and affect (usually positive) of the learner, including feedback features such as praise and judgement.

### 3.1.3 Nature

The feedback nature can be *positive* or *negative* [5, 8, 29, 30]. Positive feedback is provided when a learner has met or exceeded a specific criterion, while negative feedback is provided when the target behaviour has fallen short of the criterion [8]. A balance in the proportion of both types of feedback is very important. On the one hand, the presence of positive feedback is essential for reinforcing existing behaviours and encouraging learners, but in excess it can make learners feel not challenged, and consequently experience boredom [8]. On the other hand, negative feedback is essential to provide information for improvement [5], but in excess, it can make students to experience anxiety and frustration, and therefore feel less motivated [8].

### 3.1.4 Domain Knowledge

To be comprehensible for students and provide relevant information for improving their learning, it is essential that the feedback provides relevant information of the specific course or learning module [7, 31–33].

We call all this data *domain knowledge*, which refers, for instance, to information regarding the specific learning objects that are used (tasks, assessment exercises, etc.); the specific and general goals, and evaluation/assessment criteria of a particular learning object but also for the course; the methods and processes for performing a task; the competences or skills that students will acquire in the course or with a specific learning object; etc. It is important to note that this information must be represented in a machine-understandable format and correctly and accurately interlinked to properly construct the feedback content using a SLE.

## 3.2 Feedback Presentation or Delivery

How and when the feedback is presented to the learner may have a big impact on the effect of feedback since it can make a student pay more attention to it or perceive it and/or remember it better or worse [34].

### 3.2.1 Presentation Format

How to present feedback to students is a research matter that has been studied by several authors. For instance, in [10], the authors reported that although metacognitive feedback specific to the problems that the learners were facing was provided, a lot of learners simply did not read the feedback. The authors planned to undertake further research by using eye-tracking technology and to test different page layouts for presenting the motivational and metacognitive feedback to raise their salience. [6] analyses the efficiency, from the student's point of view, of different ways (e.g., text, video, screen-shots) of giving feedback to students' continuous assessment activities. In [27], the authors propose blogs for collecting and providing weekly and reflective ongoing feedback.

Different ways of presenting feedback can be used, e.g., in written, video, or audio format; shown in an external window independent from the learning environment interface or embedded in it; via chat dialogue with the teacher, blogs [27], text messages, emails, lofts [20], rubrics [4], etc. However, more research is still needed to understand which type of feedback presentation is likely to be more effective in each specific situation or problem.

### 3.2.2 Timing

Educational research suggests that timely feedback can have a good effect on learners and that providing feedback at the right moment is very important to allow students to improve their performance, change their strategies, or reconsider their thinking about the problem to be solved [35]. The most common moments to provide feedback are:

- *From the beginning of a task/process, anytime*: it is provided at the beginning of a task or a process and available anytime on demand. For instance, feedback in this category could be a summary of exercises performed by students from previous years with a list of mistakes and their corrections.
- *Immediate, just-in-time, or in-action* feedback [5, 20, 30, 36, 37]: it is given at the moment the problem arises to enable the learner to adapt instantly.
- *Delayed* feedback [5, 30, 36, 38]: it is provided a while after a task/process has been finished.
- *At the end of a learning task/process* or *About-action* feedback) [37]: it informs learners how they perform key skills after finishing a task and enables them to monitor their progress and adapt accordingly in subsequent tasks.

According to [5], the stage in the learning process may be also relevant in some cases. For instance, if the goal of the feedback is to encourage learners to spend more effort in order to reach success, then early stages in the learning process may be more adequate.

## 3.3  Context

That the feedback must be context-specific to be effective is nowadays an evident assumption. In [39], Gibbs affirms: *"in any particular context we need to be aware of key contextual variables that help us to identify where the pedagogic leverage is likely to be if we wish to intervene and make a difference"*. Besides Gibbs, other authors such as [4, 40] also highlight the importance of designing feedback that is context specific. The feedback recipient as well as the classroom climate are the most referred aspects in the literature to be considered as context.

### 3.3.1  Recipient

An important success for feedback is to make sure that all type of students understand it [40], therefore, the recipient of the feedback is a very important aspect to consider. Being aware of relevant information about the feedback recipient is essential to make the feedback personalized and therefore maximize its effect. The way and manner each individual interprets feedback is key to developing self-efficacy and leading to further learning [5].

The feedback recipient can be an individual or a group of learners. It is important than in the latter case, each learner still perceives the feedback as pertaining to him/her, otherwise, the feedback can be just considered as irrelevant [5]. For both, individual or group, it is important to take into account:

- *Learners' behaviour* [7, 10]: which learning actions (also known as the learning trajectory), the learner has executed to arrive to the current state and which action is currently doing.
- *Learners' cognitive/affective state* at the moment the feedback is communicated [41, 42]: The state of the learners (e.g., engaged, confused, frustrated, distracted, bored) can play a major role in their behaviour. For instance, while positive affective states can enhance learning, negative affective states can inhibit it.
- *Learners' intrinsic characteristics* [5]: self-esteem/confidence, commitment to work, culture, prior knowledge, learning style, self-efficacy, etc.

### 3.3.2  Classroom Climate

For the feedback to be used by the learners in an effective way, the environment or the climate of the classroom must also be taken into account. The classroom must leave time for feedback, encourage to seek for and use feedback to foster continual improvement, and allow learning from mistakes [1, 5, 30]. The climate of the classroom can even make feedback welcome or not welcome by students [5].

## 3.4 Use and Impact

In order to be effective, feedback must be used by the students. However, only a very small number of studies have been carried out that investigate students' actual **use of feedback** [34]. Weaver performed a study in the faculties of Business and Art and Design to explore student perceptions of feedback [33]. In this study, a majority of students said that feedback was useful for: motivation purposes; improving their coursework; reflecting on what they have learned; boosting their confidence in the case of positive comments; identifying gaps in knowledge and understanding; knowing how to improve. Ludvigsen et al. have recently identified different ways in which students use feedback in their course work [22], e.g.: to check if their learning is on track; to address difficult concepts with their peers or teacher; to adjust and focus the material to study; to identify difficult topics that they need to explore in more depth; to identify wrong applied learning strategies and adjust them. The authors state that there is a need to further investigate how students engage and work with feedback, both in lectures and in their course work [22].

Very closely related with the use that students make of feedback, it is the effect or **impact** that this feedback has on their learning. Studying this impact is crucial in order to be able to improve the feedback that is provided and make it more effective (retro feedback). To understand this effect or impact, it is important to understand how students really use the received feedback, e.g., by tracking which actions they perform in the learning environment after receiving feedback. The amount of time spent looking at feedback can also be an important predictor of the effectiveness of feedback [28]. While tracking the use of feedback or the time learners spend on processing it can be hard, time-consuming, or even impossible in a manual way, logging student behaviour in SLEs can provide a great opportunity to better understand the actual use of feedback.

## 4   Conclusions and Further Work

Although very challenging, personalized and automated feedback is essential to improve students' learning [5] and realize Smart Education. In order to help in this matter, this paper has presented a conceptual framework that compiles the most important dimensions to consider for automating feedback for learners. This framework attempts to be an analytical tool that helps researchers and educators towards the construction and delivery of high-quality and automated feedback within SLEs.

As further work, we plan to study which combinations of feedback content and presentation are recommended in the literature for each specific learning context to increase the use and the impact of feedback. Afterwards, we will investigate how the necessary information to automatically construct and deliver proper feedback following the recommended combinations can be sensed and analysed using learning analytics. Finally, we will exploit this knowledge to enrich and expand the automatic feedback provided in the learning environment *JMermaid* [9].

# References

1. Boud, D., Molloy, E.: Rethinking models of feedback for learning: the challenge of design. Assess. Eval. High. Educ. **38**(6), 1–15 (2012)
2. Spector, J.M.: Conceptualizing the emerging field of smart learning environments. Smart Learn. Environ. **1**(1), 2 (2014)
3. Hwang, G.J.: Definition, framework and research issues of smart learning environments a context-aware ubiquitous learning perspective. Smart Learn. Environ. **1**(4), 1–14 (2014)
4. Campos, D.S., Mendes, A.J., Marcelino, M.J., Ferreira, D.J., Alves, L.M.: A multinational case study on using diverse feedback types applied to introductory programming learning. In: The 42nd Frontiers in Education Conference—FIE' 12, pp. 1–6 (2012)
5. Hattie, J., Timperley, H.: The power of feedback. Rev. Educ. Res. **77**(1), 16–7 (2007)
6. Martínez-Argüelles, M.J., Plana-Erta, D., Hintzmann-Colominas, C., Badia-Miró, M., Batalla-Busquets, J.M.: Usefulness of feedback in e-Learning from the students' perspective. In: European Conference on e-Learning (ECEL), pp. 283–292 (2013)
7. Pears, A., Harland, J., Hamilton, M., Hadgraft, R.: What is feedback? connecting student perceptions to assessment practices. Learn. Teach. Comput. Eng. LaTiCE **2013**, 106–113 (2013)
8. Tanes, Zeynep, Arnold, Kimberly E., King, A.S., Remnet, M.A.: Using signals for appropriate feedback: perceptions and practices. Comput. Educ. **57**(4), 2414–2422 (2011)
9. Sedrakyan, G., Snoeck, M.: Technology-enhanced support for learning conceptual modeling. In: Enterprise, Business-Process and Information Systems Modeling, pp. 435–449 (2012)
10. Hull, A., du Boulay, B.: Motivational and metacognitive feedback in SQL-Tutor*. Comput. Sci. Educ. **25**(2), 238–256 (2015)
11. Ying, M.-H., Hong, Y.: The development of an online SQL learning system with automatic checking mechanism. In: The 7th International Conference on Networked Computing and Advanced Information Management, pp. 346–351 (2011)
12. Aljohani, N.R., Davis, H.C.: Learning analytics and formative assessment to provide immediate detailed feedback using a student centered mobile dashboard. In: International Conference on Next Generation Mobile Applications, Services, and Technologies, pp. 262–267. Information Technology and Computer Science, Jeddah-saudi Arabia (2013)
13. Gikandi, J.W., Morrow, D., Davis, N.E.: Online formative assessment in higher education: a review of the literature. Comput. Educ. **57**(4), 2333–2351 (2011)
14. Chen, D.D.: Understanding and encouraging feedback seeking behaviour: a literature review. Med. Educ. **47**(3), 232–241 (2013)
15. Hepplestone, S., Holden, G., Irwin, B., Parkin, H.J., Thorpe, L.: Using technology to encourage student engagement with feedback: a literature review. Res. Learn. Technol. **19**(2), 117–127 (2011)
16. Gabelica, C., Van Den Bossche, P., Segers, M., Gijselaers, W.: Feedback, a powerful lever in teams: a review. Educ. Res. Rev. **7**(2), 123–144 (2012)
17. Thurlings, M., Vermeulen, M., Bastiaens, T., Stijnen, S.: Understanding feedback: a learning theory perspective. Educ. Res. Rev. **9**, 1–15 (2013)

18. Evans, C.: Making sense of assessment feedback in higher education. Rev. Educ. Res. **83**(1), 70–120 (2013)
19. Ada, M.B.: MyFeedBack: an interactive mobile Web 2.0 system for formative assessment and feedback. In: International Conference on e-Learning and e-Technologies in Education, pp. 98–103 (2013)
20. Easterday, M.W., Rees-Lewis, D., Gerber, E.: Formative feedback in digital lofts: learning environments for real world innovation. In: CEUR Workshop **1009**, 1–8 (2013)
21. Loewen, S., Erlam, R.: Corrective feedback in the chatroom: an experimental study. Comput. Assist. Lang. Learn. **19**(1), 1–14 (2006)
22. Ludvigsen, K., Krumsvik, R., Furnes, B.: Creating formative feedback spaces in large lectures. Comput. Educ. **88**, 48–63 (2015)
23. Sadler, R.D.: Formative assessment and the design of instructional systems. Instr. Sci. **18**(2), 119–144 (1989)
24. Jones, A., Gaved, M., Kukulska-Hulme, E., Scanlon, A., Pearson, C., Lameras, P., Dunwell, I., Jones, J.: Creating coherent incidental learning journeys on smartphones using feedback and progress indicators. Int. J. Mob. Blended Learn. **6**(4), 75–92 (2014)
25. Burgers, C., Eden, A., van Engelenburg, M.D., Buningh, Sander: How feedback boosts motivation and play in a brain-training game. Comput. Hum. Behav. **48**, 94–103 (2015)
26. Espasa, A., Guasch, T.: Analysis of feedback processes in online group interaction : a methodological model. Digit. Educ. Rev. **23**, 59–73 (2013)
27. Stone, J.A.: Using reflective blogs for pedagogical feedback in CS1. In: The 43rd ACM Technical Symposium on Computer Science Education, pp. 259–264 (2012)
28. Miller, L.D., Soh, L.K., Nugent, G., Kupzyk, K., Masmaliyeva, L., Samal, A.: Evaluating the use of learning objects in CS1. In: The 42nd ACM Technical Symposium on Computer Science Education, pp. 57–62 (2011)
29. Mitrovic, A., Ohlsson, S., Barrow, D.: The effect of positive feedback in a constraint-based intelligent tutoring system. Comput. Educ. **60**(1), 264–272 (2013)
30. Tracey, M.W., Boiling, E.: Feedback models for learning, teaching and performance. In: Handbook of Research on Educational Communications and Technology (2014)
31. Grimalt-reynes, A., Inverno, M.: Designing educational social machines for effective feedback. In: International Conference e-Learning, pp. 239–248 (2014)
32. Iskandar, Y.H.P., Gilbert, L., Wills, G.B., Basir, N.: The development of semantic feedback for teaching and learning in physical education. In: Advanced Learning Technologies (ICALT 2013), pp. 385–387 (2013)
33. Weaver, M.R.: Do students value feedback? student perceptions of tutors' written responses. Assess. Eval. High. Educ. **31**(3), 379–394 (2006)
34. Jonsson, A.: Facilitating productive use of feedback in higher education. Act. Learn. High. Educ. **14**(1), 63–76 (2013)
35. Rasmussen, I., Lund, A., Smørdal, O.: Visualisation of trajectories of participation in a wiki: a basis for feedback and assessment? Nord. J. Digit. Literacy **2012**(1), 20–35 (2012)
36. Mathan, S., Koedinger, K.R.: Fostering the intelligent novice: learning from errors with metacognitive tutoring. Educ. Psychol. **40**(4), 257–265 (2005)
37. Van Rosmalen, P., Börner, D., Schneider, J., Petukhova, O., Van Helvert, J.: Feedback design in multimodal dialogue systems. In: 7th International Conference on Computer Supported Education, pp. 209–217 (2015)
38. Steif, P.S., Fu, L., Kara, L.B.: Computer tutors can address students learning to solve complex engineering problems. In: Frontiers in Education Conference (FIE), pp. 1–8 (2014)
39. Gibbs, G.: The importance of context in understanding teaching and learning: reflections on thirty five years of pedagogic research. In: Keynote at the International Society for the Scholarship of Teaching and Learning (2012)
40. Segedy, J., Kinnebrew, J.S., Biswas, G.: Supporting student learning using conversational agents in a teachable agent environment. In: The 10th International Conference of the Learning Sciences (ICLS 2012), pp. 251–255 (2012)

41. Grawemeyer, B., Mavrikis, M., Holmes, W., Hansen, A., Loibl, K., Guti, S.: The impact of feedback on students affective states. In: International Workshop on AMADL 2015 (2015)
42. Molloy, E.: Time to pause: giving and receiving feedback in clinical education. In: Delany, C., Molloy, E. (eds.) Clinical Education in the Health Professions, pp. 128–146. Chatswood, New South Wales, Australia (2009)

# Development of Smart-System of Distance Learning of Visually Impaired People on the Basis of the Combined of OWL Model

Galina Samigulina and Assem Shayakhmetova

**Abstract** The study describes the development of the combined model OWL (Web Ontology Language) to implement intelligent innovative technology and construction of Smart-system of distance learning visually impaired people. The proposed combination ontological model includes an ontological model of learner, learning and joint use laboratory. In the model the learner processing multidimensional data is based on neural networks, which allow the reduction of uninformative signs and selects the optimal tactics of learning. Creation model of learning adapted to the model of learner by using fuzzy logic, which defines the class as a learner vision and current knowledge. The ontological model of laboratory of joint use describes the remote access of learners to the laboratory of joint use and works on the newest expensive processing equipment for laboratory, practical works in real time. The ontological models are complementary, interconnected and allow us to determine the appropriate path learning with effective elements of the course. The combined OWL model allows for a systematic approach to construction Smart-systems based on methods of artificial intelligence and cognitive approach, facilitates writing Software, Hardware selection and helps to create effective individual process of learning visually impaired people.

**Keywords** Distance learning · Smart-system · OWL model · Visually impaired people · Cognitive approach · Artificial intelligence methods

G. Samigulina (✉)
Technical Sciences, Laboratory of the Institute of Information
and Computational Technologies, Almaty, Kazakhstan
e-mail: galinasamigulina@mail.ru

A. Shayakhmetova
Kazakh National Research Technical University named after K.I. Satpayev,
Almaty, Kazakhstan
e-mail: asemshayakhmetova@mail.ru

© Springer International Publishing Switzerland 2016
V.L. Uskov et al. (eds.), *Smart Education and e-Learning 2016*,
Smart Innovation, Systems and Technologies 59,
DOI 10.1007/978-3-319-39690-3_10

## 1 Introduction

A distinctive feature of the modern world has become a global development of information technology. Particularly effective their use in education. Today distance learning (DL) is one of the most rapidly developing areas in the global educational environment.

There are quite a number of software systems for the organization of DL. The organization of educational process in use to special information management systems (LMS—Learning Management System) [1] for performing administrative and technical support processes associated with e-learning. This system is a software platform for the deployment of e-learning in the organization and automation of educational activities. Information management systems are divided into systems with open-source and commercial [2].

Systems with open source are: eFront, Sakai, Olat (Switzerland), Ilias, Moodle (Modular Object-Oriented Dynamic Learning Environment) (Australia).

Commercial LMS systems are: WebTutor (Russia), eLearning Server, Prometheus (Russia), DLCENT (Russia), Redclass Pro (Russia), Learn eXact, Blackboard Learn (Russia).

The educational environment Moodle [3] is widely used in modern educational environment, as is free, well integrated with other information systems, supplemented with new services, support functions, reports, establishes ready or develops new additional modules. Recently, ontological models (OM) in Moodle are widely used as a basis for creation of the systematized structures.

The concept of Smart-education. It is the new approach in education providing the high level of development corresponding to tasks and opportunities of the modern world. It defines interrelation between the knowledge used by technologies and methods of teaching. Transition to Smart-education gives the chance of wide use of innovative technologies. The system of key features of Smart-education is offered in article [4]. In research [5] the technology of Smart system-education, which is connected online to virtual university is considered and is based on a web-technologies. The intellectual learning system determines the strengths and weaknesses of learners in real time. The possible level of learning and security algorithm are selected. Concepts Smart-learning [6] includes a flexible learning, consisting of a large number of sources, the maximum diversity of media (audio, video, graphics), the ability to quickly and easily adjusted to the level and needs of the learners by using mobile devices.

The basis of the Smart-systems are intelligent approaches [4] and the development of such systems DL actively apply the ontological approach. In article [7] the ontological system of Smart-class is developed. The system helps to understand relationship and to analyze existing "intelligent equipment" in a class, to define their functions, software, approaches of teaching and learning. This approach allows to organize systematized interactive courses of DL.

Especially actual problem is development of similar systems for people with special needs, including with disabilities of sight. In this article is developed the

Smart-system of DL of visually impaired people (VIP) that can be used as components of Moodle or other LMS.

The next structure of the article is proposed: the second chapter conducted a literature review of DL systems based on the methods of artificial intelligence (AI) and ontological models. The third chapter defines the aims and tasks of the research. The fourth chapter describes the construction of OWL model for the Smart-system of distance learning of VIP in ontology editor Protégé. The fifth chapter describes the results of the study. The sixth chapter is devoted to discussion of conclusions. At the end of the article there is a conclusion and given a list of used literature.

## 2  Literature Review of System of DL for VIP Based on Artificial Intelligence Methods and Ontological Models

By the development of systems of distance learning are widely used various methods of artificial intelligence [8, 9], such as neural networks (NN), fuzzy logic (FL), the neuro-fuzzy logic (NFL), genetic algorithms (GA), artificial immune systems [10] etc.

There are many publications on this topic. The research [11] proposes the implementation of a neural network model of forecasting efficiency of educational process in DL. The model is based on two parameters: the total combined score (all the modules of the learning course), and the total time spent by learners in the development of materials of all modules. In article [12] the features of use of a multilayer perceptron for automated control of knowledge are given in e-learning courses. A research objective [13] is improvement of quality of control of the acquired knowledge learnt in DL. Correctness of the response is carried out using a neural network.

In creation of DL based on intellectual methods, ontological models are actively used. In work [14] use of artificial neural networks (learning without teacher) for the solution of problems of a clustering with use of the self-organizing Kokhonen's cards is supposed. The research [15] for the construction of OM mobile learning is used approach based on the Semantic web. In article [16] the authors conclude on the need to use fuzzy logic in the design and development of intelligent subsystems for technology of DL. The algorithm and example are given an of fuzzy logic at creation of elements of expert systems for DL.

There are special tools for visual construction created by OM on computer. The most applied ontological editors: Ontolingua, OntoStudio, WebOnto, Protégé. The main criteria by which it is possible to make the comparative analysis of editors of ontologies are: openness and expansibility of architecture, language, format, access, collective network remote editing and others. All above criteria are answered by the editor of ontologies of Protégé.

Application of the AI methods and ontological approach is actual for creation of Smart-system of DL VIP. Development of these systems have their own specific features. It is especially important to consider the physiological features of perception and awareness of the proposed educational information of VIP. Often there are the following eye problems: myopia, hyperopia and astigmatism. Working capacity and visual acuity at trained with a myopia is below [17], concentration of attention and logical memory is higher in comparison with trained with normal sight. At learnt with short-sightedness endurance and working capacity is lower, in comparison with healthy contemporaries. For visually impaired memory is important. A large number of information should be stored in memory. Problems of visual impairment leads to a decrease in the perceived quality of information by learners. In learning VIP no less important to pay attention to the physiological characteristics of learners by developing their skills. Learners with low vision is developed hearing and touch. The ability as much as possible to use the acoustic analyzer is the main task in learning [18]. Different degrees of visual impairment (the nature of diseases and the degree of violations of basic visual functions) affect physiological characteristics of VIP in learning. The study of psycho-physiological characteristics of the visual system VIP gives the chance of an integrated approach to the organization of process of learning. Various techniques at different loads are applied to studying of physiological features of the visual device of people with the weakened sight [19].

## 3 Purpose and Objectives of Research

The purpose of the research is development of the combined OWL model for realization of intellectual innovative technology and creation of Smart-system distance learning of visually impaired people. For achievement of this purpose it is necessary to solve the following tasks: to develop the intellectual technology of creation of Smart-system of DL VIP based on the ontological approach; to present the ontological model of learner based on neural networks and cognitive approach; to develop the ontological model of learning based on fuzzy logic; to create the ontological model of laboratory of joint use (LJU) which describes remote access and work of VIP on the newest expensive processing equipment; to construct the combined OWL model allowing to realize system approach on the basis of methods of artificial intelligence and cognitive approach (taking into account the psycho-physiological characteristics of the knowledge obtained information).

## 4  Creation of OWL Model for Smart-System of Distance Learning of Visually Impaired People in the Editor of Ontology of Protégé

One of a ways to organize data is the method of submission information using OWL model [20]. The OWL language describes ontology for Web, which is applied at representation of objects with various complexity of structure. Ontology of OWL has three subsets of terms: OWL Lite—is designed for the solution of simple tasks. OWL DL is designed for users who need maximum degree of language features. OWL Full is designed for users who need to maximize the expressive possibilities of language and a freedom to choose the final RDF format (Resource Description Framework). In OWL identified two categories of properties: properties—objects that are linked to each other individuals (instances of classes) and properties—the values that connect individuals with the values of the data. The main elements of language are properties, classes and restrictions. In creation of the combined OWL model of Smart-system for VIP applied the ontology editor Protégé. The complete list of terms are divided on categories depending on their function in OM. The concepts (concepts or domain terms) is an object, such as "OM learner" or "Learning materials" are presented in the form of classes. Properties of classes, such as "The mode for VIP" or "Lectures", "Lab" are considered as slots. The system consists of the following basic classes such as OM of learners, OM of learning and OM of laboratory of joint use.

## 5  Results of Research

The algorithm of functioning of Smart-system of DL of VIP, which works as follows is constructed: the user must be registered for access to system. Further the model of learner of VIP taking into account their individual characteristics. The system uses a step questionnaire, which consists of two stages: a questionnaire 1 and 2. The first stage of the questionnaire is based on a cognitive approach [21, 22] to identify the physiological, mental and psycho-physiological characteristics information in which cognitive processes (perception of information, awareness, attention, memory, imagination and thinking) are considered as components of the overall process of information exchange between the learners and the environment. On the basis of processing of the questionnaire results are based vector of parameters, which is fed to the input of the NN to highlight the informative features. In the second phase questionnaire to determine the level of learning VIP answers the general questions on the chosen discipline for setting the initial level of knowledge. Further for creation of the model of education adapted to the learning model is used by FL. The fuzzy logic Mamdani type used in classifying learners by the results of the questionnaire and on the level of knowledge of VIP are selected learning class: entry, mean and continuing level.

The obtained data is recorded in the database. After studying a particular course on the chosen level, VIP passes intermediate test. According to the results of the intermediate testing is carried predicting outcomes based on learning NFL. Expeditious correction of process of learning is carried out. Further, final testing is performed. On the result of total testing the analysis of learning is performed and assessment of learning of VIP is carried out. The system sends to VIP on a new course, completion or relearning, if it is not the development of this course. After completing the course receive a certificate.

Ontological model of learner with impaired vision (Table 1), which helps to identify the intellectual, physiological and psycho-physiological peculiarities of perception VIP is designed.

**Table 1** The ontological model learner in Smart-system to visually impaired people

| Scope | Details |
|-------|---------|
| Ontological model of learner | Registration of VIP in Smart-system |
| | Application of cognitive approach for the identification of physiological, intellectual and psycho-physiological characteristics of perception and understanding of information |
| | Questionnaire of VIP for determining the individual characteristics |
| | Development of a database of individual features of VIP |
| | Data processing and preliminary data processing: normalization, filling in missing data, etc. |
| | Selection informative features on the basis of the NN |
| | Selection of the optimal tactics of learning with individual characteristics of VIP |
| | Selecting the study: normal mode, modes for VIP |
| | Changing the contrast in VIP mode |
| | Change the font size |
| | Illumination |
| | Motion |
| | Changes in the color of the screen and the color of the provided information, depending on the psycho-physiological characteristics of the learner (choice of color schemes) |
| | Location information on the monitor depending on the psycho-physiological characteristics of understanding of information and features of sight |
| | Connections of sound (For example, computer software JAWS, Job Access With Speech) |
| | Adjustment • Sound heights • Sound intensity • Timbre • Localizations of a sound |
| | Ability to connect a Braille keyboard |

Visually impaired perceive world around a little differently than people with developed sight [23]. Environment and really looks much more pale in gray and in black–white tones. The developed system considers features of VIP to reading information from the monitor screen (illumination, contrast, movement, size, color). The system provides special color schemes, which are preferred for people with different eye diseases. For increase of efficiency of perception of a learning material scoring is provided in which height adjustment, sound intensity, timbre and sound localization is possible.

Further the ontological model of learning (Table 2) adapted to the model of learner in the Smart-system of DL visually impaired people is constructed.

**Table 2** The ontological model of learning in Smart-system of DL of visually impaired people

| Scope | Details |
| --- | --- |
| Ontological model of learning | The choice of learning discipline |
| | Questionnaire of VIP to determine the level of knowledge in the learning discipline |
| | Creation of Knowledge Base |
| | Selection a learning course |
| | Choice of class-learning based on FL |
| | Selection of entry level course |
| | Selection of mean course |
| | Selection of continuing level course |
| | Selection of learning paths of VIP based on adaptation of learner model to the learning model |
| | Presentation of learning material with interactive elements of course |
| | Development of learning materials |
| | Formation of lecture materials |
| | Development of laboratory works |
| | Creation of practical tasks |
| | Preparation of independent works |
| | Creation of help system |
| | Organization of educational consultations in the form of a forum, chat |
| | Creation of service of search system |
| | Development of directory and a glossary of discipline |
| | Creation of digital libraries with electronic textbooks |
| | Intermediate control of knowledge of VIP |
| | Organization of examinations in the discipline |
| | Implementation of the intermediate tests |
| | Forecasting of results of learning on the basis of NFL |
| | Operative correction of learning process |
| | Final testing |
| | Assessment of learning of VIP |
| | Certification or re-learning |
| | Issue of a certificate |

In learning model the individual characteristics of VIP to build the adapted model for specific learners trajectory of learning are considered. Also, the model reflects the process of learning control VIP, including analysis of results of forecasting the knowledge and the adjustment of the learning process of VIP.

The ontological model of laboratories of joint use (Table 3) describes remote access of learners to the expensive processing equipment for laboratory, practical works in real time. This model defines interrelations between the main devices of computing cluster and considers connection methods of VIP to the educational equipment in LJU using virtual machine.

On the basis of research in the ontology editor Protégé was constructed the combined OWL model of Smart-system of DL visually impaired people consists of the model of learner, learning and laboratory of joint use. The proposed models complement each other and are interconnected. This combined OWL model [24] allows to analyse more deeply numerous communications between ontologic models and to consider them when developing the software of Smart-system of DL of VIP.

In software development were used: scripting language general-appointment PHP, the programming language Python, content management system open source WordPress, the web server Apache, a free database management system MySQL. The developed software can be used as MOODLE component.

**Table 3** The ontological model of laboratories of joint use in Smart-system of DL of visually impaired people

| Scope | Details |
|---|---|
| Ontological model of laboratories of joint use | Ensuring connection of VIP remote computers through Internet to LJU |
| | Organization of access to a computing cluster |
| | The use of server for management and distribution of tasks between computing nodes of a cluster |
| | Creation of virtual computer for providing remote access and installation of the necessary software for VIP |
| | The use of system of data storage for optimum operation of applications, the solution of resource-intensive technical tasks and effective data security |
| | The use of switches for ensuring continuous work of network structure and the prevention of unauthorized access to the system |
| | The use of an uninterruptible power supply for continuous operation |
| | Connection to the learning equipment (learning laboratory stands, industrial equipment, etc.) |

# 6 Discussion

The conducted researches and numerous publications on this question demonstrate the relevance of development of Smart-systems of DL based on intellectual approaches for visually impaired people, including with visual impairments.

The developed Smart-system of distance learning of visually impaired people has a number of advantages: a systematic approach to creation of Smart-system of distance learning based on AI methods, ontological and cognitive approach are applied; the OWL model in the Protégé editor is designed; the database of individual features of VIP (detection of physiological, intellectual and psycho-physiological features of perception and understanding of information of VIP) is developed; effective processing of multidimensional data and allocation of informative signs of VIP on the basis of NN is carried out; adaptation of learners model to learning model based on fuzzy logic; to forecast the results of learning on the basis of NFL; expeditious correction of process of training is made; features of VIP to reading of information from the monitor screen depending on psycho-physiological characteristics of perception of information and features of sight are considered; various color schemes of perception of information are used; possibility of work in real time for performance of laboratory, practical works learnt on the newest expensive processing equipment in laboratory of joint use.

# 7 Conclusion

Smart system of distance learning is developed for visually impaired people based on the combined OWL model allows to structure the input and output data, considers feature of functioning of the software.

The developed software received the certificate on the state registration of the rights for object of copyright and the act of introduction in the Almaty branch of the public association "Kazakh Society of Blind People" (Almaty).

The work has been performed under grant project of "Development of the Information Technology, Algorithms and Software and Hardware for Intelligent Control Systems for Complex Objects under Conditions of Parameter Uncertainty" No. GR 0215RK01472 (2015–2017).

# References

1. McIntosh, D.: Vendors of Learning Management and e-Learning Products, pp. 61–63. Trimeritus eLearning Solutions Inc. (2013)
2. Negiz, M.: Distance education implementations and the case of Russia. E-J. UDEEEWANA **1** (3), 48–60 (2015)

3. Jalobeanu, M., Naaji, A.: Using Moodle platform in distance education. In: The 7th International Scientific Conference e-Learning and Software for Education, pp. 402–409 (2011)
4. Tikhomirov, V.: The world on a way to smart education: new opportunities. Open Educ. **3**, 22–28 (2011)
5. Uskov, V., Lyamin, A., Lisitsyna, L., Sekar, B.: Smart e-Learning as a student-centered biotechnical system. Lect. Notes Inst. Comput. Sci. Soc. Inf. Telecommun. Eng. **138**, 167–175 (2014)
6. Sucheta, V., Radhika, M., Pai, M., Manohara, Pai: Modified literature based approach to identify learning styles in adaptive e-Learning. Adv. Comput. Netw. Inf. Springer **27**, 555–564 (2014)
7. Uskov, V., Bakken, J., Pandey, A.: The ontology of next generation smart classrooms. In: Smart Education and Smart e-Learning, vol. 41, pp. 3–15. Springer International Publishing Switzerland (2015)
8. Samigulina, G., Shayakhmetova, A.: The information system of distance learning for people with impaired vision on the basis of artificial intelligence approaches. In: Smart Education and Smart e-Learning, vol. 41, pp. 255–265. Springer International Publishing Switzerland (2015)
9. Rana, H., Rajiv (Prof.), L.M.: Role of artificial intelligence based technologies in e-learning. Int. J. Latest Trends Eng. Sci. Technol. **1**, 24–26 (2014)
10. Samigulina, G.: Technology Immune–Network Modeling for Intellectual Control Systems and Forecasting of the Complex Objects. Monograph, p. 172. Science Book Publishing House, USA (2015)
11. Bakhtizin, R., Gallyamov, A., Zaripov, E., Sakharov, L., Fatkullin, N., Shamshovich, V.: Enhancing feedback from trainees remotely on the basis of the associative model of phase forecasting. Open J. Syst. **1**(13), 21–23 (2015)
12. Sayed, M., Baker, F.: E-Learning optimization using supervised artificial neural-network. J. Softw. Eng. Appl. **8**, 26–34 (2015)
13. de Oliveira Neto, J.D., Nascimento, E.V.: Intelligent tutoring system for distance education. J. Inf. Syst. Technol. Manage. **9**(1), 109–122 (2012)
14. Alias, U., Ahmad, N., Hasan, S.: Student behavior analysis using self-organizing map clustering technique. ARPN J. Eng. Appl. Sci. **10**(23), 17987–17995 (2015)
15. Rodzin, S.I.: M-learning – upravlenie kontentom v kontekstno-zavisimoj mobil'noj sisteme obuchenija. Inf. Comput. Sci. Eng. Educ. **1**(16), 53–61 (2014)
16. Molchanov, A., Chvanova, M., Kiselyova, I., Hramova, M.: Using the machine theory of fuzzy sets the design of modern technologies of distance learning. Educ. Technol. Soc. **2**, 447–468 (2015)
17. Goldschmidt, E., Jacobsen, N.: Genetic and environmental effects on myopia development and progression. Sci. J. R. Coll. Ophthalmol. **28**(2), 126–133
18. Nikashina, N.: Particular of development people with confined eyesight. Vector Sci. **2**(9), 223–226 (2012)
19. Rocheva N.I.: Individual ways of teaching English to the blind. Eur. Stud. Sci. J. 1–44 (2014)
20. Shadbolt, N., Berners-Lee, T., Hall, W.: The semantic web revisited. IEEE Intell. Syst. **21**(3), 96–101 (2006)
21. Kovina, T.: Cognitive approach to learning. In: 77th International Science Engineering Conference, pp. 99–301 (2012)
22. Klingberg, T.: Training and plasticity of working memory. Trends Cogn. Sci. **14**(7), 317–324 (2010)
23. Roshchina, M.: Tifloinformational literacy as a factor of improving the quality of life of visually impaired people. J. Vestnik Lobachevsky State Univ. Nizhni Novgorod. **4**, 76–81 (2013)
24. Samigulina, G., Shayakhmetova, A.: Combined ontological model for distance education of visually impaired people. Probl. Inf. **3**, 28–36 (2015)

# Part II
# Smart Education: Research and Case Studies

# Data Mining of Students' Behaviors in Programming Exercises

Toshiyasu Kato, Yasushi Kambayashi and Yasushi Kodama

**Abstract** Programming exercises are time-consuming activities for many students. Therefore, most classes provide meticulous supports for students by employing teaching assistants (TAs). However, the programming behaviors of a particular student are quite different from other students' behavior, even though they are solving the same problem. It is hard for TAs to understand the detailed features of each student's programming behavior. We have performed data mining over the records of students' programming behaviors in order to elicit the detailed features of each student's programming behavior. The purpose of this study is to present the elicited such features for TAs so that they can provide effective assistances. We have performed data mining over the chronological records of the compilation and execution of individual students. As a result, we have found that there is a correlation between the programming activities and the duration time for problem solving. Based on the data mining, we have provided TAs some guidelines for each particular group of students. We have confirmed that our classifications and guidelines are reasonable through experiments over programming exercises. We have observed students who received appropriate guidance based on our data mining improved their programming performances.

**Keywords** Programming exercises · Teacher assistants teaching · Learning analytics

T. Kato (✉) · Y. Kambayashi
Nippon Institute of Technology, Saitama, Japan
e-mail: katoto@nit.ac.jp

Y. Kambayashi
e-mail: yasushi@nit.ac.jp

Y. Kodama
Hosei University, Tokyo, Japan
e-mail: yass@hosei.ac.jp

© Springer International Publishing Switzerland 2016
V.L. Uskov et al. (eds.), *Smart Education and e-Learning 2016*,
Smart Innovation, Systems and Technologies 59,
DOI 10.1007/978-3-319-39690-3_11

# 1    Introduction

During the programming exercises in higher educational institutes, students are individually tacking to the assigned problems. There are certain students who can easily solving the problems. On the other hand, there are many students who consume too much time to solve the same problems. For this reason, most institutes provide teaching assistant (TA) supports for programming classes. In order for TAs to effectively support the students' problem solving, TAs are expected not only to answer the questions from the students but also to grasp the particular students who really need help [1]. The TAs need to understand who need and what kind of assistances in what situations [2].

During the programming exercises, the main role of TAs is to assist students to solve the errors. The programming behavior of a particular student, which includes his or her programming style, is different from other students' behaviors. For this reason, it is hard for TAs provide appropriate guidance for each student other than error handling. The lecturers understand students' learning contexts from their teaching experiences. TAs cannot understand students' learning context because they do not have enough teaching experiences in the class.

In order to elicit the learning contexts of the students, we have performed data mining over the records of the students' behaviors in the programming classes. The purpose is to present TAs the feature of the programming behaviors of the students with difficulties. The factors of the problem solving in a programming include how much a particular students follow the programming codes, how much the particular students understand the grammar of the programming language, and how much the particular students understand the usages the compiler. This paper reports the result of our data mining that focuses the programming mode. The programming mode is the basic attitude that represents how much a particular student follows the programming codes that students are encouraged to follow in our institute. The reason why we focus the programming mode is that we can collect students' behavioral data in real-time through the programming exercise support system that we have developed [3].

# 2    Issues in the Programming Exercise

In the programming exercise, each student is individually tackling the assigned problem. The progress of the student is very different from each other depending on each skill levels [4]. The main reason why a particular student cannot progress is that his or her programming mode is not good. Such a student occasionally comes to a standstill. Such a student is often unmotivated [5]. In the programming exercise, we need to discover such students who have the difficulties in the programming in the early stages [1].

In the programming exercise, the lecturer and TAs go round students' consoles. Even such a situation, it is not easy for the lecturer and TAs to grasp which students need special assistance [5]. The lecturer and TAs check students' progresses through submitted programs and the students' screens. However, the number of TAs is limited. They are always too busy to check simple errors. We have to rely on the lecturer's teaching experience and intuition to solve problems of individual students [1].

The TA only knows how to solve particular errors. The TA only answers particular questions from students. The TA does not have enough teaching experiences to have the insight on students that the lecturers have. Therefore we need to provide TAs the detailed features of the programming behaviors of students.

## 3  Related Work

Researches on the grasping the programming behaviors include debugging training, programming tutorials, and an error analysis. Ryan has studied how to improve the debugging skills [6]. He has constructed a model from debugging logs and the development logs. The student can improve their debugging skills by using the model. The present study also analyzes the debugging logs. We also analyze other programming records.

Alex has proposed to employ tutors who support students to develop programs step by step [7]. The tutors examine students whether they take the right track. Students receive advices when their programs are incorrect. For example, the tutor gives students hint how to refactor their programs. The present study supports TAs to give students problem-solving techniques so that students can solve their own problems.

Kurasawa has implemented a system that provides students' common problems so that the students can share their problems [8]. This system infers where a particular mistake occurs from the past compile errors. The present study analyzes the programming mode.

## 4  Mining Students' Behaviors

The purpose of our mining of students' programming behaviors is to advise the TA how to guide students. The mining especially offers them the programming characteristics of the students that they are guiding.

The present study classifies the features of the students' programming behaviors in order to infer the characteristics of each student. Through the classification, we have examined the relationship between the results of questionnaire toward the programming mode and the programming behaviors. The behaviors include the number of compilations, the number of trial executions, the number of errors, and the number of repetitions of the same errors, average interval of the compilations, and average intervals of executions. The programming mode is the basic attitude for programming that how much the students follow the programming codes that they are encouraged to follow in our institute. We show the programming codes in Table 1. We have found that each student has a particular programming trait. We have also found that we can measure that trait by observing how much a particular student follows the programming codes.

## 4.1 Feature Classification of Programming Behaviors

The present study classifies the features of the programming behaviors. In order to do so, we have performed the cluster analysis over the records of the programming behaviors. Clustering is a process of grouping objects into classes of similar objects [9]. It is an unsupervised classification or partitioning of patterns (observations, data items, or feature vectors) into groups or subsets (clusters) based on their locality and connectivity within an $n$-dimensional space.

In the present study, we have performed the cluster analysis over the submitted programs that solve the assignments. The number of the subjects is eighty, and we

**Table 1** The programming codes

| 1 | When the grammar is ambiguous, examine it in the texts or manuals |
|----|----|
| 2 | Add one line of sentence, and compile |
| 3 | When compile errors appear out, deal with the first error |
| 4 | Construct programs from skeleton |
| 5 | When the execution result is not correct, trace the execution process |
| 6 | Insert spaces after keywords so that they are highlighted |
| 7 | Write output sentences first the behaviors of the program can be observed |
| 8 | Make and try several solutions to solve the errors |
| 9 | Insert spaces after commas, so that they are easily seen |
| 10 | When insert an opening parenthesis, insert corresponding closing parenthesis immediately |
| 11 | Write meaningful comments so that the semantics of the program can be understood |
| 12 | Choose meaningful variable names |
| 13 | When the usage of the instruction is ambiguous, refer the samples |
| 14 | Insert space lines so that blocks in the program clearly seen |
| 15 | Indent the codes so that the structure of the program can be clearly seen |
| 16 | Insert spaces before and after operators so that they are highlighted |
| 17 | When modifying the program, leave the old source codes as comments |
| 18 | When the program behaves strangely, print out intermediate variables |
| 19 | Write many more programs until you can write them comfortably |
| 20 | Use the patterns of program codes |

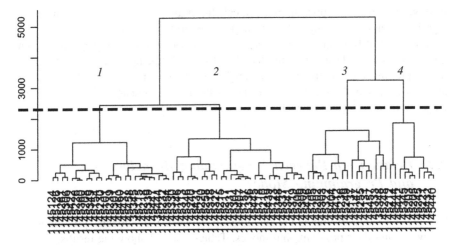

**Fig. 1** Cluster analysis by ward method

employed seven feature variables. Figures 1 and 2 show the result of the cluster analysis.

We have employed the Ward's method of a hierarchical clustering for the cluster analysis [10]. We have employed k-means for non-hierarchical clustering. Ward's

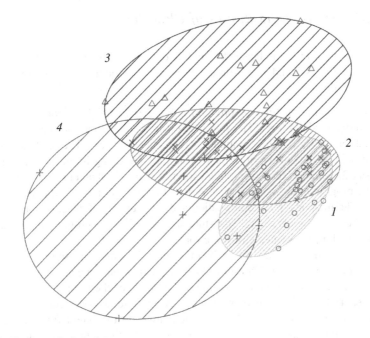

**Fig. 2** Cluster analysis by k-means

method is a criterion applied in hierarchical cluster analysis [11]. The k-means is an algorithm that clusters object based on attributes in $k$ partitions [12].

The result of k-means depends on the initial values. Therefore, we have chosen initial values as the best values produced by Pseudo-F (the number of clusters four) in our preliminary experiments. Pseudo-F is an evaluation criterion of the cluster analysis [13]. Good Pseudo-F value means the resulted clusters have little overlapping, and the density of each cluster is high.

## 4.2 Relationship Between Programming Behaviors and Programming Modes

The present study examines the relationship between the students' programming behaviors and their programming modes. Tables 2 and 3 show the results.

The numerical values in Tables 2 and 3 are mean values of the number of individuals in each cluster. The numerical values of the programming modes are the answers of four stage evaluation ($+1$ done and $-1$ not done). Moreover, we have performed the correlation analysis with the duration time for problem solving and programming mode. We have observed positive correlations (0.24) in 18 in the programming codes.

## 4.3 Discussion of the Results of Classification

The dotted line of Fig. 1 indicates the middle of the dendrogram, i.e. 2500. We can observe that Tables 2 and 3 are similar. We can conclude that we have found four clusters as follows:

- Cluster 1: Duration time of problem solving is short. The score of the programming mode is high. The intervals of the compilation and the intervals of execution are short. The compilation frequency is a few. The students understand the contents of the errors and what they are doing in the programming.
- Cluster 2: Duration time of problem solving is shorter than cluster 3. The score of the programming mode is low. A lot of errors and a lot of the same errors exist. The intervals of the compilation and the intervals of execution are shorter than those of the cluster 3. The students are doing the programming without understanding the contents of the errors.
- Cluster 3: Duration time of problem solving is long. The score of the programming mode is low. The students are repeating the same error. The students are compiling without understanding the contents of the errors.
- Cluster 4: Duration time of problem solving is long. The score of the programming mode is low. In Table 1, there are a lot of compilation frequencies and execution frequencies. The students compile frequently, and are causing

**Table 2** Breakdown of dotted line upper floor layer in Fig. 1

| Cluster | Solving time | Compilation interval | Execution interval | Compilation frequency | Execution frequency | Number of errors | Number of same errors* | Codes of programing | Number of students |
|---------|-------------|---------------------|-------------------|----------------------|--------------------|-----------------|----------------------|--------------------|--------------------|
| 1 | 528 | 209 | 267 | 7 | 5 | 1 | 1 | 10 | 25 |
| 2 | 812 | 379 | 606 | 6 | 3 | 3 | 3 | 6 | 28 |
| 3 | 1405 | 590 | 1160 | 7 | 3 | 4 | 3 | 5 | 17 |
| 4 | 1506 | 228 | 337 | 18 | 13 | 5 | 4 | 4 | 10 |

**Table 3** Cluster breakdown of Fig. 2

| Cluster | Solving time | Compilation interval | Execution interval | Compilation frequency | Execution frequency | Number of errors | Number of same errors[a] | Codes of programing | Number of students |
|---------|--------------|----------------------|--------------------|-----------------------|---------------------|------------------|--------------------------|---------------------|--------------------|
| 1 | 514 | 239 | 289 | 6 | 5 | 1 | 1 | 9 | 27 |
| 2 | 884 | 364 | 589 | 7 | 4 | 3 | 3 | 8 | 30 |
| 3 | 1398 | 599 | 1216 | 7 | 3 | 4 | 3 | 7 | 15 |
| 4 | 1694 | 225 | 378 | 20 | 14 | 6 | 5 | 5 | 8 |

[a]The same person making the same error multiple times

many errors. In Table 2, the compilation frequency and the execution frequency is a few. Surprisingly, these students are submitting the correct solutions of the problems. They have copied the correct answers from cluster 1 students.

Upon observing the results of the cluster analysis, we can conclude that the students in the first cluster, i.e. royal to the programming codes, solve the programming assignments rather quick. The features of the programming behavior appear in the difference of the programming mode.

## 5  Assessment of Effects of TA

We set up a hypothesis from the consideration of the previous section. The hypothesis is that the programming behavior appears in the programming mode.

### 5.1  Experiments

In this section, we verify the hypothesis about the programing mode and programming behavior, i.e. how much the understanding level in the programming codes corresponds to the programming behavior. Moreover, the present study verifies the differences in the results of the questionnaires about the subjective understanding level of the students before and after the experiment. The TA instructs the students based on the result of the questionnaires before the experiment. The TA guides the basic attitudes within the low numerical value for understanding level based on the questionnaires.

The verification method is to evaluate the duration time of problem solving of the experimental group and the control group. The group consists of forty students who submit their solutions late. Then we split these students into two groups at random, the control group and the experimental group. The procedures of the experiment are as follows:

1. Perform understanding level inquiry (the first time) to the programming mode. Acquire the duration time for problem solving (before the experiments).
2. First experiment (Guidance by TA).
3. Second experiment (Guidance by TA).
4. Third experiment (Guidance by TA).
5. Fourth experiment (No guidance by TA). Perform understanding level inquiry (the second time) to the programming mode. Acquire the duration time for problem solving (after the experiments).

Figure 3 shows the change in the problem solving time. We can observe the improvement of the duration time of the problem solving in the experimental group. However, there was no significant difference between the two groups.

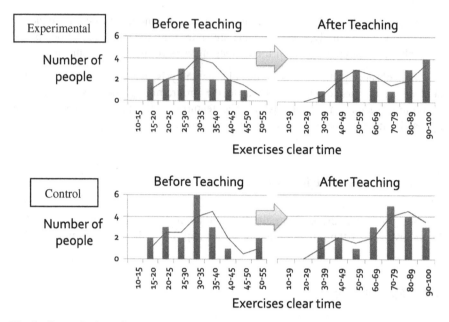

**Fig. 3** Change in time of problem solution

**Table 4** Result of the questionnaire of understanding level by student subjectivity

|  | Understanding frequency value | | | |
|---|---|---|---|---|
| Group | Before | After | Effective | None |
| Experimental | 6.66 | 9.74 46 % UP | 13 students | 4 students |
| Control | 5.84 | 8.2 41 % UP | 13 students | 6 students |

Table 4 shows the result of the understanding level inquiry. The understanding frequency value of the experimental group rose by 5 % compared with the control group. The result of the questionnaire according to the cluster is Table 5. In Table 5 the values that are in bold show where the improvements of the understanding levels are remarkable.

## 5.2 Discussion

The reason why the problem solving time has shortened may be the understanding level of the programming codes corresponds to the programming behavior. The data mining of students' behaviors enables the TAs to give appropriate guidance to the students. Because the results of the data mining presents TAs the features of the programming behaviors of the students.

**Table 5** Result of the questionnaire in each cluster

| Cluster | Matter | 1 | 2 | 3 | 4 | 5 | 6 | 7 | 8 | 9 | 10 | 11 |
|---|---|---|---|---|---|---|---|---|---|---|---|---|
| 1 | Before | 0.1 | 0.8 | 0.7 | 0.0 | 0.3 | 0.1 | 0.6 | 0.4 | -0.2 | 0.1 | 0.7 |
|  | After | -0.2 | 1.0 | **0.9** | 0.1 | 0.4 | 0.1 | 0.5 | 0.6 | **0.0** | 0.1 | 0.9 |
| 2 | Before | -0.1 | 0.3 | 0.7 | 0.2 | 0.3 | 0.3 | 0.8 | 0.4 | -0.4 | 0.6 | 0.7 |
|  | After | 0.2 | **0.7** | 0.7 | 0.2 | 0.3 | **0.7** | 0.9 | **0.9** | **0.1** | 0.6 | 0.9 |
| 3 | Before | 0.1 | 0.2 | 0.8 | -0.2 | 0.5 | 0.0 | 0.7 | 0.5 | 0.0 | 0.3 | 0.7 |
|  | After | **-0.2** | **0.6** | 0.9 | 0.1 | 0.3 | **0.6** | 0.8 | **0.8** | 0.1 | 0.5 | 0.7 |
| 4 | Before | -0.3 | 0.3 | 0.7 | -0.1 | -0.2 | 0.5 | 0.6 | 0.4 | -0.3 | 0.6 | 0.9 |
|  | After | **0.1** | 0.6 | **0.6** | **-0.3** | **0.1** | 0.6 | **0.8** | **0.3** | -0.1 | 0.6 | 0.7 |

| Cluster | Matter | 12 | 13 | 14 | 15 | 16 | 17 | 18 | 19 | 20 | Students |
|---|---|---|---|---|---|---|---|---|---|---|---|
| 1 | Before | 0.8 | 0.8 | 0.8 | 0.7 | 0.8 | -0.3 | 0.5 | -0.4 | 1.0 | 5 |
|  | After | 0.5 | 0.9 | 1.0 | 0.3 | 0.8 | -0.1 | 0.5 | **0.4** | 0.8 |  |
| 2 | Before | 0.1 | 1.0 | 0.9 | 0.3 | 0.6 | -0.2 | 0.3 | -0.3 | 0.8 | 13 |
|  | After | 0.2 | 0.7 | 0.8 | 0.3 | 0.7 | 0.0 | 0.5 | **-0.2** | 0.9 |  |
| 3 | Before | 0.3 | 0.3 | 0.8 | -0.2 | 0.3 | -0.4 | 0.1 | -0.4 | 0.7 | 11 |
|  | After | 0.3 | **0.6** | 0.8 | **0.2** | **0.6** | **0.0** | **0.4** | -0.4 | 0.9 |  |
| 4 | Before | 0.0 | 0.8 | 0.7 | 0.4 | 0.9 | -0.5 | -0.2 | -0.6 | 1.0 | 7 |
|  | After | **0.3** | 0.5 | 0.9 | 0.5 | 0.5 | **0.2** | -0.2 | -0.3 | 1.0 |  |

**Table 6** Effects of programming mode

| Cluster | Programming mode | |
| --- | --- | --- |
| | Effective | Ineffective |
| 1 | Examine the grammar<br>Write the comment<br>Make a lot of programs | None |
| 2 | Adjust the appearance<br>Write the comment | Make a lot of programs |
| 3 | Adjust the appearance | Compile for each line |
| 4 | Compile for each line<br>Deal with the first error<br>Copy the sentences that work | Examine the grammar<br>Write the output sentence first |

The reason why we did not observe significant difference between the two groups may be the existence of close connections of friends. The student shares the advice of the TA with their friend.

Guidance effects about programming mode are shown in Table 6. The data mining of students' behaviors enables TAs to give appropriate advices other than simple error corrections.

# 6 Conclusion

This paper analyzes the relation between the programming behavior and the programming mode. The result was related to the behavioral features and the programming mode. The authors have reached the hypothesis that the duration time of the problem solving is reduced as a result of the TAs instructing about the programming mode. Then, the authors performed the assessment experiments of guidance by TAs. As a result, the duration time of the problem solving of students who were taught has shortened. Therefore, the proposed technique enables TAs to give appropriate supports for students. Because TAs can obtain the deep understanding about the situation of each students.

We will perform further analysis about when a student performs what actions during his or her programming behaviors. This should enable TAs to present more effective supports about the programming mode.

**Acknowledgments** This work was supported by Japan Society for Promotion of Science (JSPS), with the basic research program (C) (No. 15K01094), Grant-in-Aid for Scientific Research, as well as Google MOOC Focused Research Award.

# References

1. Sagisaka, T., Watanabe, S.: Investigations of beginners in programming course based on learning strategies and gradual level test, and development of support-rules. J. Jpn. Soc. Inf. Syst. Edu. **26**(1), 5–15 (2009). (In Japanese)
2. Yasuda, K., Inoue, A., Ichimura, S.: Programming education system that can share problem-solving processes between students and teaching assistants. J. Inf. Process. Soc. Jpn. **53**(1), 81–89 (2012). (In Japanese)
3. Kato, T., Ishikawa, T.: Design and evaluation of support functions of course management systems for assessing learning conditions in programming practicums. ICALT **2012**, 205–207 (2012)
4. Horiguchi, S., Igaki, H., Inoue, A., et al.: Progress management metrics for programming education of HTML-based learning material. J. Inf. Process. Soc. Jpn. **53**(1), 61–71 (2012). (In Japanese)
5. Igaki, H., Saito, S., Inoue, A., et al.: Programming process visualization for supporting students in programming exercise. J. Inf. Process. Soc. Jpn. **54**, 1 (2013). (In Japanese)
6. Ryan, C., Michael, C.L.: Debugging: from novice to expert. ACM SIGCSE Bull. **36**(1), 17–21 (2004)
7. Alex, G., Johan, J., Bastiaan, H.: An interactive functional programming tutor. In: ITiCSE '12: Proceedings of the 17th. ACM, pp. 250–255 (2012)
8. Kurasawa, K., Suzuki, K., Iijima, M., Yokoyama, S., Miyadera, K.: Development of learning situation understanding support system for class instruction in programming exercises. Inst. Electron. Inf. Commun. Eng. Technol. Rep. ET, Edu. Eng. **104**(703), 19–24 (2005). (In Japanese)
9. Jain, A.K., Murty, M.N., Flynn, P.J.: Data Clustering: A review. ACM Comput. Surv. **31**(3), 264–323 (1999)
10. Kamishima, T.: A survey of recent clustering methods for data mining (part 1): try clustering! J. Jpn. Soc. Artif. Intell. **18**(1), 59–65 (2003). (In Japanese)
11. Michael, R.A.: Cluster Analysis for Applications. Academic Press, (1973)
12. MacQueen, J.: Some methods for classification and analysis of multivariate observations. In: Proceedings of the Fifth Berkeley Symposium on Mathematical Statistics and Probability. California, USA. vol. 1, pp. 281–297 (1967)
13. Calinski, T., Harabasz, J.: A Dendrite Method for Cluster Analysis. Commun. Stat. **3**, 1–27 (1974)

# Social Network Sites as a Good Support for Study Purposes

Blanka Klimova, Petra Poulova and Lucie Ptackova

**Abstract** Social network sites (SNSs) are nowadays used in all spheres of human activities, including education, thanks to their significant abilities to assist in engaging students to create, collaborate and share their learning content and outcomes. The aim of this study is to explore attitude of university students towards the SNSs with respect to their positive and negative experience, as well as their willingness to exploit SNSs as a support in their university studies and communication. Furthermore, the authors of this study summarize the main advantages and disadvantages of the use of SNSs for study purposes.

**Keywords** Social network sites · Education · Advantages · Disadvantages

## 1 Introduction

Social network sites (SNSs) are nowadays used in all spheres of human activities, including education, thanks to their significant abilities to assist in engaging students to create, collaborate and share their learning content and outcomes [1, 2]. They can be defined as web-based services that allow individuals to construct a public or semi-public profile within a bounded system; articulate a list of other users with whom they share a connection; and view and traverse their list of connections and those by other within the system [3]. They also put the main emphasis on their social aspect since these social network sites enable users to articulate and make visible their social networks. Currently, 2.5 million of people access SNSs weekly.

At present, the most popular SNS is definitely Facebook. Facebook is the biggest social platform with 1.4 billion active users, followed by QZone, the most

B. Klimova · P. Poulova (✉) · L. Ptackova
Faculty of Informatics and Management, University of Hradec Králové,
Rokitanského 62, 500 03 Hradec Králové, Czech Republic
e-mail: Petra.Poulova@uhk.cz

B. Klimova
e-mail: Blanka.Klimova@uhk.cz

© Springer International Publishing Switzerland 2016
V.L. Uskov et al. (eds.), *Smart Education and e-Learning 2016*,
Smart Innovation, Systems and Technologies 59,
DOI 10.1007/978-3-319-39690-3_12

important Chinese social network with 654 million users. Google+ is running in the third place with 300 million active "in stream" users (the number of people that see Google+ content stream each month). The fastest growing social platform is Instagram that has also reached 300 million monthly active users. Twitter is still behind with 284 million users. Sina Weibo, the Chinese Twitter clone, has 176 million monthly active users [4] (see Fig. 1 below).

Social networks can be divided into several types according to the way how they function or what their main purposes is. Among the principal categories there are social networks based on profile, content, the so-called white label networks, microblog or virtual networks [6]: *Profile based social networks*—these are the networks in which it is important to interact with people. Personal profile has its significant role in this because thanks to it, people can keep in touch with other people, share their opinions, photos or some other important events. The most popular networks of this type are undoubtedly Facebook, followed by Google+. Also the professionally oriented social network LinkedIn is among these SNs. *Content based social networks*—the personal profile is not that much important in this case. It is the content which is the key and is shared by people on the social network. This social network can include videos, music or pictures. These media are usually accessible to all, including those who are not registered on these SNs. Probably, the best known representative of these SNs is YouTube which enables users to record and share their videos. The portal which aims at photos and is worth mentioning is the mobile application Instagram. White-label social networks—these are services which offer an independent development of social network on the basis

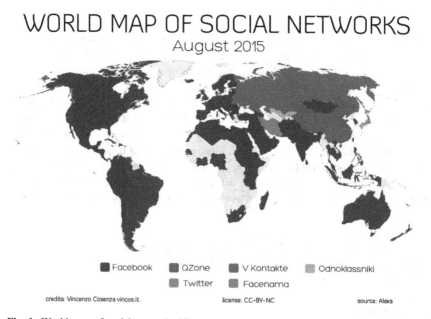

**Fig. 1** World map of social networks [5]

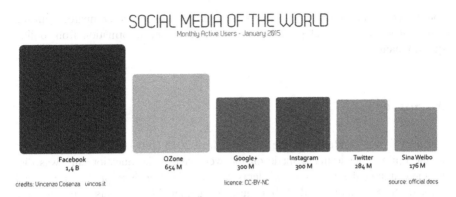

**Fig. 2** Social networks [5]

of the offered platform. Thus, users can create their own mini community, i.e. their small social network. They can make such a social network which would meet their expectations and needs. In fact, they can generate their own Facebook or Twitter in this way. Another name for this social network is Private label. The representatives of these platforms are also PeopleAggregator or Ning. *Microblog social networks*— the main feature of these SNs is publishing of short messages with a chance of adding a video or a picture. This is displayed to all subscribers of a given user. This type of social network is predominantly used on mobile phones and particularly, in case of worse text messaging. The main representative of these SNs is Twitter. This network offers a platform both for mobile phones and normal web browsers. The network is becoming quite popular in the Czech Republic. Another microblogging tool is http://Tumblr.com. *Multiuser virtual networks*—these are the categories which are borderline social networks. Their purpose is to enable users to communicate among one another, but not with the help of their profile but with the help of Avatars. Most of these networks are game servers such as World of Warcraft (see Fig. 2 below).

The aim of this study is to explore attitude of university students towards the SNSs with respect to their positive and negative experience, as well as their willingness to exploit SNSs as a support in their university studies and communication. Furthermore, the authors of this study summarize the main advantages and disadvantages of the use of SNSs for study purposes.

## 2  Methods

For the purpose of this article a method of literature review of available sources describing the use of SNSs and their advantages and disadvantages for study purposes found in the acknowledged databases such as Web of Science, Scopus, and Springer in the period of 2000–2015 was used. The findings of these literary sources are mainly discussed in the section on Discussion and compared with the results from the practical part, which are based on the information from online questionnaires.

## 3  Survey

The survey was conducted in the second half of 2015 (July–August) with the help of the online questionnaire made on the web pages. The questionnaire was distributed among the university students all over the Czech Republic and posted through SNSs, emails and web pages. Altogether 148 students submitted the questionnaire. It contained 10 questions, out of which eight and nine were open-ended questions.

### 3.1  Question 1: Do You Use SNSs?

93 % of students use SNSs, while only 7 % do not (Fig. 3).

### 3.2  Question 2: Why Don't You Use SNSs?

This question was just answered by those 7 % of students. 36 % of these students replied that they had no reason; 27 % reported that they were afraid that their personal data might be misused; 18 % preferred a personal contact; and 18 % had other reasons why they did not use SNSs.

**Fig. 3** Do you use SNSs?

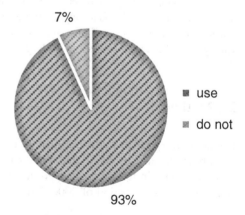

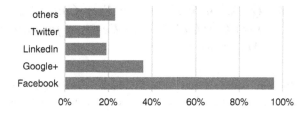

**Fig. 4** Which social networks do you use?

## 3.3  Question 3: Which Social Networks Do You Use?

As it can be expected, 96 % of students use Facebook; 36 % Google+; 19 % LinkedIn; 16 % Twitter; and 23 % of students prefer other SNSs (Fig. 4).

## 3.4  Question 4: How Often Do You Use SNSs?

83 % of students are on SNSs every day; while 11 % 3–5 times a week; and only 6 % spend less time on SNSs.

## 3.5  Question 5: How Do You Exploit SNSs?

95 % of students said that they mostly used SNSs for communication with friends; 73 % of students for following events and searching for information; 49 % for sharing data; 18 % used applications; a very small number (1 %) for other purposes such as marketing, work, or sharing materials for school (Fig. 5).

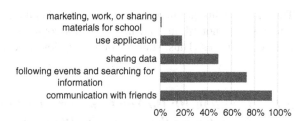

**Fig. 5** How do you exploit SNSs?

### 3.6 Question 6: Have You Ever Come Across the Teacher's Profile Which Would Be Used for Informing Students About the Subject?

46 % of students have already had such experience, while 54 % have not. In addition, most of the 46 % of students have found this profile beneficial.

### 3.7 Question 7: Do You Think that a SNS Could Be Used for Study Purposes Provided that Each Subject Had Its Own Profile?

83 % of students consider this possible; while 17 % do not.

### 3.8 Question 8: What Would the Advantages of Using Profiles of Individual Subjects Be?

83 % of students think that it would promote faster interaction between students and their teacher; 76 % of students assume that they would receive the most current data; 52 % of students could communicate without any limitations; and 45 % of students think that it would bring more clarity into the structure and content of the subject.

### 3.9 Question 9: What Would the Disadvantages of Using Profiles of Individual Subjects Be?

The replies to this question varied enormously, but the common ground was that the school should not ask its students open an account on the SNSs just for the reason of study purposes. Thus, students prefer freedom of choice in this respect.

### 3.10 Question 10: Do You Consider Effective that Individual Faculties of the University Would Use SNSs for Sharing Information and Communication with Their Students and Public?

82 % of students would be definitely for its use, while 18 % would be against it.

## 4 Discussion of the Findings

The findings from the survey show that the majority of students found SNSs as natural as breathing since they use it on a daily basis. As it might have been predicted, the most popular SNS is Facebook. 96 % of the addressed students use it since a profile on the Facebook enables individuals to raise their social presence [7]. Therefore there is a potential for its use at universities. For example, Munoz and Towner [8] in their study on the use of SNSs in education state that students prefer Facebook more than any other official online course for communication since they consider Facebook less formal than the official online course. This has been also confirmed by the study of Bueno-Delgado and Pavon-Marino [9].

Moreover, students' main reasons for the use of SNSs are communication with friends, following various events and sharing information, which could be certainly applied to the academic environment since students undoubtedly wish to integrate into a new school environment, communicate with their peers, plan events, comment or express their opinions. It has been also proved that the more students exploit social media technology for academically purposeful activities, the higher their level of engagement is [10].

Some teachers have also seen a potential of SNSs for their teaching and start to use them in teaching their subjects. The teachers who have revealing information about themselves, increase their credibility among their students who are then more willing to communicate with them. Munoz and Towner [8] report that 53 % of the students who knew that their teacher had a profile on the Facebook, visited his/her profile. This was also confirmed by the survey described above.

Klimova [11] in her study lists the main advantages and disadvantages of SNSs for university education (Table 1).

As Table 1 and the findings of the survey indicate, SNSs seem to bring a lot of benefits for university education. Furthermore, they can be very effective learning platforms that enhance students' engagement and learning experience, transforming

**Table 1** Advantages and disadvantages of SNSs for university education

| Advantages | Disadvantages |
|---|---|
| • Overcoming isolation and shyness<br>• Sharing materials from lectures, exchanging information, helping with assignments<br>• Higher students' involvement<br>• Students' cooperation and collaboration<br>• Personalization of learning<br>• Development of students' specific skills required for their future industrial occupations<br>• Development of special studying skills such as scanning and skimming skills<br>• Preparation for future employment | • Reduced reasoning and critical thinking skills<br>• Distraction from studying<br>• Misuse of SNSs, e.g., in form of cyber bullying or misuse of personal data<br>• Lack of well elaborated pedagogical methodology for the use of SNSs in engineering education<br>• Students' lack of long-term memory |

them into active learners with an increased motivation to learn while fostering big quality exchange of ideas and knowledge among participants (e.g., [12] or [13]).

Nevertheless, pedagogical principles of the implementation of SNSs into education are needed. As Wu et al. [14] state, different social media exhibited distinct and specific pedagogical features which require the involvement of a teaching assistants as facilitators of students' learning. Falah and Rosmala [15] point out that attention should be paid to the following factors when implementing SNSs into a learning process: the background and behavior of user; the university policy on the Internet access; the behavior of university communication; role and rule of social network in daily communication; and the attitude of user.

In addition, students seem to be able to retain freedom of choice in using SNSs for study purposes although they welcome its implementation at universities.

## 5   Conclusion

Overall, SNSs appear to be a good support for study purposes at the institutions of higher education. As Bloom [16] states, learning is a process helping pupils in their development from the lower forms of thinking (e.g., knowledge or understanding) to the higher forms of thinking (e.g., synthesis or evaluation). And the social network sites can assist in reaching these forms. For instance, Flickr can help in the acquisition of knowledge, Wikipedia in understanding this knowledge and You-Tube is suitable for the synthesis [17].

Moreover, since SNSs are currently exploited by nearly all university students, they represent a big potential for their use in smart learning environment, which puts emphasis on effectiveness and engagement features [18, 19].

**Acknowledgments**   This research was financially supported by the SPEV project 2105.

## References

1. Bicen, H., Cavus, N.: Social network sites usage habits of undergraduate students: case study of Facebook. Procedia—Soc. Behav. Sci. **28**(1), 943–947 (2011)
2. Uzumboylu, H., Bicen, H., Cavus, N.: The efficient virtual learning environment: a case study of web 2.0 tools and Windows Live Spaces. Comput. Educ. **56**(3), 720–726 (2011)
3. Boyd, D.M., Ellison, N.B.: Social network sites: definition, history and scholarship. J. Comput.-Mediated Commun. **13**(1), 210–230 (2008)
4. Cosenza, V.: Social media statistic, VINCOSBLOG. http://vincos.it/social-media-statistics/. Accessed 15 Jan 2016
5. Cosenza, V.: World map of social networks, VINCOSBLOG. http://vincos.it/world-map-of-social-networks/. Accessed 15 Jan 2016
6. Klimova, B., Poulova, P.: Pedagogical principles of the implementation of social networks at schools. In: Proceedings of the ICWL Workshop 2015, 5–8 Nov 2015, China (In press)

7. Poulova, P., Klimova, B.: Social networks and their potential for education. In: Proceedings of the Computational Collective Intelligence 7th International Conference, ICCCI 2015, pp. 365–366. Springer, Berlin (2015)
8. Munoz, C.L., Towner, T.: Back to the "wall": Facebook in the college classroom. First Monday 16(12) (2011)
9. Bueno-Delgado, M.V., Pavon-Marno, P.: The use of social networks as support tools to improve teaching/learning process in universities. In: Proceedings of EDULEARN12: 4th International Conference on Education and New Learning Technologies, pp. 35–44. IATED, Spain (2012)
10. Center for Community College Student Engagement (CCCSE): Making Connections: Dimensions of Student Engagement (2009 CCSSE Findings). The University of Texas at Austin, Community College Leadership Program, Austin (2009)
11. Klimova, B.: Benefits and limitations of social network sites for engineering education—a review study. In: Proceedings of the IEEE Global Engineering Education Conference (EDUCON) 2016 (In press)
12. Selwyn, N.: Faceworking: exploring students' education-related use of Facebook. Learn. Media Technol. 34(2), 157–174 (2009)
13. Wang, Q., Woo, H.L., Quek, C.L., Yang, Y., Liu, M.: Using the Facebook group as a learning management system: an exploratory study. Br. J. Educ. Technol. 43(3), 428–438 (2012)
14. Wu, W.H., Yan, W.C., Wang, W.Y., Li, S.L., Hy, K.: Comparison of varied social media in assisting student learning. Int. J. Eng. Educ. 31(2), 567–573 (2015)
15. Falahah, D.: Rosmala, study of social networking usage in higher education environment. Procedia—Soc. Behav. Sci. 67(1), 156–166 (2012)
16. Bloom, B.: Taxonomy of Educational Objectives, Handbook 1: The Cognitive Domain. David McKay, New York (1956)
17. Klimova, B., Poulova, P.: Social networks in education. In: Proceedings of the 12th Cognition and Exploratory Learning in the Digital Age (CELDA 2015), pp. 240–246. Maynooth, Ireland (2015)
18. Klimova, B.: Assessment in smart learning environment—a case study approach. In: Uskov, V., Howlett, R.J., Jai, L.C. (eds.) Smart Innovation, Systems and Technologies, vol. 41, pp. 15–24 (2015)
19. Spector, J.M.: Conceptualizing the emerging field of smart learning environments. Smart Learn. Environ. 1, 2 (2014)

# Design and Application of MOOC "Methods and Algorithms of Graph Theory" on National Platform of Open Education of Russian Federation

Lyubov Lisitsyna and Eugeniy Efimchik

**Abstract** The paper describes results of development and practical application of MOOC "Methods and algorithms of graph theory" in 2015 on National Platform of Open Education of the Russian Federation https://openedu.ru. The structure and content of the course are presented, as well as evaluation tools for monitoring learning outcomes. The paper describes features of the implementation of interactive practical exercises in the course by means of RLCP-compatible virtual laboratories which are SMART objects of the course. An example of a virtual stand of such laboratory and an example of criteria used in checking of solutions of the students are given. Analysis of the practical application of the course showed its effectiveness. Although only 9.2 % of 2605 registered students of the course went for certification, nearly 40 % of them gained a certificate of successful completion of the course, every fourth gained certificate with honors and 4.2 % of active students achieved the maximum score.

**Keywords** MOOC · Graph theory · National platform of open education of russian federation · Virtual laboratory · SMART object · Online exam

## 1 Introduction

Massive Open Online Courses (MOOC) are an important trend in the development of modern education in the world [1]. A MOOC project "National Platform of Open Education" (https://openedu.ru) [2, 3] was launched in Russian Federation in 2015 by association of eight leading Russian universities the ITMO University (Saint-Petersburg). The mission of this project is to develop and promote open education to assist improving the quality and accessibility of higher education in

L. Lisitsyna (✉) · E. Efimchik (✉)
ITMO University, Kronvrkskiy pr. 49, Saint Petersburg 197101, Russia
e-mail: lisizina@mail.ifmo.ru

E. Efimchik
e-mail: efimchick@cde.ifmo.ru

© Springer International Publishing Switzerland 2016
V.L. Uskov et al. (eds.), *Smart Education and e-Learning 2016*,
Smart Innovation, Systems and Technologies 59,
DOI 10.1007/978-3-319-39690-3_13

Russia. There were 56 MOOCs of basic disciplines of bachelor and specialist level in Russian developed by leading professors of these universities that have already been applied at the platform, and it is planned to further expand the list of online courses. Among applied courses there are the most popular ones from ITMO University—"Web-programming" by associate professor Pershin [4] and "Methods and algorithms of graph theory" by professor L. S. Lisitsyna. Special attention is paid to the efficiency and quality of the courses, their interactivity and procedures of evaluation of learning outcomes. User authentication is provided by the proctoring procedures and biometric technologies. Student's learning outcomes of the online courses on this platform can be recognized in related disciplines in all universities of Russia. This paper describes features of the development of the online course "Methods and algorithms of graph theory" and the first results of its application on the National Platform of Open Education of the Russian Federation in the autumn of 2015.

## 2  Structure and Content of the Course

The online course is focused on the study of methods and algorithms of graph theory and their application in practice. The goal of the course is to develop basic knowledge and skills to solve the most important and frequently encountered in practice graph problems. Scientific methodology that is used in design of the course is based on the modeling of the educational process to achieve expected learning outcomes [5]. Learning outcomes in this course are: LO1—a readiness to demonstrate a basic knowledge of mathematics (graph theory); LO2—the ability to apply in practice effective methods and algorithms for solving typical graph problems. The course is an educational module of discipline "Discrete Mathematics" which is a part of the basic educational programs for bachelors in various areas of training at universities in the Russian Federation.

The course duration is 10 weeks, and its labor intensity is three credit units. The average weekly load of student is 10 h. For the theoretical training course uses video lectures. It was experimentally proved that the length of video lectures in general is three times less than the duration of a similar lecture at the traditional training. Therefore, each video lecture lasts about 15 min. After completion of each video lecture an online quiz is conducted to verify the ability to apply this knowledge in practice. For theoretical study of methods and algorithms in the course interactive demonstrations are used. Those demonstrations include not only a description, but also examples of their use. After video lectures students are invited to explore practical exercise to test skills of solving relevant standard graph problems. Each practical exercise is implemented as a virtual laboratory task [6, 7]. For each student an individual variant of task is automatically created. His solution is checked step by step according to learned algorithm. Table 1 describes the structure and content of the course, and Table 2—description of its practical exercises and maximum score that a student can get for completing them.

**Table 1** Structure and content of course

| Number | Topic | Amount of video lectures with quizzes | Amount of practical exercises |
|---|---|---|---|
| 1 | Basics of graph theory | 5 | 0 |
| 2 | Connectivity of graphs | 6 | 2 |
| 3 | Cycles in graphs | 5 | 1 |
| 4 | Trees | 6 | 2 |
| 5 | Optimization on graphs | 7 | 3 |
| 6 | Bipartite graphs | 6 | 1 |
| 7 | The isomorphism and the homeomorphism of graphs | 2 | 1 |
| 8 | Planar graphs | 4 | 1 |
| | **Total** | **41** | **11** |

**Table 2** Practical exercises of the course

| Number | Typical graph problem | Algorithm | Max score |
|---|---|---|---|
| 1 | Search shortest route | Lee algorithm | 5 |
| 2 | Search route with minimal weight | Bellman-Ford algorithm | 6.5 |
| 3 | Search for Hamilton loops | Roberts-Flores algorithm | 6.5 |
| 4 | Search for minimum spanning tree | Prim algorithm | 6.5 |
| 5 | Search for minimum spanning tree | Kruskal algorithm | 6.5 |
| 6 | Search for largest empty subgraphs | Magu-Weismann algorithm | 6.5 |
| 7 | Minimum vertex coloring of graph | Method based on Magu-Weisman algorithm | 6.5 |
| 8 | Minimum vertex coloring of graph | Greedy heuristic algorithm | 6.5 |
| 9 | Search perfect matching in a bipartite graph | Hungarian algorithm | 6.5 |
| 10 | Detecting of isomorphism of two graphs | Algorithm based on ISD method | 6.,5 |
| 11 | Graph planarization | Gamma-algorithm | 6.5 |
| | | **Total** | **70** |

Topic 5—"Optimization on graphs" (Table 1) consists of two parts and lasts two weeks. The rest of topics last one week. Upon completion of the course online exam is provided, which is held on the 10th week. Exam consists of two parts—A and B, and time limit to complete exam is one hour. Part A is held as an extended online quiz, and part B is a practical exercise. General characteristics of the examination work is shown in Table 3.

**Table 3** Structure and content of online exam

| Part | Characteristic of part of online exam | Content elements being checked (numbers of topics) | Number of tasks | Max score |
|------|---------------------------------------|---------------------------------------------------|-----------------|-----------|
| A | Checking the ability to apply knowledge in practice | 1, 2, 5 | 12 | 18 |
| B | Checking skills of solving typical problems applying studied graph algorithms | 3, 4, 6, 7, 8 | 1 | 12 |
| | | **Total** | **11** | **30** |

Result of evaluation learning outcomes is scaled to 100 points in accordance with designed criteria. Each student can get maximum of 70 points for practical exercises (Table 2) and 30 points for online exam (Table 3). Threshold level for each evaluated kind of work is 60 %.

## 3 Virtual Laboratories Applied in the Online Course

Practical exercises (Table 2) are important interactive elements of the online course, they form and evaluate skills of learning outcomes. RLCP-compatible virtual laboratories are used for their implementation [6]. Such laboratories are SMART—objects of the course and they consist of two independent modules—virtual stand and RLCP-server. Virtual stand is responsible for visual representation of task variant details and provision student with tools to form and edit intermediate results and final answer. RLCP-server is a TCP-server, which provides an RLCP interoperability and protection from unauthorized access [8]. RLCP-server performs by request several types of operations: building task variant, calculation of intermediate results, evaluation of student's solution. These two modules are not aware of each other and interact through a special control environment of RLCP-compatible virtual laboratories (hereinafter - environment), which automatically controls the assignment of the listener. Sequence diagram of interaction of a student, virtual laboratory modules and the environment is shown in Fig. 1. When the environment receives request for test it requests RLCP-server for test variant, prepares and sends test frame with embedded virtual stand to student. The environment controls the progress of the solution and allows student to freely move to previous solution stages and edit intermediate results. When student finishes test he initiates procedure of evaluation of his solution by RLCP-server. After that RLCP-server returns evaluation result with all needed comments to the student.

Since such environment is absent within National Platform of Open Education, applied virtual laboratories are placed in AcademicNT, which is learning

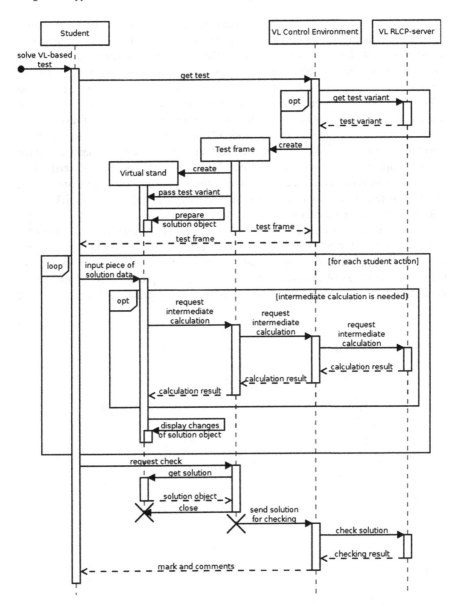

**Fig. 1** Sequence diagram of interaction of student, modules of RLCP-compatible virtual laboratories and control environment

management system at ITMO University [7]. AcademicNT performs as external evaluating service for the platform. The interaction of the systems have been implemented at the level that is sufficient to ensure the seamless student experience and registration of the evaluation results in both systems.

Virtual laboratories applied in the online course are designed to form and control skills of solving tasks using known algorithms, that require rigid sequence of actions and logical methods of solution. On the one hand, this type of tasks requires strict control of the variability including control of final complexity of generated variants. Because of that the procedure of variant preparation for each virtual laboratory is based on specially designed algorithm that builds variant taking into account given complexity class [9, 10]. On the other hand, this type of tasks allows to check all intermediate results of student's solution in details using results of reference algorithm. It allows to detect exact stage where student made a mistake and tell him about it in related comment. In addition, this method of verification allows to automatically estimate and assess the proportion of correct answers in student's solution. That is much better than binary "true/false" rule, which leaves student no room for mistakes.

Figure 2 shows an example of the virtual stand with completed solution of a virtual laboratory for practical exercises #2 (Table 2). The initial data of test variant are visualized as an interactive weighted graph with the indicated vertex of start (vertex f) and vertex of end (vertex b) of the route. Complexity class of test variant depends on the number of vertices and edges of the graph and the number of

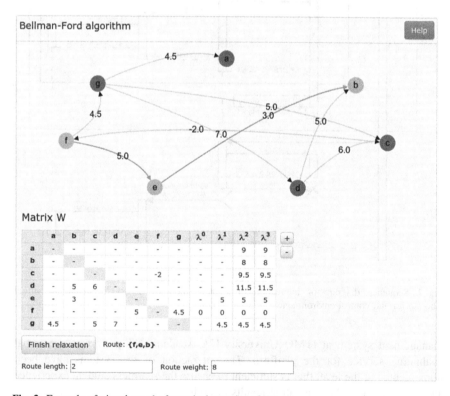

**Fig. 2** Example of virtual stand of practical exercise of the course

**Table 4** Evaluation criteria of practical exercise "Bellman-Ford algorithm"

| Number | Criteria | Score |
|---|---|---|
| 1 | Errors in building adjacency matrix | 0 |
| 2 | Errors in determining amount of relaxing iterations | 10 |
| 3 | Errors in relaxing | 20 |
| 4 | Errors in determining amount of steps of the route | 40 |
| 5 | Errors in determining the route | 50 |
| 6 | Errors in determining the weight of the route | 60 |
| 7 | No errors | 100 |

iterations needed to perform relaxation of the graph. The algorithm for generating such variants is implemented on the side of RLCP-server and takes into account all of the above parameters. It also prevents generating graph with some undesirable properties, for example, disconnected graphs or graphs with cycles having negative weight. Instructions for filling of intermediate results and final answer are located in the tab "Help". When filling out the weight matrix W student can use the interactive graph. When he hovers over one of the vertices of the graph all images of this vertex become highlighted, as well as related arc, along with their weights.

Once student fills the matrix W, he proceeds to the filling of λ-labeled column of the table. Button "+" allows to create another λ-labeled column of the table, and the button "−"allows to remove the last of such columns. After completion of the graph relaxation student proceeds to detecting of vertices of the required route using λ-columns of the table. While student points vertices at this stage the route step by step emerges on the graph. Figure 2 shows the stand, which is filled out with all the intermediate results and the final answer. After that, the student gives the command "Answer is ready", and the protocol of his solution is sent to be evaluated on RLCP-server.

The criteria that are used to evaluate the student's solution of this practical exercise are given in Table 4. Due to fact that all intermediate results are based on previously calculated ones evaluation stops at first commited error and the student gets corresponding score.

# 4 Results of Application of the Course

The course "Methods and algorithms of graph theory" on the National Platform of Open Education of the Russian Federation was started at September 28, 2015 for the first time. The number of registered students was 2605 persons, but only 239 students actually started to study (it is the number of students who have scored a non-zero evaluation results), i.e. 9.2 % of the registered students. Table 5 provides

**Table 5** Evaluation results of students of the course

| Number | Category of active students | Amount of students | Percentage (%) |
|---|---|---|---|
| 1 | All students that passed certification | 239 | 100 |
| 2 | Students who gained more than 20 points | 138 | 57.7 |
| **3** | **Students who gained more than 60 points** | **94** | **39.3** |
| 4 | Students who gained more than 70 points | 80 | 33.5 |
| 5 | Students who gained more than 80 points | 71 | 29.7 |
| 6 | Students who gained more than 90 points | 61 | 25.5 |
| 7 | Students who gained 100 points | 10 | 4.2 |

information about the evaluation results of course participants. Students who gained more than 60 points received certificates. There were 39.3 % of all active students who gained such result. Each forth active student gained certificate with honors (90 points or more), while 4.2 % of active students scored the maximum score of the course.

Table 6 describes information on the results of practical exercises of the course. For each exercise student had maximum of five attempts, including attempts to

**Table 6** Practical exercises results

| Number | Related algorithm | Amount of students | Percentage of students with non-zero result (%) | Percentage of students with max result (%) | Average amount of attempts |
|---|---|---|---|---|---|
| 1 | Lee algorithm | 256 | 90 | 86.3 | 2.55 |
| 2 | Bellman-Ford algorithm | 212 | 95.3 | 81 | 2.33 |
| 3 | Roberts-Flores algorithm | 163 | 96.3 | 79.1 | 3 |
| 4 | Prim algorithm | 150 | 98 | 94 | 1.56 |
| 5 | Kruskal algorithm | 139 | 98.5 | 97 | 1.51 |
| 6 | Magu-Weismann algorithm | 128 | 92.9 | 78 | 3.23 |
| 7 | Method based on Magu-Weisman algorithm | 110 | 96.3 | 71 | 2.41 |
| 8 | Greedy heuristic algorithm | 104 | 100 | 88.4 | 3.1 |
| 9 | Hungarian algorithm | 112 | 94.6 | 88.4 | 3.38 |
| 10 | Algorithm based on ISD method | 103 | 94.1 | 88.3 | 2.23 |
| 11 | Gamma-algorithm | 102 | 98 | 93.1 | 3.41 |
| **Average** | | | **96** | **85.4** | **2.61** |

improve the previous results. As it can be seen from Table 6, the students spent much less than five attempts in average, 96.0 % of students in average coped with the task and 85.4 % of students in average gained maximum score, which demonstrates the effectiveness of application of virtual laboratories in the course.

It is known that online communication between students and authors of the course is very valuable in terms of increasing effectiveness of online courses. While applying our online course there were 79 topics started, which contain 535 messages with questions, answers and suggestions, and 97 of messages were the answers from authors of the course. There were 14 suggestions about improvement of the course at the forums, organized at the end of the course, as well as words of gratitude and contentment.

# 5  Conclusions

Course "Methods and algorithms of graph theory" applied at the National Platform of Open Education of the Russian Federation includes videolectures supported by quizzes, interactive demonstrations and practical exercises for 8 topics. The course duration is 10 weeks, and its labor intensity is three credit units. The online exam is provided upon completion of the course. Expected learning outcomes of the course are: LO1—a readiness to demonstrate a basic knowledge of mathematics (graph theory); LO2—the ability to apply in practice effective methods and algorithms for solving typical graph problems. The course is an educational module of the discipline "Discrete Mathematics" which is a part of the basic educational programs for bachelors in various areas of training at universities in the Russian Federation. It can be said that the course has shown good practical results in 2015. And although the proportion of active students was only 9.2 % (this proportion does not exceed 5 % in average for MOOCs [1]), strong results have been obtained, confirming its efficiency: nearly 40 % of active students gained certificates of successful completion of the course, every fourth student gained a certificate with honors, and 4.2 % of students achieved the maximum score for the course. An important result of the practical application of the course was to confirm the efficiency of interactive practical exercises implemented with help of RLCP-compatible virtual laboratories. Almost all the students cope with the task of these exercises, and 85.4 % of students achieved the maximum score of the exercise. Perhaps the most important proposal of students was the desire to have in this course practical exercises for the development and programming algorithms for solving problems. It has defined new objectives for the improvement of the course, and first results of that improvements will be demonstrated to participants of conference in our report.

**Acknowledgment** This paper is supported by Government of Russian Federation (grant 074-U01).

# References

1. Vasiliev, V., Stafeev, S., Lisitsyna, L., Ol`shevskaya, A.: From traditional distance learning to mass online open courses. Sci. Tech. J. Inf. Technol. Mech. Opt. **89,** 199–205 (2014). (in Russian)
2. Edutainme. http://www.edutainme.ru/post/russian-national-mooc (in Russian)
3. MK.RU. http://www.mk.ru/social/2015/09/22/vosem-universitetov-obrazovali-nacionalnuyu-otkrytuyu-platformu-onlaynobrazovaniya.html (in Russian)
4. Lisitsyna, L.S., Pershin, A.A., Kazakov, M.A.: Game mechanics used for achieving better results of massive Online, In: Smart Education and Smart e-Learning, pp. 183–193 (2015)
5. Lisitsyna, L., Lyamin, A.: Approach to development of effective e-learning courses. In: Frontiers in Artificial Intelligence and Applications, vol. 262, pp. 732–738 (2014)
6. Lyamin, A.V., Efimchik, E.A.: RLCP-compatible virtual laboratories, In: E-Learning and E-Technologies in Education, pp. 59–64 (2012)
7. Kopylov, D.S., Fedoreeva, M.K., Lyamin, A.V.: Game-based approach for retaining student's knowledge and learning outcomes. In: Application of Information and Communication Technologies, pp. 609–613 (2015)
8. Silnov, D.S., Tarakanov, O.V.: Assessing the stability of antivirus software and data protection means against erroneous outcomes. Int. J. Appl. Eng. Res. **10**(19), 40342–40349 (2015)
9. Efimchik, E.A., Chezhin, M.S., Lyamin, A.V., Rusak, A.V.: Using automaton model to determine the complexity of algorithmic problems for virtual laboratories. In: Application of Information and Communication Technologies, pp. 541–545 (2015)
10. Chezhin, M.S., Efimchik, E.A., Lyamin, A.V.: Automation of variant preparation and solving estimation of algorithmic tasks for virtual laboratories based on automata model, In: E-Learning, E-Education, and Online Training, pp. 35–43 (2015)

# A Framework for Human Learning Ability Study Using Simultaneous EEG/fNIRS and Portable EEG for Learning and Teaching Development

Boonserm Kaewkamnerdpong

**Abstract** In ageing society, it is not only that we need better care and treatment to maintain the quality of life of elderly population but also need better ways to strengthen the development of our children so that we could live in a healthy ageing society. This study proposed a framework for human learning ability study by using multimodal neuroimaging through simultaneous EEG/fNIRS measurement and neuroinformatics to understand target learning ability in laboratory and using portable EEG device to monitor real-time brain state for evaluating the learning/teaching methods introduced for improving the target learning ability in learning environment. By incorporating neuroscience approach in both laboratory and learning environment, not only scientific findings from neuroscience studies could give implications to educators to improve students' learning, but also the developed learning/teaching methods could be assessed their effectiveness in classroom practice. This framework may help contribute in bridging the gap between neuroscience and education, which is the aim of educational neuroscience. Toward smart education, the proposed conceptual framework deploys truly brain-based learning/teaching approach to enhance students' learning to better acquire the knowledge and cognitive skills.

**Keywords** Educational neuroscience · Electroencephalography (EEG) · Functional Near-infrared spectroscopy (fNIRS) · Multimodal neuroimaging · Neuroinformatics · Simultaneous EEG/fNIRS · Smart education

B. Kaewkamnerdpong (✉)
Biological Engineering Program, Faculty of Engineering, King Mongkut's University of Technology Thonburi, Bangkok, Thailand
e-mail: boonserm.kae@kmutt.ac.th

© Springer International Publishing Switzerland 2016
V.L. Uskov et al. (eds.), *Smart Education and e-Learning 2016*,
Smart Innovation, Systems and Technologies 59,
DOI 10.1007/978-3-319-39690-3_14

# 1   Introduction

With the current trend of the increasing life expectancy and decreasing birth rate, the impacts of ageing society on our everyday life are inevitable. In many countries, there are great numbers of elderly population. Moreover, the ratio between the elderly population and the child population has been continuingly increasing. These children will grow and become the country's crucial workforce. They will work not only to support themselves but also to support the elders. For some countries like Japan, the problem seems even more immense. Hence, not only that we should maintain the quality of life of elderly population but also we need to strengthen the development of our children so that we could live in a healthy ageing society.

Education plays an important role in sustainable development. Learners with good knowledge and skills will be crucial workforces in the society. Hence, education will take part in shaping the future of the society in both the direction and quality of development toward sustainability. Nevertheless, our society is complex and dynamics; the characteristics and behaviors of people in the society change through time and are different from one region to another. One example of such changes is that, nowadays, to effectively solve a real-world problem it requires the integration of multiple disciplinary related to the problem. Hence, systems thinking and interpersonal skills become attractive competences. For the change on the learners, the learners in 21st century who grew up with computers and smart technologies could learn what they want to know from anywhere any time through ICT technologies and communication networks. They could better understand visual communication, better learn from self-discovery than listening and give quickly response. On the downside, they could hardly concentrate on one thing for a long time; they quickly change their interests [1]. Consequently, learning and teaching methods in education should also adapt appropriately according to the need of the society.

For tackling complex problems that require multidisciplinary approach, problem-based learning (PBL) seems to be an appropriate pedagogy. Groups of PBL Learners learn the subject through identifying problems and hands on solving problems [2]. PBL projects student-centered learning and promotes lifelong learning. By interviewing learners in undergraduate- and graduate-level PBL class in computer engineering [3], most learners agreed that PBL approach allowed them to deeply understand the subject and develop suitable skills for self-directed learning. However, it was found that some learners required more time to adjust from lecture-based learning to PBL approach and more time to develop necessary skills than the others. Those who could not keep up with the group felt pressured and retracted themselves from participating in class discussion. On the other hand, those who could quickly understand the knowledge and develop relevant skills felt bored sometimes. It has evidentially shown that PBL can help learners obtain knowledge and effectively develop problem solving skill, collaboration skill as well as communication skill. Unfortunately, the evaluations for the effectiveness for learning approaches are usually performed qualitatively. It is difficult to utilize such results in improving learning/teaching methods.

In cognitive neuroscience, researchers have been studying through cognitive experimental tasks and analyzing brain activity data to understand how brains work in cognition. Findings in research studies in cognitive neuroscience could be useful for developing learning and teaching methods effectively. In the field of educational neuroscience, researchers in cognitive neuroscience, educational psychology, developmental neuroscience, educational technology and other education-related disciplines are invited to conduct research to understand neural mechanisms in human learning abilities better and to provide scientific implications for educators so that they could structure appropriate educational materials, learning environments, learning and teaching methods and curriculum development in order to improve education to be more effective and more suitable for learners in this generation.

For example, it has been identified that the construction of meaning is the key to understand and remember information [4]. When we encounter new information, the information become memorable to us once we could make meaning of it by linking this new information to our existing knowledge. After the meaning is constructed, neural activity in the left inferior prefrontal cortex is increased. Such scientific implications could be useful for smart education in the design and development of smart educational materials with active contents so that the related knowledge previously learnt could be included along with new information to provide easier meaning construction and, in turn, faster knowledge delivery to students.

Although the concept of educational neuroscience seems valid, two possible problems could arise. On one hand, since neuroscience findings are usually obtained from studying the target function in isolation from other functions in controlled environment in laboratory for simplicity, it may be doubted whether the findings could be truly applicable in real complex learning environment. On the other hand, neuroscience findings can take on many forms in pedagogical practices. Translating those neuroscience findings into pedagogical procedures requires knowledge and experience on human neurophysiology and behaviors. Incorrect translation could lead to another neuromyth. Educators could design some pedagogical procedures, but there is not yet a way to scientifically correlate pedagogical results back to neuroscience findings.

This paper proposed a framework for studying human learning abilities through bringing neuroscience methodology that is used in laboratory into pedagogical practices in classroom/learning environment as well. With this framework, the constructive collaboration between neuroscientists and educators is needed to develop effective learning/teaching methods for students to better/faster achieve educational outcomes in terms of knowledge contents and cognitive skills. However, both parties could benefit from the collaboration. Educators could develop pedagogical methods more effectively. Neuroscientists could learn how applicable the neuroscience findings could be in real environment and realize practical research questions. In Sect. 2, neuroimaging and neuroinformatics used in methodology for educational neuroscience studies are described. In Sect. 3, the proposed framework is described and discussed with current implementation. The framework is concluded in Sect. 4.

## 2  Neuroimaging and Neuroinformatics for Educational Neuroscience

To gain better understanding of the human brain, neuroimaging modalities were introduced to non-invasively (or less invasively) obtain neuronal activity data to identify brain regions and their interconnection corresponding to a certain function. In neuroinformatics, techniques in information sciences are used to analyze neuroimaging data for discovering new knowledge on brain and nervous system. Non-invasive neuroimaging modalities that are currently used can be divided into two groups: electrophysiological and hemodynamic principles [5].

Based on electrophysiological principles, electroencephalography (EEG) records neuronal electrical activity from human scalp surface. EEG provides electrophysiological data with high temporal resolution but poor spatial resolution as it is a summation of activity from neurons within as well as adjacent area of the measurement. Magneto-encephalography (MEG) detects magnetic flow caused by current flow from neuronal electrical activity in the area; hence, MEG sensors have to be placed tangentially to the surface and can detect only current flow in the tangential orientation [5]. Nevertheless, MEG is not affected by shunting effect causing distortion in EEG [6].

Based on hemodynamic principles, positron emission tomography (PET) and single-photon emission computed tomography (SPECT) require an injection of radioactive tracer intravenously. After injection, the scanner detects the emitted radioisotopes at the hemodynamically or metabolically activated region and constructs signal data into images. These techniques can still be considered as one of the non-invasive neuroimaging techniques [5]. Nevertheless, PET and SPECT have low levels in both temporal and spatial resolution compared with other modalities.

Functional magnetic resonance imaging or fMRI can give us the structural images of the brain and the measurement of hemodynamic response (change in blood flow) related to neural activity in the brain; it cannot, however, register changes in neural activity fast enough to reveal the sequence of activation. fMRI provides images with good spatial resolution but inadequate temporal resolution. Moreover, it is not recommended to operate on infants, toddlers and patients with inserted metallic implants. For similar purposes, functional near-infrared spectroscopy (fNIRS), which is infant- and toddler-friendly technology optically measuring the changes of concentrations of oxyhemoglobin and deoxyhemoglobin in the tissue to support neuronal activity, offers better temporal resolution and adequate spatial resolution.

Each of the existing neuroimaging technologies has its own characteristics in terms of temporal and spatial resolution. Nevertheless, many experimental studies [7–11] have shown that there exists the correlation between hemodynamic response and electrophysiological activity. Such coupling between hemodynamic response and neuronal electrical activity, or neurovascular coupling, analyzed from data collected from two or more neuroimaging modalities simultaneously, or *multimodal neuroimaging*, could provide further information and reveal underlying mechanisms on how human brain functions. Moreover, novel discoveries in multimodal neuroimaging could bring about several potential benefits including obtaining neuronal activity with high level in both temporal and spatial resolution as well as revealing complex relationships of neurophysiology to cognitive functions.

Neuroimaging modalities are used to collect brain functional data during designed cognitive experiments. The collected brain data could be analyzed by using statistical and information processing techniques for prominent features representing target cognitive functions. For example, Dennis and Cabeza [12] studied different impacts of age range in learning vocabulary via memory. Hammer et al. [13] investigated the difference between errorless and errorful learning in face-and-name associative learning.

Research findings from neuroscience studies often report the difference on how brains respond between the target cognitive function and control function or between one type of learning and the other. Many studies identify brain regions corresponding to such difference. The implications from research findings to educators are sometimes also given. However, if not given, findings must be translated into pedagogical implications. The educators can develop teaching and learning methods based on guiding implications in their practices in traditional, e-learning and smart educations. Educational implementation for neuroscience implications can take on many forms. Thus, it could be questionable which implementation would be more effective. Due to biological variation, the effectiveness of any implementation for different learners may vary.

In today's education system, the assessment for students' learning is done based on the end results. We assess whether students have obtained the knowledge contents, have acquired target cognitive skills or could extend their knowledge and skills further on more complex cases. The assessment for students' learning itself seems missing. The sooner the educators acknowledge students' learning performance, the faster they could make change on their current teaching method to accommodate to students' learning ability. In traditional education, the learning process could still be indirectly evaluated based on students' behavioral-based performance in the classroom. In smart education where the learning process could occur elsewhere, the assessment for students' learning process is unaccounted for. It would be better if the role of neuroscience could extend to assist in evaluating the effectiveness of teaching and learning methods during the learning process. If we measure neural activity in the brain during learning, we could monitor how learning process executes and assess whether it is good learning. The ability to assess students' learning could allow us to develop learning/teaching methods for learners in new generation more effectively.

## 3   Proposed Framework for Human Learning Ability Study

Bringing neuroscience methods into action in education development, this study proposed the framework for human learning ability study using simultaneous EEG/fNIRS to collect brain functional data in laboratory to characterise the mechanisms of target learning ability and using portable EEG to assess the performance of different educational approaches in classroom/learning environment. The proposed framework illustrated in Fig. 1 can be divided into two parts: in laboratory and in learning environment.

The target learning ability must firstly be identified. As learning is a complex function involving several cognitive tasks, the cognitive experiment should be carefully designed so that it includes two or more cases with different subject responses for the purpose of comparison in order to effectively capture the neural process of the target learning ability. While the cognitive experiment is conducted in laboratory or controlled environment on one or more groups of subjects, EEG and fNIRS are simultaneously measured at shared locations based on the international 10/20 system.

EEG and fNIRS are chosen to simultaneously measure brain functional data in a cognitive experiment to understand how brain neuronal connectivity during learning. Although it was the simultaneous use of EEG and fMRI that become attractive and popular in multimodal neuroimaging as the integration of high temporal resolution from EEG and high spatial resolution from fMRI, fNIRS was selected instead of fMRI for the following reasons:

1. The measurement of hemodynamic response in term of blood oxygen level dependent or BOLD in fMRI techniques allows high spatial resolution to identify the activated brain region, but the measurement values are collected and averaged in a period of time; hence, it is unclear that the detected hemodynamic response is truly corresponding to the neuronal activity. In fMRI study, subjects are required to concentrate and repetitively perform the experimental tasks in tight, noisy environment in MR scanner; it might be, however, difficult for subjects to concentrate on the tasks and get good data as fMRI has low temporal resolution. Although real-time fMRI recently introduced [14] can provide information with high spatial resolution and improved temporal resolution, fNIRS has higher temporal resolution than real-time fMRI and better spatial solution than EEG.
2. It is not recommended for using fMRI in infant and small children [15]; there is no known effect on using fNIRS in infants. Although the measurement may be limited to cortical cerebral region, fNIRS can be recorded concurrently with EEG in a more natural manner for a longer time than fMRI [16] with a variety of subjecting ranging from infants to elderly.
3. For the purpose of cognitive research in learning ability, it is better to have a system that could allow the experiment to resemble real learning environment. The MR scanner is inappropriate for using in real learning environment.

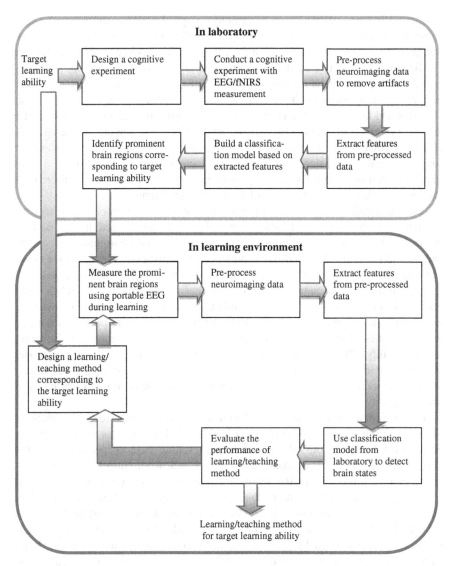

**Fig. 1** The proposed framework for human learning ability study using simultaneous EEG/fNIRS and portable EEG

The identified brain region from fMRI study may not be able to monitor in real learning environment. On the other hand, fNIRS can be performed in everyday life conditions, which allow considerable freedom in the design of experimental tasks [17]. Moreover, there is a portable version for both EEG and fNIRS so it is more promising that they can be used for monitoring learning ability in real learning environment.

Using EEG and fNIRS, both electrophysiological activity and hemodynamic response reflecting neuronal activity can be recorded with high temporal resolution and good spatial resolution. Simultaneous EEG-fNIRS has been employed toward the diagnosis of epilepsy by detecting epileptic activity [18]. Fazli et al. [19] integrated NIRS data to enhance the performance of EEG analysis for brain computer interfacing. Simultaneous EEG-fNIRS has been chosen for language studies as it can be performed in everyday life conditions, which allow considerable freedom in the design of experimental tasks [14]. Compared to EEG-fMRI, the integrative analysis of EEG-fNIRS for understanding the underlying mechanisms in neurovascular coupling is still in its infancy; it has, however, great potential to improved overall spatiotemporal resolution for the purpose of human brain mapping, cognitive neuroscience studies and development of neuroimaging diagnostic system.

After data collection, the neuroimaging data must be pre-processed to remove internal artifacts from eye blinking, breathing and cardiac activity as well as external artifacts from noise from the instrument. Several neuroinformatics techniques could be applied to identify prominent characteristics or features of the target learning ability in collected brain functional data. These features could be analyzed in terms of time and frequency domains as well as spatial domain. The relationship between EEG and fNIRS in the area could also be features. Statistical and machine learning methods could be applied to extract features. It is more often that not one feature but a combination of several features could be used to represent the target learning ability. The extracted features would be included in a mathematical model that can classify the learning ability based on different types of subject responses. Such model called classification model can be used for detection different cases of learning ability.

The identified features usually lead to neuroscience findings (e.g. identification of brain regions corresponding to target learning ability) serve as implications for educators. Educators would design a learning/teaching method specifically for the target learning ability. The classification model built from collected data in laboratory can be employed in the learning environment as a detection system for evaluating the learning/teaching method designed for the target learning ability. Not all measured brain regions may be necessary for real-time evaluation; the prominent areas and/or patterns should be identified and used. In the real learning environment, a portable-version EEG recording device could be used to measure brain responses during learning. The data must be pre-processed and computed for the previously identified features. The resulting classification model from the cognitive experiment can be used to detect the brain states and/or patterns of brain connectivity which can indicate the performance of learning/teaching method and use for further improving the design of learning/teaching approach to enhance students' learning more effectively.

According to the proposed framework, for example, we would like to study about learning via visual working memory [20–22]. The cognitive experiment would be performed through visual event-related potentials (vERP) approach with simultaneous EEG-fNIRS recordings. During the experiment, subjects would be asked to view the screen and give response through mouse clicking. Indoor/outdoor scenes are used as tools for learning. The indoor/outdoor scenes are randomly shown for a limited time period. After that, the subject needs to decide whether he/she would like to *remember* or *forget*, and response accordingly through buttons. The left button is for remembering case while the right button is for forgetting case. There are 250 indoor/outdoor scenes included in the experiment. After the measurement, the subjects rest for 10 min and perform a scene recognition test by identifying which of the 500 scenes they had seen in the experiment. The scenes in recognition test comprise of 250 scenes from the experiment and 250 new scenes from the database.

After data collection, both EEG and NIRS data must be pre-processed to remove cyclic artifacts, cardio activity and respiration artifacts, and noise. Machine learning methods may be performed to identify features that can distinguish between brain activity corresponding to remember and forget cases. Prominent features/patterns are selected and used to build a mathematical model for classification the brain signal data to detect for the brain state corresponding to good visual working memory and the signs of good learning via visual working memory at their earliest.

Let us consider, for example, the case that we develop smart educational materials with active contents. The knowledge contents are shown in pictures on the screen. We design a teaching/learning method through storytelling with accompanying pictures. Two sets of pictures are included: pictures of the new knowledge contents and pictures of related knowledge previously learnt. The model could be used with portable EEG device for monitoring the brain state for signs of good learning via visual working memory in learning environment. We may design the teaching/learning method with adaptive contents based on the current brain state of students; when the brain state does not show signs for good learning via visual working memory, the teaching/learning method will select the previously learnt contents that are related to the new contents with greater proportion than the new contents. Once the brain state shows signs for good learning via visual working memory, the teaching/learning method will illustrate the new contents with greater proportion. With suitable teaching/learning method, the knowledge could be delivered to students faster. If the current teaching/learning method is not yet effective enough, the method can be modified and assessed with brain-based learning model again. With the proposed framework, neuroscience evidence in learning environment could be provided and used for further effective pedagogical development toward smart education.

# 4 Conclusion

Through the use of neuroimaging modalities to collect brain functional data and the use of statistical and information processing techniques to extract brain characteristics that describe how brains work, neuroscience studies can provide useful findings. Educators could develop pedagogical practices based on neuroscience findings to improve education. However, the role of neuroscience could be extended to help assess the performance of educational approaches as well. The proposed framework in this paper introduced the use of neuroscience methodology in both research laboratory for neuroscience findings and learning environment for neuroscience evidence of how effective the developed pedagogical practices could be. With the proposed framework, the proactive collaboration between neuroscientists and educators could plausibly drive education development forward more effectively toward smart education. Portable EEG devices may become another crucial technology for smart education; portable EEG devices with classification model can not only serve as tool for indicating the performance of learning/teaching methods but also can be used to develop smart educational materials for better educational outcomes.

**Acknowledgements** The author is grateful for all financial supports for research projects conducted with this framework including (1) MRG5680144 by Thailand Research Fund, Office of the Higher Education Commission, and King Mongkut's University of Technology Thonburi, (2) Hitachi Research Fellowship HSF-R136 by Hitachi Scholarship, the Hitachi Global Foundation, and (3) Research Strengthening Project of the Faculty of Engineering, King Mongkut's University of Technology Thonburi.

# References

1. Oblinger, D., Oblinger, J.: Is it age or it: first steps toward understanding the net generation. Educating the Net Generation. EDUCAUSE (2005)
2. Moore, S., Barrett, T.: New Approaches to Problem-based Learning: Revitalising Your Practice in Higher Education. Routledge, New York (2010)
3. Kaewkamnerdpong, B.: Problem-based learning approach for bio-inspired artificial intelligence: a case study. In: Proceedings of International Consortium for Educational Development (ICED), pp. 327–330, Bangkok, Thailand, 23-25 July 2012
4. Neuroscience and Education: Issues and Opportunities. Teaching and Learning Research Programme, The Economic and Social Research Council (2007)
5. Shibasaki, H.: Human brain mapping: hemodynamic response and electrophysiology. Clin. Neurophysiol. **119**, 731–743 (2008)
6. Shibasaki, H., Ikeda, A., Nagamine, T.: Use of magnetoencephalography in the presurgical evaluation of epilepsy patients. Clin. Neurophysiol. **118**, 1438–1448 (2007)
7. Logothetis, N.K., Pauls, J., Augath, M., Trinath, T., Oeltermann, A.: Neuro-physiological investigation of the basis of the fMRI signal. Nature **412**, 150–157 (2001)
8. Devor, A., Ulbert, I., Dunn, A.K., Narayanan, S.N., Jones, S.R., Andermann, M.L., Boas, D. A., Dale, A.M.: Coupling of the cortical hemodynamic response to cortical and thalamic neuronal activity. Proc. Natl. Acad. Sci. U.S.A. **102**, 3822–3827 (2005)
9. Arthurs, O.J., Boniface, S.J.: How well do we understand the neural origins of the fMRI BOLD signal? TRENDS Neuroscience. **25**, 27–31 (2002)
10. Arthurs, O.J., Boniface, S.J.: What aspect of the fMRI BOLD signal best reflects the underlying electrophysiology in human somatosensory cortex. Clin. Neurophysiol. **114**, 1203–1209 (2003)
11. Arthurs, O.J., Donovan, T., Spiegelhalter, D.J., Pickard, J.D., Boniface, S.J.: Intracortically distributed neurovascular coupling relationships within and between human somatosensory cortices. Cereb. Cortex **17**, 661–668 (2007)
12. Dennis, N.A., Cabeza, R.: Age-related dedifferentiation of learning systems: an fMRI study of implicit and explicit learning. Neurobiol. Aging **32**, 2318.e17–2318.e30 (2011)

13. Hammer, A., Tempelmann, C., Münte, T.F.: Recognition of face-name associations after errorless and errorful learning: an fMRI study. BMC Neurosci. **14**, 30 (2013)
14. Weiskopf, N.: Real-time fMRI and its application to neurofeedback. NeuroImage **62**, 682–692 (2012)
15. Seghier, M.L., Hüppi, P.S.: The role of functional magnetic resonance imaging in the study of brain development, injury, and recovery in the newborn. Semin. Perinatol. **34**, 79–86 (2010)
16. Wallois, F., Patil, A., Héberlé, C., Grebe, R.: EEG-NIRS in epilepsy in children and neonates. Neurophysiol. Clin./Clin. Neurophysiol. **40**, 281–292 (2010)
17. Wallois, F., Mahmoudzadeh, M., Patil, A., Grebe, R.: Usefulness of simultaneous EEG–NIRS recording in language studies. Brain Lang. **121**, 110–123 (2012)
18. Machado, A., Lina, J.M., Tremblay, J., Lassonde, M., Nguyen, D.K., Lesage, F., Grova, C.: Detection of hemodynamic responses to epileptic activity using simultaneous Electro-EncephaloGraphy (EEG)/Near Infra Red Spectroscopy (NIRS) acquisitions. NeuroImage **56**, 114–125 (2011)
19. Fazli, S., Mehnert, J., Steinbrink, J., Curio, G., Villringer, A., Müller, K.R., Blankertz, B.: Enhanced performance by a hybrid NIRS–EEG brain computer interface. NeuroImage **59**, 519–529 (2012)
20. Yoo, J.J., Hinds, O., Ofen, N., Thompson, T.W., Whitfield-Gabrieli, S., Triantafyllou, C., Gabrieli, J.D.E.: When the brain is prepared to learn: enhancing human learning using real-time fMRI. NeuroImage **59**, 846–852 (2012)
21. Angsuwatanakul, T., Iramina, K., Keawkamnerdpong, B.: Multi-scale sample entropy as a feature for working memory study. In: The Seventh Biomedical Engineering International Conference (BMEiCON), pp. 1–5, Fukuoka, Japan, 26–28 Nov 2014
22. Angsuwatanakul, T., Iramina, K., Keawkamnerdpong, B.: Brain complexity analysis of functional near infrared spectroscopy for working memory study. In: The Eighth Biomedical Engineering International Conference (BMEiCON), pp. 1–5, Pattaya, Thailand, 25–27 Nov 2015

# A Conceptual Framework for Knowledge Creation Based on Constructed Meanings Within Mentor-Learner Conversations

Farshad Badie

**Abstract** Constructivism is a learning philosophy and an educational theory of learning. In the framework of constructivism, a human being with insights based on her/his pre-structured knowledge and on background knowings could actively participate in an interaction with another human being. The central focus of this article is on construction of conceptual knowledge and its development. This research localises the constructivist learning in the context of mentor-learner interactions. It will analyse meaning construction relying on my own conceptual framework that represents a semantic loop. The learner and the mentor as intentional participants move through this semantic loop and organise their personal constructed conceptions in order to construct meanings and produce their meaningful comprehensions. This research is concerned with definitions, linguistic expressions and meanings within the developmental processes of personal world constructions. It proposes a scheme for interpretation based on semantics and in conversations. I shall conclude that the outcomes of this research could be able to support smart education and smart learning environments.

## 1 Introduction

The dialogues and the conversational exchanges between mentors and their learners ask questions and give answers concerning their individual conceptions and realisations. I shall stress that what I will use and express under the label of 'concept' aims at providing a comprehensible characteristics of human being's conceptions and conceptualisations. In the section 'related works and proposal' I will focus on the works related to realisations of concepts. *Constructivism* is a learning philosophy and an educational theory of learning that can be recognised as a model (and theory) of knowing with roots in philosophy as well as in psychology and cybernetics, see [6]. A constructivist conversation could be seen as a radical

F. Badie (✉)
Center for Linguistics, Aalborg University, Rendsburggade 14, 9000 Aalborg, Denmark
e-mail: badie@id.aau.dk

© Springer International Publishing Switzerland 2016      167
V.L. Uskov et al. (eds.), *Smart Education and e-Learning 2016*,
Smart Innovation, Systems and Technologies 59,
DOI 10.1007/978-3-319-39690-3_15

constructivist account of the learner's and the mentor's comprehensions, realisations and understandings, see [17]. This account is capable of enabling the mentor and the learner to develop their own understandings of the specified concepts with regard to their understandings of the more general concepts. Producing the understanding of a world (a universe of discourse) and developing it during the conversation could be said to be the most valuable product of a constructivist conversation. When a mentor and her/his learner start a conversation, then, based on their personal knowings and on personal pre-structured knowledge, they attain deeper realisations of the world. Conversation supports them to develop their understandings of the world (and of each other and of themselves). Jean Piaget, the originator of constructivism, believed that human being's mental objects (or schemata) gradually develop into more abstract (= general) and conceptual mental entities, see [12, 13, 18]. In fact, according to constructivist learning the human being's mentality manifests itself in the form of schemata [3], and I will explain some details in the next section. Considering learning in the framework of constructivism with reference to Conversation Theory (CT), which is conceived and elaborated by Gordon Pask, the enterprise begins with the negotiation of an agreement between the learner and the mentor to converse about a given domain and to learn/train about some particular topics and skills in that domain. It could work as an explanatory, heuristic and developmental framework, see [15, 16]. The CT is fundamentally an explanatory ontology combined with an epistemology, which has wide implications for psychology and educational technologies. Pask's main premise is that the reliable knowledge exists, is produced, and evolves in action-grounded conversations, see [4].

In this research I will work on a conceptual framework that represents the process of meaning construction in constructivist conversational exchanges between mentors and learners. My main reference is my own semantics-based framework [1, 2] and this research attempts to develop my framework for analysing 'conceptual knowledge creation based on constructed meanings within conversational exchanges between mentor and learner'. According to [1, 2], the learner and the mentor as intentional participants move through a semantic loop and organise their personal constructed conceptions in order to construct meanings, to improve the constructed meanings, and to produce their individual meaningful comprehensions. What could be offered by conceptual knowledge construction in the developed framework is 'a body of thought' and 'a semantic model to account for the emergence of the domain of the learner's and of the mentor's conceptual knowledge'. It can express how the produced meanings based on human being's constructed concepts could support her/him in constructing the personal worlds and in creating her/his conceptual knowledge.

## 2 Related Works and Proposal

The most momentous building block of this research is 'concept', thus I shall start with the realisation of concepts. The reference [14] identifies concepts as the furnitures of human beings' minds. According to this realisation, a well furnished mind can be a source of successful knowledge acquisition (and learning). According to [5], a concept is a linkage between the human being's linguistic expressions and the mental images (of all kinds of perceptions) that s(he) may have in her/his mind. Regarding [5] and taking [3] into consideration, the human being may represent a concept as a 'thing' under a specified label. Consequently, the labels are the constituents of propositions that mediate between thought, language, and referents. My conceptual approach has relied on this realisation and grasp of concepts with regard to [3, 5]. I assume that a concept can also be seen as an 'idea' that corresponds to a distinct entity or to its essential features, attributes, characteristics and properties. As mentioned, concepts are the conjunctions between the human being's linguistic expressions and her/his mental images. Focusing on conversational exchanges between mentors and learners, the mental images may be seen as the representation(s) of the aspects of the world (a universe of discourse). Taking this realisation into consideration, any conception of a leaner/mentor within a conversation can be identified as her/his act of visualising (in a broad sense) different concepts by linking her/his expressions (and specifications) to her/his own mental images [that have been visualised over various schemata). Let me be more specific. First, I focus on clarifying schema and schema-based knowledge constructions, and consequently, I will draw your attention to the interrelationships between 'specified characteristics of concepts in my approach' and 'the Kantian philosophy and Kantian account of schemata'. In my opinion, knowledge can actively be constructed over human being's constructed concepts and conceptions. I believe that knowledge is built up (created) based upon the learner's/mentor's comprehensions of concept meanings with regard to their definitions. I shall stress that the definitions are highly influenced by the learner's/mentor's world descriptions based on their linguistic expressions. Also, the linguistic expressions have been assigned to their mental images of phenomena within the world (the universe of discourse). Reconsidering the introductory section, when a mentor and her/his learner start a conversation, then, based on their personal knowings and on personal pre-structured knowledge, they attain deeper realisations of the world. Conversation supports them to develop their understandings of the world [and of each other and of themselves]. As mentioned, in constructivist learning a human being's mentality manifests itself in the form of schemata. It's important to say that the schemata demonstrate the human being's realisation of the world. They conceptually represent the constituents of the learner's/mentor's thoughts for knowledge acquisition [and learning/mentoring] with regard to their realisations of the world. Now, let me focus on the interrelationships between specified characteristics of concepts in my approach and the Kantian account of schemata. Regarding [7, 8], Kant defines a non-empirical (or pure) concept as a category. According to Kantian philosophy, schemata are the procedural rules by which a category is associated with a sense impression. Kant claimed that the schema provides a reference to intuition in a way similar to the manner of empirical concepts. According to the Kantian account of schemata, empirical concepts[1] are the most fundamental types of concepts that employ schemata. For instance, the concept *Liquid* can be explained by a rule according to which a human being's imagination can visualise a general figure of a description like "*A state and a distinct form that matter takes on*" without being restricted and closed to any particular and specific shape produced by human being's experience. The empirical concepts provide the origin of what I have brought under the label of concepts. In fact, a human being proposes various linguistic expressions in order to describe her/his mental images of a phenomenon. This description is directly dependent on her/his own schemata, which have been designed and shaped over her/his experiences. Moreover, regarding the Kantian account of schemata, the human being is concerned with 'pure concepts of the understanding[2]. According to a human being's realisation and grasp of the pure concepts of understanding, s(he) focuses on characteristics, features, attributes, qualities and properties of an object, that are also other objects in general or as such. They all support her/him in producing better understanding of the things/phenomena in the world. At this point I shall emphasise that my conceptual analysis may look a bit like Wittgenstein I [19] but I will make no use of Wittgenstein's approach.

## 3  Conversational Learning Theory

According to [10, 11], Laurillard's framework[3] can be interpreted as a learning theory and as a practical framework for designing educational and pedagogical environments. This framework includes four important components: (*i*) Mentor's concepts (and conceptions), (*ii*) Mentor's constructed learning environment [that produces her/his constructed world], (*iii*) Learner's concepts [and conceptions] and (*iv*) Learner's specific actions [that construct her/his world]. Regarding Laurillard's framework, we need to consider different forms of communication and associated mental activities. The main constructors of the associated mental activities are (1) discussion, (2) adaptation, (3) interaction and (4) reflection. In fact, relying on Pask's Conversation Theory and focusing on Laurillard's framework, there are four kinds of human being's activities that take place in different kinds of flow between the components of Laurillard's framework:

- Discussion between the mentor and the learner on their conceptions, descriptions, realisations and reasonings.

---

[1]http://kantwesley.com/Kant/EmpiricalConcepts.html.

[2]http://userpages.bright.net/~jclarke/kant/concept1.html.

[3]http://edutechwiki.unige.ch/en/Laurillard_conversational_framework.

**Fig. 1** Conversational learning framework

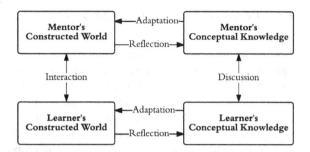

- Adaptation of the learner's actions and of the mentor's constructed environment. There is a type of adaptation of the learner's world and of the mentor's world.
- Interactions between the learner and the environment which is defined by the mentor.
- Reflection of the learner's performance by both the mentor and the learner.

I have illustrated the above-mentioned characteristics in the framework represented in Fig. 1. For instance, relying on the conversational learning framework and reconsidering the concept *Liquid*, the existed knowledge about the object *liquid* is strongly dependent on the interrelationships between learner and mentor (i.e., 'learner → mentor → learner'). In fact, the learner incorporates the internalised mentor, and the mentor incorporates the internalised learner. Thus, the interactions and the discussions about 'liquid' (initially, over the empirical concept of *Liquid*) starts. Accordingly, regarding the pure concepts of understanding, learner and mentor focus on multiple characteristics, predicates, attributes, qualities and properties of the object *liquid*. Consequently, the learner and the mentor converse and exchange their conceptions in order to construct and develop their conceptual knowledge.

## 4 Meaning Construction

I have assumed that the definitions and meanings based on a human being's conceptions strongly support her/him in knowledge construction processes in the framework of constructivist learning (mentoring) and in the context of mentor-learner conversations. I also assume that an explanation is the actual act of explicating definitions and meanings. I shall emphasise that the main objective of the learner's (and the mentor's) explanations are to shed light on the produced personal comprehensions, realisations and understandings. In this section I focus on definitions and meanings, which are the main building blocks of my framework.

## 4.1   From Definitions to Meanings

In a logical and semantics-based system a definition can be figured out as a kind of equivalency (and, semantically, as a kind of equality,) whose left-hand side is a concept (the concept that is going to be defined) and whose right-hand side is a description in the form of a number of expressions. For instance, a person defines Liquid by '*Liquid ≐ A state and a distinct form that matter takes on. Also the matter in this state maintains a fixed volume, but has a variable shape that adapts to fit its container*'. I shall stress that this definition has proposed a concept definition (a type of equivalence between the right-hand side and the left-hand side). It also presents a semantical equivalence between the right-hand side and the left-hand side. This semantical equivalence represents two concurrent implications; one from the left-hand side into the right-hand side and one from the right-hand side into the left-hand side. In a constructivist conversation, any of the agents may define a concept based upon her/his individual conception. Subsequently, regarding the feedbacks of the conversant, s(he) modifies and updates her/his definition. I have called it 'definition updating'. Let me point to the fact that that the learner and the mentor interact with each other in order to adapt and to conform their 'personally constructed worlds'. They also discuss in order to exchange their personal conceptual knowledge (their constructed knowledge over concepts and conceptions) and to develop it. Consequently, any proposed definition provides a supportive backbone for performing more developed and more well-organised concept descriptions. Additionally, a more developed concept description supports both the agent and the interlocutor in providing the more understandable explanations of meanings. At this point considering the specified realisation of definitions, I focus on the expression 'meaning'. I have assumed the following descriptions to be comprehensible in the context and in my framework. Linguistically, meaning is— according to one approach—realised as a context-update function, see [9]. So, meaning is regarded as a function from a context into a context. Any context comprises different types of (general and specified) concepts. Therefore, I describe a meaning as a function from a concept into its updated form. In fact, in my conceptual and logical approach a meaning is seen as a 'concept-update function'.

## 4.2   A Framework for Meaning Construction

In [1, 2] I have analysed a (conceptual and semantic) loop that the learner and the mentor move through in order to organise their personal constructed concepts, in order to describe their definitions, in order to construct and formulate the meanings and in order to produce their individual meaningful comprehensions; see the red sections of Fig. 2. The proposed semantic loop is self-organising and can promote itself on higher conceptual levels and on higher levels of interaction and conversation. It has proposed a scheme for interpretation based on semantics and on interactions. The most primary building block of learning in the framework of constructivist conversational learning can be built up over an asked question. Regarding empirical concepts and the human being's experimental conceptions, the one who undertakes to learn/train something and to produce her/his (deeper and deeper) understanding within a constructivist conversation, primarily gets concerned with various general concepts related to that thing. Later on, s(he) focuses on main characteristics, attributes, qualities and properties of that thing in order to be concerned with realisation of that thing and to recognise its state and its condition within the world. More specifically, s(he) focuses on forming[4] [and reforming] concepts. A mentor who has defined the concept *Liquid* as "*a state and a distinct form that matter takes on. Also, the matter in this state maintains a fixed volume, but has a variable shape that adapts to fit its container*" is directly and indirectly concerned with a number of general concepts (e.g., *State, Form, Matter, Volume, Shape, Container*). Therefore, the learner needs to conceptualise these notions and to discuss her/his conceptions with the mentor. The preliminary understanding supports the learner in building various patterns in her/his mind, and any of these patterns describes her/his thought over a conception. These patterns could be seen as the mental structures (consisting of mental objects or schemata). Accordingly, the learner relies on her/his mental structures to organise her/his conceptual knowledge, and to provide strong backbones for her/his interpretations and explanations. In fact, the collection of the rules and processes that manage various linguistic expressions (and, respectively, definitions) based upon logical foundations do not, and cannot, have any meaning until the non-logical words[5] of the language are given interpretations. And the human being's mental structures support her/his personal interpretations. The collection of (1) concept formation, (2) concept transformation (from the mentor's mind into the learner's mind and vice versa) and (3) concept reformation is the most significant matter in the development of the concept constructions within constructivist conversations. I have identified the process '1 → 2 → 3' as the main basis of 'Concept Construction' in my framework. Taking the constructed concepts into consideration, the reflection of the prior knowledge (what has been acquired or created before participating in the conversation) and the new knowledge (what is being acquired or created during the conversation) and the initial definitions must all be negotiated. So, the sequence "Prior Knowledge → Definition → New Knowledge" supports the learner in searching for the appropriate initial meanings of the class/classes of her/his constructed concepts. This phase is highly affected by 'interpretations'. In fact, the initial meanings are being interpreted in order to be balanced. The interpretation of a defined constructed concept is a function from a 'definition' into a 'meaning'. Logically, we may have iterative loops between 'definitions' and 'meanings'

---

[4]http://teachinghistory.org/teaching-materials/teaching-guides/25184.

[5]The words like, e.g., *and, or, not, since, then, so, all, every, any*, have logical consequences and are identified as the logical words in a natural language with regard to classical symbolic logic and predicate logic..

(consisting of functions from definitions into meanings and inverse functions from meanings into definitions). The agent could be able to balance and adjust the initial meanings based on the interrelationships between 'interpretation' and 'the inverse of the interpretation' (by comparing the subject of interpretation before and after being interpreted). The conclusions make an appropriate background for verifying the personally found meanings. Therefore, a meaning would be given a more appropriate shape after checking the balanced definitions based on the personal constructed concepts. The conclusions will support the agent in formulating the balanced meanings. Any formulated meaning is a basis for providing a supportive conceptual structure of meaning production. These conceptual structures are all personally formulated over personal constructed concepts and definitions. On the other hand, they induce new formulated meanings on higher conceptual levels and on higher levels of conversation. Finally, the supportive conceptual structures support the agent in producing meanings. Note that the produced meanings reinforce the meaningful comprehensions. I shall emphasise that the produced meanings support the construction of the individual mental worlds. Any produced meaning has been designed, shaped, balanced, formulated and produced based upon the learner's/mentor's formatted concepts, constructed concepts, expressed definitions and interpreted definitions. Any produced meaning reflects in the constructor's self and supports her/him in re-shaping and developing her/his schemata on the next levels of her/his conversation. Thus, the learner's and the mentor's produced meanings are employed in the developmental processes of personal world construction. As you may have realised, at this point I have entered Laurillard's conversational learning framework. The dashed arrows in Fig. 2 show that my framework is getting connected with the simplified Laurillard's framework. As

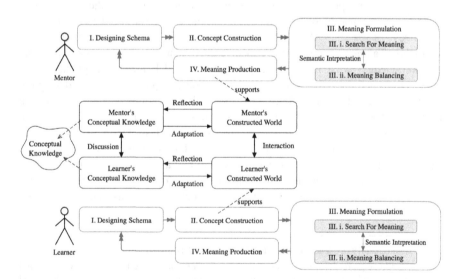

**Fig. 2** Constructivist onversation: meaning construction and knowledge creation

mentioned, regarding Piaget's developmental theory of learning, constructivist learning is concerned with how the individual human being goes about the construction of knowledge in her/his own cognitive apparatus. Therefore, the most important output of my 'meaning construction' framework is an input for the developmental processes of personal world construction.

## 5 The Outcomes Within Smart Learning Systems

I shall draw your attention to the relation between the main objectives of this research and the main concepts of smart education and smart learning. Considering the following items, I might claim that this research has proposed a conceptual analysis of a constructivist paradigm within smart learning systems. (i) The smart learning approaches[6] motivate higher levels of learners' understandings. Similarly, this research has focused on conceptual analysis of knowledge creation [based on produced understandings] in the framework of constructivism and in the context of mentor-learner conversations. Let me say that this research focuses on higher [and deeper] levels of learners' understandings with regard to their produced meanings and their generated meaningful comprehensions. (ii) Through the various processes within the framework of smart learning, the learner initially activates background knowledge, identifies goals for her/his personal learning, focuses on information processing and focuses on self regulations. In a corresponding manner, this research is structured over the learner's [and the mentor's] background knowledge. It sees learning as the process of construction over personal background knowledge. It focuses on a human being's realisation of her/his objectives and specifies a self-organising process on the part of the learner. (iii) Regarding the framework of smart learning, the learner transforms learning into demonstrations of understanding, and accordingly reflects on her/his own learning. In a very similar way, the most significant matter in this research has been 'transforming learning phenomena into knowledge construction that has been achieved over constructed concepts and conceptions'. Also, this research focuses on transforming learning into the learner's/mentor's comprehensions of concepts' meanings with regard to their definitions. These two important matters could support learners in reflecting on their learnings [and on themselves].

## 6 Conclusions and Future Work

This research has focused on constructivist conversational learning. A constructivist conversation has been seen as a radical constructivist account of the learner's and the mentor's comprehensions and understandings. Regarding Piaget's

---

[6]http://connectedteachers.weebly.com/what-is-smart-learning.html.

developmental theory of learning, constructivist learning is concerned with how the individual human being goes about the construction of knowledge in her/his own cognitive apparatus. I have proposed a conceptual framework that represents the process of meaning construction in constructivist conversational exchanges between mentors and learners. The proposed framework represents the processes of conceptual knowledge creation based on constructed meanings within conversational exchanges between mentor and learner. I have assumed that the definitions and meanings based on human being's conceptions strongly support her/him in the process of conceptual knowledge creation. I have also assumed the explanation as the actual explaining of definitions and meanings. The backbone of this research is my proposed semantic loop. The learner and the mentor move through this semantic loop in order to organise their personal constructed concepts. Thus they will be supported in constructing meanings, to improve the constructed meanings, and to produce their individual meaningful comprehensions. This article's proposed framework is 'a body of thought' and 'a semantic model to account for the emergence of the domain of the learner's (mentor's) conceptual knowledge'. It expresses how the produced meanings based on the constructed concepts could support personal world constructions. The personal worlds are employed by Laurillard's conversational learning framework that is a learning theory and a practical framework for designing educational and pedagogical environments. It values discussions, adaptations, interactions and reflections. In future research I will focus on logical and semantic analysis of conceptual knowledge creation based on constructed meanings within mentor-learner conversations relying on First-Order Predicate Logic and Description Logics.

# References

1. Badie, F.: A semantic basis for meaning construction in constructivist interactions. In: International Conference on Cognition and Exploratory Learning in Digital Age (CELDA). Dublin, Ireland (2015)
2. Badie, F.: Towards a semantics-based framework for meaning construction in constructivist interactions. In: The 8th Annual International Conference of Education, Research and Innovation (ICERI2015). Seville, Spain (2015)
3. Bartlett, F.C.: A Study in Experimental and Social Psychology. Cambridge University Press (1932)
4. Boyd, G.M.: Conversation Theory. Handbook of Research on Educational Communications and Technology (2004)
5. Götzsche, Hans: Deviational Syntactic Structures. Bloomsbury Academic, London/New Delhi/New York/Sydney (2013)
6. Husen, T., Postlethwaite, T.N.: Constructivism in education. The International Encyclopaedia of Education, Supplement vol.1, pp. 162–163. Pergamon Press, Oxford/New York (1989)
7. Kant, I.: Theoretical Philosophy after 1781—First Section Of the Scope of the Theoretico-Dogmatic Use of Pure Reason. The Cambridge Edition of the Works of Immanuel Kant (1999)
8. Kant, I.: Critique of Judgment—the Unity of Kant's Thought in His Philosophy of Corporeal Nature. Hackett publishing company Indianapolis/Cambridge (1790)

9. Larsson, S.: Formal Semantics for Perception. Workshop on Language, Action and Perception (APL) (2012)
10. Laurillard, D.M.: Rethinking University Teaching: A Framework for the Effective Use of Educational Technology. Routledge, London (1993)
11. Laurillard, D.: Rethinking University Teaching. A conversational framework for the effective use of learning technologies. London: Routledge (2002). ISBN 0415256798
12. McGawand Peterson, P.: Constructivism And Learning. International Encyclopaedia of Education, 3rd Edn. Elsevier, Oxford (2007)
13. Moallem, Mahnaz: Applying constructivist and objectivist learning theories in the design of a web-based course: Implications for practice. Educ. Technol. Soc. **3**, 113–125 (2001)
14. Parker, W.C.: Pluto's Demotion and Deep Conceptual Learning in Social Studies. Social Studies Review (2008)
15. Pask, Gordon: Conversation, Cognition and Learning: A Cybernetic Theory and Methodology. Elsevier Publishing Company, New York (1975)
16. Pask, G.: Developments in conversation theory (part 1). Int. J. Man-Mach. Stud., Elsevier Publishers (1980)
17. Scott, B.: Conversation theory: a constructivist, dialogical approach to educational technology. Cybernetics and Human Knowing (2001)
18. Spiro R., Feltovich P., Michael J., Richard C.: Cognitive Flexibility, Constructivism, and Hypertext. Random access instruction for advanced knowledge acquisition in ill-structured domains. Edu. Technol. pp. 24–33 (1991)
19. Wittgenstein, L.: Tractatus Logico-Philosophicus. Routledge (1922)

# Inductive Teaching and Problem-Based Learning as Significant Training Tools in Electrical Engineering

Fredy Martínez, Holman Montiel and Edwar Jacinto

**Abstract** Professional training in engineering is traditionally deductive. Under this approach, curriculum and its development are designed to start with a first stage of theoretical foundation, then, and progressively interact with the application of that theory. However, given the high labor requirements in terms of specific knowledge and skills, and the diversity of students and learning styles, a training with inductive approach may allow a more autonomous growth of the student and greater proximity to solving real problems. In the inductive approach, the contents are introduced by presenting study cases or problems whose solution involves the analysis and discovery of the theories involved. This paper presents an evaluation in a comparative way, in terms of meaningful learning, for electronics courses, on a technique of inductive learning (the problem-based learning (PBL)), compared to previous experiences in traditional training for students of Technology and Electrical Engineering at the District University of Colombia. This strategy of problem-based learning is complemented by the use of technology tools to structure a system of smart pedagogy. In this sense, the use of smart robotic systems becomes a key element in the smart learning, given the development of skills involving intelligence and adaptability. In the end, an analysis of the effectiveness of these two approaches is made from the perspective of learning outcomes.

**Keywords** Electric engineering · Inductive teaching · Problem-based learning · Significant learning

F. Martínez (✉) · H. Montiel · E. Jacinto
District University Francisco José de Caldas, Bogotá D.C., Colombia
e-mail: fhmartinezs@udistrital.edu.co
URL: http://www.udistrital.edu.co

H. Montiel
e-mail: hmontiela@udistrital.edu.co

E. Jacinto
e-mail: ejacintog@udistrital.edu.co

© Springer International Publishing Switzerland 2016
V.L. Uskov et al. (eds.), *Smart Education and e-Learning 2016*,
Smart Innovation, Systems and Technologies 59,
DOI 10.1007/978-3-319-39690-3_16

# 1 Introduction

Traditional teaching techniques of engineering training tend to be selfish on autonomy and student participation, focusing the process in the teacher [15]. The teacher should organizing the knowledge, selecting the topics and develop the necessary strategies to enable the student learning processes. The student is generally limited to collect and copy information, with an emphasis on content. Such techniques adversely affects the performance of students and even at labor level, given the sedative effect that applies, and particularly for not to show their connection with the real world. Some researchers and experts believe that unless a new pedagogy is consolidated in the process of training, students will lose more and more motivation gradually reducing their performance, and teachers will increase their level of stress, also negatively impacting on the teaching [5].

The new pedagogies required in current training processes require changes in relationships between teachers and students, in teaching and learning strategies, and how the learning process is evaluated. The student profile has changed drastically in recent years, and the relationships and strategies must be integrated into a different world. The youth of this generation have grown up in a world in which computers, hand-held devices and smart phones are integrated in their existence. Access to all kinds of information, along with a technology that is directly related to the need to be socially connected, has shaped the youth in a unique manner. This new social group has a huge potential that goes along with current trends in industrial and commercial development [11, 16, 19].

Is here where it is postulated that an inductive approach can positively impact as motivator and academic performance booster. Different implementations of inductive methods, such as problem-based learning (PBL), usually involve active learning methods and collaborative work that tends to actively engage students. Many studies have documented the positive effects of these strategies in terms of meaningful learning outcomes [8, 14].

The PBL is considered an inquiry-based learning method, which is a pedagogical method that have roots in constructivist philosophy, particularly the work of Piaget, Dewey, and Vygotsky [3, 6, 17]. The PBL is characterized as a dynamic strategy in which students face a structured real-world problem (strongly structured or weakly structured, depending on the cognitive processes to develop), for which they should be organized in groups or teams to identify learning needs and develop a viable solution [12]. Teachers act as facilitators rather than primary sources of information, accompanying the whole process [1, 4, 18].

A key element in the PBL is the student freedom for decision making. This is a key motivating element, students have greater motivation for their learning planning, and organizing their own research in solving real-world problems, particularly when they are working in groups [2]. In parallel, the pertinence by the occupation and the training institution is encouraged.

This paper shows the experiences over three years of work with undergraduates students of Technology and Electrical Engineering at District University Francisco

José de Caldas (Colombia). We observe the change of a deductive methodology to a PBL approach in a controlled electronic course.

Regarding the evaluation of the strategy, it is appropriate to clarify that the conclusions of a particular study may depend on many variables, such as the topics covered, the teacher, the student population, processes of knowledge acquisition, the curriculum, the attitude, etc. This does not make it easy to analyze results between inductive and deductive methods, already, in fact, it is difficult to compare two implementations of the same teaching strategy [13]. However, in this paper we focus on meaningful learning outcomes demonstrated by different groups of students with similar population characteristics, in order to try to establish the real contribution of the method.

The paper is organized as follows. In Sect. 2 objectives pursued with the use of PBL are discussed, Sect. 3 describes the learning strategies used with the groups of students. Section 4 shows the autonomous robots as smart systems used in the learning process. Section 5 introduces some results obtained with different groups of students. Section 6 shows some considerations about the validity of the results and future work. Finally, conclusion and discussion are presented in Sect. 7.

## 2 Objectives

An engineering curriculum requires that its future professionals should be fully prepared to accomplish a variety of tasks and activities involving not only theoretical concepts, but also skills. These skills are typically acquired by students through practical activities. In this sense, to increase the performance of electronics courses coordinated by the research group ARMOS in the programs of Technology and Electrical Engineering at the District University (Colombia), the research group proposed the use of methods that integrate technology into learning, together with problem-based learning (PBL) as a teaching strategy of active learning.

In Colombia becoming less young initiate engineering studies. While in other countries the number of students is increasing year by year (for example, 26 % per year in China, 17 % in India or 10 % in Brazil), in Colombia a fall of 5 % occurs. Every year it is observed as there is less interest in studying engineering. This lack of interest is partly due to a saturation of professionals appeared in the country a few years ago, which made engineering less attractive for young people, because labor competition increased and salaries were reduced.

Although labor conditions have now changed, selflessness still persists for these careers, which is beginning to affect the industrial development. This trend goes hand in hand with high student desertion rates. In the case of the Electric program of the District University these percentages have exceeded the 70 %. That is why we have raised the need to, on the one hand increase the competitive level of the students in meaningful learning optimizing real-world problems, and secondly increase motivation and interest in the study through the implementation of PBL. The study was specifically applied in electronics training, area coordinated by the research group.

Electronic courses will help students to gain the knowledge and skills necessary for professional practice in the industry. The practical laboratory work helps students to study the problem (a real problem), search and study of related theory, nomination of possible solutions, evaluation of strategies, solution design, prototype implementation and evaluation of results. The students are presented with a artificial scenario, quite similar to the real world, where they are required to perform as they would in a real life situation.

## 3   Learning and Teaching Strategy

Courses of the programs of Technology and Electrical Engineering at the District University francisco Joé de Caldas are traditionally deductive. During the last three years it has changed the learning strategies in the area of electronics to incorporate an inductive approach. Among the strategies introduced the use of PBL and computer support (cloud shared repository with documentation on IPython Notebook) is contemplated to form a system of smart pedagogy.

At the beginning of the course the student body is organized in small teams of three individuals. The learning objectives are defined and explained to the students. After explaining the game rules of the course, a real world problem weakly structured is proposed, with the characteristics required in the solution. With this introduction, we asked students to create solution strategies that achieve the objectives of both the problem and the course.

The problem that they receive is selected to acquire own specific knowledge of the course. This selection is quite complex, since it is often not possible to imagine the creativity of students, which should be encouraged with the problem. In most cases we contemplate concepts including instrumentation, measurement, sensors, calibration, monitoring, automation, signal conditioning, filtering, searching, identification and communication.

The design of each tool used (problem selection, monitoring online platform and results evaluation system) included the following elements: (1) Application to real problems whose solution would require creativity and critical thinking, (2) encourage exploration and search, (3) teamwork, (4) documentation and dissemination of advances, (5) independent work and responsibility, and (6) encourage the planning and coordination of activities. Each tool used is subjected to a process of continuous adjustment according to feedback with students and teachers (Fig. 1).

Since students are grouped into small teams, each student participation, as well as the self-organization and interaction between teams is encouraged. Each team must present to his teammates the progress every two weeks. These advances are also recorded in the online shared platform to encourage criticism and fed back to the other teams. The evaluation considers the success of the proposed solution, the conceptual domain achieved, the demonstrated social skills and the concept of the other students in the course.

**Fig. 1** Workflow summary
of the formation dynamics
and adjustment of strategies

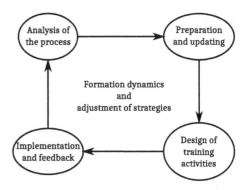

In most cases the solution of problems requires the development of both hardware and software prototypes [9]. The accompaniment by the teacher is supported by a software tool that raises issues to be developed in the solution, and presents some examples. This tool is mounted on a cloud storage system where each student team has its own folder. The tool is developed entirely in IPython Notebook, and has support for NumPy and SciPy for numerical operations, Matplotlib for plotting, and scikit-learn that integrates machine learning algorithms [7]. IPython Notebook is an interactive computational environment in which it is possible to combine code execution, rich text, mathematics, plots and rich media [10]. The problems assigned to students include self-balancing systems, robots with specific tasks of navigation, sensing or other activities in response to environmental conditions, and identification systems based on image processing (Figs. 2 and 3).

**Fig. 2** Self-balancing
system prototype built by
students

```
%pylab inline
```

```
Populating the interactive namespace from numpy and matplotlib
```

```
WARNING: pylab import has clobbered these variables: ['clf']
`%matplotlib` prevents importing * from pylab and numpy
```

```
x_index = 12
y_index = 5
```

```
# this formatter will label the colorbar with the correct target names
formatter = plt.FuncFormatter(lambda i, *args: data.DESCR[int(i)])
```

```
plt.scatter(data.data[:, x_index], data.data[:, y_index], c=data.target)
plt.colorbar(ticks=[0, 1, 2,3,4,5,6,7,8,9,10,11,12], format=formatter)
plt.xlabel(data.feature_names[x_index])
plt.ylabel(data.feature_names[y_index])
```

```
<matplotlib.text.Text at 0x12bbd0f0>
```

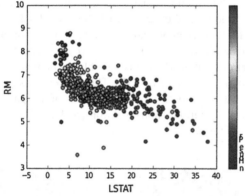

**Fig. 3** Part of a report developed by a group of students in IPython Notebook for a classification problem

## 4   Smart Systems for Learning: Autonomous Robots

There are many smart systems developed for education. Some of them focus in the classroom, and others in the development of smart environments. In the latter case takes great importance the use of prototypes of smart systems, systems in which specific processes are strengthened. In electronics, these processes are related to sensing activities, information processing, communication, management of actuators, and especially with smart answers and adaptability. The work with our electricity students has benefited from the development of some smart prototypes, in particular, small autonomous robots.

**Fig. 4** Intelligent system prototype (quadricopter) used in learning

With the help of students, we have developed a large number of prototypes, from differential wheeled platforms (Fig. 2) until aerial vehicles (Fig. 4). These systems are provided with processing units (embedded systems that may be small 8-bit microcontrollers until platforms with graphic OS as Raspberry Pi), a comprehensive set of sensors and actuators, and communication units. The problems posed to students are related to path planing in different kinds of environments, promoting the development of smart algorithms and schemes.

The smart education system is designed and developed as a smart student-centred autonomous robotic system with certain features of smart systems (sensing, data processing, transmission, activation of actuators) and some degree of intelligence and autonomy (modelling, inferring, learning, adaptation, and self-organization).

# 5  Results

The aim of the migration to an inductive education is the development of individual responsibility of students in their training process. The educational tools are complemented by continuous assessment and feedback to enable reflection and self-criticism. The final report is based on their own achievements and observations derived from the process. The oral presentation to their classmates and the teacher is fundamental, because it allows continued progress of students, increases their responsibility and confidence in their work, in addition to feed back to the other students.

As part of their final assessment, students answer a follow-up survey. This survey aims to obtain statistical information related to the student's perception against the new teaching strategy. One of the questions relates to the impact of the methodology on the student, the question was: Does the development of the project increases the interest and dedication to the course compared to other courses? Fig. 5 shows the results, where, at least in perception of students, wide acceptance of the teaching strategy (over 70 %) is observed.

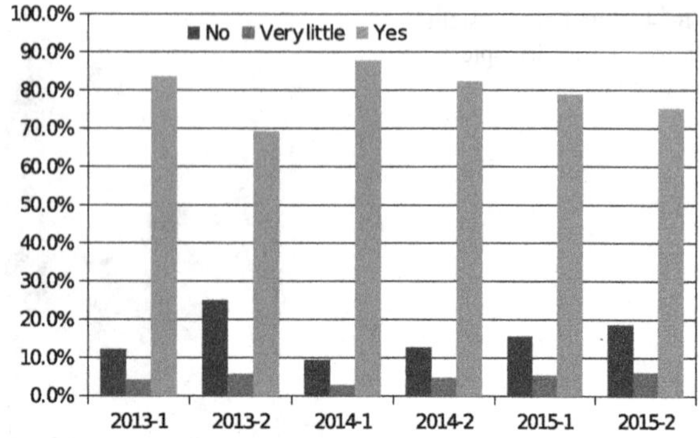

**Fig. 5** Student response to question: Does the development of the project increases the interest and dedication to the course compared to other courses?

**Fig. 6** Semester average obtained by students in their final examinations

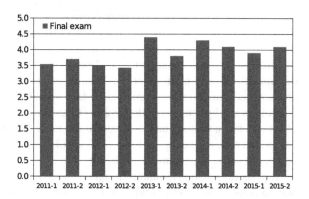

Another parameter used to assess the impact of the new methodological strategy is the academic performance of students in the written tests. This parameter has remained unchanged, and historically has been designed to determine the specific competencies of each course. Figure 6 shows the average values of the courses in the last five years. The evaluation scale is from zero (minimum value) to five (maximum value). Students pass the written test when they have a value exceeding three. During the years 2011 and 2012, students develop a traditional deductive process, the new inductive process was applied from 2013.

The figure shows from 2013 onwards a significant increase in academic performance (about 16 %), which is accompanied by a reduction in academic mortality. It is also worth noting that cancellations of courses by students were reduced from 15 to 2 %.

## 6 Limitations and Future Research

We must be mindful of the methodological limitations when considering the results of the study. The most critical variable is that the different methodological techniques were applied to different groups of students in different academic periods. Although the student community is more or less homogeneous, and widely characterized by the District University, learning styles and the dynamics that exist between students are unique. To increase the validity of the results, we propose to apply the same techniques to many more groups of students to analyze statistically the results. Further research is needed to show the benefits to students as a result of inquiry-based methods to justify a long-term commitment to PBL, and a possible overall curriculum redesign.

# 7 Conclusions

A change to an inductive approach in electronic area courses in an Electrical Engineering programme is presented in this paper. Students are directed to design the solution to a real problem involving theoretical concepts of the course. Monitoring, support and documentation is done on a cloud shared repository with documentation on IPython Notebook. The new proposed methodology combines theoretical concepts with their practical application. It is intended that students to construct their own knowledge by analyzing the problem and finding a workable solution. The goal is to allow the consolidation of specialized knowledge and professional practice that encourages student activity. Other objectives relate to the development of communication skills, teamwork and activities management.

Students are organized into small teams of three students, and after assign them the problem (the same for each team), are asked to self-organize to develop the solution. This cooperative work approach imposes a very different dynamic in the classroom. Each team is responsible for weekly advance, but also each student is individually responsible for the performance of their activity since the final evaluation depends on everyone. PBL builds critical thinking, practical skills, activity management, understanding, learning how to learn, and cooperative work.

The benefits of PBL in the electronic area courses are visible in the students' academic performance. In general, students recognize the potential of these tools, and they are noted with very high motivation. When they were asked if they see more interest in the course as a result of the methodology, about 70 % expressed a positive impact, this over three years of activity. In terms of improving in academic performance over previous strategies (deductive strategies), we compare the average results in the final written exam of each year, observing a significant increase in academic performance (about 16 %), which is accompanied by a reduction in academic mortality. These results gave a very positive feedback for the use of PBL and Inductive teaching to promote understanding of contents, theoretical concepts and skills development.

**Acknowledgments** This work was supported by the District University Francisco Jos de Caldas, in part through CIDC, and partly by the Technological Faculty. The views expressed in this paper are not necessarily endorsed by District University. The authors thank the research groups DIGITI and ARMOS for the evaluation carried out on prototypes of ideas and strategies.

# References

1. Beichner, R.: History and evolution of active learning spaces. New Dir. Teach. Learn. **2014**(137), 9–16 (2014). doi:10.1002/tl.20081
2. Bell, T., Urhahne, D., Schanze, S., Ploetzner, R.: Collaborative inquiry learning: models, tools, and challenges. Int. J. Sci. Educ. **32**(3), 349–377 (2010). doi:10.1080/09500690802582241
3. Dewey, J. (ed.): How We Think. D. C. HEATH & CO. (1910)
4. Dole, S., Bloom, L., Kowalske, K.: Transforming pedagogy: changing perspectives from teacher-centered to learner-centered. Interdisc. J. Probl. Based Learn. **10**(1), 1–15 (2016). doi:10.7771/1541-5015.1538
5. Fullan, M., Langworthy, M.: Towards a new end: new pedagogies for deep learning, Jun 2013. www.newpedagogies.org
6. Ginsburg, H., Opper, S.: Piaget's Theory of Intellectual Development, 3rd edn. Pearson (1987)
7. Jovic, A., Brkic, K., Bogunovic, N.: An overview of free software tools for general data mining. In: 37th International Convention on Information and Communication Technology, Electronics and Microelectronics (MIPRO 2014), pp. 1112–1117 (2014). doi:10.1109/MIPRO.2014. 6859735
8. Lacuesta, R., Palacios, G., Fernandez, L.: Active learning through problem based learning methodology in engineering education. In: 39th IEEE Frontiers in Education Conference (FIE '09), pp. 1–6 (2009). doi:10.1109/FIE.2009.5350502
9. López, C.: Evaluación de desempeño de dos técnicas de optimización bio-inspiradas: Algoritmos genéticos y enjambre de partículas. Tekhnê **11**(1), 49–58 (2014). ISSN: 1692-8407
10. Lovejoy, M., Wickert, M.: Using the ipython notebook as the computing platform for signals and systems courses. In: IEEE Signal Processing and Signal Processing Education Workshop (SP/SPE 2015), pp. 289–294 (2015). doi:10.1109/DSP-SPE.2015.7369568
11. Martínez, F.: Editorial. Tekhnê **10**(1), 3–5 (2013). ISSN: 1692-8407
12. Mohd-Yusof, K., Jamaludin, M.Z., Harun, N.F.: Cooperative problem-based learning (cpbl): a practical pbl model for engineering courses. In: IEEE Global Engineering Education Conference (EDUCON 2011), pp. 366–373 (2011). doi:10.1109/EDUCON.2011.5773162
13. Prince, J., Felder, R.: Inductive teaching and learning methods: definitions, comparisons, and research bases. J. Eng. Educ. **95**(2), 123–138 (2013). doi:10.1002/j.2168-9830.2006. tb00884.x
14. Prince, M.: Does active learning work? a review of the research. J. Eng. Educ. **93**(3), 223–231 (2004). doi:10.1002/j.2168-9830.2004.tb00809.x
15. Prince, M., Felder, R.: The many faces of inductive teaching and learning. J. Coll. Sci. Teach. **36**(5), 14–20 (2007)
16. Selvi, S., Kaleel, D., Chinnaiah, V.: Applying problem based learning approach on e-learning system in cloud. In: International Conference on Recent Trends In Information Technology (ICRTIT 2012), pp. 244–249 (2012). doi:10.1109/ICRTIT.2012.6206814
17. Vygotsky, L.: Thought and Language. MIT Press (1962)
18. Weiss, R.: Designing problems to promote higher-order thinking. New Dir. Teach. Learn. **2003**(95), 25–31 (2003). doi:10.1002/tl.109
19. Yu-Lin, J., Tien-Chi, H., Chia-Chen, C., Yu, S., Yong-Ming, H.: Developing a mobile instant messaging system for problem-based learning activity. In: International Conference on Interactive Collaborative Learning (ICL 2015), pp. 313–316 (2015). doi:10.1109/ICL.2015.7318044

# A Supporting Service in Teacher Training: Virtual Newspaper

Nuray Zan and Burcu Umut Zan

**Abstract** This study focuses on establishing communication via virtual newspaper in institutions where pedagogical education is provided. By considering the traditional newspapers, the role of virtual newspapers, published in the internet, in the improvement of teacher candidates, its significance on his/her interaction with the institution and the usage of this communication channel are dealt with in this study. In 2015–2016 academic year, 35 teacher candidates took part in the study. In this study qualitative research methods were used both in collecting and evaluating data. The virtual newspapers created by teacher candidates were analyzed by content analysis method and the viewpoints of the teachers to the study were researched. Obtained results of the study were as follows: Teacher candidates performed meaningful learning on educational subjects, established a link between their fields of training and areas of expertise and had the advantage of using various techniques and methods, However, teacher candidates, who were below the average in using information technologies, had some difficulties in preparing a virtual newspaper and they indicated that working on a virtual newspaper had a disadvantage as it consumed too much of their time.

**Keywords** Virtual newspaper · Teacher training · The internet · Learner · Teaching science

N. Zan (✉)
Department of Educational Sciences, Çankırı Karatekin University,
Çankırı, Turkey
e-mail: nurayyoruk@gmail.com

B.U. Zan
Department of Information and Records Management, Çankırı Karatekin University,
Çankırı, Turkey
e-mail: burcumut@gmail.com

© Springer International Publishing Switzerland 2016
V.L. Uskov et al. (eds.), *Smart Education and e-Learning 2016*,
Smart Innovation, Systems and Technologies 59,
DOI 10.1007/978-3-319-39690-3_17

# 1 Introduction

Even in the early 1990's, academicians were waiting for weeks to have an access to the research articles published in academic journals. However, in our modern day we carry out our studies by using a computer and we can have an instant access to the full version of the articles. New opportunities stemming from the scientific and technological developments fastened the social change. Besides, change has been inevitable in many institutions as we are heading towards an information society. The change in the field of education, forces the "traditional" education system and correspondingly the roles of teachers to also experience a change.

## 1.1 Social Change and Changing Teachers

In our modern day, it is known that many new professions emerged due to fast changing technology and accumulation in information production. However in this knowledge-based society, it is not possible to be an expert on a field and to fully comprehend all necessary qualifications in a profession without getting a proper formal training. In other words, in order to guarantee a sustainable competition and to have universal professional qualities, individuals should go through the steps of a good education. What we need in the global knowledge-based society where we want to reach to the levels of civilized societies, are fully equipped schools and qualified teachers to be employed in these schools. In this period as a profession, teachers should have the largest proportion in investments and they should have a far more technological knowledge when compared with an average individual. The vision of a teacher will help the students to create their visions in the future. Teachers are giving life to the fully equipped schools that we need and they are the ones making the schools endurable and powerful. It is necessary to educate trainers who are capable to use every technological tool and who will carry out their professions in the most appropriate way which is providing the fastest and best learning environment for students. Application of information technology to education provides savings in time, cost and labor in relation with the principles of economy and it also increases the efficiency. No longer schools have a duty as to make pupils capable to live in the digital world so that teachers have to deal with the new generations and the technologies they use, encouraging a greater educational use during course [1]. That is why teachers must be able to integrate the use of technology and technology standards for students into the curriculum. Teachers must know where, when, and how to use technology for classroom activities and presentations. Also teachers must know basic hardware and software operations, as well as productivity applications software, a web browser, communications software, presentation software, and management applications. Teachers must have the technological skill and knowledge of Web resources necessary to use technology to acquire additional subject matter and pedagogical knowledge in support of teachers' own professional [2].

## 1.2  Teacher and Technology

In the middle of 1990's the number of families having internet connection were rising day by day and it is known that benefiting from the internet in teaching and learning environments creates a mutual sharing environment and a place to discuss ideas for individuals. Together with the internet, the increase in using the information technologies for educational purposes is in a close relation with the teachers' awareness and proficiency in the field. To use the internet in-class teaching practices, teachers should be competent in how to adapt this technology to teaching programmes. For this reason how to use information technologies for educational purposes should be systematically presented to teacher candidates. Acquiring the usage of information technologies for educational purposes and transferring the wide range of documents, animations, experiments and innovations that they benefited to students to help them also benefit from this global knowledge net has been a necessity for teachers [3]. There is a debate about technology using in classroom teaching. Some researchers believe that technology prepares students by improving their skills with information and communicative technologies, [4, 5]. At the same time researchers are in the same opinion with increasing academic achievement and improving students' interactions in classroom [6–9]. But on the other hand, it has been arguing that technology in itself does not teach but it is an effective tool when placed in the hands of skilled teachers [10]. Whether children use technology in their lessons depends largely on the preferences of their teachers' interests and skills [11].

Individuals studying in the field of education should carry out their studies by focusing on the areas widely used by the students and therefore it will be easier for them to achieve their intended goals. The aim of using educational technology and media is to increase the efficiency in learning. To do these materials should be carefully chosen and used within a context. The contribution of educational technology and media to learning depends on how they are used by the teachers. Trainings for launching a virtual newspaper in the field of education and producing it as a material to supplement teacher candidates and publishing it for the access of the target group will guide those studying in the field and contribute them in training themselves. One of the most significant functions of mass media is to be instructional and by this way the virtual newspaper aims to educate contemporary and qualified individuals with being used by the teacher candidates.

## 1.3  Virtual Newspaper

People have always felt the necessity of communicating with others who are living in the same time period with them. This is a natural outcome of life in being socialized. In other words, considering the conditions of our are, individuals are trying to do this by using the most familiar means of mass media. In the past, the most important and functional means of mass media was the newspapers [12].

Newspapers have also started to appear in the internet throughout the years. Generally newspapers have four main functions; to inform, to entertain, to be instructional and to provide service to the readers by the advertisements. Media tools are quite efficient in gaining knowledge and skills that are difficult to gain via formal education [13]. Individuals internalize the information gained through media tools more easily and they do not struggle to adapt this information to their daily lives [14]. As one of the media tools, the newspaper has a high motivational strength [15] and it is quite effective in drawing attention to science matters in articles [16].

The instructional function: It is effective in developing the personality of individuals, in helping them to be aware of their different areas of interest and in facilitating the period of socialization.

The informative function: It has a role in informing the individuals about local, national and universal developments and providing him to be aware of daily/weekly or monthly news. When the individuals are informed their level of awareness increases and they may be directed to thinking.

The entertaining function: Newspapers will carry away the individuals from their problems and will temporarily entertain them with enabling them not to be too much occupied with mental processes devoted to the problems.

The role of providing service by advertisements: With the help of newspapers, individuals will instantly learn about interesting events, celebrities and new places and products.

As it is mentioned above newspapers informs the readers, instructs them and they have roles in entertaining and advertisements. By taking these main functions of newspapers as a basis, the experience of presenting field information to the students within this framework forms the subject matter of our study.

The objective of this study is to analyze the viewpoints of the teacher candidates, who are studying in pedagogical formation certificate programme, about the method of forming a virtual newspaper.

## 2  Method

Content analysis, which is among the qualitative research methods in the processes of collecting data and discussing the results, is used in this study. Similar data are organized and interpreted within the framework of specific concepts and themes. Therefore the study was read and categorized by the researcher and then the same process was repeated by a second researcher.

### 2.1  Sampling

The sample group of the study was formed by the students who attended to pedagogical formation certificate programme in the University of Çankırı Karatekin in

**Table 1** The distribution of students to the departments

| Faculty | Department | Number of students |
|---|---|---|
| Faculty of Science | Physics | 9 |
| Faculty of Science | Chemistry | 3 |
| Faculty of Science | Biology | 5 |
| Faculty of Engineering | Food Engineering | 5 |
| Faculty of Health Science | Department of Nursing | 13 |

2015–2016 Academic year, and who took Special Teaching Methods course in the Spring semester. All the 35 students in the sample group are teacher candidates graduated from Faculty of Science, Faculty of Engineering and Faculty of Health science. The distribution of students to the departments is given in Table 1.

## 2.2 Data Collecting

After the study interviews were made with teacher candidates and data were collected by using the "Teaching Method Evaluation Form".

**Application**
In the class of 35 teacher candidates who took part in the study, major area courses within the content of special teaching method were thought both theoretically and practically for seven weeks. After this teaching process, teacher candidates were asked to prepare a newspaper for educational purposes about their undergraduate fields. All the teacher candidates were informed that structural teaching approach was to be taken as a basis in the study. Moreover, the areas to take part in the newspaper and the length of it (four pages) were also given to the teacher candidates. A sample template was given in Fig. 1.

The general template of the virtual newspaper was prepared. In this template, the name of the newspaper and the subjects to take part on the pages were identified. The newspaper was decided to be 4 pages. On the first page, the main subject based on the unit took place. On the second page, when possible, science people that contributed to scientific studies from the past and the recent time were provided.

On the third page, information on science related activities that took part in classrooms throughout the school year and their images, and activities related to trips and observations were provided.

On the fourth page cross word puzzles related to scientific information and suggested publications from the library took place. And also in this page, information related to national or religious holidays and important dates were provided which was in the interval of this issue. And also in last page, with the preferable of

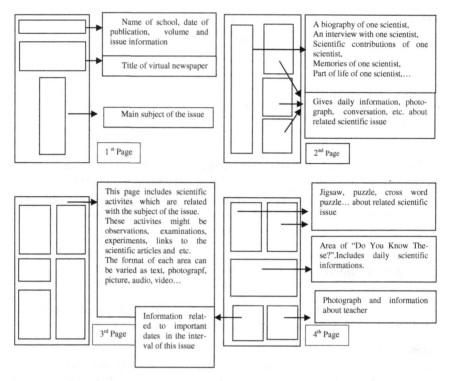

**Fig. 1** A sample template

teacher or teachers, who constitutes the committee of this virtual newspaper, should give their names, photographs and information about themselves. But this is not a mandatory template for teacher candidates. This is just a proposal for scientific virtual newspaper. This template can be converted to different formats but the content of this template should be taken and developed by teacher candidates.

A-two week time was given to teacher candidates to prepare their newspapers. Every teacher candidate presented his/her virtual newspaper to his/her peers and discussed the efficiency of the study. Furthermore, they indicated the effects of the application to teacher candidates studying in other fields. A sample of four paged virtual newspaper prepared by teacher candidate is given in Fig. 2.

## 3   Findings

Virtual newspapers of teacher candidates were assessed by two researchers. Assessment was done by using basic criteria. Basic criteria are given in Table 2.

90 % of virtual newspapers were found successfully in giving main subject. In those newspapers, main scientific subject of the issue was given clearly with

**Fig. 2** An original sample of virtual newspaper

striking titles and attractive content, limitations of subject was sufficient for the reader student group. 30 % of the virtual newspapers did not provide the issue with the presentation of one scientist. 80 % of the virtual newspapers supported with connections of scientific issues with daily life problems. These parts mostly provided within the corner of "Do you know these?" 60 % of virtual newspapers were lack of giving scientific activities. But the newspapers provided with scientific activities were very successful. 70 % of these were supported with both visual and audio-visual content and 80 % of these were supported with links directed to the scientific publications and experiment videos on youtube. 80 % of the virtual newspapers were enriched with entertainment materials. 50 % of the teacher candidates were used more than one entertainment material. Mostly used material were respectively, cartoons and crossword puzzles. To the aim of not to tiring eyes 20 % of virtual newspapers gave too much empty space. This situation cause presenting

**Table 2**  Criterias for assessment

| Criteria | Details |
|---|---|
| Main subject of issue is sufficient | Main subject is explained with an understandable language |
| | Limitations of main subject is sufficient |
| A presentation of one scientist is sufficient | Main topic is given clearly (E.g. Newton and "law of gravity") |
| Daily information about scientific issue is sufficient | Is it supported with;<br>　　Photographs or pictures,<br>　　Videos,<br>　　Observations,<br>　　Links to scientific publications... |
| Scientific activity about related issue is sufficient | Activities are understandable |
| | Easy activities are selected to courage students to do it |
| | Activities are supported with visual or audio-visual materials |
| Related scientific issue is supported with qualified entertainments | Is this part includes, Comics, cartoons, jokes, anecdotes, jigsaw, puzzle, cross word puzzle... |
| Visual qualifications of virtual newspaper | Does it make eyes tired? |
| | Visual materials, audio-visual materials and texts are used in a balance? |

weak contents in the newspaper. The texts and figures are compressed 20 % of them, to give augmented context within 4 pages. Sample of virtual newspaper, made by teacher candidate, was given in Appendix.

In this study which is carried out with the goal of analyzing the viewpoints of teacher candidates in preparing a virtual newspaper, the findings and gathered data are presented below.

*Views of teacher candidates about preparing a virtual newspaper*

> It provided us to put forth abstract data in transferring the techniques and methods which we learnt in Educational Sciences to our own fields. A study like this makes education more attractive.

> It's a method which does not force the individual to rote learning, rather than that individuals are expected to be curious and thoughtful. Moreover it's a cost effective method through which a beneficial product is achieved by spending less money.

> With the help of this method we learnt how to adapt the subject matters of our fields to the levels of the students and we also had ideas about how to teach.

> This is not only a teaching method, studying on it takes time but it provides me to imagine the subjects of my field and developed different approaches. Moreover it also provides met o make researches.

> I took lots of courses throughout the certificate programme but only in this I was able to materialize a subject matter in my field. Whatever was taught in the class was very far from my area of expertise and I did not know how to establish a link. With the help of this study, I managed to relate the subjects of Physics with education.

It gives us the opportunity to use puzzles, cartoons, experiments photographs, poems, information about daily life and etc. simultaneously. Therefore we can address to multiple intelligence areas and this makes learning and comprehension easier.

Besides, many of the teacher candidates mentioned the advantages of this electronically produced study. These advantages are specified below.

1. If this was a printed newspaper, we would not be able show the experiments in video format.
2. The resources can be easily reached simply by clicking on the links given on the virtual newspaper. Hence students can study the subject matter without spending time in the internet.
3. If this was not an online Newspaper, we should bear in mind the cost of printed newspaper. Whereas, I only printed one or two copies of the newspaper just for the purpose of introduction which I made to students.
4. In a virtual newspaper, students can study the visuals in detail by enlarging them.
5. In a virtual newspaper, the group which I reach is not limited with the class. It gives us the chance to reach to more crowded groups. For example my virtual newspaper can be an educational material for many other teachers dealing with the same subject in Turkey.
6. The virtual newspaper application enables the students to access to the course material and to attend the activities of the course with their smart phones or mobile devices regardless of time and place. Hence, excuses like tearing, losing or deformation of the course material are packed away.

On the other hand some teacher candidates indicated that this electronically produced study has some disadvantages. Teacher candidates who have an insufficient level of computer literacy and who are not good at technology mentioned that they had some difficulties and needed help. An interesting view about this is as follows:

If this had not been a virtual newspaper but a printed one, I would have searched various resources and picked the ones that I needed. Both the visuals and the necessary information could be photocopied and pasted to the printed newspaper. Moreover I would have prepared the puzzles and pictures manually. I would also have had the chance to give the steps of the experiments and spend less time and benefited from the opportunities which I had.

When Table 3 below is analyzed, it is seen that teacher candidates have generally positive attitudes about the study. However two of the views indicate that this method is boring and time consuming. Teacher candidates mentioned that this approach has been more effective than other courses in the field of education and they have reached too many educational attainments with the help of this method.

**Table 3** Content analysis of the views of teacher candidates about virtual newspaper

| Propositions prepared in accordance with the views of teacher candidates | F | % |
|---|---|---|
| The method of preparing a virtual newspaper enables us to integrate the subjects of our fields with education | 20 | 57.1 |
| The method of preparing a virtual newspaper enables us to materialise the subjects in the field of education | 17 | 48.6 |
| The method of preparing a virtual newspaper enables us to carry out the applications of our fields in education | 25 | 71.4 |
| The method of preparing a virtual newspaper is not expensive | 23 | 65.7 |
| The method of preparing a virtual newspaper has too many advantages when compared with printed newspapers | 25 | 71.4 |
| The method of preparing a virtual newspaper will be my one of the first applications to use when I become a teacher | 16 | 45.7 |
| The method of preparing a virtual newspaper enables us to use many methods and techniques together | 22 | 62.9 |
| The method of preparing a virtual newspaper causes met o spend too much time with the computer and in the internet | 9 | 25.7 |
| The method of preparing a virtual newspaper is boring. The interested students can already find these information | 5 | 14.3 |

# 4  Discussion and Suggestions

In this chapter, both the advantages and disadvantages of producing a virtual newspaper is discussed by taking the obtained data into consideration.

The persistence of knowledge is the result of meaningful learning. As for Ausubel, new learning is meaningful when it is associated with previous learnings [17]. Transferring the learnt data to other realms increases the persistency of learning. Carrying a method or technique in the field of education to the environment of a virtual newspaper in relation with a subject in science, enables the teacher candidates to use all his/her knowledge which s/he acquired with formation courses. This means that meaningful learning is realized.

Another advantage of this method is that teacher candidates who remained silent in a lesson atmosphere and did not take active part in the classes were seen to be involved in the lessons while presenting their virtual newspapers and emphasizing up-to-date links about the areas of expertise.

Some of the teacher candidates within the sample group emphasized that they were not good at information technologies and they had some difficulties while using this method. However, all the students that they will confront will be demographically more capable in information technologies. For example today's high school or university students met with the internet and mobile devices when they were really very young and learnt how to use them. For a generation like that, preferring the traditional methods in teaching will distract them and make them bored. Therefore we should speak the same language with the students, that is to say we should also follow the popular modes of communication and entertainment and use these modes effectively in teaching.

## 4.1 Suggestions to Teacher Candidates

Teaching is a profession which is only done with passion. Therefore, most of the time teachers use various teaching methods and techniques with spending time and effort to obtain a complete learning process and to reach to the expected learning outcomes. While preparing the methods to be used to reach the students, it is necessary for teachers to prepare themselves before the lessons.

When the principle of economy is taken as a basis in education, it should be noted that our study can be used every following year with making some improvements. It should always be kept in mind that a material prepared about a specific subject can be used by making some necessary updates whenever the same subject will be taught. With this method we can form our material archives. Because of these preparing materials before the lessons is very crucial for teachers.

Sharing our studies in a virtual platform transfers professional knowledge, draws attention to the events happening around, provides a close contact between student and lesson, establishes links between science, technology, society and environment, presents cultural knowledge, enables the mechanisms of criticism and self criticism of the system, reinforces the sense of belonging to a group, provides the connection, interaction and communication between learner-teacher and learner-learner.

Teacher training programmes should increase the efficiency in education by using the advancements of technology in every educational activity. For this reason, when necessary teacher candidates should take an exam evaluating their proficiencies in information technologies and they should be more competent than an average user in information technologies. The method of producing a virtual newspaper which forms the basis of our study was carried out by teacher candidates of math and science. It is thought that carrying out the same study in a group of teachers in social sciences can have similar but also different results. This virtual newspaper application which is prepared to present a specific subject matter to students can also be prepared practically for how to teach the teaching methods and techniques in the field. For example the already used versions of teaching methods and techniques in the fields such as using concept cartoons in teaching physics, experience-based teaching applications in chemistry, teaching physics about analogies can be organized as issues of the virtual newspaper.

## 4.2 Suggestions Related to Technological Approaches to Virtual Newspaper

The dissemination of the virtual newspaper can be done by using different methods. One of the options that can be used in dissemination is QR (Quick Response) application. QR application is used by smart phones which have digital cameras and this kind of phones is widely used in our age. With the QR application the users can be directed to web sites, to e-mail addresses, phone numbers, and contact information, SMS or MMS or to geographical location data. By QR applications students can have an easier access to virtual newspaper. Moreover, students' e-mail accounts face book accounts, tweeter accounts can also be used as a means in disseminating the new issues of virtual newspaper in classes where it is used. However the borders of the virtual newspaper should not be limited with the classes and it should be noted that it can be disseminated to various web-addresses with the help of reposts.

There may also be a group who want to have an access to the new issues of the virtual newspaper and who want to follow it continuously, so these people can be subscribed to the virtual newspaper by which they will always be aware of the new issues. Furthermore subscriptions can be saved on a database and an appropriate place can be given in virtual newspaper for e-mail addresses of those who want to subscribe to virtual newspaper.

It is known that the internet is growing very fast and especially the students are dragged into a crowd of knowledge. Reliable and verified publications are needed on the internet. Therefore, if publications like the virtual newspaper are consistent and followed by a group of people in the area of interest, this will eventually pave the way to various sources of income. In this point, the popularity of the virtual newspaper in the internet is very significant and can be the subject of another study.

Authors can also prepare virtual newspapers about their books. They can add a QR code inside their books to direct both the students and teachers to the relevant web-site. With this application subjects can be presented theoretically in the book and by the virtual newspaper application, the connection between science, technology and environment, applications for multiple intelligences, relevant documentaries, videos of experiments, contemporary and popular texts and occupational information about the subject, can be given.

As a result of the increase in online culture, traditional paradigms will lose their power. It is foreseen that the boundaries between disciplines will be removed. In the following years online communities will form a new culture by uniting web-sites databases and online chat rooms with traditional media communication forms like television, music and film. In time, the existence of school and curriculum as it is today will vanish and current school system will be a virtual field.

New developing information technologies will remove the school of the industrial age, the printed material which is the main component of this school, teachers, the content and organization of the curriculum and school buildings [18]. The culture of tomorrow can only arise from the output of mass means of communication. A new type of teaching will have to take that into account [19]. To what extent these will be realized is not clear but the importance of education will increase in the following years and both the content and methods will not remain the same.

# References

1. Ghislandi, P., Facci, M.: Schools in the digital age: teachers' training role in the innovative use of the interactive whiteboard. J. Theor. Res. Educ. (Ricerche Di Pedagogia E Didattica) **8**(1), 61–78 (2013)
2. UNESCO: ICT competency standards for teachers. Competency Standards Modules. United Nations Educational, Scientific and Cultural Organization, Parigi (2008)
3. Mustafa, M.Q., Şahin, S.: Öğretim elemanlarının öğretim amaçlı internet kullanımları Eğitim Teknolojisi Kuram ve Uygulama: Educ. Technol. Theory Pract. **3**(1), 1–12 (2013)
4. Johnson, L., Adams, S., Cummins, M.: NMC Horizon Report: 2012 K–12 Edition. The New Media Consortium, Austin, TX (2012)
5. National Educational Technology Standards: The standards for learning, leading, and teaching in the digital world. International Society for Technology in Education, Washington, DC (2012). http://www.iste.org/standards. Accessed 2014
6. O'Brien, D., Beach, R., Scharber, C.: "Struggling" middle schoolers: engagement and literate competence in a reading writing intervention class. Reading Psychol. **28**(1), 51–73 (2007)
7. Scherer, M.: Transforming education with technology: a conversation with Karen Cator. Educ. Leadersh. **68**(5), 17–21 (2011)
8. Storz, M., Hoffman, A.: Examining response to a one-to-one computer initiative: student and teacher voices. Res. Middle Level Educ. **36**(6), 1–18 (2013)
9. Wendt, J.: Combating the crisis in adolescent literacy: exploring literacy in the secondary classroom. Am. Second. Educ. **41**(2), 38–48 (2013)
10. Philip, T., Garcia, A.: The importance of still teaching the generation: new technologies and the centrality of pedagogy. Harvard Educ. Rev. **83**(2), 300–319 (2013)
11. McDermott, P., Gormley, K.A.: Teachers' use of technology in elementary reading lessons. Reading Psychol. **37**(1), 121–146 (2016)
12. Demiray, U., Gürcan, H.: 4–8 Mayıs, Uzaktan Eğitimde bir öğrenci destek hizmeti olarak gazetenin önemi ve rolü Türkiye İkinci Uluslarası Uzaktan Eğitim Sempozyumu ss-197–207 (1998)
13. Özay, K.E.: Using newspaper articles in biology education. Theoret. Educ. Sci. **1**(2), 84–89 (2008)
14. İlkörücü, G.Ş.: Primary School 6th Students' Association Levels Between Information Gained in Biology Classes and Daily Issues. Uludağ University, Institute of Social Sciences, Bursa (2007)
15. Dee Garrett, S.: Improving comprehension with newspapers. In: 2008 NIE Week Teacher's Guide. Newspaper Association of America Foundation (2008). http://www.naa.org/docs/Foundation/NIE%20Week%2008%20teacher%20guide.pdf, Accessed 05 Aug 2014
16. Halkia, K, Mantzouridis, D.: Students' views and attitudes towards the communication code used in press articles about science. Int. J. Sci. Educ. **27**(12), 1395–1411(2005)
17. Ausubel, D.P.: Educational Psychology, A Cognitive View. Holt Rinehart and Winston Inc., New York (1968)
18. Marshall, J.D., Sears, J.T., Allen, L.A., Roberts, P.A., Schubert, W.H.: Turning Points in Curriculum: A Contemporary American Memoir, 2nd edn. Pearson, Merril Prentice Hall, Ohio (2007)
19. Use of newspaper as a distance teaching medium. A case study. Prepared by the Educational Research and Development Department of the college Maria- Victoria, Montreal, Quebec, Canada (1983)

# Participation in State R&D Projects Jointly with Industrial Enterprises: Factor in Employability Improving of University Graduates

Alexey Margun, Artem Kremlev, Aleksandr Shchukin,
Konstantin Zimenko and Dmitry Bazylev

**Abstract** The paper describes an impact of participation in state R&D projects jointed with industrial companies on quality increasing of educational process. The potential impact of such R&D projects is based on the experience of state project between ITMO University and AlkorBio Group biotechnological company. The results of the project can be used as a basis for development of educational programs, which have a direct interaction with real sector of economy and industry. The development of such R&D projects and educational programs can contribute to a sharp increase of employability of students after graduation.

**Keywords** State R&D projects · Employability · Complex technical objects · Quality of educational process · Competence-based approach

## 1 Introduction

Enhancing the prestige of the engineering profession is one of the priorities in modernization of the Russian education system. This priority is chosen due to the fact that technological re-equipment of the country is essentially depends on level of qualification of engineers, technologists and designers. To achieve the goal it is necessary to

A. Margun · A. Kremlev · A. Shchukin · K. Zimenko (✉) · D. Bazylev
Department of Control Systems and Informatics, ITMO University,
49 Kronverkskiy av., 197101 Saint Petersburg, Russia
e-mail: kostyazimenko@gmail.com

A. Margun
e-mail: alexeimargun@gmail.com

A. Kremlev
e-mail: kremlev_artem@mail.ru

A. Shchukin
e-mail: shhukinaleksandrsd1990@yandex.ru

D. Bazylev
e-mail: bazylevd@mail.ru

form a new generation of engineers, able to create, maintain and develop innovative technological solutions. One of the main tasks of engineering education is to prepare not only professionally educated and self-dependent specialists, but also their preparation for successful entry into the labor market. It is also required to develop an active social position and ability of self-education of future engineers in order to form a competent person [1].

However, the absence of state order for training of engineers, non-participation of government authorities in career guidance process have led to the fact that engineering professions are not popular for Russians in our days. For example, about 20 % of Russians suppose that the most prestigious profession is legal profession, and economists are in second place. Engineers and scientists rating is about 1–3 % [2]. Besides, young engineers without experience at the beginning of their careers do not have big chances to get good job in their professional field. This is confirmed by the data of the Federal State Statistics Service, namely, growth of unemployment among university graduates has reached 21.9 % [2].

Nowadays, the Government of the Russian Federation conducts an effective policy on the development of cooperation of Russian higher educational institutions and industrial enterprises [3]. In this regard, new projects and programs of financial support aimed for development of high-tech industry and stimulation of innovations in the Russian economy are created. Particularly, special grants (so-called Megagrants) are established for development of cooperation of Russian higher educational institutions and industry organizations for integrated projects implementation and creation of high-tech production. The program of Megagrant projects is founded according to the decision #218 of the Russian Federation Government of April 9, 2010 "Measures of state support for development of cooperation between Russian universities, research organizations and companies which implement complex projects for high-tech production".

A large amount of funding for such projects (about 1–2 mln. euro per year) has an impact on the increasing of popularity and competitiveness of engineering specialties. Moreover, these projects have an significant effect on improving the quality of education process and increase student interest in their own profession. The present paper describes and analyzes the Megagrant project influence on improving the quality of educational process, changes in educational structures, etc. This paper is based on the experience of ITMO University which is the participant of the Megagrant project. The R&D work in ITMO University is devoted to development of analytical robotic complex for clinical laboratory tests using nano-reagents and is conducted in cooperation with AlkorBio Group biotechnological company.

## 2 Formation of General and Professional Competencies in the Innovative Teaching Technologies

The ongoing reform of higher education in Russian Federation is associated with the transition to competence-based approach. Competitive education programs are aimed to developing of a new integrated characteristics of graduates, namely, employ-

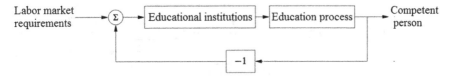

**Fig. 1** Feedback control of educational process

ability. Employability of graduates means presence of skills that allow them to be flexible workers who can easily operate in very different environments [4, 5].

In accordance with this, potential job place in the chosen professional field is a unique landmark in the formation of specific competences. Thus, the external environment corrects the education system by increasing its practice-oriented nature. Roughly speaking, this correction process can be presented as a well-known two-dimensional control process with negative feedback (Fig. 1).

In this case "Educational process" is the process to be controlled, "Educational institutions" is the control block, "Labor market requirements" forms the reference requirements and competent person, who is needed for the labor market, is presented as the process output.

In this case labor market requirements contain the following competences:

- autonomy in knowledge acquisition;
- disciplined and concentrated working;
- communicability and social interactivity;
- competitiveness and adaptability of the labor market;
- cognitive interest, self-development and ongoing training;
- etc.

Considering the educational process in more detail, we can conclude that this process is quite complex and pursues multiple objectives. To fulfill the requirements for the educational process mentioned above the common for competence-based educational scheme and modular system are used. The structural scheme of the educational process is shown in Fig. 2 [2], construction principles of modular system are presented in Fig. 3 [6].

Presented education system is accepted and has been successfully used around the world [1, 6]. However, this system has several drawbacks. For example, the system allows to prepare a professional who is capable of self-development and has a number of mentioned competencies. But, this specialist usually does not have in-depth knowledge. Instead of working at the new job the specialist has to learn again to meet the requirements of the employer. In this situation, the employer receives a "pupil" instead of the employee, who solves different engineering challenges. In response to this, when hiring an employee, employers demand work experience [5, 7].

Thus, the system receives the following contradiction: on the one hand, the education system is aimed at training the specialists needed for the labor market, on the other hand, employers do not want to hire university graduates. A number of articles

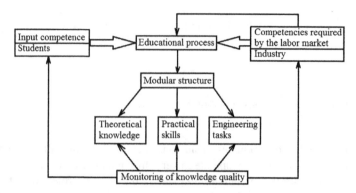

**Fig. 2** Structural scheme of the educational process

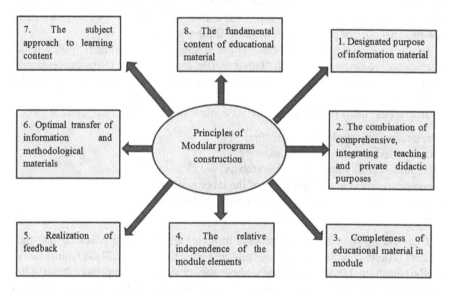

**Fig. 3** Construction principles of modular system

is devoted to the solution of this problem (see, for example, [4, 5, 8–10]). However, most of these articles describe only partial (as in [10–12]) or local (as in [5, 8]) solution.

An interesting fact that in Russia one of the solutions for this situation was the establishment of Megagrants, that are primarily aimed at an entirely different purpose. Firstly, these grants are aimed to support cooperation between Russian universities, research organizations and companies. During the execution of one of these grants in ITMO University it is turned out that Megagrant has an significant impact on improving the quality of education process, increases student interest in their own profession and improves their competitiveness.

## 3    Megagrant Framework

This section describes the framework and objectives of the Megagrant under decree of Russian Government no. 218 [13].

At 9th April, 2010 the Government of Russian Federation has approved decision "Measures of state support for development of cooperation between Russian universities, research organizations and companies which implement complex projects for high-tech production". Decree provides the possibility of allocating subsidies to industrial enterprises for a period from 1 to 3 years with volume of financing up to 2 million of Euro per year for supporting of high-tech production organizing in cooperation between industry companies and universities. The volume of contributed to the project funds of industrial enterprises should be not less than 100 % of the subsidy amount, and should be adequate to complete the organization of a new high-tech production. The subsidy is allocated to the production enterprises to guarantee the demand of scientific results from university and its further using for the organization of a new high-tech production. Also, within the framework of the grant there are a lot of students to be involved to research and engineering work.

In 2014 ITMO University became a participant of several of such grants. One of them is devoted to the development of analytical robotic complex for clinical laboratory tests using nano-reagents and it is conducted in cooperation of ITMO University with AlkorBio Group biotechnological company. The following section describes how the grant allows to improve education process in ITMO University and increase employability of graduates.

## 4    R&D Project as Factor in Quality Increasing of Educational Process

One of the key conditions for the grant is the involvement of students, PhD students, young scientists and specialists into the research group [13]. Namely, half of the participants in the projects should be young Russian scientists under the age of 35 years. In other words, realization of the Megagrant provides attractiveness of the engineering science for wider ranges of young people.

World experience shows that the student participation in R&D works contribute considerably to the students' positive attitude to this form of education (see, for example, [14–16]). The enhancement of student motivation for active participation in projects by involvement of an industrial partners is considered in [14, 16]. The cooperative research for both industry and universities allows to apply the powerful educational and research tool [16, 17].

Within the framework of the grant at ITMO University a lot of students were attracted to scientific and engineering work. At the current stage, in the framework of the grant a working sample of analytical robotic complex for clinical laboratory tests (Fig. 4) is developed. Most of engineering and research works are carried out

**Fig. 4** Construction principles of modular system

with the participation of students. Involved students have received themes of their diploma works and master's theses, which are related to various aspects of theoretical and practical research of the project. Practical and approbation parts of PhD student theses are also related to project thematic. Due to the fact that the project is quite extensive, and incorporates a number of scientific and engineering areas (Fig. 5), students have chosen the most suitable and interesting themes for themselves. Based on this fact, participation of students allows to obtain the following results related to educational activity:

- interest of students to their profession is significantly increased;
- inclusion of students into the work, which corresponds to their future professions;
- determination of an individual learning program;
- due to the fact that the project incorporates a number of scientific and engineering areas, students get a lot of knowledge, not only in their scientific field but also in a number of related areas;
- high academic progress of students, who are involved to the project (improved by 7 % compared with the previous results);
- improvement of student competencies, such as communicability, ability to work in team, ability to work with the customer and the employer, etc.

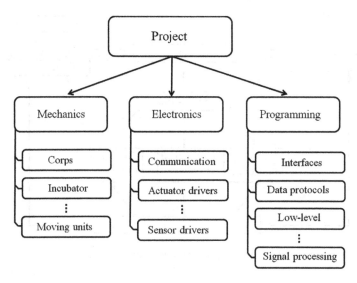

**Fig. 5** Scientific and engineering areas of the project

- since the project involves the collaboration of the ITMO University with AlkorBio Group company, students get the first work experience in industry;
- etc.

A very most important thing is that AlkorBio Group company is interested in employment of students after their graduation. In its turn, the students are interested to get more knowledge and experience as the project could be the first step in their career. This fact motivates future decisions of students and promotes synergies in their work.

Thus, embedding of industry elements to the educational process does not destroy the established system of education with the competence approach, but on the contrary, it allows to fill gaps in knowledges. Namely, the project complements traditional engineering education process that often lacks focus in fields of critical need for industry sector. In general, such R&D project allows to restructure and complete structure of the educational process presented in Fig. 2. In the modified scheme industry elements are not only form necessary competencies, but also have direct effect on the educational process and become part of this process (Fig. 6). The main advantage of this scheme is to increase the employability of students after graduation: over 60 % of the project team consists of bachelor, master and PhD students, and among these, about 40 % intend to continue careers in AlkorBio company. The another quantitative result is that performance of involved students is improved on average by 9.2 %.

Of course, this result is based only on the basis of the grant between ITMO University and AlkorBio Group company and applied only to the students involved in the project. However, this example can serve as a good impetus for the development of such grants as Megagrant under decree of Russian Government. The experience

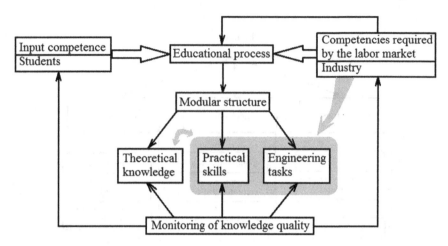

**Fig. 6** Changed structural scheme of the educational process

accumulated during project implementation can serve as a basis for the development of educational programs, which have a direct interaction with real sector of economy and industry.

# 5    Conclusion

This paper presents the results of implementation of joint projects between universities and engineering companies and describes its impact on the educational process of students and employability improvement of university graduates. The results are based on the example of Megagrant under decree of Russian Government between ITMO University and AlkorBio Group company. It is shown that these grants influence positively on the education system in general and enhance the employability of students after graduation. The results of the project between ITMO University and AlkorBio Group company can have an impact in development of educational programs, which have a direct interaction with real sector of economy and industry. Such educational programs and projects allow to get benefits for education system and make it more directed at the industry. Namely, these benefits are in complementation of traditional engineering education process that often lacks focus in fields of critical need for the industry.

**Acknowledgments**  The paper was written in accordance with R&D work "Development of analytical robotic complex for clinical laboratory tests using nano-reagents" in ITMO University with financial support from the Russian Federation Ministry of Education and Science, according to the decision of the Russian Federation Government of April 9, 2010 #218 "Measures of state support for development of cooperation between Russian universities, research organizations and companies which implement complex projects for high-tech production".

# References

1. Tikhomirov, V., Dneprovskaya, N., Yankovskaya, E.: Three dimensions of smart education. In: 2nd International KES Conference on Smart Education and Smart e-Learning, SEEL 2015. vol. 41, pp 47–56 (2015)
2. Matushkin, N.N., Stolbova, I.D.: Formation of the list of professional competencies of graduates. High. Educ. Today **11**, 28–30 (2007)
3. Kremlev, A.S., Bazylev, D.N., Zimenko, K.A., Margun, A.A.: Involvement of students and postgraduates into megagrant laboratory under supervision of leading scientist Romeo Ortega at ITMO University. In: Proceedings ERPA International Congresses on Education, in press (2015)
4. Finch, D.J., Peacock, M., Levallet, N., Foster, W.: A dynamic capabilities view of employability: exploring the drivers of competitive advantage for university graduates. Educ. Training **58**(1), 61–81 (2016)
5. Hameed, S., Nileena, G.S.: IEEE student quality improvement program: to improve the employability rate of students. In: IEEE International Conference on MOOC, Innovation and Technology in Education (MITE), pp. 219–222 (2014)
6. Kremlev, A.S., Bazylev, D.N., Margun, A.A., Zimenko, K.A.: Transition of the Russian Federation to new educational standard: independent work of students as a factor in the quality of educational process. In: Proceedings ERPA International Congresses on Education, in press (2015)
7. Barte, G.B., Yeap, G.H.: Problem-based learning approach in enhancing engineering graduates' employability. In: IEEE Colloquium on Humanities, Science and Engineering (CHUSER), pp. 771–775 (2011)
8. Jiangqin, X., Zhiwen, G.: Employability development of graduates in China. In: International Conference on Management and Service Science (MASS), pp. 1–4 (2011)
9. Zou, X., Ye, l.: Study on structure dimensions of ability to work for university graduates based on the employability. In: 2nd International Conference on Artificial Intelligence, Management Science and Electronic Commerce (AIMSEC), pp. 662–665 (2011)
10. Bhardwaj, J.: Evaluation of the lasting impacts on employability of co-operative serious game-playing by first year computing students: an exploratory analysis. In: IEEE Frontiers in Education Conference (FIE), pp. 1–9 (2014)
11. Bazylev, D.N., Margun, A.A., Zimenko, K.A., Shchukin, A.N., Kremlev, A.S.: Active learning method in System Analysis and Control area. In: Proceedings—Frontiers in Education Conference (FIE 2014) (2015). doi:10.1109/FIE.2014.7044339
12. Zimenko, K.A., Bazylev, D.N., Margun, A.A., Kremlev, A.A.: Application of innovative mechatronic systems in automation and robotics learning. In: Proceedings of the 16th International Conference on Mechatronics, Mechatronika 2014, pp. 437–441 (2014)
13. Decree of Russian Government no. 218. http://p218.ru/ (in Russian)
14. Donoval, D., Hajtas, D.: Team project-an effective tool for application of knowledge and deriving engineering competencies. In: Proceedings of IEEE International Conference on Microelectronic Systems Education, pp. 55–56 (2003)
15. Soares, F., Leao, C.P., Carvalho, V., Vasconcelos, R.M., Costa, S.: Automation & control remote laboratory: evaluating a cooperative methodology. In: 6th IEEE International Conference on e-Learning in Industrial Electronics (ICELIE), pp. 34–39 (2012)
16. Hosier, W.J.: University/industry co-operation in research. In: Proceedings of International Professional Communication Conference, IPCC'91, vol. 2, pp. 394–398 (1991)
17. Bayless, D.J., Pawliger, R.I.: American electric power's project probe/sup SM/-enhancing power engineering education through industrial-academic cooperation. In: 28th Annual Frontiers in Education Conference, FIE'98, vol. 3, pp. 1230–1235 (1992)

# On-Line Formative Assessment: Teaching Python in French CPGE

**François Kany**

**Abstract** This paper describes the design and implementation of a website devoted to automatic assessment of programming in Python. This server uses many technologies (Linux, Nginx, PostgreSQL, Django, PySandBox). We report a 3-month experimentation of this site within the learning process of computing as set in France in the syllabus for the first and second years of studies after the "Baccalauréat". This kind of formative assessment enables the students to be active and to create emulation among them.

**Keywords** Computer based assessment · E-marking · Python

## 1 Introduction

Formative assessment is a kind of grading process which deeply modifies the statute of the mistake and the involvement of the student in the process. The point of this kind of assessment has been made by numerous authors [1].

First point: making mistakes has to occur during the learning process (a student who would never make any mistake would either be a genius or a cheater). A student—contrary to a graduate engineer—must be allowed to make mistakes.

Second point (which derives from the first one): as students know that they are allowed to make mistakes, they are not afraid of failing; they can learn from their mistakes; they find confidence and measure their own progress.

Formative assessment reaches a higher dimension when the assessing is made either by the students themselves [2] or by the way of instant and automatic assessment. The students, thus, become the masters of their own learning process.

F. Kany (✉)
ISEN-Bretagne [Institut Supérieur d'Electonique et du Numérique],
20 Rue Cuirassé Bretagne, CS 42807, 29228 Brest Cedex 2, France
e-mail: kanyfrancois@hotmail.com

F. Kany
La Croix Rouge La Salle, 2 Rue Mirabeau, 29200 Brest, France

© Springer International Publishing Switzerland 2016                213
V.L. Uskov et al. (eds.), *Smart Education and e-Learning 2016*,
Smart Innovation, Systems and Technologies 59,
DOI 10.1007/978-3-319-39690-3_19

The new French CPGE[1] syllabus [3] re-introduces real computer science courses in the common core. Even if the academical curriculum imposes the contents, the teacher is entitled by the law [4] to a certain extent of pedagogical freedom. According to this, computer science has been introduced, in six forms, using the automatic formative assessment.

## 2 Examples of On-Line Assessment Web-Site

Many web-sites offer on-line computer science self-teaching through solving algorithms or mathematical problems. These sites—*in fine*—are based on the same learning process as the formative assessment as described in the introduction.

Among the most popular sites let us quote: France-IOI [5]; Sphere Online Judge [6], Project Euler [7], Code Academy [8],....

The e-learning site France-IOI [5] is an excellent one for pupils of secondary schools. Nevertheless it is not quite suitable for CPGE students. Even if, from levels 3 and 4, exercises start to be interesting, the moving forward in levels 1 and 2 is a bit slow. But, on the contrary, for levels 5 and 6, time restraints are really strict for an interpreted language such as Python. Exercises, then, require to code in Python perfectly. The CPGE syllabus is clear and does not require to train specialists in Python (able to re-define the `input` and the `print` functions to gain a few hundredths of a second). The aim is to understand algorithmics, Python being only the support language.

## 3 The ISEN-La Croix Rouge Web-Site

We have set up a server very similar to the one of France-IOI but including problems complying with the CPGE syllabus. Unlike France-IOI, in order to avoid cheating, the exercises put on site are blind-assessed: the students do not know with which kind of input parameters the algorithms are tested; they do not know either what the answers of their own algorithms might have been. (On France-IOI site, there is a guidance for students; the server shows the differences between the values computed by the submitted code and the values expected by the server. The students then may cheat by asking their own programs to show the input parameters given by the server. As the server also shows the expected values, the students are able to write a program designed as follows: "for such input parameters, print these results" without programming the general algorithm which would answer the problem). However, like on France-IOI, the server indicates errors that happened while interpreting the Python code in order to guide students for the debugging.

---

[1]CPGE ("Classes Préparatoires aux Grandes Écoles") is a 2-year-intensive program in math and physics after the "Baccalauréat" (A Level). Most of these classes prepare students for competitive exams for the entrance in the top-ranking higher education establishments. Some of these classes (called CPGE-intégrées) have the same syllabus but do not prepare for competitive exams.

**Table 1** Technical details of the project

| Functionality | Technology |
| --- | --- |
| Internet access provider | OVH |
| Virtual Private Server (VPS) | OpenStack (Python) |
| Operating System | Debian 8 |
| Database | PostgreSQL |
| WSGI interface | Gunicorn |
| Process Control System | Supervisor |
| HTTP server | Nginx |
| Language | Python 2.7 |
| Framework | Django (Python) |
| "Safely" run untrusted code | PySandBox |

# 4 Architecture of the Project

The technical details of the project are summarized here after in Table 1.

Choosing Django [9], a very popular framework, seemed obvious for many reasons: it is written in Python; once the basis learnt, it enables to code quickly; the developer does not need to take heed of the fastidious tasks (user's authentication, passwords management, creation of the administration interface,...) and can concentrate on the main task. The project represents 1,400 lines of Python; Django provides automatically the major part of it; the core of the project represents 500 lines of Python.

The codes submitted by the students are assessed in a "sandbox" [10]. Even if this solution is not perfectly secured [11], it allows to limit access to the Python external libraries and prevent a student from trying to introduce a malicious code.

Moreover, PySandBox helps to force off the assessment of a program after 1 or 2 s in order to avoid that the code of a student might run endlessly and paralyze the server.

# 5 Database

## 5.1 Architecture of the Database

The architecture of the database is shown on Fig. 1.

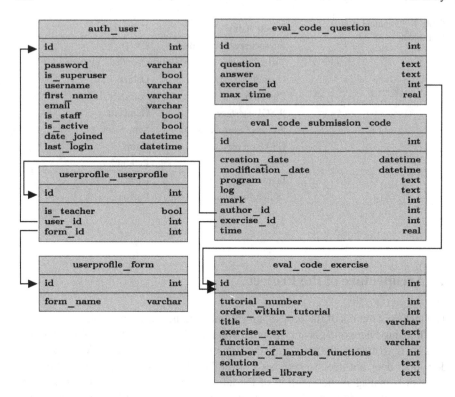

**Fig. 1** Architecture of the database

---

**Alg. 1** Example of a lambda function.

Lambda function is not a serious limitation as many functions can be rewritten as lambda functions. Here an example, for the mastermind board game, of a function returning the number of pegs of a correct "color" in (correct, wrong) positions. The function can only be called 10 times.

```
>>> ask=lambda attempt,MaxColor=10,answer="1234",
MaxQues=iter(range(10,-1,-1)): [(correct, common-correct) for correct in
[sum([1 for c1,c2 in zip(attempt,answer) if c1==c2])] for common in
[sum([min(attempt.count(str(i)),answer.count(str(i))) for i in
range(MaxColor)])]][0] if MaxQues.__length_hint__() and
next(MaxQues) else None # invisible for students
>>> ask("1320") # students only know the lambda function name
(1,2) # "1" is in correct position; "2" and "3" are in wrong positions
```

---

## 5.2 Content of the Database

The database contents 188 students divided into 6 forms: 1 first-year CPGE, 2 first-year CPGE-intégrées[3], 1 second-year CPGE, 2 second-year CPGE-intégrées. It entails 125 original exercises at the moment. The "algorithmic" orientated exercises are shared out among thematic lists: variables, tests, loops, lists, strings, numerical recipes, sorting algorithms, trees,... The "numerical simulation" oriented exercises are shared out among other lists: arithmetics, probabilities, electricity, mechanics,... In each list, the exercises are sorted in increasing order of difficulty. Each exercise compels the student to code a function with a certain name, a certain number of input (with their types) and a certain number of output (with their types). All types are possible (integer, float, list, string,...). It is also possible to set a lambda-function as an input (see example Algorithm 1); this is particularly useful for numerical recipes: for the Newton algorithm, it is possible to enter a function and its derivative.

The database contents 530 tests (on the average, it will be 4.2 tests per exercise).

The tests are static: the questions (input parameters) and the answers (output of the functions) are stored in the database. This solution was preferred to dynamical tests (where questions would have been generated randomly and the answers compared with the computation of a solution algorithm). Thus, the computing time can be minimized and the access time for the users can be reduced. Moreover, static tests are carefully designed to check if the code is effective in all particular cases. In order to make the comparison easier with the answers in the database, the students are asked to give float results rounded to the fifth decimal, to give string results in uppercase or to give list results according to a certain order. Thus, whether the asked result is of integer, float, list or string type, the server only verifies the strict equality between the student's answer and the database's answer.

As soon as a student submits a code to solve one exercise, his or her previous code is erased[2] and his or her score recalculated according to the number of passed tests. If a student tries the same exercise many times, only the last attempt is taken into account. According to the principles of formative assessment, multiple tryings are not sanctioned.

For each exercise, the student can see the list of students who have already solved (or tried to solve) the exercise. The list is sorted by decreasing score and chronologically creasing.

## 5.3 Use Cases

The CPGE students officially have 1-h tutorial course per week. The on-line assessment site is used then. The procedure may change from a session to another.

- From September to Halloween, the second-year students use it freely to revise.
- During some sessions of tutorial, there is a mandatory subject: all the students must solve the same exercises.
- From time to time, I organize "sprint competition". I put an exercise on-line at a defined time and the students must solve it as quickly as possible. Unlike during

---

[2]Tracks of the successive tryings might have been kept to make statistics (for example: number of tryings per exercise) or to make comparisons between different versions of the same code; but this might have over-loaded the database.

**Table 2** Example of a sprint competition

| Name | Date | Grade |
|------|------|-------|
| OB | Nov. 13, 2015, 4:20 p.m. | 20 |
| FP | Nov. 13, 2015, 4:21 p.m. | 19 |
| RJ | Nov. 13, 2015, 4:28 p.m. | 18 |
| BF | Nov. 13, 2015, 4:35 p.m. | 17 |
| GLT | Nov. 13, 2015, 4:39 p.m. | 16 |
| LD | Nov. 13, 2015, 4:40 p.m. | 15 |
| MLG | Nov. 13, 2015, 4:45 p.m. | 14 |
| VG | Nov. 13, 2015, 4:47 p.m. | 13 |
| ... | | |
| FC | Nov. 14, 2015, 5:48 p.m. | |

*Start* Nov. 13, 2015, 3:30 p.m.

a supervised test, I answer questions; I help in debugging syntax errors; I regularly give clues to the entire form (each 20 or 30 min). The grading is simple: the first student who validates all the tests gets 20 out 20. An example of "sprint competition" is given in Table 2. There one can see that the subject was given at 3.30 p.m., the first entirely correct submission was posted at 4.20 p.m., the second one at 4.21 p.m. (there were in fact less than 20 s between the fist and the second one). The last student posted her solution the next day at 5.48 p.m.

## 6   Analysis

### 6.1   Figures

On a 3-month span, with 1 h a week of tutorial, 3,767 submissions have been made, that is to say an average of 20.0 exercises par student. 16,132 tests were passed, that is to say an average of 85.8 tests per student. This average does not exactly match the average number of exercises (20.0) multiplied by the average of tests per exercise (4.2) because the second-year students concentrated on review exercises (with a little more tests per exercise) whereas the first-year student started with basic exercises (with slightly fewer tests per exercise).

The Django administration interface allows to get an easy access to the database if the teacher wishes to get further statistics.

## 6.2   Pedagogical Point of View

Without any doubt, this method of formative assessment is efficient.

Students are no longer passive; they cannot pretend to be working; they have to be productive at each tutorial. Sometimes, I use the video-projector to show the general ranking with the total amount of points (the students have only access to the ranking per exercise).

As the correction is done automatically, it prevents any objection; the students learn to be rigorous: a meer space, a meer lowercase instead of uppercase letter, a meer typing error,... and the machine won't obey!

Displaying a ranking for each exercise creates emulation (even among students of the ISEN and of the Croix-Rouge who do not know each other). Two second-year students (FP at the ISEN and POHB at the Croix-Rouge) compete sharply to be the first one to solve new exercises. This experience confirms that gamification (points, bonuses, ranking,...) is a powerful incentive to motivate students.

The best students do not hesitate to carry on their work long after the tutorials set in their time-table (see Table 3).

This emulation also worked among the four CPGE-intégrées[3] forms whereas usually (in mathematics, physics and engineering science) there is no real fighting spirit in those classes because they are not preparing for a competitive exam.

The fighting spirit is towering during the "sprint sessions". The very good students who compete for the first ranks would really code in a feverish mood. When the server returns its verdict and validates all the tests of a student, the sound of a big "Yes!" would burst out in the classroom as in a sports arena.

There is little cheating (previously, when I used to work with France-IOI, students would exchange codes via a Dropbox). I think that ranking now each exercise brings about much change: a student who spent much time developing by himself a code does not wish it to be disclosed. He understands that if this practice was to be spread, this would work against him on another exercise. He could be overtaken in the ranking by a plagiarist. It happened thus that without my setting of rules, the students will keep their programs preciously.

**Table 3**  Example of dates of submission

| Name | Date | Exercise |
|------|------|----------|
| FP | Oct. 21, 2015, 10:48 p.m. | Horner-1 |
| FP | Oct. 21, 2015, 11:09 p.m. | Horner-2 |
| POHB | Oct. 16, 2015, 11:41 p.m. | Rational fraction |
| POHB | Oct. 18, 2015, 12:40 a.m. | Partition of an integer |
| FT | Dec. 2, 2015, 1:59 a.m. | Grocery list |
| FT | Dec. 2, 2015, 2:37 a.m. | Back to the entrance of the maze |

*FT* is a first-year student; *FP* and *POHB* are second-year students

Displaying the ranking for each exercise is also a good incentive for weaker students. When a student sees that a problem has already been solved by a hundred of students, he considers that the exercise is manageable and he works over it. (The Euler site works using a similar motivation but on another scale: one can see that the first exercise has been solved by more than 500,000 people). For any set exercise, as the students are allowed to propose as many submissions as they wished without any penalty, everybody is incited to submit his or her code. By changing the statute of mistakes, formative assessment urges the students to be active.

The choice of the set of exercises is important too. Some algorithms are playful on purpose (eg. Pledge to get out of a maze, solving Sudokus, mastermind board game,...) so that students may appropriate the problem and be really interested in finding the solution.

For example, when a student, after many working sessions, eventually finds his or her way out of the maze there is a real burst of joy in the classroom. Another relevance of gamification....

## 7 In and Out the Classroom

### 7.1 In the Classroom

The tutorial sessions are very dynamic both for students and the teacher. It happens sometimes that ten hands would rise at the same time asking for help; it is up to the teacher then to react quickly. It is essential that the teacher may be sufficiently trained to spot and solve syntax errors in less than 10 s (otherwise student number ten asking for help starts to be impatient). It is also important that the teacher remembers the 125 algorithms perfectly to find more serious mistakes rapidly. When I realize that a mistake in logics is made three or four times in a row, I stop the session and take the opportunity to review the point on the blackboard for the whole form.

### 7.2 Out of the Classroom

The students' enthusiasm has a setback for the teacher. Students would not hesitate to email me seven days a week (and nearly 24 h a day) asking for help (see Table 4).

In these exchanges of emails, the point is not to give the student the answer to the problem but rather to give him or her clues to solve it or to show counter-example so that he or she may understand why his or her algorithm is wrong.

**Table 4** Example of email exchange with POHB

| Sender | Date |
|--------|------|
| POHB | 21/09/2015 19:50 |
| kany | 21/09/2015 19:54 |
| POHB | 21/09/2015 21:49 |
| kany | 21/09/2015 22:25 |
| POHB | 24/09/2015 21:14 |
| kany | 25/09/2015 07:51 |
| POHB | 27/09/2015 12:09 |
| kany | 27/09/2015 12:36 |
| POHB | 16/10/2015 23:52 |
| kany | 17/10/2015 06:25 |
| POHB | 17/10/2015 19:36 |
| kany | 18/10/2015 00:16 |
| POHB | 18/10/2015 09:36 |
| kany | 18/10/2015 11:28 |
| POHB | 20/10/2015 14:48 |
| kany | 20/10/2015 15:37 |
| POHB | 20/10/2015 18:54 |
| POHB | 20/10/2015 19:08 |
| POHB | 20/10/2015 19:10 |
| POHB | 21/10/2015 18:23 |
| kany | 21/10/2015 20:03 |
| POHB | 22/10/2015 17:03 |
| kany | 22/10/2015 19:06 |
| POHB | 22/10/2015 23:32 |
| kany | 23/10/2015 08:08 |
| POHB | 23/10/2015 11:47 |
| POHB | 08/11/2015 12:41 |
| POHB | 08/11/2015 13:30 |
| kany | 11/11/2015 23:52 |
| ... | |

## 8  Possible Evolutions

On the short term, the first evolution consists in giving a wider range of exercises in order to scan the whole syllabus of CPGE point by point.

At the moment, when a student submits a program which loops endlessly, the assessment process is brutally stopped after a few seconds. If many students submit, at the same time, programs with infinite `while`, the latency for other users may rise up to 10 s. It would be possible to force students to test their codes on their own PC before submitting any to the server by blocking for a minute the IP address of a student submitting an infinite loop. The point is not at all to change the spirit of formative assessment: mistakes are part of the learning process, it is out of question to penalize a student who goes the wrong way. The point is rather to encourage students to perform a minimal check before submitting.

The ISEN-Bretagne is an engineering school belonging to a federation (HEI-ISA-ISEN) gathering ten schools, located on the French territory, schooling some 18,000 students.

It could be considered to generalize this formative assessment for the whole of the students. Nevertheless at the moment, the website is hosted on a unique virtual server and the code evaluation is run synchronously. As there are, at most, 30 or 60 students on-line at a given time, the server is not too busy. If we were to extend the system on a wider scale, we would have to bring two simple modifications. First, the code evaluation should be done asynchronously (which is quite simple to do in Python thanks to the `threading` library). Second, the computation should be dispatched on several virtual servers (which is perfectly possible thanks to the OpenStack technology that allows to create servers—on the fly—as soon as one server starts being too busy).

## 9  Conclusion

The Nginx/PostgreSQL/Django/PySandBox technologies allowed us to set up a system of automatic assessment of Python codes which is simple, flexible and scalable.

According to the rules of formative assessment, the students are allowed to make mistakes without being penalized. This change of the mistake statute has positive consequences.

Students are active; they do not cheat; the ranking per exercise encourages better and also weaker students; students go on working long after tutorial sessions. There is emulation inside the form and between forms that do not know each other. As the students ask for help, it shows that they are motivated to solve problems by themselves.

The "sprint sessions" prepare students to their future jobs as engineers; they learn how to code in an efficient way under time pressure.

Moreover, setting up a smart e-learning system, with real time feedback, completely change the educational environment. Whereas CPGE courses are usually very formal, directive and knowledge-based, the Python course is now quite informal, auto-didactic and competence-based. The number of standard "formal" sessions is reduced; theoretical contents are taught using an agile approach [12]: information are given when needed to solve a specific problem (in a kanban way). This new environment is more adapted to the cognitive needs of the students.

# References

1. Crooks, T.: The validity of formative assessments. In: British Educational Research Association Annual Conference, University of Leeds (2001)
2. Nunziati, G.: Pour construire un dispositif d'évaluation formatrice. Cahiers Pédagogiques: Apprendre **280**, 48–64 (1990)
3. Bulletin officiel spécial n° 3 du 30 mai 2013. http://www.education.gouv.fr/pid25535/bulletin_officiel.html?cid_bo=71586
4. Article L912-1-1 du Code de l'Education: Loi n° 2005-380 du 23 avril 2005 - art. 48 JORF 24 avril 2005. http://www.legifrance.gouv.fr/affichCodeArticle.do?cidTexte=LEGITEXT000006071191&idArticle=LEGIARTI000006525569
5. The International Olympiad in Informatics (IOI) is an annual competitive programming competition for secondary school students. http://www.france-ioi.org/
6. Sphere Online Judge is an online judge system with over 315,000 registered users and over 20,000 problems. http://www.spoj.com/
7. Project Euler is a series of challenging mathematical/computer programming problems. https://projecteuler.net/
8. Codecademy is an online interactive platform that offers free coding classes. https://www.codecademy.com/
9. Django is a free and open source web application framework which follows the model-view-controller (MVC) architectural pattern. https://www.djangoproject.com/
10. PySandBox is a Python sandbox: By default, untrusted code executed in the sandbox cannot modify the environment. https://pypi.python.org/pypi/pysandbox/
11. The PySandBox project is broken. https://mail.python.org/pipermail/python-dev/2013-November/130132.html
12. Vermeulen, M., Fleury, A., Fronton, K., Laval, J.: Approches agiles pour l'enseignement supérieur. Colloque Questions de Pédagogie pour l'Enseignement Supérieur (QPES), pp. 243–248. Brest (2015)

# PageRank Algorithm to Improve the Peer-Led Team Learning Pedagogical Approach

Nazar Zaki

**Abstract** Peer-Led Team Learning model has been introduced into several undergraduate courses and it showed improvement in students' learning performance. Despite its success, little attention has been given to further enhance the model for academic benefits particularly in terms of suitable peer leaders selection, homogenous groups forming and the process of disseminating the educational knowledge. In this paper, we utilized PageRank Algorithm to improve the PLTL pedagogical approach. Unlike the traditional way, we introduced social interaction analysis as a way to create natural groups of students, and select the best peer leaders in classrooms who can disseminate the educational knowledge in an efficient and smooth fashion. The new proposed approach was tested on a dataset of 16 students in an Operating System course offered at the College of Information Technology (CIT), United Arab Emirates University. The improvement in students' performance achieved is encouraging evidence in favor of the proposed method.

**Keywords** Peer-Led team learning · Google PageRank algorithm · Peer leader · Learning improvement · Social network analysis · Cooperative learning · Group learning

## 1 Introduction

Over the past two decades, several leading education organizations have recognized the importance of group activity in improving students' conceptual understanding and learning in classrooms. This has led to a major paradigm shift from traditional, instructor-centered classrooms to student-centered classrooms; providing students opportunities to be actively engaged and contribute to their learning process. This has triggered the birth of the Peer-Led Team Learning (PLTL) pedagogical

N. Zaki (✉)
College of Information Technology United Arab Emirates University (UAEU),
Al Ain P.O. Box 15551 UAE
e-mail: nzaki@uaeu.ac.ae

© Springer International Publishing Switzerland 2016                           225
V.L. Uskov et al. (eds.), *Smart Education and e-Learning 2016*,
Smart Innovation, Systems and Technologies 59,
DOI 10.1007/978-3-319-39690-3_20

approach which provides small group of students' instruction without ignoring or neglecting the large lecture component of the undergraduate courses that has become so deeply entrenched in college and university systems. Peer leaders work collaboratively with the course instructor to facilitate small group following intensive trainings in learning theory, pedagogical methods, and the conceptual content of the course [1]. Since its conception [2], the PLTL model has been introduced into several undergraduate courses, including organic chemistry, biology, anatomy and physiology [3–5] and showed noticeable improvement in students' performance. However, little attention has been given to further enhance the model for academic benefits particularly for the peer leaders selection, group forming and the entire learning process. The important questions that could potentially improve the PLTL experience are yet to be addresses. This study is intended to address the following three questions:

- How to form homogenous group of students with the right chemistry to work together?
- How the best peer leader is selected?
- How the knowledge is disseminated in efficient and smooth manner?

Some of the intuitive and traditional attributes of a peer leader selection are excellent interpersonal skills, interactive, communicative, supportive, positive, respectful of others, responsible, availability, experience with PLTL and very good knowledge of the course materials. These attributes are often judged by the student's performance in the PLTL course, recommendations of the peer leader, interviews, personal invitations, etc. It is always difficult to find such quality in one student beside the success of the process is not guarantee. The experience is often hampered by personal issues, class work load, commitment, emotional issues, out of control behavior, not being liked by peers, etc.

The knowledge is often disseminated through frequent formal sessions/ workshops where the peer leader is expected to effectively run the sessions, keeping student attendance, coordinating with the course instructor, creating exercises, help prepare other class activities, etc. The PLTL is a voluntary student academic resource, however, the expectation is that the student should be committed to attend all of the sessions/workshops. Many students are often reluctant to participate in any additional academic or extra-curricular activities particularly if they don't feel comfortable around the peers and the peer leader.

In this paper we introduced a novel approach of student group forming and peer leader selection. We introduce graph knowledge and social interactions as a way to create natural and homogenous groups of students. Social interactions between students are a major and under-explored part of undergraduate education. Understanding how learning relationships form in undergraduate classrooms, as well as the impacts these relationships have on the learning outcomes, can inform educators in unique ways and improve the educational reform [6].

The peer leader or here we refer to them as "most influential students" in the classroom is selected by applying Google PageRank algorithm. Google PageRank algorithm [7–9] is one of the most successful algorithms used to quantify and rank the importance of web pages. The PageRank algorithm is utilized to quantify the importance of each student in the class based on the social relationships with other classmates. Once all students in the class network are ranked, we select the most influential ones and then use the hub/spoke model to identify the suitable students group. The course instructor has to work with a limited number of students (peer leaders) and the knowledge is expected to be naturally disseminated via social links without formal sessions or workshops.

## 2　Method

### 2.1　Data Collection and Network Construction

The relational and nodal attribute data is collected using short surveys. The students were informed that we are interested in investigating how in-class study networks formed in large undergraduate classes and how the network could improve the PLTL pedagogical approach in terms of group forming, peer leader selection and dissemination of knowledge of the course. The students are then handed out a survey includes a check-box next to each student name in the class says "Socially Linked" for each student to evaluate whether they fit the description. If no one fits these descriptions, the student can simply write "none". The term socially linked is explained in a way that it works for the benefit of the course such as studying together, exchange knowledge of the course, solve homework together, explain materials to each other, etc. Once this data is collected the network can simply be constructed and the Social Network Analysis (SNA) can simply be done based on the network topological information which includes nodes ranking. In this case SNA could assist us to understand how relationships between the student form, what kinds of relational structures emerge from the building blocks of individual relationships between pairs of students, and what, if any, the impacts are of these relationships on other students in the network.

### 2.2　Peer-Leader Selection

To identify potential peer leaders ("most influential students") in the classroom, we utilize Google PageRank Algorithm as a way to rank and quantify the importance of each student in the class based on the social links with other classmates.

To formulate the method, consider a network of $n$ students indexed by integers from 1 to $n$. This network is represented by the directed graph $G = (V,)$. Here, $V: = \{1, 2, ..., n\}$ is the set of nodes (students) and $E$ is a set of edges (links) among the students in a class. A student $i$ is connected to a student $j$ by an edge, i.e.,

$(i, j) \in E$, if student $i$ interacts (links) with $j$. Once an adjacency matrix of links is calculated, we utilized Google PageRank algorithm [7–10] to provide some measure of importance to each student in the network. The ranking value of student $i \in V$ is a a real number in $[0, 1]$; we denote this by $x_i$. The values of $x$ are ordered such that $xi > xj$ implies that the student $i$ is more important than the student $j$ based on the PageRank algorithm. The value of a student is determined as a sum of the contributions from all other students that have links to him/her. In particular, the value $x_i$ of student $i$ is defined as:

$$x_i = \sum_{j \in \tau_i} S^*_{i,j} \cdot x_j \tag{1}$$

where $\tau_i : = j: (j, i) \in E$, i.e., this is the index set of students linking to student $i$, and $n_j$ is the number of outgoing links of student $j$. It is customary to normalize the total of all values so that $\sum_{i=1}^{n} x_i = 1$.

Let the values of $x$ be in the vector form where $x \in [1, 0]^n$. Then, from Eq. 1, the PageRank algorithm can be rewritten as:

$$x = S^* x, \; x \in [1, 0]^n, \; \sum_{i=1}^{n} x_i = 1 \tag{2}$$

where $S^*$ is the adjacency matrix.

Note that the vector $x$ is a non-negative eigenvector corresponding to the eigenvalue 1 of the nonnegative matrix $S^*$. In general, for this eigenvector to exist and to be unique, it is critical that the student links network is strongly connected. To find the eigenvector corresponding to the eigenvalue 1 a modified version of the values has been introduced in [7] as follows:

Let $m$ be a parameter such that $m \in (0, 1)$, and let the modified interaction matrix $M \in \mathfrak{R}^{n \times n}$ be defined by:

$$M : = (1 - m) \times S^* + \frac{m}{n} \times 1 \tag{3}$$

where $1$ is a $n \times n$ matrix with all elements equal to 1.

In the original algorithm in [7], a typical value for $m$ is reported to be 0.15; The value of $m$ is too small and it does not really influence the results. It is rather important in avoiding convergence problems. Notice that M is a positive stochastic matrix. Thus, according to Perron theorem [11], this matrix is primitive; in particular, $|\lambda| = 1$ is the unique maximum eigenvalue. To find corresponding eigenvector $x$ we apply the following formula:

$$x(k + 1) = M_x(k) = (1 - m) S^* \times x(k) + \frac{m}{n} \times 1 \tag{4}$$

where $x(k) \in \mathfrak{R}^{n \times 1}$ and the initial vector $x(0) \in \mathfrak{R}^{n \times 1}$ is a probability vector.

Now let us expand on the convergence rate of this scheme. Let $\lambda_1(M)$ and $\lambda_2(M)$ be the largest and the second largest eigenvalues of M in magnitude. Then, for the power method applied to M, the asymptotic rate of convergence is exponential and depends on the ratio $|\lambda_2(M) / \lambda_1(M)|$. Since M is a positive stochastic matrix, we have $\lambda_1(M) = 1$ and $|\lambda_2(M)| > 1$ Furthermore, it is shown in [10] that the structure of the link matrix M leads us to the bound:

$$|\lambda_2(M)| \leq 1 - m \qquad (5)$$

Figure 1 shows a simplified network with four nodes (students) that illustrates the ranking step .

The link matrix $S^*$ can easily be constructed as:

$$S^* = \begin{bmatrix} 0 & 0.5 & 1 & 0.5 \\ 0.333 & 0 & 0 & 0.5 \\ 0.333 & 0 & 0 & 0 \\ 0.333 & 0.5 & 0 & 0 \end{bmatrix},$$

$$M = \begin{bmatrix} 0.038 & 0.463 & 0.888 & 0.463 \\ 0.321 & 0.038 & 0.038 & 0.463 \\ 0.321 & 0.038 & 0.038 & 0.038 \\ 0.321 & 0.463 & 0.038 & 0.038 \end{bmatrix}$$

Using the power method with $m = 0.15$, the eigenvector $x$ can be computed as $x = [0.367, 0.246, 0.141, 0.246]^T$. Notice that the student number 1 has the largest value since he/she links with three other students. On the other hand, student 3 appears to have the lower ranking score since he/she links with only one student.

**Fig. 1** Example of a simple network which includes 4 nodes (students) and 4 social links

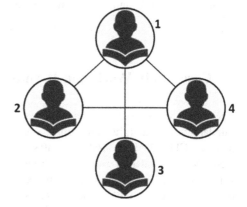

**Fig. 2** A hypothetical network of 9 nodes (students) and 11 social links to illustrate the selection of 2 peer-leaders and the formation of 2 groups, accordingly

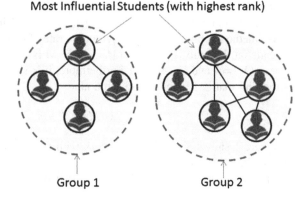

## 2.3 Groups Formation

To form homogenous groups we use the hub/spoke model. The spoke-hub distribution paradigm (or model or network) is a system of connections arranged like a wire wheel, in which all traffic moves along spokes connected to the hub at the center. For a network of $n$ nodes, only $n - 1$ routes are necessary to connect all nodes. In this case, the peer leaders, are consecutively considered by their decreasing ranking order and each of them is pulled from the interaction network along with his/her socially linked neighbors (directly connected neighbors and if necessarily the second or third levels neighbors) to form a homogenous group of students. Here, each student can belong to one group only. The homogeneity and links density of a group can easily be visualized. This step is illustrated in Fig. 2. In this case the student number S5 is the highly ranked student and can be pulled out of the network along with his/her socially linked neighbors S6, S7, S8 and S9 to form group 1. Similarly, for the remaining nodes in the network, the node S1 is the highly ranked node and it can similarly be pulled out along with nodes S2, S3 and S4 to form group 2.

## 3 Case Study Work and Results

The case study is conducted on a dataset of students in ITBP315 Operating System Fundamentals undergraduate course offered at the College of Information Technology (CIT), United Arab Emirates University. The ITBP315 course is a 3 credit hours mandatory course which taught twice a week for 14 weeks. The case was conducted on a group learning based on topic entitled "CPU Scheduling". Here the notion of the CPU scheduling, which is the basis for multi-programmed operating systems is introduced. The students experience working with various CPU-scheduling algorithms and evaluate the appropriate CPU-scheduling

**Table 1** Classroom network details

| Number of nodes | 16 |
|---|---|
| Number of links | 18 |
| Network density | 0.15 |
| Number of isolated nodes | 0 |
| clustering coefficient | 0.2625 |
| Network radius | 2 |
| Network centralization | 0.21 |
| Average number of neighbors | 2.25 |

algorithm for a particular system. These important practical concepts are covered in around 4 weeks of lectures and labs (8 lectures and 2 lab sessions). The data which was recorded in the Fall 2015 consists of 16 female students. The students were asked to list classmates with whom they share knowledge and materials of the course. This step resulted in a network of 16 nodes and the details of the network are summarized in Table 1.

Following the network construction, as shown in Fig. 3, the PageRank algorithm was utilized to identify the influential students in the class.

The obvious observation is that the class mainly consists of 2 natural sub-networks based on two batches of students who joined the college in 2012 and 2013, respectively. Following the construction of the network, we apply PageRank

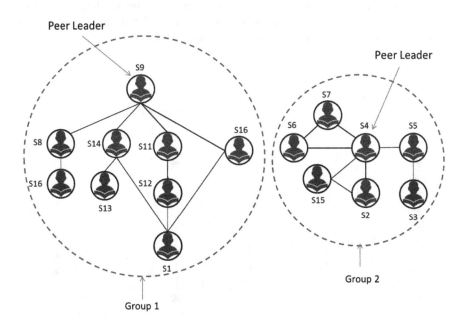

**Fig. 3** The construction of 16 students in one classroom. Following the construction of the network, PageRank algorithm was applied to rank all the student in the class based on their social links. In this case two peer leaders (node S4 and node S9) were identified

**Table 2** PageRank scores of all 16 students

| 1 | 2 | 3 | 4 |
|---|---|---|---|
| 0.0792 | 0.0792 | 0.0352 | **0.1266** |
| 5 | 6 | 7 | 8 |
| 0.0608 | 0.0537 | 0.0537 | 0.0623 |
| 9 | 10 | 11 | 12 |
| **0.1057** | 0.0543 | 0.0553 | 0.0553 |
| 13 | 14 | 15 | 16 |
| 0.0326 | 0.0820 | 0.0537 | 0.0359 |

algorithm to rank all the student in the class based on their social links. All the 16 nodes along with their corresponding ranking score are listed in Table 2.

In this case two peer leaders (node 4 and node 9) who were ranked top were selected and 2 groups of sizes 7 and 9 students, respectively were formed as explained in Sect. 2. To test the improved PLTL pedagogical approach, 2 tests with the same number of questions, similar score (out of 20), similar level of difficulties and related to the "CPU scheduling" topic were given to the student before and after the adoption of the proposed method. Following the first test, more attention and additional sessions/meetings were conducted to improve the performance of the two peer leaders (node 4 and note 9). The students' performances in test 1 and test 2 before and after the adoption of the proposed method is shown in Fig. 4. Except student number 12 all the other 14 students have significantly benefited from the adoption of the proposed method. Student number 15 achieved similar score in both tests. It is worth to notice the performance improvement of the peer leaders (student number 4 and 9). The performance of the peer leader number 4 has increased from 60 to 80 %, while the student number 9 has improved from 65 to 90 %.

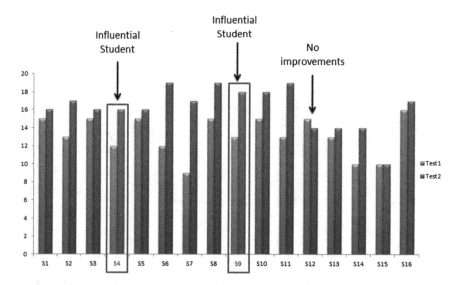

**Fig. 4** The performance of the 16 students in test 1 (*blue*) and test 2 (*red*), students 4 and 9 were ranked as influential students in the classroom

# 4  Conclusion and Discussion

In this study, we have utilized PageRank Algorithm to improve the PLTL peda-gogical approach. Unlike the traditional way, we introduced graph knowledge and social interactions analysis as a way to create natural groups of students and select the best peer leaders in classrooms. The new approach was evaluated on a data set of 16 students in a classroom and the improvement in students' performance is great evidence in favor of the proposed method. This is the second introduction following Grunspan et al. efforts [6] of the power and complexity of educational research aims that might benefit from SNA. We showed that network visualization and analysis are a relatively simple but quite powerful way of looking at the small and vital communities in classrooms. We hope this primer helps to guide educators into a highly promising field that can simply help to investigate classroom-scale hypotheses and improve the educational process and learning.

This study could also leads to ways in which the e-Learning process could be improved. It could be a continuation of the effort done by Edwards and Bone [12]. The authors investigated interfacing Peer Assisted Learning (PAL) and e-Learning and showed that the concept provides an important context for re-positioning the ways in which tutorials and lectures could be used as a basis for collaborative learning between students and lecturers alike.

Furthermore, the method has to be further tested on a classroom data of more than 30 students. The method should be tested on more than one classroom and course. The performance of the students should be monitored over a longer period and over several assessment processes (tests, quizzes, exams, etc.). Issues related to groups sizes should also be investigated further. It is also worth to investigate the sub-group notion which can be defined as a functional group which is a subset of larger functional groups [13]. A simple tool to assist the instructor to automatically rank the students in the class, select peer leaders and form groups should be developed.

**Acknowledgment**  The author would like to thank the students of section 51, ITBP315 who were registered in the Fall of 2015 for their volunteering to provide the social data which made this study a success. Special thanks to the College of Information Technology and the United Arab Emirates University for facilitating the study.

# References

1. Gosser, D., Roth, V., Gafney, L., Kampmeier, J., Strozak, V., Varma-Nelson, P.: Workshop chemistry: overcoming the barriers to student success. J. Chem. Educ. **1**, 1–17 (1996)
2. Woodward, A., Gosser, D.K., Weiner, M.: Problem solving workshops in general chemistry. J. Chem. Educ. **8**(70), 651–652 (1993)
3. Hewlett, J.: In search of synergy: combining peer-led team learning with the case study method. J. Coll. Sci. Teach. **33**(4), 28–31 (2004)

4. Tenney, A., Houck, B.: Peer-led team learning in introductory biology and chemistry courses: A parallel approach. J. Math. Sci.: Collab. Explor. **6**, 11–20 (2003)

5. Wamser, C.C.: Peer-led team learning in organic chemistry: effects on student performance, success, and persistence in the course. J. Chem. Educ. **83**(10), 1562–1566 (2006)

6. Grunspan, D.Z., Wiggins, B.L., Goodreau, S.M.: Understanding classrooms through social network analysis: a primer for social network analysis in education research. CBE-Life Sci. Educ. **3**, 167–178 (2014)

7. Brin, S., Page, L.: The anatomy of a large-scale hypertextual web search engine. Comput. Networks ISDN Syst. **30**, 107–117 (1998)

8. Bryan, K., Leise, T.: $25,000,000,000 eigenvector: the linear algebra behind google. SIAM Rev. **48**(3), 569–581 (2006)

9. Ishii, H., Tempo, R.: A distributed randomized approach for the pagerank computation: Part 1. In: 47th IEEE Conference on Decision and Control, pp. 3523–3528, Yokohama, Japan (2008)

10. Zaki, N.M., Berengueres, J., Efimov, D.: Detection of protein complexes using a protein ranking algorithm. Proteins: Struct. Funct. Bioinf. **80**(10), 2459–2468 (2012)

11. Meyer, C.M., Langville, A.N., Meyer, C.: Matrix Analysis. Princeton University Press (2006)

12. Edwards, S., Bone, J.: Integrating peer assisted learning and eLearning: using innovative pedagogies to support learning and teaching in higher education settings. Aust. J. Teach. Educ. **37**(5) (2012)

13. Zaki, N.M., Mora, A.: A comparative analysis of computational approaches and algorithms for protein subcomplex identification. Sci. Rep. **4**, 4262 (2014)

# Developing Computational Thinking Abilities Instead of Digital Literacy in Primary and Secondary School Students

Eduardo Segredo, Gara Miranda, Coromoto León and Anthea Santos

**Abstract** Core subjects by field of knowledge for official University studies have been established in Annex II of the RD 1393/2007. Computer Science appears only in Engineering and Architecture Degrees. It is therefore necessary that the training received by high school students in the Computer field is not limited only to the intrinsic knowledge of current digital technologies and their immediate practical uses. It is crucial that the training also focuses on the development of skills that enable students to adapt to new technologies that might emerge in the future, especially, in the field of smart education and next generation smart classrooms. Whereas *computational thinking* may be the most appropriate for developing such skills, in this work, a particular proposal for measuring the development of computational thinking abilities in students is described, together with the results obtained in an experiment carried out during the practicum of the Master's Degree in Secondary School Teaching from the Universidad de La Laguna.

**Keywords** Computational thinking · Algorithmic thinking · Scratch · Visual programming languages · Practicum

E. Segredo (✉) · G. Miranda · C. León
Dpto. Ingeniería. Informática y de Sistemas, Universidad de La Laguna,
Avda. Astrofísico Fco. Sánchez s/n, 38071 La Laguna, Spain
e-mail: esegredo@ull.edu.es

G. Miranda
e-mail: gmiranda@ull.edu.es

C. León
e-mail: cleon@ull.edu.es

A. Santos
Facultad de Psicología, Dpto. de Psicología Cognitiva Social y Organizacional,
Universidad de La Laguna, Campus de Guajara, 38071 La Laguna, Spain
e-mail: alu0100719888@ull.edu.es

© Springer International Publishing Switzerland 2016                                235
V.L. Uskov et al. (eds.), *Smart Education and e-Learning 2016*,
Smart Innovation, Systems and Technologies 59,
DOI 10.1007/978-3-319-39690-3_21

# 1  Introduction

Most current computer courses are focused on office suites and specific applications such as web browsers, for instance. To a lesser extent, they are also focused on hardware and/or the different existing operating systems. However, computer science is not only limited to specific applications and computer architectures. It also includes the set of scientific and technical knowledge that enables automatic data processing, allowing its implementation through computers. In this regard, it is essential to study the key concepts of computer science, such as abstraction, algorithms, and simulation, among others. Given the importance of technology and computers in our daily lives, those concepts should be handled by everybody, and not only by those specifically devoted to information technologies. Undoubtedly, maths and languages are essential to start any type of study, and from our point of view, the same should happen with computer science since it is a key tool for all areas of our society [1]. That is the reason why a new approach to education is being developed currently—at all education levels—for including *computational thinking* as an essential element of the curricula.

In order to achieve the above, *smart education*, and particularly, next generation *smart classrooms*, will play a key role. During last years, the smart education market, as well as the market of hardware and software for smart classrooms and smart universities, have significantly increased, and will continue increasing for the next decade, thus showing their relevance [2]. Bearing in mind that the main goals of next generation smart classrooms are, on the one hand, to show maturity on different smartness levels, like sensing or self-learning, among others, and on the other hand, to be based on modern hardware and software, the environment provided by these types of classrooms will allow computational thinking to be introduced in a faster way. In the same way, the introduction of computational thinking in the curricula will allow students to better take advantage of next generation smart classrooms, and consequently, of smart education.

Computational thinking could be described as the thought processes involved in problem formulation and solutions representation, so that these solutions can be implemented by a processing information agent (either a human, a computer or combinations of both). This term became famous thanks to Wing [3], who introduced computational thinking as a procedure that allows problem solving, designing systems and understanding human behaviour by the use of fundamental concepts of computing. Since then, computational thinking has attracted attention in the context of primary and secondary education, especially in English-speaking countries. However, there is still no consensus on the definition of computational thinking, thus having multiple variants [4–6]. For instance, the *International Society for Technology in Education* (ISTE) and the *Computer Science Teachers Association* (CSTA) define computational thinking as a process for problem solving which includes at least the following dimensions:

- Formulate problems to allow the use of computers to solve them.
- Organise and analyse data logically.
- Represent data through abstractions, models and simulations.
- Automate solutions through algorithmic thinking, i.e. through a series of orderly steps that achieve those solutions.
- Identify, analyse and implement possible solutions in order to find the more efficient and effective combination of steps and resources.
- Generalise the process of problem solving to wide range of problems.

On the other hand, the *National Research Council* (NRC) of the United States recommends mathematics and computational thinking as one of the eight main practices in the Science, Technology, Engineering, and Mathematics (STEM) fields [7]. Some of the definitions of computational thinking are based on the belief that students make use of computational thinking even when they do not use any kind of software. Conversely, programming itself implies that students make use of computational thinking through the construction of artefacts [8, 9]. This paper analyses the development of computational thinking through the use of programming. For this purpose, we will focus on the definition proposed by Brennan and Resnick [5], and considering *Scratch*[1] as the programming language. Scratch is a visual programming language widely used, especially in primary and secondary schools [10–13].

Scratch shares some characteristics with other modern visual programming languages: it is easy to learn, provides visual feedback of the developed programs (in the form of animated objects), and allows students to create interactive media, such as animations and games. Scratch is probably the most suitable visual programming language to develop computational thinking abilities through programming in the field of primary and secondary education. Considering Scratch, three dimensions where proposed in [5] for computational thinking: *computational concepts*, *computational practices* and *computational perspectives*. Table 1 shows a description and some examples for each of these three dimensions.

These three dimensions allow to understand how students from primary and secondary school address the learning of programming. Such dimensions also allow to check if the students are familiar with Scratch and its main precursor, *Logo*.[2] Logo first appeared in 1967 and was created by Seymour Papert. The knowledge of the programming language involves the syntactic, semantic and schematic knowledge (computational concepts) and also the strategic knowledge (computing practices). In 1982, Harold Abelson stated that "Logo is the name for a philosophy of education and a continually evolving family of programming languages that aid in its realisation". Logo programming environments that have been developed over the last decades are all based on a constructionist philosophy, which is based on the constructivist education theory, and therefore, they have been designed to support constructionist learning. Constructionism argues that human knowledge is generated from an interaction between individuals and the world around themselves. In fact,

---

[1]The reader is referred to https://scratch.mit.edu/ for more information.

[2]The reader is referred to http://el.media.mit.edu/logo-foundation/index.html.

**Table 1** Computational thinking dimensions

| Dimension | Description | Examples |
|---|---|---|
| Concepts | Concepts used by programmers | Variables, statements, etc. |
| Practices | Problem solving practices that arise during programming tasks | Be incremental and iterative |
| | | Testing and debugging |
| | | Reusability |
| | | Abstraction |
| | | Modularity |
| Perspectives | Students' knowledge about themselves, their relationships with equals, and the technological world that surrounds them | Express and question ideas about technology |

such an assumption coincides with the third of the proposed computational dimensions in [5] (Table 1). In this paper, we propose a general framework to analyse the impact of using a visual programming language, like Scratch, to develop computational thinking abilities. We also present some preliminary data that have been collected using mechanisms for measuring computational thinking in the context of the proposed framework.

## 2 Methodology

Both, the ISTE and the CSTA, consider "*algorithmic thinking*" as one of the dimensions of computational thinking. An *algorithm* is a method which consists of a sequence of precise instructions for solving a given problem [14]. Algorithmic thinking is considered one of the key concepts that allows people to be fluent in the use of information technologies. Thus, the NRC describes it as a set of concepts that includes, among others: functional decomposition, repetition (iteration and/or recursion), organisation of basic data (structures, registers, matrix, list, etc.), generalisation and parameterisation, algorithms versus programs, top-down design, and refinement. According to Futschek [14], algorithmic thinking includes the following capabilities or competencies:

1. Analyse given problems.
2. Specify or represent a problem accurately.
3. Find the basic and appropriate operations (instructions) to solve a given problem.
4. Build an algorithm to solve the problem following the given sequence of actions.
5. Think about all possible cases (special or not) of a given problem.
6. Improve the efficiency of an algorithm.

Lye and Koh [1] proposed that the presence of computational thinking should be increased in primary and secondary classes. Specifically, this presence should focus on practical and computational perspectives, two of the dimensions of computational thinking (see Table 1). To achieve this, a problem-based learning environment, including information processing activities, scaffolding and reflection to develop both, computational practices and perspectives, could be designed. This work is intended to continue with the previous proposal, but with the aim of measuring the impact that a visual programming language like Scratch can have on the development of computational thinking skills. To do so, three phases are identified in this study:

1. **Application of measuring instruments for computational thinking**. The objective of this first phase is to collect data which measure somehow the degree of use of computational thinking by students before carrying out specific activities for developing computational thinking skills.
2. **Development of computational thinking in a constructionist learning environment based on problems**. Following the recommendations provided in [1], this phase consists of a battery of information processing activities, scaffolding and reflection by students, using Scratch, as well as other instruments. This phase should be applied for a significant period of time, such as a semester or a full academic year.
3. **Application of measuring instruments for computational thinking**. In this last phase the measuring instruments are applied again in order to analyse the evolution of students regarding computational thinking skills.

As measuring instruments for computational thinking—applied in phases 1 and 3—we propose the use of the instruments defined in [15]. These instruments consist of a total of five activities that allow students to work four major computational thinking skills. These skills are: logical thinking, abstraction, algorithmic thinking, and cognitive planning. In this study we obtained data for the five activities, although only the results for two of the activities are shown: "*Organise and draw the objects*" and "*Win points!*", which have been selected since they demand greater extent of abstraction and algorithmic thinking skills.

The activity "*Draw and order objects*" is a variant of the one proposed by Morra [16] and is mainly focused on the ability of abstraction. The activity consists of some objects (a ball, a pencil, a book, and a bear) that have to be drawn into a rectangle following a set of instructions. Two different sets of instructions are given to the students, so the result consists of two schemes with different spatial distribution of the four objects. A possible solution to this activity is shown in Fig. 1. Students must interpret the set of instructions in order to establish the spatial relationship between the objects to distribute them in the rectangle. The instructions are based on the "*Test of Space Description*" proposed by Ehrlich and Johnson-Laird [17]. The spatial relationship between objects is represented by sequential statements of the form "A is below B", "B is to the left of C" and "D is to the right of C". In this activity, the first scheme uses a set of semi-continuous instructions: A–C, C–D, and A–B. The second scheme uses a set of discontinuous instructions: D–B, C–A,

**Fig. 1** Possible solution for the activity "*Organise and draw the objects*" considering the first scheme (*left-hand side*) and the second one (*right-hand side*)

and A–D, i.e. two different objects in the first instruction, two different objects in the second instruction, and finally, an object from the first instruction and an object from the third instruction. According to [17], the youngest students have difficulties processing discontinuous instructions due to the amount of information they have to memorise. In order to assess the level of performance of students when undertaking this activity, it is not only necessary to attend to the number of instructions properly processed and included in the scheme, but also to analyse the procedure used to successfully solve both schemes using knowledge obtained from the first scheme in order to solve the second one. The rubric used to assess the performance of students involves three levels: basic (students withhold information or confuse the relationships between objects), medium (students understand the composition of both schemes and are able to transfer knowledge from the first scheme to the second one), and advanced (students understand all the information given in the instructions and are able to properly build both schemes).

On the other hand, the activity "*Win points!*" determines the ability of students to understand and use control structures, especially conditional statements. Students should fulfil two conditions: find words with the last three letters equal to those of another word and search for words beginning with "p" and ending with "a". Figure 2 shows a potential solution to this activity. As in the previous activity, for the assessment of students, three levels of performance (basic, medium, and advanced) are considered.

For the second stage of the study, focused on the development of computational thinking abilities, we have used two different activities, which have to be implemented in Scratch by students. The main goal is twofold herein. Firstly, to introduce the language syntax and the graphic interface provided by Scratch to students. Secondly, to present some computational concepts and terminology that they will use afterwards in other kinds of activities. In this case, the scaffolding example consists of a story in which some friends are planning a trip. Students early realise that a character called Billy disappear from the Scratch scene, since Billy does not agree

| Pastorcita | Points |
|---|---|
| Pastorcita perdió sus ovejas | 1+5+5+5_____ |
| ¡y quién sabe por dónde andarán! | 5+5+5_____ |
| No te enfades, que oyeron tus quejas | 5+5_____ |
| y ellas mismas bien pronto vendrán. | 5+5+5+5_____ |
| | |
| Y no vendrán solas, que traerán sus colas, | 5+5+5+5+5+5_ |
| Y ovejas y colas gran fiesta darán. | 5+5+5_____ |
| Pastorcita se queda dormida, | 1+5+5_____ |
| y soñando las oye balar. | 5_____ |
| | |
| Se despierta y las llama enseguida, | 5+5_____ |
| y engañada se tiende a llorar. | 5_____ |
| No llores, pastora, que niña que llora | 1+5+5+5_____ |
| bien pronto la oímos reir y cantar. | 5+5_____ |
| | |
| Total points: | 163_____ |

**Fig. 2** Rafael Pombo's poem "Pastorcita" used for the activity "Win points!"

with the idea his friends have proposed for spending the day. During the activity's progress, students are guided by teachers in the usage of a resolution procedure with the aim of helping them to discover and solve these problems. For instance, teachers could help students to implement a new scene where the characters accept the idea proposed by Billy to spend the day. These modifications would include the replacement or addition of new Scratch blocks to control the movement of the characters, what they say, or even, whether they would be present at the scene or not. Before implementing the solution with Scratch, however, students must fill a problem analysis template that allows them to identify the characters, the scenes of the story, the problems that could potentially arise, the aspects they would change to solve those problems, as well as the Scratch blocks they would use to implement their solution. This template contains four sections, each one focused on one of the four first abilities proposed in [14], i.e. the problem definition, the expected results, the available data, the problem constraints, and the pseudocode that implements a possible solution. The other two abilities, i.e. thinking in all possible cases of a problem, and improving the efficiency of an algorithm, are not considered in the template, because of their complexity for primary and secondary school students. This analysis, which has to be carried out before implementing a particular solution in any programming language, like Scratch, involves the development of cognitive tasks that include skills such as planning, abstraction, and linguistic comprehension, among others. All of these abilities have been identified in computational thinking studies proposed by Wing [3], the ISTE, and the CSTA as well.

In the second activity, students have to interact with a story by the usage of a Lego™ inclination sensor. A character called Holly moves around her bedroom trying to pick up different items she have to store in her backpack. Students have to control Holly's movement with the inclination sensor, and they early realise that Holly only can go up to the top or down to the bottom of the scene, and that for solving the problem, they should modify different conditional blocks in their programs which would allow Holly to move left and right. Following the same procedure explained for the former activity, students have to gather all possible information to fill the problem analysis template before implementing their solution in Scratch, while teachers have to collect the highest amount of evidences about the activity's progress. For doing that, for instance, teachers might automatically record all class activities by using different components of smart classrooms, such as different video cameras installed to capture movements, discussions, expressions or gestures [2].

## 3   Results and Discussion

The proposed study was carried out during the practicum of one of the authors before obtaining his Master's Degree in Secondary School Teaching from the Universidad de La Laguna. The practicum was performed at the Primary School *"Tomé Cano"* and the Secondary School *"Domingo Pérez Minik"*, both located at Tenerife, Spain. In these centres there exist a large variety of students: high performance students, students who are not able to achieve the proposed aims, foreign students, students who are dependent due to impaired or reduced mobility, students with special educational needs, students with enough resources, and students without them. The above is, without any doubt, a clear example of the diverse environment that surrounds us. Particularly, we applied this study to 54 students who belong to three different groups and levels:

- **Sixth course of the primary school** (SPS). Students with ages between 11 and 13 years old (19 students who are 11 years old, 9 students who are 12 years old, and 2 students who are 13 years old).
- **Fourth course of a curricular diversification program** (FCDP). Students with ages between 15 and 19 years old (2 students who are 15 years old, 3 students who are 16 years old, 8 students who are 17 years old, 3 students who are 18 years old, and a only one student who is 19 years old).
- **First course of basic vocational training in telecommunications and informatics** (FBVT). Students with ages between 16 and 18 years old (4 students who are 16 years old, 2 students who are 17 years old, and a only one student who is 18 years old).

Due to the features and length of the practicum, it was possible to apply the first two stages of the study. The third stage, whose aim is to globally measure and evaluate the computational thinking development produced in the students, was not carried out, since it would have involved a larger amount of time to perform the whole set of programming and problem solving activities in Scratch. Bearing the above in mind, in this section we present the results obtained in the first stage of the study for the particular activities *"Organise and draw the objects"* and *"Win points!"*, which are shown on Figs. 3 and 4, respectively.

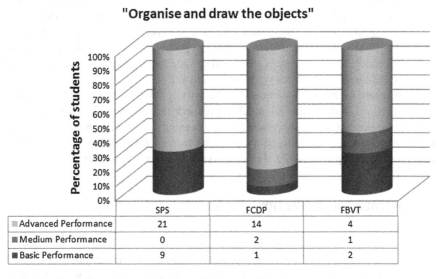

| "Organise and draw the objects" | SPS | FCDP | FBVT |
| --- | --- | --- | --- |
| Advanced Performance | 21 | 14 | 4 |
| Medium Performance | 0 | 2 | 1 |
| Basic Performance | 9 | 1 | 2 |

**Fig. 3** Results for the activity *"Organise and draw the objects"*

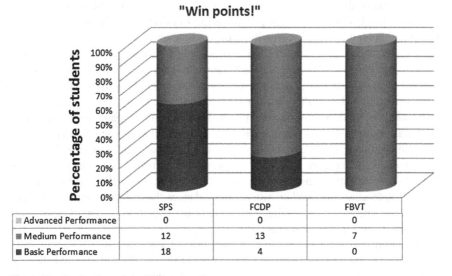

| "Win points!" | SPS | FCDP | FBVT |
| --- | --- | --- | --- |
| Advanced Performance | 0 | 0 | 0 |
| Medium Performance | 12 | 13 | 7 |
| Basic Performance | 18 | 4 | 0 |

**Fig. 4** Results for the activity *"Win points!"*

In the case of the activity *"Organise and draw the objects"*, we should note that, in general, most students who belong to the different groups analysed showed an advanced performance level. That means that students were able to understand and integrate all the information provided in the instructions into a single scheme. Consequently, students demonstrated a high development of abilities related to the abstraction, planning, and parallelism. Taking into account the different performance levels independently achieved by each group, it is worth pointing out that the groups SPS and FCDP presented a higher percentage of students with an advanced performance level than the group FBVT. The reason why this happened is the shortage of personal motivation that students belonging to the group FBVT demonstrated. At the same time, differences between groups FCDP and SPS are likely related to the age of their members, although differences were not significant.

With regard to the results obtained in the activity *"Win points!"*, the first conclusion is that, no student showed an advanced performance level, i.e. students were not able to successfully apply conditional statements. The second conclusion is that, in this case, the age seems to be a significant factor for students when they try to understand and apply conditional or other types of control statements.

## 4   Conclusions and Future Work

Resnick stated that when people program, they learn to solve problems, design projects and communicate ideas, in addition to understand computational and mathematical concepts [8]. People do not only program for solving problems, but also program for learning. Programming is like writing, and consequently, not only computer scientists should learn to program, just like not only novelists should learn to write. People can use Scratch programming to develop and implement interactive media with the aim of expressing their ideas. That is the reason why one of the main goals of this work was the design of a study for measuring the impact that a visual programming language, such as Scratch, has over the development of computational thinking abilities in primary and secondary school students. The proposed study consists of three different stages, where different tools to boost and measure computational thinking abilities are utilised. First and third stages focus on obtaining quantitative data, while the second one, applied in a problem-based constructionism learn environment, is more focused on obtaining qualitative information. For doing that, students use a problem analysis template, among other mechanisms for gathering information, before implementing a particular solution for their problem. In the same way, teachers also gather data about evidences that arise during the development of the activities, and without any doubt, smart education and smart classrooms will allow to better carry out these tasks.

Practicum was not large enough to carry out the whole study. As a result, only the first stage and part of the second one, were performed. Results conclude that, in general, primary and secondary school students showed and advanced performance level related to abilities like abstraction, planning, and parallelism. Students did not show, however, the same high performance level in tasks related to the usage of control statements. In this sense, one of the main research lines for the future will be the application of the remaining stages belonging to the proposed study. Hence, qualitative information that supplement quantitative data already collected could be obtained. Additionally, we could investigate whether the use of visual programming languages, like Scratch, significantly influences the improvement of computational thinking abilities.

# References

1. Lye, S.Y., Koh, J.H.L.: Review on teaching and learning of computational thinking through programming: what is next for K-12? Comput. Human Behav. **41**, 51–61 (2014)
2. Uskov, V.L., Bakken, J.P., Pandey, A.: The ontology of next generation smart classrooms. In: Smart Education and Smart e-Learning, pp. 3–14. Springer International Publishing, Cham (2015)
3. Wing, J.: Computational thinking. Commun. ACM CACM **3**, 33–35 (2006)
4. Barr, V., Stephenson, C.: Bringing computational thinking to K-12: what is involved and what is the role of the computer science education community? ACM Inroads **2**, 48–54 (2011)
5. Brennan, K., Resnick, M.: New frameworks for studying and assessing the development of computational thinking. In: Annual American Educational Research Association Meeting, Vancouver, BC, Canada (2012)
6. Grover, S., Pea, R.: Computational thinking in K-12: a review of the state of the field. Educ. Res. **42**, 38–43 (2013)
7. National Research Council: A Framework for K-12 Science Education: Practices, Crosscutting Concepts, and Core Ideas. The National Academies Press, Washington, D.C., USA (2012)
8. Resnick, M., Maloney, J., Monroy-Hernández, A., Rusk, N., Eastmond, E., Brennan, K., Millner, A., Rosenbaum, E., Silver, J., Silverman, B., Kafai, Y.: Scratch: programming for all. Commun. ACM **52**, 60–67 (2009)
9. Kafai, Y.B., Burke, Q.: Computer programming goes back to school. Phi Delta Kappan **95**, 61–65 (2013)
10. Kafai, Y.B., Fields, D.A., Burke, W.Q.: Entering the clubhouse: case studies of young programmers joining the online scratch communities. J. Organ. End User Comput. (JOEUC) **22**, 21–35 (2011)
11. Tangney, B., Oldham, E., Conneely, C., Barrett, S., Lawlor, J.: Pedagogy and processes for a computer programming outreach workshop: the bridge to college model. IEEE Trans. Educ. **53**, 53–60 (2010)
12. Theodorou, C., Kordaki, M.: Super Mario: a collaborative game for the learning of variables in programming. Int. J. Acad. Res. **2**, 111–118 (2010)
13. Baytak, A., Land, S.: An investigation of the artifacts and process of constructing computers games about environmental science in a fifth grade classroom. Educ. Technol. Res. Dev. **59**, 765–782 (2011)
14. Futschek, G.: Algorithmic thinking: the key for understanding computer science. In: Mittermeir, R. (ed.) Informatics Education—The Bridge between Using and Understanding Computers. Lecture Notes in Computer Science, vol. 4226, pp. 159–168. Springer, Berlin, Heidelberg (2006)
15. López, J.C.: Impacto de Scratch en el desarrollo del pensamiento algorítmico. Master's thesis, Universidad Icesi (2014)
16. Morra, S.: On the information-processing demands of spatial reasoning. Think. Reason. **7**, 347–365 (2001)
17. Ehrlich, K., Johnson-Laird, P.: Spatial descriptions and referential continuity. J. Verbal Learn. Verbal Behav. **21**, 296–306 (1982)

# Innovations in Subjects Knowledge Technologies

Petra Poulova and Ivana Simonova

**Abstract** Reflecting the call for new knowledge and skills required from the graduates of the Faculty of Informatics and Management, University of Hradec Kralove, the topic of smart and knowledge technology was implemented in the curriculum. To ensure the quality of highly professional training programme, the faculty started cooperation with companies. The aim of the relationship is to provide students with theoretical and practical professional skills and prepare them for new positions that have recently been created on the labour market. Since smart phones were launched, the mobile market has rapidly changed from feature phones to smart phones. As a result, the field of education is concerned with delivery of knowledge through smart devices.

**Keywords** Smart · Knowledge management · e-Learning · LMS · Efficiency · Success rate · Visit rate · Tracking

## 1 Introduction

One of the main objectives of the Faculty of Informatics and Management, University of Hradec Kralove (FIM) is to facilitate the integration of our students in the national and international labour markets by equipping them with the latest theoretical and applied investigation tools for the IT area.

That is why the current study programmes are analyzed in co-operation with specialists from the practical environment to detect the disaccord with needs required on the labour market.

P. Poulova (✉) · I. Simonova
Faculty of Informatics and Management, University of Hradec Králové,
Rokitanského 62, 500 03 Hradec Králové, Czech Republic
e-mail: Petra.Poulova@uhk.cz

I. Simonova
e-mail: Ivana.Simonova@uhk.cz

© Springer International Publishing Switzerland 2016
V.L. Uskov et al. (eds.), *Smart Education and e-Learning 2016*,
Smart Innovation, Systems and Technologies 59,
DOI 10.1007/978-3-319-39690-3_22

Reflecting the results of analyses, the curricula of study programmes are innovated so that they are of a modular structure and graduates gain appropriate professional competences to be easily employable. The proposal of a new design is assessed by the HIT Cluster (Hradec IT Cluster), i.e. by the society of important IT companies in the region, and by the University Study Programme Board. Its main objective is to create a transparent set of multidisciplinary courses, seminars and online practical exercises which give students the opportunity to gain both theoretical knowledge and practical skills as well as to develop their key competences. Subjects in each study programme are structured into five groups—data engineering, computer networks, software engineering, enterprise informatics and knowledge technologies.

## 2 Innovations in Knowledge Management Subjects

Currently, smart and knowledge technologies are of increasing importance. That is why students of the Computer Science study programmes at FIM are expected to be well-prepared in this field. Most companies focusing on software development expect the graduates to have both good theoretical knowledge and practical experience in this field. That is the reason why new competences have been included in the curricula and required from the FIM graduates.

The Knowledge technology courses were included in the curricula of bachelor and master study programmes of Information Management and Applied Informatics more than 15 years ago.

Following subjects have been embraced in the group of Knowledge technology:

- Theory of systems I, II
- Introduction into artificial intelligence
- Knowledge technology I, II, III, IV
- Computational intelligence I, II
- Knowledge management: seminar
- Cognitive science

This proposal was designed in co-operation with companies GMC Software, UNIEPOS, ORTEX, GIST and DERS.

### 2.1 Company Requirements on Graduates

New competences required for FIM graduates were set by the expert group consisting from 11 members of the Hradec IT Cluster. The experts considered the original state and proposed a new concept subjects supported by online courses in the Learning Management System (LMS) Blackboard. All the study materials were created and uploaded there. Each online course was equipped with instructions, study materials, tests and communication tools [1–4].

The expert team headed by the vice-dean for study affair set 20 roles of IT specialists (e.g. Analyst, Business Analyst, Designer, Flash Developer, Internet Marketer PPC/SEO, Programmer, IT Manager, IT Sales, IT Administrator etc.). For each role the description of the working position was defined and consequent requirements on the position described. For example, the description of the IT systems implementer position includes following tasks:

- provides concrete solutions to customers (mainly abroad), either remote, or on the place;
- independently secures expert support to internal and external customers;
- proposes configuration of appropriate solution reflecting requirements pre-defined by the customer;
- installs the product which meets customer's requirements;
- sets the system so that it met the customer's requirements;
- trains customer's staff;
- makes documentation of installations and configurations.

Consequently, following knowledge and skills are included in the Requirements on the position:

- advanced knowledge of English, both written and oral, experience in IT field, in project management, analytic skills, negotiating skills, communication and presentation skills, willingness to travel abroad for longer time-periods;
- experience in economic, production and logistic process;
- general knowledge of IT environment, good knowledge of office SW, SQL databases, experience in enterprise IS;
- legal and economic minimum;
- communication skills, assertivity;
- time flexibility, resistance to stress;
- driving license.

Reflecting the expert opinions, the general knowledge comprises following fields:

- Problem analysis.
- System approach.
- Result orientation.
- Precision.
- Consistency.

## 2.2 Innovative Contents in Single Subjects

In the below listed subjects following innovative design and content was included.

## Systems Theory I, II

The subject is designed in such a way which reflects the knowledge development, students' abilities and skills, as defined with single IT roles—these have been worked out in co-operation with the HIT Cluster. Particularly, the roles of IT analyst, business analyst. IS implementer, coder/programmer, IT consultant and IT manager.

- In the role of Analyst and Business Analyst the knowledge of creating/designing models and general understanding of the model approach is required.
- In the role of IS implementer the ability to perceive customers' problems and requirements is necessary; these skills are supported by the development of system thinking.
- In the role of Coder/Programmer the system thinking is even on the top of the list of skills defined by the HIT Cluster.
- IT consultants and IT managers should be good at discovering correlations of various types within the process of IS management, which is considered to be the basis of the system approach.

## Introduction into Artificial Intelligence

This subject provides basic outline and motivation to studying the field of artificial intelligence. The provided information cover the introduction to studying other subjects which follow up in other years of university study. Their aim is to introduce to students selected topics within the field of traditional and new artificial intelligence, important personalities, crucial problems to be solved, supported by samples of good practice.

## Knowledge Technologies I, II, III, IV

These subjects focus on problems related to knowledge presentation through ontological modelling, to the field of automated intelligent manipulation and exploitation of information through knowledge modelling via TopicMaps and to problems related to multi-agents systems.

## Computational Intelligence I, II

These subjects aim at acquiring basic methods of computational intelligence, their background, theoretical bases and application potential. After the courses students should be able to identify problems in practice, to solve them by applying the mentioned methods in a more efficient way compared to standard computational procedures, including the exploitation of appropriate software.

## Knowledge Management: Seminar

The subject of knowledge management introduces a rather new concept whose multi-disciplinary character touches many fields of human activities, not only in the sphere of business and entrepreneurship. The main objective of the subject is to apply the active seminar form to introduce students the core characteristics of knowledge society and knowledge economy, the knowledge management system environment on the managerial level, as well as theoretical outcomes of knowledge management, practical examples and ways of implementation, and to outline and explain basic principles, procedures, methods, techniques and tools which can be exploited in knowledge management in practice.

**Cognitive Science**

In this subject students meet the problems of human thinking, reasoning and the impact of cognitive processes on the decision-making process, resp. planning. The emphasis is paid on the holistic and multi-dimensional approach of the cognitive science to mutual impact and inter-relations of the cognitive science and artificial intelligence. Instead of others, the subject should serve as the basis for further studies, particularly of neuron networks.

## 2.3   E-Learning in the Courses

The whole programme is supported by e-learning lessons. All the study materials are created and uploaded in the FIM LMS BlackBoard [5–7].

Each module in the e-learning course is equipped with its own guiding instructions, study materials and modified tests with comprehensive sets of questions. Study materials were developed not only as standard presentations; they contain audio-visual materials, animations and instructional video-recordings. Students' attendance in the courses and lectures is monitored in order to evaluate not only the results of the individual tests but also to detect the fields of students' interest.

Since smart phones were launched, the mobile market has rapidly changed. As a result, the field of education is concerned with delivery of knowledge through smart devices (Smart Education). Smart Education mechanism can be seen as an integrated educational environment in which cooperative, interactive, participative, sharing, and intelligent learning are available through new forms of teaching—learning content, environment, and ICT.

## 2.4   Expert Evaluation of Subjects

As in the above described activities, the evaluation process also resulted from the cooperation of the HIT Cluster—GMC SW, UNIEPOS, ORTEX, GIST and DERS. Their experts had complete information on innovated subjects, including the access to e-subjects in the LMS Blackboard.

The evaluation was made by questionnaires containing 25 statements. Their dis/agreement was expressed on four-level scale: completely agree—partially agree —partially disagree—completely disagree. Moreover, experts also commented the evaluated subjects in the form of open answers.

The group of knowledge management subjects was evaluated in the written form by companies of the HIT Cluster. Totally 30 evaluation reports were provided which can be summarized as follows:

- The knowledge objectives in single subjects are defined in measurable outputs, they reflect requirements on knowledge and skills in the field of computer networks. The information on objectives is available to students within several tools in the e-subjects.
- The learning content reflects the declared objectives. It is structured into appropriate modules and presented in the logical order. Multi-media elements are appropriately integrated in various parts of the subjects. Students are expected to have a deeper understanding of the learning content which is explained in examples and/or models.
- The material for students' guidance through courses in included in each course which makes the process of learning easier for them.
- Moreover, clearly defined evaluation criteria are available to students, e.g. sample project works, semestral works etc. Each testing of knowledge clearly relates to pre-defined objective.
- Various methods of assessment are exploited (test questions of various types, projects, discussions etc.) and students have self-evaluation activities and tools available to get a constructive feedback and correct the knowledge.

Experts also provided oral evaluation. Selected samples are presented below:

*This group of subjects is difficult to be evaluated by the expert from the sphere of practical entrepreneurship. It is the most academic of all, which is the main contribution—the university studies should be of 'academic' character, as well as scientific basis exceeding the common IT practice. Otherwise, it would be a vocational school only.*

*Theoretical knowledge is well proportioned with practical ones where students can try less common procedures and tools (e.g. fuzzy logics, neuron networks, expert systems, TopicMaps, Matlab etc.), which are probably to appear in practice more and more frequently. The set of subjects and their learning objectives is appropriate in my opinion, I would recommend to provide a more detailed description of syllabi.*

ORTEX

*In every respect, the innovation is considered a step toward. Within the primary and secondary schooling the education in the field of analytic thinking, thinking skills development, understanding the reality, problem solving and thoughts defining is omitted, despite it is crucial for managers. That is why we consider important for the quality of FIM graduates the teachers participated in defining curricula to eliminate this handicap.*

*The original learning content of subjects of this group rather aimed at theoretical knowledge than gaining practical skills. The current content is proportionally weighted from this view. In our opinion, in the innovated subjects students get appropriate theoretical knowledge to understand the field and practical skills to apply the on real problems of the IT profession.*

*Most of our prioritized requirements have been accepted in the innovated subject content. For the future we would recommend to conduct seminars on soft skills—communication, both in the team and out of the team, particularly with users. We agree the development of soft skills have been included in seminars but it should be done in wider extent. Low level of communication skills stands behind most problems in IT projects, which is why the graduate should be able to communicate on high level.*

*We propose to include the subject Cognitive Science in the second semester—it is the theoretical basis of other subjects, it is late to study it in the 8th semester. Moreover, the number of credits should be re-considered.*

GIST

*The innovation in this group of subjects is mostly (but not fully, see recommendations) in accord with our requirements and needs of real practice. We appreciate particularly following:*

- *The learning content in subjects Computational Intelligence and Knowledge technologies (neuron networks, semantic webs, TopisMaps, MAS).*
- *The implementation of CLIPS language into the learning content.*
- *The exploitation of advanced tools within the instruction (MATLAB in Computational Intelligence, Stella in modelling system dynamics, Protege in OWL modelling).*
- *Creating thematic explanatory dictionaries.*
- *The emphasis on strengthening the ability of abstraction, discovering correlations, criticism, creativity innovativeness etc.*
- *Restructuration of selected subjects with respect on specifics and differences in study programmes and requirements of real practice (HIT Cluster)*
- *Increasing the number of students' independent work on projects.*

*With respect to the above mentioned, we sum up our recommendations as follows:*

- *We strongly appreciate the entire existence of the subject Knowledge Science (KOGN), but recommend to teach it in earlier semesters, as it serves the theoretical basis of follow-up ones.*
- *We also appreciate the trend towards team co-operation which is clearly visible in summaries of innovations in single subjects (e.g. those from the group of Communication technologies). However, we recommend to add other independent blocks focusing on techniques of training "soft skills", particularly (but not only) in the Information Management study programme. Appropriate topics (in the form of practical role-playing) are e.g. effective communication, team co-operation, targeting, discovering the core of the problem, time management, agile approaches in thinking, motivation, self-fulfillment, self-assessment etc.*
- *The involvement of wikinomic principles of co-operation and building knowledge bases is also helpful, as well as exploitation of knowledge in building knowledge systems for education.*

DERS

*In subjects Theory of Systems we appreciate the emphasis on practical experience, criticism and ability of abstraction; these are very useful in practice.*

GMC

# 3   Conclusion

Fast development of latest technologies is reflected in new requirements for graduates' knowledge and skills. This was the main reason, why the co-operation started between the Faculty of Informatics and Management, University of Hradec Kralove (FIM UHK), and companies. In the data analysis and their proper usage the IT companies see a great potential. What they are concerned about, however, is the lack of qualified professionals who would be able to handle these data. Consequently, this was the main reason why the FIM curricula were adjusted to latest requirements.

Reflecting the expert recommendations, large changes were made in the evaluated subjects. To illustrate the process, two examples are presented—Cognitive Science and Knowledge Technologies I.

In the subject of Cognitive Science the content of lectures was updated to reflect latest knowledge on human thinking, methods of its research and application of results within the artificial intelligence. Detailed thematic dictionaries were created, list of links to important web sources, video-library and the case study on the research of cognitive and mental maps on FIM.

In the subject of Knowledge Technologies I, reflecting the volume of Hradec IT Cluster recommendations, not all comments could be implemented. However, the whole concept of the subject was changed. In the original version, the subject was oriented on the technical side of designing ontologies and thematically limited to ontologies only. In the new concept, knowledge technologies are included in the context of other disciplines, i.e. knowledge technologies applications in companies, knowledge societies and artificial intelligence. Lectures are not limited to ontologies only; within the practical application students' work in pairs is required, and emphasis is paid on communication and negotiation skills. In the seminar project the analysis and proposal of knowledge application is required, as well as implementation as it used to be in the original concept.

**Acknowledgement** This research was financially supported by the SPEV.

# References

1. Poulova, P., Simonova, I.: Flexible e-learning: online courses tailored to student's needs. In: Capay, M, Mesarosova, M, Palmarova, V (eds.) DIVAI 2012: 9th International Scientific Conference on Distance Learning in Applied Informatics: Conference Proceedings, pp. 251 –260. UKF, Nitra (2012)
2. Tomaskova, H., Nemcova, Z., Simkova, M.: Usage of virtual communication in university environment. Procedia Soc. Behav. Sci. **28** (2011)
3. Zentel, P., Bett, K., Meiter, D.M., Rinn, U., Wedekind, J.: A changing process at German universities—innovation through information and communication technologies? Electron. J. e-Learn. **2**(1), 237–246 (2004)
4. Kotzian, J., Konecny, J., Krejcar, O.: User perspective adaptation enhancement using autonomous mobile devices. In: Lecture Notes in Artificial Intelligence. vol. 6592, pp. 462 –471 (2011)
5. Bradford, P., Porciello, M., Balkon, N., Backus, D.: The blackboard learning system. J. Technol. Syst. **35**, 301–314 (2007)
6. Stepanek, J., Simkova, M.: Design and implementation of simple interactive e-learning system. Procedia Soc. Behav. Sci. **83**, 413–416 (2013)
7. Poulova, P., Simonova, I., Manenova, M.: Which one, or another? Comparative analysis of selected LMS. Procedia Soc. Behav. Sci. (2015)

# Economic Aspects of the Introduction of the SMART Technology into Kindergartens and Primary Schools—Czech National and Local Scene

**Libuse Svobodova and Miloslava Cerna**

**Abstract** SMART technology and the use of other advanced technologies are becoming increasingly important. Ten years ago universities dominated in using technology to support teaching/learning process but in recent years advanced technologies have been intentionally and intensely included also into pre-primary, primary and secondary education and even have become a part of educational program framework. While some students and teachers and a part of the professional public can see great potential in them, there are a lot of opponents on the Czech scene, as well. Even so, it seems that the trend in the implementation of smart technologies will indisputably follow and that individual schools will try or even will be made to introduce interactive whiteboards and tablets into their education. Economic aspect is an integral part of each project. The paper focuses on one fragment of the economic issue, on mapping the costs on purchase of SMART technology equipment per one class.

**Keywords** Benefits · Costs · Education · Implementation · SMART technologies

## 1 Introduction

The technology is literally all around. The technology assists in both business and personal life where education is no exception. While a computer, laptop or tablet pc, data projectors and online education are a standard issue in most universities in the Czech Republic, the situation in kindergartens, primary and secondary schools is different, in spite of the fact that a significant shift has been made. Still it can't be

L. Svobodova (✉) · M. Cerna
Faculty of Informatics and Management, University of Hradec Kralove,
Rokitanskeho 62, Hradec Kralove, Czech Republic
e-mail: libuse.svobodova@uhk.cz

M. Cerna
e-mail: miloslava.cerna@uhk.cz

© Springer International Publishing Switzerland 2016
V.L. Uskov et al. (eds.), *Smart Education and e-Learning 2016*,
Smart Innovation, Systems and Technologies 59,
DOI 10.1007/978-3-319-39690-3_23

stated that each kindergarten, primary or secondary school is equipped with at least one interactive white board which could be used by children and students during their classes or stay in the kindergarten. As for tablets the situation is even worse. But it can be expected that changes are coming. Ministry of Education headed by the ex-minister of Education himself Mr. Chládek announced their plan to introduce tablet devices into schools; this plan is currently being promoted [1]. There is a project worth highlighting which preceded this plan; it is called "Tablet devices into schools - a tool for teachers in the world of digital education - CZ.1.07/3.1.00/51.0002". The project was run last academic year 2014/2015. The project involved three regions—Hradec Kralove, Pardubice and South Moravia and is considered as a starting line for a massive implementation of touch devices into the teaching/learning process.

"The objective of the project is the implementation of ICT into the curricula of primary and secondary schools and increase of ICT competences of teaching staff at schools when using touch devices in the educational process both in didactics and in classes. Another goal is to increase the competencies of the top management of schools involved in the issue of selection procedures and related legislation, planning of school ICT development, planning of organization development by School Profile 21. Managers will be trained so that they could effectively use potential of School Profile 21 for the development planning of their organizations.

An important goal is the assistance in the professional development of teachers provided by mentors and coaches. Teachers will be given chance to contact a trained expert or get the remoted online assistance when using ICT. One ICT expert will be systematically retrained in selected schools. This expert is supposed to solve emerging potential IT problems which might arise at their schools and be able to provide sufficient support to all colleagues in the use of ICT in the process of education" [2].

In total about 900 teachers from all three regions were supposed to take part in the project and be given technical support throughout the project.

There are two distinctive groups with differ in perceiving smart technologies and their benefits in preprimary and primary education. Firstly there are a lot of experts and general public who consider technology as an important component of life and learning, and so they fully encourage implementation of advanced technologies into education. Beside academia papers there can be found a long list of links on the Internet to YouTube websites, blogs or various educational centers discussing enthusiastic positive experience of practitioners. E.g. "The Interactive Smart Board in the kindergarten. Reality or a dream?" [3] contribution written by the director of a village kindergarten which was published on the websites of a preprimary center brings just positive field experience; higher competitiveness among other kinder-gartens, affection by preschool children, etc. The other illustrative example is taken from YouTube "We are playing and we are learning: Spring" [4] showing chil-dren's work with IWB in the kindergarten with the interactive Colored pebbles. Enthusiasm of the teacher gets reflected in the video but the opposite rather skeptical approach to the issue of Smart technologies can be found immediately in the comments. So the other group, group of opponents in the Czech Republic who are against utilization of IWB (interactive white boards) and tablet devices in kindergartens, primary and secondary schools is plentiful and represented by academia experts, as well. E.g. Neumajer [5] states eight objections. Komárek [6] discusses in his works Digital Dementia. Klaus [7] also addresses problems with the introduction of table devices into schools.

However, as mentioned above, it is without doubt expected that interactive whiteboards and tablets (or smartphones) will be widely supported and used in primary and secondary schools in the Czech Republic. As for kindergartens, installation and use of interactive whiteboards can be expected, especially in classes of preschool children. That is why; the authors have focused in their article on calculation of costs associated with the introduction, maintenance and use and of these technologies.

## 1.1 Literature Review

Usage of digital devices is a global phenomenon. At the end of the introductory part a shortlist of local authors was presented. This chapter brings a deeper insight into the issue coming from various corners of the globe covering various aspects like rules in the usage of tablet devices [8], Positive and negative aspects of the IWB and tablet computers [9], creating of teaching materials [10, 11], ongoing classroom presence for the hospitalized children [12], middle school students' perceptions regarding the integration of tablets into the learning process [13], digital pencil [14] and digital writing tool versus traditional writing tool [14, 15] effectiveness of touchscreen devices [16].

Selected resources refer predominantly to utilization of digital devices by children attending primary school. A Singaporean study published last year [8] with the sample exceeding one hundred children in their second year of primary school presented following findings: all children have access to mobile phone, nearly all of them also have home computer access but just 57 % have access to tablet PC. This study [8] discussed an access and usage of digital devices at home.

The following qualitative study takes readers to primary school setting to the first grade classes. Pros and contras of interactive whiteboard and tablet PC were explored by means of a multi-perspective approach: by teachers, the pupils and an independent observer. Main positives connected with the use of interactive whiteboard were pupils' attention and motivation. As for negatives; technical problems were identified in both IWB and tablets. Teacher's lesser control over pupils' work was identified as another distinctive negative in the use of tablets [9].

Utilization of smartphones, tablets for study purposes at the primary school setting and for free time is discussed in the study [10] covering national and international scene.

Next area of utilization of digital media is specific and worth highlighting. The core of the paper stems from the needs of children with long-term bad health conditions whose school attendance is low or disrupted. Even if the research sample consisting of hospitalized children is small the outcomes of the research based on specially designed Presence App run on a mobile tablet computer are promising [12]. Presence App enabled absent children to maintain a social presence in the classroom [12].

Plan to introduce tablets into schools isn't only a current topic of the Czech Ministry of Education. A large research with nearly one thousand respondents was conducted in Turkey called "Tablets in Education". The objective of the research [13] was to find out how the integration of tablets into the learning process is perceived by students.

A deep literature review was provided on one segment of digital devices utilization, on early writing outcomes among first writers where "revolutionary" digital writing tools like tablets were compared with "traditional" writing tools. Studies were categorized and their methodological quality was analyzed, key problems were identified in the review findings [15].

The last selected paper on digital devices doesn't refers to children but to teachers to their readiness to create and use electronic educational material. Methodological support of teacher training together with the system of assessment were elaborated and consequently presented in the paper [11].

## 2   Materials and Methods

The research procedure followed two phases:

- The first phase deals with introduction to the topic of implementation of technology into schools and into the process of education.
- The second phase discusses selected economic aspects of introducing technologies in schools.

Primary and secondary sources were used. As for secondary sources, they comprised websites of selected kindergartens and primary schools in Hradec Králové and Pardubice, websites on digital devices and education, e.g. EDUIN, technical literature, information gathered from professional journals, discussions or participation at professional seminars or conferences. Then it was necessary to select, categorize and update available relevant information from the collected published material. Work in the field followed: top management of selected institutions was contacted, and then appointments and meetings with directors and teachers from three kindergartens and two primary schools were organized. Information on costs of purchase and implementation of technology, its maintenance and teachers' training was collected in individual kindergartens and schools. Gained data were processed and presented in an anonymous form to avoid possible problems and misunderstandings or legal pitfalls.

The amounts in calculations are listed in both CZK and EUR. The calculations were based on an exchange rate when 1 EUR equaled 27 CZK.

## 3   Economic Aspects of Implementation of Tablets to School

### 3.1   Budget and Costs

At the preparatory stage of the introduction of technology into schools it is necessary to calculate the budget with expenses that are related to the implementation of technology.

Interactive whiteboards, tablets, laptops, netbooks, smart mobiles are defined as digital devices as technology which supports the learning/teaching process.

The budget will be divided into 2 sections. The first one will be kindergartens. Their main equipment will be interactive whiteboard and one tablet for the teacher. The second group will consist of primary and secondary schools. Each classroom will be equipped with one interactive whiteboard and one tablet of higher quality for the teacher and 30 tablets for students.

Calculations are based on the data from the statistics of the Ministry of Education [17], with 23.4 children in a kindergarten class in 2014/15. Average amount of pupils in the primary school was 19.7 per class. Calculation of tablets for learning/teaching will be based on these data from the statistics.

### 3.2   Equipment of One Class in the Kindergarten

Calculation is based on the assumption that one classroom in the kindergarten will be equipped with interactive whiteboards and tablet for teachers. There is a trend to equip especially preschool classes with this digital device.

Even though the technological readiness of teachers in the Czech Republic is at a good level and that teachers have already gained good literacy competence it can be expected that they will enroll for a training course, where they will learn how to work properly with a new equipment and how to use special software.

It is possible that within a new programming period of funds from the European Union free seminars will be announced. E.g. EDULAB organization [18] already offers applications for nursery and primary schools. Each application costs 1.500 CZK (56 EUR). Training is currently free. That is why the interest in training is enormous and at this moment 3 months ahead are fully booked.

Calculation will be based on the fact that there are two teachers in the class. According to statistics there are 12.6 children per one fulltime preprimary teacher. Stated figures come from the current prices of preprimary teacher trainings.

Training and preparation of technical personnel will not be incorporated into the budget since these workers aren't affordable for kindergartens. If necessary, those needed services will be provided within the device installation and delivery. In the future, if teachers themselves will not be able to provide proper maintenance, these services will be outsourced, if necessary. At the same time it is expected that the

**Table 1** Non-recurring expenses—kindergarten

| Costs | Price CZK (CZK) | Item | Total CZK (CZK) | Total EUR (EUR) |
|---|---|---|---|---|
| Training of managers | 3,000 | 1 | 3,000 | 111 |
| Training of teachers | 3,500 | 2 | 7,000 | 259 |
| Purchase of the IWB and its accessories (holder, computer, etc.) | 60,000 | 1 | 60,000 | 2,222 |
| Purchase of the educational applications | 1,500 | 20 | 30,000 | 1,111 |
| Purchase of the tablet for the teacher | 15,000 | 1 | 15,000 | 556 |

kindergartens will have to update used educational applications. It means that further costs will have to be spent on the purchase of new interactive device applications that will be used with the children in the class in the future.

Basic technical knowledge should be ensured for managers of individual educational institution, as well. Kindergartens are currently connected to the Internet and therefore there is no need to calculate the cost of the introduction of the Internet into the budget. In some cases, however, it will be desirable to expand and intensify cable network or Wi-Fi.

Charges for repetitive yearly trainings and updates of materials will belong to the category of regular charges. Charges for the Internet or servicing equipment will be excluded from the calculation because the kindergartens would have the Internet connection even without Interactive whiteboards. These charges are estimated to be 8,000 CZK per year (296 EUR). Updating of materials will likely be the most costly part. It is possible that these payments could rise depending on the number of applications.

Nonrecurring expenses (Table 1) associated with the introduction of SMART Technologies into the process of instruction in kindergartens can be calculated at about 115,000 CZK (4,259 EUR) per class.

In case that the kindergarten management decides not to buy just those 20 applications but the recommended whole package of 4,672 materials with valid license for 1 year, they will have to pay 89,900 CZK (3,330 EUR). This application can be used on multiple devices within the same school. In total the onetime costs will rise to 174,900 CZK (6,478 EUR).

Other extra costs might be related to the purchase of mobile whiteboard. This kind of IWB is practical but much more costly than fixed IWB. Its price is currently in the Czech Republic about 120,000 CZK (4,444 EUR).

## 3.3    Equipment of One Class in the Primary School

Calculation is based on the assumption that one classroom in the primary school will be equipped with an interactive whiteboard, and tablets for the teacher and pupils. The teacher will be given a more powerful tablet than pupils.

The tablet for the teaching/learning purposes will be selected from the wide current tablet offerings. The screen size should be 10.1″ as recommended in [19]. Depending on the size of the interactive whiteboard, the holder and other equipment is supposed to be greater and more robust than equipment in kindergarten.

Also in the case of primary school teachers it can be expected that they will participate in a technical training where they will be instructed how to use new devices and their software properly. It is possible that within a new programming period of funds from the European Union free seminars will be announced. Primary school teachers like preprimary teachers can also attend free seminars organized by EDULAB [18] and buy special educational applications from them.

The calculation is based on one teacher in the class as is usual in the Czech Republic. A lesser amount of money from the budget will go on training and preparation of the technical staff. It is evident that new applications will have to be purchased into the interactive device which will be used during classes. Managers are supposed to have or gain basic competence in the use of advanced technology, as well. Primary schools have access to the internet the significant obstacle is that very often not all areas of school are covered with the signal. That is why there might arise initial nonrecurring expenses on implementation of stronger Internet access with wider network coverage depending on the size of the school to some 50,000 CZK (1,852 EUR).

Charges for regular yearly trainings together with charges for updates of materials will belong to the category of regular charges. (Charges for the Internet or servicing equipment will be excluded from the calculation because the school would the Internet connection even without Smart technologies. These charges are estimated at 8,000 CZK (296 EUR) per year. Updating of materials will likely be the most costly part. It is possible that these payments could rise depending on the number of applications.

In Table 2 there are described the non-recurring expenses expected for primary schools.

Nonrecurring expenses associated with the introduction of SMART Technologies into the process of instruction in primary school can be calculated at about

**Table 2** Non-recurring expenses—primary school

| Costs | Price (CZK) | Item | Total CZK (CZK) | Total EUR (EUR) |
|---|---|---|---|---|
| Training of managers | 3,000 | 1 | 3,000 | 111 |
| Training of teachers | 3,500 | 1 | 3,000 | 111 |
| Training of technical staff | 3,500 | 1 | 3,500 | 130 |
| Purchase of the IWB and its accessories (holder, computer, etc.) | 80,000 | 1 | 80,000 | 2,963 |
| Purchase of the educational applications | 1,500 | 20 | 30,000 | 1,111 |
| Purchase of the tablet for the teacher | 15,000 | 1 | 15,000 | 556 |
| Purchase of the tablets for pupils | 8,000 | 20 | 160,000 | 5,926 |

265,000 CZK (9,815 EUR) per class in case that twenty educational applications will be purchased.

The costs in nonrecurring expenses might increase mostly in the purchase of apps licenses. E.g. EDULAB [17] offers whole packages, where Mathematics for primary schools with 5498 educational materials costs 69,900 CZK (2,589 EUR). Then it offers a variety of opportunities to learn the English language from over 19,900 CZK (737 EUR) to 39,900 CZK (1,478 EUR). The last major area that is currently being offered is science subjects (physics, chemistry, biology and natural history). Applications and implementation of technology is more financially advantageous for greater schools because school licenses may be used throughout the school. The more classes will be involved, thus decreasing the cost of purchase applications per student.

Another fruitful source of electronic materials is web of DUMY (Digital material for educational purposes) [20]. Materials are free. But their form differs from materials created by EDULAB [17]. Free materials are less interactive than materials from EDULAB.

## 3.4  New Projects from European Union for Schools (Operational Program for Research and Development and Integrated Regional Operational Program)

Currently new possibilities of applying for projects are announced [21]. The key principle of the Operational Program Research, development and education is the development of human resources for the knowledge economy and sustainable development in a socially cohesive society and is supported by interventions in the context of several priority axes.

The highlighted issue in the new period is mainly improvement of the educational system in the Czech Republic. Compared to the period 2007–2013 greater focus will be on universities, linking modernization of research infrastructure (i.e. a hard activity) and the development of research teams and their capacity (i.e. soft activities).

Supported areas include e.g. equal access to quality preprimary, primary and secondary education and technical assistance.

## 4  Findings

In the Czech Republic there is a wide span in the approach, acceptance and utilization of SMART technology in the pre-primary and primary education. Nevertheless the strong trend into the implementation of SMART technology is undeniable which might be simply expressed as follows in two levels, at the level of government plan in the educational sector and the level of natural development where technology forms a natural part of our lives.

The number of kindergartens and primary schools which have acquired SMART technology equipment from the projects and donations currently prevails over those which do not have this kind of equipment, at all. There are great differences in utilization of already gained devices; some use these devices actively, some of them only sporadically. A considerable part of kindergartens and primary schools are still unaffected. The same is true for teachers in schools. There are still teachers who do not have any experience with SMART technology.

School top management is currently waiting for the new projects launched by the European Union for the program period 2014–2020. Many schools have problems with the finances and planning. For this reason, we decided to work on this paper which is focused on the economic aspects of the financing of introducing new technologies into the educational process. The article elaborated presumed costs, which are connected with introduction of new SMART technology into schools.

The cost of investment of SMART Technologies into one class may be around 115,000 CZK (4,259 EUR). As for primary schools, the costs will rise by eventual training of technical staff and purchase of tablets for students.

## 5  Discussion and Conclusion

Necessity of the introduction of tablets and interactive devices into kindergartens and primary schools in the coming years has been and will be a hot topic in the Czech Republic also due to the present situation; currently, according to data from the Czech Statistical Office there were 4.8 computers per 100 children in kindergartens in 2014 and 22.4 computers per 100 pupils in the first grade of the primary school and (39.2 computers per 100 pupils in secondary schools).

Out of these numbers just 1. 9 computers were available to pre-primary children and 16.7 to primary school pupils (27.3 to secondary school students).

If we focus on Computers with an Internet connection available to children in the kindergartens the number is even lower; there were 1.4 computers available. In the primary school the statistics shows 16.3 available computers (and 27 in the secondary schools). For this reason, we can see great potential in introducing modern technologies not only in kindergartens but also in primary schools.

Thanks to recently announced projects, an increase in wider implementation of SMART technology into schools and preparation of their staff can be expected.

The anticipated trend will be that there will be an enormous increase in creation of new teaching materials and that even more new companies dealing with digital media products will enter the business so that consequently this kind of products might become less costly.

It can be expected that if costs of the technology aren't covered by the project, parents should contribute to the tablet purchase and in this case they will seek the least costly option. It means that expenditure on the purchase of a tablet would be reduced even by half. Parents will likely seek the tablet price of about 3,000 CZK (111 EUR) + holster.

It can be presumed that if the school pays the technology from its own budget, it will be much more interested not only in search of affordable prices, but also in the systematic use of bought devices. On the Czech market there is available special equipment marked EDU which currently dominates the market. These products are usually more expensive than other competitive products because they are considered to be highly professional due to sophisticated marketing and promotion. Other offers may deliver products at lower prices and provide even higher quality. This does not apply to every offer. Prices of interactive boards and their equipment range from about 30,000 CZK up to 165,000 CZK (1,111–6,111 EUR). It depends on the parameters of the selected technologies used.

There is an issue which might raise a discussion; is there dependency between sources and spending? In case that the school will fund new equipment from the project or gift and not from own budget, it could be expected that the school will not consider the price of the equipment as a major criterion and we can assume that the project costs will rise.

Another question stems from the hypothetical situation when the kindergarten or school dares to invest more into the technical equipment than is necessarily required, e.g. into more powerful Internet to cover all institution premises. Will the costs finally increase or decrease? Could higher investments lead to lower operational costs?

**Acknowledgments** This work was supported by the Specific project at the Faculty of Informatics and Management from University of Hradec Kralove no. 2105/2016.

# References

1. Kopecky, J.: Free tablet for first-graders in the primary school and first graders in the secondary school again, dreams Chládek. http://zpravy.idnes.cz/tablet-pro-prvnacky-zadarmo-a-na-druhem-stupni-znovu-sni-chladek-pxj-/domaci.aspx?c=A150415_134634_domaci_kop. Accessed Jan 10 2016
2. Tablets to schools. http://www.csystem.cz/tablety-do-skol. Accessed Jan 10 2016
3. Key to education, Interactive Board in Kindergarten: Dream or Reality? https://www.klickevzdelani.cz/Veřejnost/Než-zazvoní/Pod-pokličkou-mateřských-škol/ID/22330/Interaktivni-tabule-v-materske-skole-Skutecnost-nebo-sen (2013). Accessed Jan 10 2016
4. Interactive Board in the kindergarten: We are playing and we are learning. https://www.youtube.com/watch?v=DAXn6-c95xg (2012). Accessed Jan 10 2016
5. Neumajer, O.: Tabooed tablets to schools. http://spomocnik.rvp.cz/clanek/18827/TABUIZOVANE-TABLETY-DO-SKOL.html. Accessed Jan 10 2016
6. Komarek, O.: Are our children threatened by digital dementia? The second part of the discussion: who really wants tablets? http://www.literarky.cz/politika/domaci/17332-hrozi-naim-dtem-digitalni-demence-druhy-dil-diskuse-kdo-o-tablety-vbec-stoji. Accessed Jan 10 2016

7. Klaus, V.: Free tablets to schools? And for pupils a pony with a ribbon—Vaclav Klaus jr. http://www.novinky.cz/komentare/368619-komentar-tablety-zdarma-do-skol-a-pro-zakyne-ponika-s-masli-vaclav-klaus-ml.html. Accessed Jan 10 2016

8. Goh, W.W.L., Bay, S., Chen, V.H.H.: Young school children's use of digital devices and parental rules. Telematics Inf. **32**(4), 787–795 (2015)

9. Fekonja-Peklaj, U., Marjanovic-Umek, L.: Positive and negative aspects of the IWB and tablet computers in the first grade of primary school: a multiple-perspective approach. Early child development and care **185**(6), 996–1015 (2015)

10. Manena, V., Rybenska, K., Spilka, R.: Research of mobile technologies used by students in primary school as a basis for the creation of teaching materials. In: Proceedings from International Conference on Advanced Educational Technology and Information Engineering, (AETIE 2015), Beijing, pp. 330–335 (2015)

11. Lapenok, M.V., Lapenok, O.M., Simonova, A.A.: Preparation and evaluation of teachers' readiness for creation and usage of electronic educational resources in school's educational environment. Smart Innov. Syst. Technol. **41**, 299–308 (2015)

12. Hopkins, L., Wadley, G., Vetere, F., Fong, M., Green, J.: Utilising technology to connect the hospital and the classroom: maintaining connections using tablet computers and a 'Presence' App. Aust. J. Educ. **58**(3), 278–296 (2014)

13. Gorhan, M.F., Oncu, S., Senturk, A.: Tablets in education: outcome expectancy and anxiety of middle school students. Kuram ve uygulamada egitim bilimleri **14**(6), 2259–2271 (2014)

14. Van Thienen, D., Sajjadi, P., De Troyer, O.: Smart study: pen and paper-based e-learning. Smart Innov. Syst. Technol. **41**, 93–103 (2015)

15. Wollscheid, S., Sjaastad, Jø., Tømte, C.: The impact of digital devices vs. Pen(cil) and paper on primary school students' writing skills—a research review. Comput. Educ. **95**, 19–35 (2016)

16. Srivastava, A., Yammiyavar, P.: Effectiveness of tangible and tablet devices as learning mediums for primary school children in India. Smart Innov. Syst. Technol. **35**, 353–363 (2015)

17. The annual report on the status and development of education in the Czech Republic in 2014. http://www.msmt.cz/file/36121/. Accessed Jan 10 2016

18. EDULAB. http://www.edulabcr.cz/. Accessed Jan 10 2016

19. Test of tablets 2016. http://www.testkvality.eu/test-tabletov/. Accessed Jan 10 2016

20. DUMY. http://dumy.cz/. Accessed Jan 10 2016

21. New challenges OP VVV and IROP. http://www.itveskole.cz/2016/01/10/nove-vyzvy-op-vvv-a-irop/. Accessed Jan 10 2016

# Part III
# Smart e-Learning

# An Integrative Approach of E-Learning: From Consumer to Prosumer

**Adriana Schiopoiu Burlea and Dumitru Dan Burdescu**

**Abstract** This article analyses the challenges of the e-learning platform Tesys for both students and teachers in a Smart environment. In our research, two theoretical approaches are used to develop the hypotheses related to the relationships between user and technology: the relationship user-prosumer and the relationship principal-agent. Therefore, our quantitative and qualitative research put into value a strong relationship between user and technology. We arrived at the conclusion that the entrepreneurial behaviour of the agent is positively related to the Tesys features and also that the agent's satisfaction is positively related to the principal feedback. We find that the principal-agent theory offers an integrated approach of the e-learning system in a Smart environment, and that a lack of agents control coupled with the lack of the principal's motivation generated artificial barriers difficult to overcome even with Smart technology.

**Keywords** E-learning platform Tesys · Agency theory · Prosumer · User's satisfaction · Smart environment

## 1 Introduction

The relationship between principal and agent in e-learning is oriented towards increasing the user's satisfaction and it is related to the structure of the e-learning platform Tesys which transforms the learning process in an instrument able to develop the entrepreneurial competencies of the users.

A.S. Burlea (✉)
Department of Management, Marketing and Business Administration,
University of Craiova, Craiova, Romania
e-mail: adriana_burlea@yahoo.com

D.D. Burdescu
Department of Computers and Information Technology,
University of Craiova, Craiova, Romania
e-mail: dburdescu@yahoo.com

© Springer International Publishing Switzerland 2016
V.L. Uskov et al. (eds.), *Smart Education and e-Learning 2016*,
Smart Innovation, Systems and Technologies 59,
DOI 10.1007/978-3-319-39690-3_24

Many studies are based on the evaluation of the digital content for the e-learning system and the importance of the principal-agent interaction in terms of acquiring new competencies is neglected [1, 2].

The principal-agent relationship in the e-learning system leads to a new challenge: to use Smart technology to achieve a motivational learning climate and to develop, at the same time, an individualized and group learning activities. Meanwhile, the principal is able to provide a positive individualized feedback only if he/she transforms the platform in a "friend".

The e-learning platform Tesys is realised by the member of the Research Centre and it is used by the students from the University of Craiova. Since 2005 the platform has continuously improved and its facilities generate challenges for the students and the teachers [3–6]. Therefore, our quantitative and qualitative research put into value the relationship between user and technology.

In our research, two theoretical approaches are used to develop the hypotheses related to the relationships between user and technology.

First, the relationship user-prosumer is used as a basis for a model that predicts the role of the platform Tesys as a tool to promote innovation and entrepreneurial skills. We argued that the platform Tesys is a mechanism used for managing learning process characterised by high levels of interdependence and interconnection. Second, the relationship principal-agent will be sometimes relayed on teachers' level of expertise and sometimes relayed on user/students' competencies. The results indicate that the platform has limited feature for developing entrepreneurial skills to users. Therefore, it is necessary to use additional technology for developing and improving entrepreneurial skills.

We will analyse the relationship between the proactive e-learning strategy and the e-learning platform Tesys integration when there are some obstacles that hinder the principal-agent relationship.

## 2 How to Transform the Challenges in Opportunities in the Smart Environment

In 1976, Jensen and Meckling defined the principal-agent relationship as a contract between the principal and the agent. In this contract 'the principal(s) engages another person (the agent) to perform some service on their behalf which involves delegating some decision making authority to the agent' [7, p. 308].

In the e-learning system the relationship between principal and agent presents some particularities:

- The teacher (principal) engages the students (agents) to perform the knowledge and to achieve satisfactory results in their learning process.
- The interactions between principal and agent are mediated by the platform Tesys.

- The relationship between principal and agent is formalised in order to clearly define the agent's needs and to put into value the principal's efforts to meet these needs.
- The e-learning activity has the role to create agents for helping the principal to share the knowledge and to deal with the inadequacies of the system.
- Taking agency theory and the technology (e-learning platform Tesys) as points of departure, we propose some solutions that explain the relationship between user and technology, and certain aspects of the user's (as agent) and the teacher's (as principal) strategic behaviour.

## 2.1 The Role of the Prosumer in Smart Environment

In her work about prosumer, Adriana Schiopoiu Burlea [8, p. 3] tried to find the answer to the dilemma if the prosumer is a consumer or a stakeholder for the organisation and she arrived at the conclusion that the prosumer 'is a new customer that is very well informed for realising that, on the market, there are no products of technical level that satisfy his/her needs. Therefore, the prosumer used his/her abilities and competencies to produce a personalised product and pushes the organisations to elaborate a mass prosumption strategy that includes a mass-customization strategy...' and as a result, the prosumer 'is a promoter of the change in organisations'.

In this context, considering the e-learning system particularities, the role of the prosumer is taken, alternatively, by both the principal and the agent.

*The principal* is motivated by the 'desire to gain control over the product (especially, over the Internet technology and web sites)' and to develop personal skills of the agent [8, p. 3]. For these reasons, the principal becomes a prosumer for knowledge that contributes to the learning process, and for some modules of the platform Tesys, including software and other resources that shape the Tesys platform.

*The agent* being motivated by a variety of desire (e.g. desire for information, desire to acquire new knowledge, desire to save time and money, desire to gain greater confidence in the product that is being used—[8, p. 3]) is directly interested to develop his/her entrepreneurial competencies.

The platform Tesys features play an important role in developing the entrepreneurial behaviour of the agent (e.g. content of the courses, the architecture of some modules of the e-learning platform Tesys, self assessment, and forum). If the principal is motivated to be a prosumer for courses content (production of knowledge), the agent is more interested to become a prosumer and to use his/her creativity for improvement of the platform (improvement of the technology features).

**Hypothesis 1** The entrepreneurial behaviour of the agent is positively related to the platform Tesys features.

## 2.2 The Role of Opportunism in Building a Strong System of Knowledge

We argue that when the agents are not monitored and supported by the principal, the evaluation may not be the norm and agents may prefer to abandon the e-learning system. In some cases, overconfident agents sometimes obtain very good results and tend to overinvest in their knowledge to consolidate their reputation as good agents and to reinforce the principal trust. This is a critical point when the technology can be used as a control mechanism that should reduce the agent's opportunism [9, 10].

In the e-learning literature there is a dispute between the actors (e.g. student/agent or teacher/principal) and their role in developing creative resources [11, 12]. The e-learning platform has the role to develop a type of 'agent control' that provides principals with the trust to spread knowledge for the benefit of the agent. The efficiency of the agent's control is measured by the timely feedback provided by the principal to the agent. In addition, the agent has been given control over knowledge and the outcomes for the agent's control are competencies-based. The agent's control should be related to the individual's objectives to have a positive impact on his/her entrepreneurial behaviour. Therefore, the university should develop a proactive e-learning strategy and take into account the barriers that hinder the development of such strategy. The barriers, such as the complexity of the e-learning platform Tesys and the lack of principal-agent communication are difficult to overcome. The agents' perception of barriers is different from the principal's perception and this perception depends on the e-learning strategy of the university. The agents have problems related to the lack of learning resources, whereas the principal struggles with the lack of the agents' interest.

In this context, the principal has to transform the e-learning environment from a rigid one (characterised by lack of possibility for the agent to become a prosumer at two levels: technological and production of knowledge) into a flexible environment (e.g. a Personal Learning Environment—PLE that promotes a prosuming behaviour) and to provide an efficient feedback to agents for improving their satisfaction.

**Hypothesis 2** The agent's satisfaction is positively related to the principal feedback.

We have used the agency theory as a method to explain the importance of the agent's control in the e-learning process and to increase the principal's involvement in the knowledge transfer.

The role of the principal is to avoid the opportunistic behaviour of the agent through technology, and at the same time, to promote the prosuming behaviour of the agent (see Fig. 1).

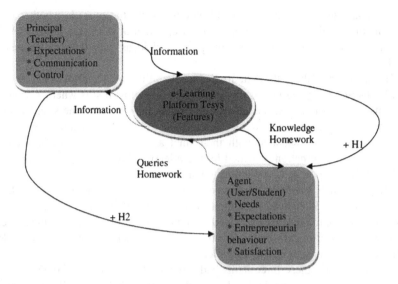

**Fig. 1** The model of the principal-agent relationship in the e-learning system

## 3 Methods

### 3.1 Sample

The sample, composed of 206 students from the Faculty of Economics and Business Administration, was used to cross-validate the results. Two hundred and six students enrolled in the University of Craiova volunteered to fill the questionnaire.

The structure of the sample of 206 platform users, most respondents 67 % (138) were females, and 33 % (68) were males. The age of the respondents was between 20 and 29 years of age and the mean was 23.33 years of age.

### 3.2 Measures

All ratings of the items were made on a five-point Likert scale (1 = strongly disagree to 5 = strongly agree).

*The features of the e-learning platform Tesys* were measured by three items: one is about the access on the platform; the second is about the flexibility of the platform; and the last is whether or not the students are feeling lost in the navigation process on the e-learning platform Tesys.

These three items had a reliability coefficient of 0.93.

*The entrepreneurial behaviour of the agent* was measured by the following items: one is whether or not the platform provide the students with the opportunity to communicate their opinions and ideas effectively; the other is whether or not students find motivation to develop their entrepreneurial skills; the third is whether or not students can exchange the entrepreneurial ideas with their colleagues and teachers.

Taken together, these three items had a reliability coefficient of 0.94.

*The agent's satisfaction* was measured by three items: first is whether or not the students are satisfied with the information provided by the teachers and with the quality of the interactions between them and the teachers; the second is whether or not the students are satisfied with their final assessment; the third is about the individualised information provided by the e-learning system to the students.

These three items had a reliability coefficient of 0.93.

*The principal's feedback* was measured by the following items: one is whether or not the feedback received from the teacher was timely for the students; the other is whether or not the feedback received from the teacher was useful for the students; the last is whether or not the involvement of the teacher in providing the students with valuable information was tangible.

Taken together, these three items had a reliability coefficient of 0.95.

Cronbach's alphas are calculated for each construct and the results are over 0.90 (0.932 and 0.946, respectively). This suggests a strong reliability and validity of the study, exceeding the threshold value of 0.70 recommended by Nunnally [13].

## 3.3   Results, Discussion and Findings

The evaluation of the measurement model was realised by a confirmatory factor analysis (CFA) and we arrived at the conclusion that the model fit is good and the root mean squared an error of approximation [RMSEA] = 0.07; confirmatory fit index [CFI] = 0.98 [14].

Table 1 presents the means, standard deviations, and correlations among the variables related to the e-learning platform Tesys in the principal-agent relationship.

As shown in Table 1, all correlations are greater than 0.8 and this supports our model. Therefore, the practical model had a good fit and all of the items had statistically significant ($p \leq 0.01$) loadings on their intended construct.

**Table 1** Descriptive statistics and correlations

| Variable | Mean | SD | 1 | 2 | 3 | 4 |
|---|---|---|---|---|---|---|
| Features of e-learning platform Tesys | 4.06 | 0.46 | (0.93) | | | |
| Entrepreneurial behaviour of the agent | 3.78 | 0.51 | 0.84 | (0.94) | | |
| Agent's satisfaction | 3.85 | 0.47 | 0.87 | 0.86 | (0.93) | |
| Principal' feedback | 3.96 | 0.56 | 0.85 | 0.87 | 0.84 | (0.95) |

[a]Numbers in parentheses on the diagonal are the Cronbach's alpha in coefficients of the composite scales

The e-learning system involves a strong principal-agent relationship that positively guides the ongoing development of entrepreneurial behaviour, based on creativity, confidentiality and trust [15].

An agency relation between user and teacher in e-learning can build both theoretical and empirical foundation for the development of the learning organisational strategy. The teacher is in relationship with the students and with the technical team of the e-learning platform Tesys and these relationships are important to the basic e-learning process.

In agency relation, the agent (student) has the obligation to serve the interest of the principal (teacher), and also the principal the duty to satisfy the needs of the agent [16].

**Features of the e-learning platform Tesys**. The agents which participated in this study had a mean score of 4.06 (SD = 0.46). Based on the descriptive analysis, the agents' efficiency is influenced by the features of the e-learning platform Tesys. The agents which have little experience in using the platform are interested in developing their skills and try to transform their gap in a motivational factor for personal development.

**The agent's entrepreneurial behaviour**. The agents had a mean score of 3.78 (SD = 0.51) and the three items specific means ranged from 3.53 to 3.92 and suggest that the agents' entrepreneurial behaviour was moderately good to good. The agents mostly agreed with the following statement: the information found on the e-learning platform Tesys provided me motivation to develop my entrepreneurial skills (M = 3.92, SD = 0.697).

The multiple regression analysis conducted examined whether the entrepreneurial behaviour of the agent is positively related to the platform Tesys features we arrived at the conclusion that the features of the e-learning platform Tesys made a significant contribution to the prediction of the agents' entrepreneurial behaviour ($\beta = 0.46$, $p < 0.001$).

**Hypothesis 1** Predicts that the entrepreneurial behaviour of the agent is positively related to the platform Tesys features. We found support for Hypothesis 1($\beta = 0.46$, $p < 0.001$).

The entrepreneurial behaviour of agents depends on the effect of knowledge on their needs and expectations, on the collaboration of the principal, and on the features of the e-learning platform Tesys. Therefore, the agent should avoid the opportunistic behaviour and establish a trusty relationship with the principal [17].

In this situation, the e-learning platform Tesys offers to the principal greater amounts of control over the agent learning. Paradoxically, this technological control provides agents with the freedom to selecting knowledge.

This approach is empirically-oriented analysis of the principal-agent relationship, because both the principal and the agent try to perform tasks desired by each other [18]. In the e-learning case, the efficiency is measured by the final evaluation received by the agent from the principal, and by the principal from the agent.

**Agent's satisfaction**. The agents which participated in this study had a mean score of 3.85 (SD = 0.47) and the three items specific means ranged from 3.65 to

4.14 and suggest that the agents' satisfaction was good to very good. The agents were very satisfied with their final assessment (M = 4.14, SD = 0.349) which means that the entire e-learning system had a motivational effect on the agents' learning behaviour, even if the following statement: the e-learning system developed on platform Tesys provided the individualised information was good (M = 3.65, SD = 0.588).

**Principal's feedback**. The agents had a mean score of 3.96 (SD = 0.56) and the three items specific means ranged from 3.73 to 4.09 and suggest that the feedback received from the teacher was good to very good. The agents considered that the following statements were very good: the feedback received from the teacher was useful for me (M = 4.09, SD = 0.598) and the involvement of the teacher in providing me with valuable information was tangible (M = 4.06, SD = 0.630). The main problem was registered in relation with the delay in the feedback, because the agents considered that this attribute of the e-learning system should be improved (M = 3.73, SD = 0.634).

Overcoming the problem of information asymmetry by the principal represents a positive point for the agency theory and develops the agents' trust that reconsiders the mechanism designed to control their behaviour. Therefore, the two parties to the relation are assumed to hold learning and the organisational interests and to reduce the conflict of interest between the principal and the agent such that the principal's feedback positively impacts the agent's satisfaction. Sometimes, if a feedback is not timely could lead to the misunderstanding of the message and the agents could be lost and lack the motivation. Therefore, a timely and useful feedback could build a strong relationship between the principal and the agent and reinforce the agents' motivation (the principal rewards the agent's effort) in developing their learning and entrepreneurial behaviour.

The multiple regression analysis conducted examined whether the principal's feedback positively influences the agent's satisfaction and we arrived at the conclusion that the timely feedback of the principal made a significant contribution to the increasing satisfaction of the agent ($\beta = 0.49$, $p < 0.001$).

**Hypothesis 2** Predicts that the agent's satisfaction is positively related to the principal's feedback. We found support for Hypothesis 2 ($\beta = 0.49$, $p < 0.001$).

The performance of the agent is poor and he/she is self-interested to maximize his/her results without regards for the other (e.g. principal, colleagues). The agent builds his/her relationship on competition and his/her criterion of success is measured in final evaluation note. Sometimes, the agent expects the worst from the principal for justifying his/her unsuccessful behaviour [19].

As a conclusion, the e-learning system in Smart environment involves human and technological interactions based on dual agency relationship. The e-learning platform Tesys is a tool that helps the principal and the agent to start making decisions. This tool has software that transforms it in a technological agent that does some activities on behalf of the teacher or of the student. Therefore, the e-learning platform becomes a technological agent.

In the e-learning system, the particularity consists of the student that acts as trustworthy agent for the teacher and has the moral obligation to behave in the principal's best interest and also in his/her own interest. This particularity sometimes has a negative influence on the agency relationship and generates a principal-agent problem.

## 4 The Practical Implication of the Research

We find that the principal-agent theory offers an integrated approach of the e-learning system, and that a lack of agents control coupled with the lack of the principal's motivation generated artificial barriers difficult to overcome.

In a Smart environment, the control of the agent on the platform Tesys reduces the role of the principal to the simple moderator/instructor. Therefore, the agency theory is useful to highlight the systemic relationship between impacts of the principal's efforts on the agent's needs.

In the e-learning process it is very important to create and to develop the motivation of both the agent and the principal to involve and to maintain an agency relationship.

The platform Tesys controls the results of the agent and the role of the principal is reduced to the moderator/instructor. Therefore, we have identified tactics that the principal can employ to solve the problems that can arise in the agency relationship during the e-learning process, as following:

- To encourage the creativity of the agent based on the e-learning platform features.
- To create and share knowledge and ideas through the e-learning platform and to make these knowledge available on the other multimedia devices.
- To focus the main e-learning activities on the agent's needs and to provide a timely feedback.
- To collaborate with the agent for translating the learning initiatives into strategic alignment for other users.
- To use the e-learning platform leverage system for developing a scalable evaluation system for the agent
- To reduce the pressure of the time and to improve the flexibility in time spent on courses by the agent

The principal-agent approach of the e-learning system in Smart environment brings a new light on the role that the principal plays in the creative and innovative process at organisational level. The principal being aware that whenever he/she can change the place with the agent, he/she will be pushed to become a prosumer and to adapt the answers to the agent's needs and expectations. Instead, the agent must be aware of the principal's effort and not consider this involvement of the principal as something habitual.

The agent, through the e-learning platform Tesys, might evolve from the exploratory stage to the stage of the entrepreneur prosumer and later to the stage of the virtual world prosumer [8, p. 4], [20, 21]. The usability of the e-learning platform in a Smart environment is related to both the main goal of the Smart educational research and to the business environment and agents manifests their opportunist behaviour with aversion to risk.

The maximization of the learning process efficiency allows the control of the agent's performance and the entrepreneurial behaviour is highly motivated by the field of control and programmability of the learning tasks performed by the agent and by some degree of uncertainty of results.

## 5 Conclusions

We contend that this approach of the e-learning in Smart environment focuses on the principal-agent relationship could be compatible with the value of the universities.

The principal-agent relationship does not fail to acknowledge the e-learning context in which the Tesys platform provides not a barrier but an opportunity for extending the understanding of the agency theory to a variety of learning contexts. Thus, even if agents could differ in their needs and competencies, sometimes these differences may result in their entrepreneurial behaviour. The principal plays a delicate role in this mechanism, because he/she never knows the full extent to which the knowledge converge or diverge with technology (e-learning platform Tesys), and when the opportunism of the agent occurs in the dynamic learning process.

We suggest that the principal-agent relationship in Smart environment is not a restrictive one and future research could be oriented to the impact of imperfect asymmetric information on developing of the entrepreneurial behaviour of the agent and on building trust of the principal.

## References

1. Politis, D.: The process of entrepreneurial learning: a conceptual framework. Entrepreneurship Theory Pract. 29(4), 399–423 (2005)
2. Tacu, M.G., Mihaescu, M.C., Burdescu, D.D.: Building professor's mental model of student's activity in on-line educational system. In: The Third International Conference on Cognitonics, (Cognit-2013), Ljubljana, Slovenia (2013)
3. Burdescu, D.D.: TESYS: an e-Learning system. In: The 9th International Scientific Conference Quality and Efficiency in e-Learning (2013)
4. Burdescu, D.D., Mihaescu, C.M.: Building intelligent e-learning systems by activity monitoring and analysis. In: Tsihrintzis, G., Jain, L. (eds.) Multimedia Services in

Intelligent Environments, Smart Innovation, Systems and Technologies, vol. 3, pp. 153–174 (2010)

5. Burlea Schiopoiu, A., Badica, A., Radu, C.: The evolution of e-learning platform TESYS user preferences during the training processes. In: ECEL2011, 11–12 Novembre, Brighton, UK, pp. 754–761 (2011)

6. Burlea Schiopoiu, A.: The complexity of an e-learning system: a paradigm for the human factor. In: The Inter-Networked World: ISD Theory, Practice and Education, vol. 2, pp. 267–278. Springer (2008)

7. Jensen, M.C., Meckling, W.H.: Theory of the firm: managerial behavior, agency cost and ownership structure. J. Financ. Econ. **3**, 305–360 (1976)

8. Burlea Schiopoiu, A.: The challenges of the prosumer as entrepreneur in IT. In: Pankowska, M. (ed.) Frameworks of IT Prosumtion for Business Development, pp. 1–15. IGI-Global (2014)

9. Bartikowski, B., Walsh, G.: Attitude contagion in consumer opinion platforms: posters and lurkers. Electron Markets **24**, 207–217 (2014)

10. DeRouin, R.E., Fritzsche B.A., Salas, E.: E-learning in organizations. J. Manage. **31**(6), 920–940 (2005)

11. Henry, P.: E-learning technology, content and services. Education + Training **43**, 249–255 (2001)

12. Norman, D.A., Spohrer, J.C.: Learner-centered education. Commun. ACM **39**(4), 24–27 (1996)

13. Nunnally, J.C.: Psychometric Theory, 3rd edn. McGraw-Hill, New York (1994)

14. Anderson, J.C., Gerbing, D.W.: Structural equation modeling in practice: a review and recommended two-step approach. Psychol. Bull. **103**(3), 411–423 (1988)

15. Burgess, J.R.D., Russell, J.E.A.: The effectiveness of distance learning initiatives in organizations. J. Vocat. Behav. **63**, 289–303 (2003)

16. Heath, J.: The uses and abuses of agency theory. Bus. Ethics Q. **19**(4), 497–528 (2009)

17. Caers, R., Du Bois, C., Jegers, M., De Gieter, S., Schepers, C., Pepermans, R.: Principal-agent relationships on the stewardship-agency axis. Nonprofit Manage. Leadersh. **17**(1), 25–47 (2006)

18. Wiseman, R.M., Cuevas-Rodríguez, G., Gomez-Mejia, L.R.: Towards a social theory of agency. J. Manage. Stud. **49**, 202–222 (2012)

19. Ghoshal, S.: Bad management theories are destroying good management practices. Acad. Manage. Learn. Educ. **4**(1), 75–91 (2005)

20. Malaby, T.M.: Making Virtual Worlds: Linden Lab and Second Life. Cornell University Press, Ithaca (2009)

21. Zwick, D., Bonsu, S.K., Darmody, A.: Putting consumers to work: co-creation and new marketing govern-mentality. J. Consum. Cult. **8**, 163–196 (2008)

# Using E-Learning in Teaching Economics at Universities of the Czech Republic

Roman Svoboda, Martina Jarkovská, Karel Šrédl, Lucie Severová and Lenka Kopecká

**Abstract** The rapid development of information and communication technologies (ICT) allows their use in all areas of human activity. An increasing number of university students is also reflected in the conditions of the universities in the Czech Republic. However, current financial possibilities of the educational and research capacities of departments do not suffice to a rapid increase in the number of students. A solution can be the introduction of ICT into a teaching process; computers save time for both the teachers and students while enabling intensive use of space and simplifying the administration of tests. The paper aims to express the impact of using ICT on level of knowledge, respectively classification of university students enrolled in course Economics on the example of e-learning system Moodle. A partial aim of the study is the generalization of experience with learning management systems and the expression of their positives and negatives. As shown by our research the introduction of smart education from 2006 to 2015 first led to a drop and subsequently to an improvement in the quality of the students' performance.

**Keywords** e-learning education universities · ICT · Moodle

R. Svoboda (✉) · M. Jarkovská · K. Šrédl · L. Severová · L. Kopecká
Faculty of Economics and Management, Czech University of Life
Sciences in Prague, Prague, Czech Republic
e-mail: svobodar@pef.czu.cz

M. Jarkovská
e-mail: jarkovska@pef.czu.cz

K. Šrédl
e-mail: sredl@pef.czu.cz

L. Severová
e-mail: severova@pef.czu.cz

L. Kopecká
e-mail: kopeckal@pef.czu.cz

© Springer International Publishing Switzerland 2016
V.L. Uskov et al. (eds.), *Smart Education and e-Learning 2016*,
Smart Innovation, Systems and Technologies 59,
DOI 10.1007/978-3-319-39690-3_25

# 1 Introduction

It is necessary to change the school's focus on traditional knowledge transfer methods to dealing with problem of how are information processed and used. Recently are gaining popularity comprehensive e-learning systems as a public ICT projects in field of education, which includes both the teaching part and the subsequent verification of students' knowledge [2].

The rapid development of information and communication technologies (ICT) allows their use in all areas of human activity. Contents of the knowledge acquired during the studies of individual branches must be regularly upgraded, due to the emergence of new knowledge. It is necessary to transfer such information to end users as quickly as possible so they can continue to use it. Therefore, the implementation of ICT and information systems into learning has become an important priority throughout school systems in developed countries as well as the need of corporate education.

According to OECD statistics, in the last 10 years the number of university graduates in the Czech population increased about 3 percentage points to 14 %; citizens of developed European countries, however, are twice as educated. If the current share of university educated people in the population of developed market economies amounts to around 28 % (e.g. Finland has already reported a 75 % share of tertiary education in a completed education of the population), not even the Czech economy can avoid this important transformation in human resources [3]. A growing number of students in all forms of study at FEM CULS Prague relates to the above-mentioned efforts to increase the number of university students and, consequently, a higher proportion of university graduates among the entire Czech population. It's not just this quantitative aspect but also the preservation of quality education at a growing number of university graduates.

# 2 Materials and Methods

## 2.1 Literature Overview and Aim of the Paper

Despite the growing number of students, the required level of teaching must be maintained; it is therefore necessary to deal with new ways of teaching and learning. A suitable solution that can reduce the need for human resources appears to be the introduction of new ICT in the educational process and the related new methods of checking students' knowledge [10].

Issues of efficiency of public projects in field of education are also dealt by Ochrana in his works "Theory and Practice of the Public Sector Savings: the Case of Czech Regions" [8] a Starting Points for Creating a Transparent and Non-corruption Public Procurement System [9].

The paper aims to express the impact of using ICT on level of knowledge, respectively classification of university students enrolled in course Economics on the example of e-learning system Moodle. A partial aim of the study is the generalization of experience with learning management systems and the expression of their positives and negatives.

## 2.2  Trend in the Area of E-Learning

E-learning is developing at a platform called Learning Management System (LMS). "LMS is a web-based platform to support the management of the educational process" [12]. Functionally LMS enables the creation and manipulation of the content of courses, study monitoring and evaluation of results. Development and operation of courses takes place precisely in this environment. Among the best-known LMS belong Canvas, Blackboard, Moodle, Sakai, SharePoint, Fronter. "All the LMS support the creation of instruction materials, tests, organization of discussion groups, assessment and grading, and the flexibility to manipulate the content of the courses. The main difference among LMS is in the availability of individual functions" [6].

Czech University of Life Sciences Prague (CULS) and its largest faculty, the Faculty of Economics and Management (FEM), are by their nature of courses a suitable environment for the implementation of the LMS. The faculty has a significant number of students to the existing number of teachers. The faculty annually offers basic (core) courses which, when taught by a classical method of teaching and assessment, put very high demands on the teachers time-wise. Therefore, in 2009, the CULS Prague launched LMS "Moodle" [5].

In the first period, the use of the system mainly focused on the development of Moodle supported exam modules for most core courses (including Economics). The aim was to speed up and objectify the examination process during the examination period. The second stage of the development of the courses on the Web platform began in 2011 and is still ongoing.

Moodle is a software package designed to support full-time and distance learning through online courses available on a website [7]. It is a comprehensive e-learning software for universities offering a variety of functions with the possibility of its customization for the needs of a particular school. Moodle allows for example the insertion of an outline for a course, use of the keys to enroll in a course, pasting files into the course, dividing students into groups according to a course teacher, submitting tasks using a Web interface, entering students' surveys, attendance, establishing forums and, finally, it offers the creation of tests and their subsequent generating to be used by students within their written exam.

The main advantage of **smart education** lies in its accessibility. A Smart, multi-disciplinary student-centric education system—linked across schools, tertiary institutions and workforce training, using adaptive learning programs and learning portfolios for students; collaborative technologies and digital learning resources for teachers and students; computerised administration, monitoring and reporting to keep teachers in the classroom; better information on our learners and online learning resources for students everywhere [11].

Our system can be logged into from wherever there is a computer with the Internet access. Hence its most typical use in teaching the students of distance learning. Teachers can communicate with their students in the form of public statements within the course, by establishing a forum or sending private messages. In addition to the actual subject curriculum and basic information, the course may be enriched by a variety of other learning materials, including lectures and model examples. This lowers requirements for regular commuting to school either by a student or teacher and no individual communication between a student and teacher via e-mail is necessary [5].

While preparing courses in Moodle may seem to teachers superfluous and redundant in connection with the standard instruction based on participation in lectures and seminars, the use of Moodle as a tool for checking students' knowledge is a great automation and simplification of teachers' work. In the preparation and application of exam tests, the function of the teacher is narrowed to the problem of making an examination database. Moodle itself generates tests for students from the database and evaluates them. A distinct advantage of Moodle is its variability of examination questions formation. Moodle offers an extensive range of task types. The most common task is a multiple choice question; with such questions it is possible to select what percentage of the total number of points per question a student receives for individual answers. A practical type of task, such as in mathematical tasks, is a question enabling students to write a particular calculation result as an answer. As a question parameter, the creator of the question enters an interval in which the specified outcome is still correct. Frequently used are also additional or top-up questions, where students are asked to write a word or a whole sentence. It is in these cases where the application of the percentage of total points per questions determined by the quality of the answers is welcome. The disadvantage is the impossibility of using negative points. However, this Moodle imperfection can be avoided by moving the overall marking boundary. For example, if we wanted to allow students to leave test questions unanswered (i.e. in essence the answer "I do not know"), in the system of negative points we would set up a zero number of points for such an answer. In Moodle, however, half the number of points is set up for such answers and the boundary of marking is shifted from its half [5].

Moodle stores the results of students' tests in the database from where they can be recalled at any time. It also allows exporting a complete list of results into Excel spreadsheets for statistical purposes. Moodle itself calculates the average of the test results and processes the statistics of specific successful answers to some questions [5].

# 3 Results and Discussion

## 3.1 Preparation of Students' Knowledge Verification Through ICT and Sample Formulation of the Test

Based on the analysis of the range of teaching materials, the number of qualified teachers for the testing and the extent of working hours and availability of classrooms the Department of Economic Theories has at its disposal, the following findings were elicited [2]:

- The range of teaching material offered approximately corresponds to a standard course accredited with the 1 A classification of the faculty, similarly to all faculties of the University of Economics, Prague. The scope of the curriculum thus cannot be reduced. On the other hand, the number of lessons and students' homework—the equivalent of one semester, corresponds to the half of the capacity of teaching hours, unlike at the University of Economics, Prague.
- The difficulty of the computer tests is set by ourselves; technology only makes work easier. And if all of the curriculum cannot be converted into a computer (software) form, a remaining part of the knowledge can be verified by oral examination.
- The system of students' knowledge verification by examinations had already been previously introduced by other departments with convincing positive results. That a test can be prepared even from economic theories is confirmed by the experience of developed countries; e.g. in the United States at Oregon State University (Corvallis), where a test of knowledge on the subject of economics was thus introduced in the previous century, and this was not by far the first case in the United States.
- In the course of microeconomics a test was created not only covering the entire material taught, but also allocating it into 13 chapters. The 13 chapters, randomly generated by the computer, contain 8 quiz questions, 3 examples and 2 graphs. Each test variant in the text is assessed 8 points per 4 answers, while individual questions are posed in the form of the expected yes—no—I do not know answer assessed by two points. An illustrative example of a sub-task is as follows:
Yields from variable input:

(a) Refer to a long-term period

   – not correct

(b) are declining in a situation where production is growing faster than input

   – not correct

(c) are declining in a situation where production is growing slower than input

   – correct

(d)  refer to a short-term period

    – correct

• System speed and time savings associated with it. Students learn the result immediately after the computer test, and do not wait long for the announcement of the result, which better reflects the validity of performance in a follow-up oral exam with actual knowledge. The computer itself informs about the result: failed, passed with certain classification and percentage of correct answers. The examiner then performs real intellectual work he or she is destined for; an activity the student finds both evolving and motivating while being examined.

If we evaluate the results of tests processed using information technology, we can say that in terms of content and extent of the materials taught in a course, using this type of the test seems adequate and fully appropriate to the content of the course.

## 3.2  Comparison of Test Results and Analysis of ICT Impact

Comparison of test results from macroeconomics in three observed school years (2006/2007, 2007/2008 and 2014/2015) clearly shows the difference in testing students using conventional written tests and the use of modern ICT for distance learning. The test took place in the summer semester (Fig. 1, Table 1).

In 2006/2007 the total of 307 distance learning students registered for the subject of Economics; 216 students attended the test. After three attempts at the exam an average grade 2.86 was detected for the course.

In 2007/2008 the total of 330 distance learning students registered for the subject of Economics; 196 students attended the test. After three attempts at the exam an average grade 3.25 was detected for the course.

**Fig. 1** Classification from Economics (own data and calculations, 2016)

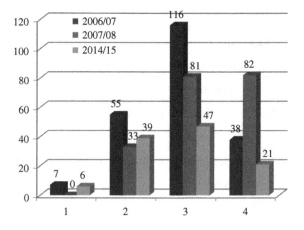

**Table 1** Classification from Economics (own data and calculations, 2016)

| Classification | | 1 | 2 | 3 | 4 | Students in total |
|---|---|---|---|---|---|---|
| 1st attempt | Written 2006/07 | 7 | 34 | 66 | 109 | 216 |
| | Computer-based 2007/08 | 0 | 19 | 47 | 130 | 196 |
| | Computer-based 2014/15 | 4 | 27 | 26 | 56 | 113 |
| 2nd attempt | Written 2006/07 | 0 | 17 | 37 | 28 | 82 |
| | Computer-based 2007/08 | 0 | 10 | 25 | 60 | 95 |
| | Computer-based 2014/15 | 2 | 8 | 13 | 26 | 49 |
| 3rd attempt | Written 2006/07 | 0 | 4 | 13 | 2 | 19 |
| | Computer-based 2007/08 | 0 | 4 | 9 | 29 | 42 |
| | Computer-based 2014/15 | 0 | 4 | 8 | 7 | 19 |
| Final mark | Written 2006/07 | 7 | 55 | 116 | 38 | 216 |
| | Computer-based 2007/08 | 0 | 33 | 81 | 82 | 196 |
| | Computer-based 2014/15 | 6 | 39 | 47 | 21 | 113 |
| Average mark | Written 2006/07 | 2.86 | | | | |
| | Computer-based 2007/08 | 3.25 | | | | |
| | Computer-based 2014/15 | 2.73 | | | | |

In 2014/2015 the total of 165 distance learning students registered for the subject of Economics; 113 students attended the test. After three attempts at the exam an average grade 2.73 was detected for the course.

As is clear from comparing the above-mentioned mark diameters—2.86 before the introduction of ICT into the students' knowledge testing and 3.25 after the introduction—a significant difference in classification was found, which can confirm previous reflections on higher objectivity when testing students using ICT. Similar test results were reported in the following school year.

After seven years of using the e-learning system, the average grade improved significantly from 3.25 to 2.73, which was even ahead of the average grade of the written tests. It is obvious that the students adapt to a new way of testing and began to prepare better.

In terms of the number of teachers involved in student testing, this saves time especially during corrections, a manual activity which employed a substantial portion of teachers. The transition to the new knowledge testing does not place demands on space, on the contrary, it allows for an efficient use of computer laboratories during the exam period.

The experience with the introduction of this form of testing is similarly positively evaluated at educational institutions. Experience from universities in developed countries, particularly the United States, point at an overall backwardness of some Eastern European countries, not only with the introduction of ICT, but especially with its implementation in practical activities. It is estimated that those EU countries face up to 15 years of delay in contrast to the USA. Such a negative trend must be reversed, and even in our mentioned application.

### 3.3  Advantages and Disadvantages of E-Learning Systems

The trend of e-learning systems is currently growing stronger as well as the development of new ICT and computer skills among the population of developed countries. The use of e-learning systems suggests a number of important advantages:

- *The absence of the physical presence of a student or teacher in a university building.* This reduces the cost of commuting to school either by the student or the teacher and also the costs of the university rent and energy.
- *The possibility to study anywhere and at any time of day and night.* Students can enter the e-learning course regardless of public holidays or the time of day, and anywhere. They are only limited by the Internet access.
- *Self-paced learning. Students can customize their learning pace to fit their needs.* If necessary, they can spend more time studying complex lessons or study them again.
- *Interactive form of teaching and learning.* The course can be enriched by various modules involving students in active learning. This may be for example a survey providing a necessary feedback to the teacher.
- *Diverse communication channels for contact with the teacher and other students.* If the teacher sets the possibility of forum or chat in the course, students can receive quick responses not only from the teacher, but also from other students.
- *A quality, already created course can be reused with minimal demands for its adjustment.*
- *Saving the teacher's time at testing the student's knowledge.* Computer-generated tests and automatic correction of tests will allow the instructor to monitor the achievements of students without the instructor's own involvement in the process of students' knowledge verification.
- *Simple update of the course content.* Unlike for example printed scripts, e-learning text can be easily adjusted to fit the current status of the studied issue. The same possibility arises in the case of additions to the examination database.

However, disadvantages of using e-learning systems can be also specified:

- *Considerable time demands on the initial creation of the course.* Creating a quality course requires reflection on the overall content of the course, subsequent implementation of teaching modules and particularly the transfer of learning materials to HTML and other languages.
- *Lack of direct contact with the teacher.* If necessary, the student can consult with the teacher via any of the communication channels, which is part of the e-learning system, however, subsequent feedback from the teacher often lacks a significant impact or is delayed in time compared to traditional consulting at the workplace of the faculty.

- *Problem to persuade teachers to use e-learning systems.* Very often the effort encounters a routine teaching process of a teacher who is only little willing or even refuses to change the current teaching style and start using ICT.
- *Need for adequate ICT and the Internet connection.* In developed countries, this disadvantage has lately lost its significance. According to Eurostat, the European Statistical Office, "on average 74 % European dwellings had a computer and 70 % households had the Internet connection" [4].

As is evident from the above-mentioned comparison, the advantages of e-learning systems greatly outweigh their disadvantages. What is more, with growing computer literacy of the population the last two mentioned disadvantages will increasingly lose their importance.

## 4 Conclusion

An increasing number of university students, which should enable a qualitative growth of human resources in developed countries, is also reflected in the conditions of the universities in the Czech Republic. However, current financial possibilities of the educational and research capacities of departments do not suffice to a rapid increase in the number of students. A solution can be the introduction of ICT into a teaching process; computers save time for both the teachers and students while enabling intensive use of space and simplifying the administration of tests.

As the study by Andrius [1] claims, when creating an e-learning course there are three levels of difficulty, while at the basic level it is sufficient for the course creator to have basic knowledge only. It is therefore obvious that the creation of an e-learning course does not require a computer expert. The main obstacle seems to be abandoning the established practice of teaching and reserving enough time for the initial creation of a quality course, which can then be updated and used in future years. As the authors of the paper can confirm from their experience with the system Moodle, currently the advantages of using e-learning greatly outweigh its disadvantages.

The implementation of the new system is received positively not only by teachers but also students; long and stressful waiting for the result of the written examination is thus removed. The adoption of the ICT system (Moodle) can be recommended for use at other educational sites, where it can save the trouble of verifying students' knowledge in the form of written tests.

Moodle can also be recommended for use in company courses and for testing employees' knowledge in companies. A significant contribution consists in flexible study of staff who can access the course and sit for a test (verify knowledge) at any time suitable for them.

As shown by our research the introduction of smart education from 2006 to 2015 first led to a drop and subsequently to an improvement in the quality of the students' performance. Randomly generated tests by computer made it impossible for students to copy and it motivate them to be more diligent when studying the course. Automation and test processing speed allow teachers to devote their time for gaining new knowledge and advance their intellectual skills.

**Acknowledgements** This research project is supported by Faculty of Economics and Management, Czech University of Life Sciences in Prague, grant number 20161003—"Price Discrimination of Monopolistic Companies on Food Commodity Markets".

# References

1. Andrius, J.: Learning Management Systems: A Teacher´s Perspective. Australian Flexible Learning Community, http://community.flexiblelearning.net.au/TeachingTrainingLearners/content/article_4840.htm (2003)
2. Brčák, J., Šrédl, K.K.: problematice projektů ICT v ekonomickém vzdělávání. Media4u Mag. **8**(2), 40–44 (2011). ISSN: 1214-9187
3. České školství v mezinárodním srovnání, vybrané ukazatele publikace OECD education at a glance 2008, http://www.oecd.org/education/skills-beyond-school/41284038.pdf (2008)
4. Czech News Agency, Počet domácností s počítačem neustále roste, http://www.finance.cz/zpravy/finance/304550-pocet-domacnosti-s-pocitacem-neustale-roste/ (2011)
5. Dittrich, O.L., Svoboda, R.: Možnosti e-learningových systémů v praxi. Media4u Mag. **9**(4), 94–98 (2012). ISSN: 1214-9187
6. Instructure, Compare Canvas, http://www.instructure.com/compare-canvas/ (2012)
7. Moodle.cz, http://moodle.cz (2012)
8. Ochrana, F., Maaytova, A.: Starting points for creating a transparent and non-corruption public procurement system. Ekon. Cas. **8**(7), 732–745 (2012)
9. Pucek, M., Ochrana, F.: Theory and practise of the public sector savings: the case of Czech Regions. Transylv. Rev. Admin. Sci. **9**(42), 203–224 (2014). ISSN: 1454-1378
10. Severová, L.: Hodnocení kvality vzdělávání ve vysokém školství. Media4u Mag. **7**(2), 4–6 (2010). ISSN: 1214-9187
11. Smart Education, (2016) https://www.ibm.com/smarterplanet/global/files/au__en_uk__cities__ibm_smarter_education_now.pdf
12. Wang, M., Ran, W., Liao, J., Yang, S.J.H.: A performance-oriented approach to e-learning in the workplace. J. Educ. Technol. Soc. **13**(4), 167–179 (2010). ISSN: 1436-4522

# The Gamification Model for E-Learning Participants Engagement

**Danguole Rutkauskiene, Daina Gudoniene, Rytis Maskeliunas and Tomas Blazauskas**

**Abstract** The active use of Information and Communication Technologies (ICT) caused the creation of new ICT application models in various sectors. Due to that, education is getting more and more interactive as more concepts of smart education are implementing into learning process. The new smart ways of learning bring new challenges to manage the learning processes and make them as much evolving as possible at the same time opening new self-learning opportunities through gamification as an effective engaging learning method. The aim of this article is to present the gamification model for e-learning participants' engagement. The objectives of the paper: (1) Overview the existing approaches and models of gamification; (2) Present the design of gamification model for participants' engagement; (3) Present the results of implementation of the gamification model.

**Keywords** ICT · Gamification · Pedagogical aspects · Technical aspects · Engagement

## 1 Introduction

Gamification is a relatively novel technology in educational and working area. Gamification is defined as the application of typical elements of game playing (rules of play, point scoring and competition with others) to other areas of activity,

D. Rutkauskiene (✉) · D. Gudoniene · R. Maskeliunas · T. Blazauskas
Kaunas University of Technology, Kaunas, Lithuania
e-mail: danguole.rutkauskiene@ktu.lt

D. Gudoniene
e-mail: daina.gudoniene@ktu.lt

R. Maskeliunas
e-mail: rytis.maskeliunas@ktu.lt

T. Blazauskas
e-mail: tomas.blazauskas@ktu.lt

© Springer International Publishing Switzerland 2016                                      291
V.L. Uskov et al. (eds.), *Smart Education and e-Learning 2016*,
Smart Innovation, Systems and Technologies 59,
DOI 10.1007/978-3-319-39690-3_26

especially to engage users in problem solving [1]. For the purpose of the gamification techniques and according to organizational goals various models and elements are used in practice. In this paper will be presented Lithuanian and worldwide scientists' experiences and ideas, which can be used in developing gamification area. The scientific practice presents the innovations brought by gamification implementation in conjunction with technological solutions in educational context.

From the global perspective of gamification, the concept of training is analysed very widely. As an illustration, scientists [2] analysed the impact of gamification elements in the "brainstorming" system. They proved that the use of game mechanics can contribute to promoting the involvement of "flow" status [3], but also highlighted the importance the quality of gamification.

Student motivation is always something to keep in mind when planning lessons, as a teacher. By applying gamification to the classroom, students could be motivated to learn in new ways or enjoy otherwise tedious tasks. According to Crawford [4], the gamification can also be used to improve student behaviour in the classroom. Students can be rewarded when they are engaged in examples of good behaviour. The resulting rewards can be specific badges, which is a tangible symbol of all their work. The game provides real-time feedback from the teacher, which helps with positively reinforcing appropriate behaviour.

Hill, Crok and Wickramasekera [5] analysed students' involvement to the lectures and their attendance. Kapp [6] analyses games engagement—the cornerstone of any positive learning experience. "With the growing popularity of digital games and game-based interfaces, it is essential that gamification be part of every learning professional's toolbox. In this comprehensive resource, Kapp reveals the value of game-based mechanics to create meaningful learning experiences".

Educational Gamification is not to be confused with Game-based Learning, Simulation, or Serious Games. This is the direct opposite of educational gamification, which seeks to add game-like concepts to a learning process. Reiners and Wood [7] suggest there are six elements which make up the 'recipe' for meaningful gamification. These elements are:

- Play—Setting the boundaries for play and allowing exploration and failure within;
- Exposition—Creating metaphors which integrate with real world settings;
- Choice—Putting the power of choosing in the user's hands;
- Information—Using engagement through game design and display to create concepts which enhance learning process;
- Engagement—Encouraging collaborative learning with other users who have similar real-world interests;
- Reflection—Assisting participants to find interests and past experiences which can enrich the learning process with deeper engagement.

The aim of this article is to present the gamification model for e-learning participants' engagement. The objectives of the paper: (1) Overview the existing approaches and models of gamification; (2) Present the design of gamification model for participants' engagement; (3) Present the results of implementation of the gamification model.

## 2 Review on Existing Practices

### 2.1 Overview on the Gamification Models

Possibility to learn or work, while person is playing seems so useful, that this idea spread all around the world exclusively fast. Researchers have their opinion about the association of gaming and problem solving skills such as deduction power, dimensional thinking (in addition to linear thinking) and proof based decision making [6, 8]. These researchers believe, that gaming allows one to exercise his/her imagination, to fantasize about aspirational roles (a roller coaster tycoon!). Below it is seen Miller's enhanced gamification model [9] (Fig. 1). The biggest benefit from the gamification, as noted in Miller's written literature, that learners while playing reach the leaning outcomes.

The main goal of the game is the long-term goal of completing the game. Then there is the medium-term goal of completing the levels in the game, and the short-term goal of completing the game missions in the level. The main requirement of each goal "layer" in a game is to get increasingly harder as you move from short-term to long-term goals. These goals "layers" let players learn and practise skills, which are needed for the games, prior to mandatory requirement to demonstrate mastery of those skills in the most challenging parts of the game. Researcher [10] created game goals structure, which is used in designing eLearning material (see Fig. 2).

The skills of the participant must grow when the challenge of an experience rises. The "Flow Channel" illustrates an optimal user experience (see Fig. 3). The squiggly line in the Fig. 3 shows the experience described above where a user is challenged to a high degree with new experiences, and then given an opportunity to

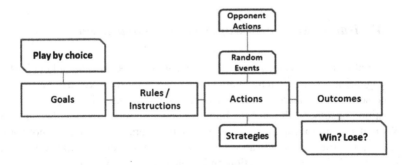

**Fig. 1** Miller gamification model [9]

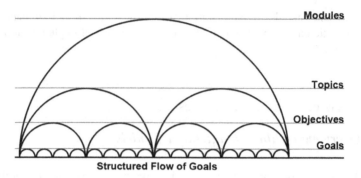

**Structured Flow of Goals**

**Fig. 2** e-Learning material designed on the model of linear flow of goals in gamification [10]

**Fig. 3** The flow channel [10]

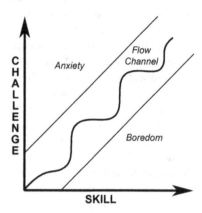

demonstrate and master the skill of that experience, before given a completely new challenge to conquer.

Aseriskis and Damasevicius [11] research focused on building, evaluating gamification solutions for software engineering teams. Furthermore, they focused on modelling of gamified systems and discovering patterns in gamified applications [11].

## 2.2 Motivation and Encouragement in Gamification

Engagement is the key factor for an effective learning process. Chatfield [12] states that engagement occurs when the brain is rewarded, it must evoke positive emotions in a person. When learners are engaged with learning they become active information-consumers who take the initiative and push forwards their development.

When it comes to engagement in games there are two types of motivation: intrinsic and extrinsic. Gamification desires to combine intrinsic motivation with extrinsic in order to raise motivation and engagement (see Fig. 4).

**Fig. 4** Motivation types

Intrinsic motivations come from within, the user/actor decides whether to make an action or not, some examples are: altruism, competition, cooperation, sense of belonging, love or aggression [13].

Extrinsic motivations, on the other hand, occur when something or someone determines the user to make an action, for example: classifications, levels, points, badges, awards, missions [14]. In the learning sphere, extrinsic motivation always comes with a promise of a pay rise at the end of action or process. There are four types of extrinsic motivators [15]:

- **Achievements**: awards, trophies, badges;
- **Progress**: level, scores, points;
- **Content**: quests, missions, virtual goods;
- **Reputation**: leader boards, ranking, rating.

According to Muntean [16] there are some key components that need to be taken into consideration and that build up a coherent overview of the entire functionality (utility) of an application/website. Game mechanics and features are comprised in the game design in order to create gameplay. Game mechanics are a set of rules and feedback loops that create the gameplay. They represent the fundamentals of any gamified context. Each game mechanic is characterized by three attributes:

- **Game mechanics type**: Progression, Feedback, Behavioural
- **Benefits**: engagement, loyalty, time spent, influence, fun, SEO, UGC, Vitality
- **Personality types**: explorers, achievers, socializers and killers.

Banyte and Gadeikiene [17] were searching how consumer motivation to play video games affects their engagement with video game-playing. They found out three types of motivation, i.e. intrinsic motivation, extrinsic motivation and experiential motivation [4, 18], describe general consumer motivation to play video games both on personal and unipersonal (game) level.

Kostecka and Davidaviciene [19] were solving the problem of motivation of employees, who are working with information system and whose work environment is full of monotonous, boring and repetitive tasks, is analysed [20, 21]. Theoretical aspects of work motivation are analysed and it is suggested to use gamification in order to solve this problem [22]. After all, it is suggested to use model which joins main aspects of employee needs and gamification [23, 24]. Based on the results of the research, opportunities of motivating accounting specialists through gamification of information system are evaluated.

# 3    Design of the Gamification Model

## 3.1    The Pedagogical Aspects of the Gamification Model

Creating the new gamification model it is essential to think of a valid pedagogical model to deliver the content. The motivational factors cannot be forgotten while planning the learning process as well. Having in mind the main motivation boosts and principles of gamification in education, the new model of gamification was created and directed to the increase of learners' engagement to learning content through gamified elements in e-learning.

The created gamification model involves most of main educational gamification principles. The model is based on the pedagogical view that guides to the games mechanics for extrinsic motivation [25] (see Table 1). The new gamification model including various game mechanics create a multi-layer environment where every learner is motivated to learn for number of reasons: to unlock next level, to gain more points; to raise up his/her status, to learn to code and many more.

The gamification model is based on three levels of mastery: basic, second and third levels. Each of the learner can move through all the levels. However the primary skills in coding will be tested by taking the test. This ensure the relatively customized learning content. The move through levels are based on pointing system which describes the certain rules of how many points or achievements you need to collect for unlocking next level. In this case points and achievements shows the mastery and skills (the more points you have, the higher level you are in).

**Table 1** Educational gamification design principles [25]

| Design principles | Game mechanics |
|---|---|
| *Goals:* specific, clear, moderately difficult, immediate goals | |
| *Challenges and quests:* clear, concrete, actionable learning tasks with increased complexity | |
| *Customization:* personalized experiences, adaptive difficulty; challenges that are perfectly tailored to the player's skill level, increasing the difficulty as the player's skill expands | Multiple game pathways |
| *Progress:* visible progression to mastery | Points, levels |
| *Feedback:* immediate feedback or shorten feedback cycles; immediate rewards instead of vague long-term benefits | |
| *Competition and cooperation/social engagement loops* | Badges, leader boards, levels |
| *Visible status:* reputation, social credibility and recognition | Points, badges, leader boards |
| *Access/unlocking content* | |
| *Time restriction* | Countdown clock |

## 3.2 Technological Aspects of the Gamification Model

Some of gamification techniques strive to leverage people's natural desires for socializing, learning, competition or achievement.

The gamification model is designed on presentation of rules, practise guide, knowledge enhancement and deepen learning experience levels (see Fig. 5). Each of the level includes one or more activities (tools) to implement it in the platform. As an example, presenting basic rules level includes animation mode. That means that animated interactive content helps learner to understand the rules of the game. Guided practising level includes Feedback mode. That means learners gain feedbacks as results for their actions in the game. To stimulate the knowledge quiz and formula modes are also implemented into the game as it trains the mastery into specific subject.

One of the most important level is the deepening learning experience process. It is a complex process including competition, tactical, collaboration and cooperation modes. These modes are made of three levels of independency where learners can

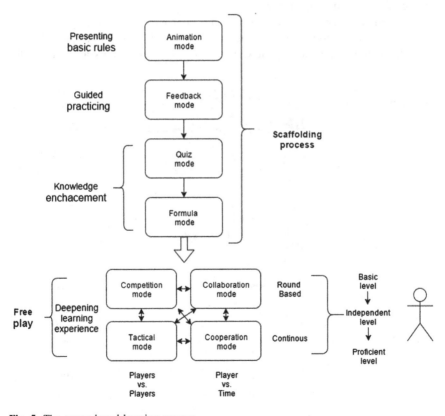

**Fig. 5** The games based learning process

demonstrate their mastery in the subject. As it was mentioned before, competition mode includes leader boards and pointing system which stimulates learners to compete and win. Tactical mode is about the strategic thinking of how to gain as more points as possible with less effort. To solve the tasks learners join the groups on social networks. There learners can share the insights and help each other to solve the difficult parts of tasks. This fulfil the collaboration mode. The last mode is cooperation which is similar but restricted by timing (the certain amount of time is given for solving each of tasks).

## 4 The Model Integration to the Learning Platform and Exploitation Results

Informic environment was created for pupils from Lithuania schools to participate in online programming learning contest (see Fig. 6). Informic environment offered students access to tutorials and exams. Tutorials were not a mandatory if user wanted to participate in the contest. Exams were required to be done for students to get a certificate issued by university.

Each finished task gives additional points. For example, finishing exams and tutorials gives users points (0–50 for tutorial and 0–100 for exam) displayed in top 10-leader board and a ratings view, which showed user points and badges for

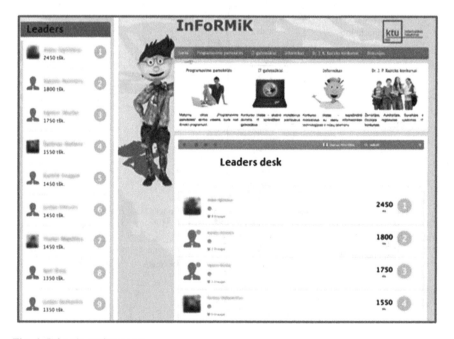

**Fig. 6** Informic environment

**Fig. 7** The levels of games based education

everyone (see Fig. 7). To make it more complex, multiple types of badges where awarded for students as ladders to next level which means the student needed to collect as many rewards as possible to get to the next level (and raise up in the leader board).

Gamification model is based on "shopping" awarding/stimulation methodology, where each student can gain points out of main (compulsory) tasks as well as of additional free to choose tasks. These points one can spend later in several ways:

- to open further (compulsory and free to choose) quizzes and tasks;
- to buy additional grade points for the current (or upcoming) compulsory tasks which form final evaluation grade for the subject (final grade of the semester).

The model exploitation results revealed that this gamified learning model based on points encouraged pupils to complete more tasks as they get more points for that. The students were more active and got better grades in exams. Also, students were more determined to learn new subjects as that could raise them up to the next level. However, it was noticed that, some students complete tasks for speed not for quality. Some of the students wanted to gain as more points and badges as possible and did not care about the quality of the solutions. This shows that model still distinct the objective quality evaluation criteria to motivate students to solve tasks as good as they can. Also, some of students lost their interest in gamified content after a while and stopped learning which shows the lack of multi-perspective motivational tools and methods to monitor student's performance and boost the motivation when the first signs of demotivation appears. The results will be used for the further improvement of the gamification model and e-learning platform to create a unique learning environment with effective motivational tools.

## 5 Conclusions

Gamification models can be designed differently which opens new perspectives for creativity. However, the difference of gamification models has one similarity which is included by all models—the motivation. Motivation is the key factor for the success of gamification models. Various tools to boost motivation stimulate all learners and each of them personally to be engaged into the game and master the subject. Creating the effective gamified model educator must think of the target learners and their needs for learning as well as the methods to boost their engagement to the learning content.

The created gamification model is based on the extrinsic motivational tools. The method is directed to create a pack of tools which would be effective for learning online and would keep learners engaged. The method involve pointing system, social competition and achievements to boost learner's engagement to the learning content.

The method for participants' engagement was implemented into the learning platform and tested with pupils. The results of the pilot testing revealed that tools used for engagement stimulation was chosen right. Most of students are motivated to learn by themselves. However, the part of students does not show the expected interest which indicates the problems with motivation boosts or the platform. The method requires further development.

**Acknowledgements** This paper has been partially supported by Educational content modernization using new methods and tools for competences development of educators.

# References

1. Oxford Online Dictionary, http://www.oxforddictionaries.com/
2. Yuizono, T., Xing, Q., Furukawa, H.: Effects of gamification on electronic brainstorming systems collaboration technologies and social computing. Commun. Comput. Inf. Sci. **460**, 54–56 (2014)
3. Witt, M., Scheiner, C., Robra-Bissantz, S.: Gamification of online idea competitions: insights from an explorative case. Informatik schafft Communities (2011)
4. Crawford, C.: Chris Crawford on Game, pp. 5–10, 1st edn. New Riders Publishing, Thousand Oaks, USA (2003)
5. Hill, C., Cronk, T., Wickramasekera, R.: Global Business Today: An AsiaPacific Perspective. McGraw-Hill, North Ryde, Australia (2008)
6. Kapp, K.M.: The Gamification of Learning and Instruction: Game-Based Methods and Strategies for Training and Education (2012). ISBN: 978-1-118-09634-5
7. Reiners, T., Wood, L.C.: Gamification in Education and Business (2015). ISBN: 978-3-319-10208-5
8. Blair, L., Mesch, R.: Virtualized shared protection capacity. Ciena Corporation, US8456984 B2 (2013)
9. Miller, C.: The Gamification of Education. ABSEL (2013)
10. Mahle, D.A., Anderson, P.E., DelRaso, N.J., Raymer, M.L., Neuforth, A.E., Reo, N.V.: A generalized model for metabolomics analyses: application to dose and time dependent toxicity. Metabolomics 7, 206–216 (2011). doi:10.1007/s11306-010-0246-3
11. Aserskis, D., Damasevicius, R.: Gamification patterns for gamification applications. In: 6th International conference on Intelligent Human Computer Interaction, IHCI (2014)
12. Chatfield, T.: Fun Inc: Why games are the 21st century's most serious business. Virgin Books (2010)
13. Marczewski, A.: A Simple Introduction: Tips, Advice and Thoughts on Gamification (2013)

14. Viola, B.F.: Genome-wide association study identifies susceptibility loci for IgA nephropathy. (2011). doi:10.1038/ng.787
15. Sudan, J.: Extrinsic Rewards, Intrinsic Rewards: The Gamification of Motivation (2013)
16. Muntean, C.I.:. Raising engagement in e-learning through gamification. In: Proceedings of 6th International Conference on Virtual Learning ICVL, pp. 323–329 (2011)
17. Banyte, J., Gadeikiene, A.: The effect of consumer motivation to play games on video game-playing engagement (2015). doi:10.1016/S2212-5671(15)00880-1
18. Torsten Reiners, T., Teräs, H., Chang, V., Wood, L.C., Gregory, S., Gibson, D., Petter, N., Teräs, M.: Authentic, immersive, and emotional experience in virtual learning environments: the fear of dying as an important learning experience in a simulation. In: Teaching and Learning Forum 2014 (2014)
19. Kostecka, J., Davidaviciene, V.: Model of employees motivation through gamification of information system. In: Science—future of Lithuania/mokslas—lietuvos ateitis, 7(2) (2015)
20. Erenli, K.: Gamification and law: on the legal implications of using gamified elements. In: Gamification in Education and Business, pp 535–552 (2014)
21. Zichermann, G., Cunningham, C.: Gamification by Design: Implementing Game Mechanics in Web and Mobile Apps (2011)
22. Bruneel, H., Kim, B.G.: Discrete—time Models for Communication systems Including ATM. Springer Science + Business Media, LLC (1993)
23. Apostol, S., Zaharescu, L., Alexe, I.: Gamification of learning and educational games. eLearn. Software Educ. 2, 67–72 (2013)
24. Engler, R.: Serious Games—Gamification of Education. Vrije Universiteit, Amsterdam (2012)
25. Dichev, C., Dicheva, D., Agre, G., Angelova, G.: Trends and opportunities in computer science OER development. Cybern. Inf. Technol. (2015)

# RLCP-Compatible Virtual Laboratories in Computer Science

Evgeniy A. Efimchik, Elena N. Cherepovskaya
and Andrey V. Lyamin

**Abstract** Education, and especially e-Learning, is rapidly changing and more electronic instruments are becoming available. One of the most frequently applied electronic environments is virtual laboratory. Virtual laboratories provide many benefits as one of their main aims is developing practical skills. This paper presents three virtual laboratories that are applied in the online course of Computer Science and describes RLCP protocol, developed in ITMO University, that supports interaction between virtual stand and a server.

**Keywords** e-Learning · Computer science · Virtual laboratories · RLCP

## 1 Introduction

In the fast changing environment, requirements for the staff in different working spheres, particularly in engineering and computer science, are rapidly increasing. One of the basic requirements is that a person should possess practical skills that would be useful for successful working activity. e-Learning occupies one of the major places in preparing high-qualified students, so that they can prove all necessary practical skills they obtained. It supports different learning schemes, e.g. massive open online courses (MOOCs) [1], despite the traditional ones. As well as large learning material databases, e-Learning systems provide a possibility of adding different tools and environments that help to develop practical skills, i.e. virtual laboratories that are frequently used in MOOCs' platforms and learning management systems (LMS) [2].

E.A. Efimchik · E.N. Cherepovskaya · A.V. Lyamin (✉)
ITMO University, Saint Petersburg, Russia
e-mail: lyamin@mail.ifmo.ru

E.A. Efimchik
e-mail: efimchick@cde.ifmo.ru

E.N. Cherepovskaya
e-mail: cherepovskaya@cde.ifmo.ru

© Springer International Publishing Switzerland 2016
V.L. Uskov et al. (eds.), *Smart Education and e-Learning 2016*,
Smart Innovation, Systems and Technologies 59,
DOI 10.1007/978-3-319-39690-3_27

Virtual laboratory (Lab) is a special programming tool embedded in e-Learning systems that provides an opportunity to use virtual environment for investigating real-world complex systems. Main aims of the virtual laboratory are developing practical skills, assessing skills by checking student's actions while dealing with virtual stand and, hence, enhancing learning outcomes of the learning program, where Lab is applied. All virtual laboratories have different architecture and can be divided into three classification groups according to their main characteristics:

- Autonomous [3] or client-server [4, 5] realizations of virtual laboratories;
- Virtual laboratories with an access to physical equipment [6, 7] and Labs based on mathematical models [8];
- Virtual laboratories that provide automatic results assessment [5, 9].

Each of the presented categories has its benefits and shortcomings. Autonomous virtual laboratories provide full set of functions as well as client-server realizations. However, actions performed by a person using autonomous Lab and his results could not be stored in the database or automatically checked by server. Considering another realization: virtual laboratory can be provided to a client as a virtual stand embedded in e-Learning system, and after the assignment is completed, student's response is sent to the server, processed and results are stored in system's database. Another benefit of such realization is that for some virtual laboratories an assignment could automatically be generated by server [9] and inserted in a virtual stand. The next category assumes modelling type of the virtual laboratories. The first type infers virtual environment with the remote access to some physical equipment. This type of Lab does not require complex programming realization; however, it leads to expensive support of the applied mechanism. On the contrary, Lab based on the mathematical models requires complex programming realization, though having many advantages, e.g. low-cost support and possible addition of extensions to the existing programmed algorithms in order to provide a full-function virtual environment. Virtual laboratories related to the last category are client-server based realizations that provide an opportunity to check person's answer or a sequence of actions on the server automatically. As information technologies had improved, virtual laboratories became perfect instruments for developing person's practical skills and assessing learning outcomes.

In this paper we propose three virtual laboratories used for the training and assessment purpose in the discipline of Computer Science in ITMO University. The second section is related to the Remote Laboratory Control Protocol (RLCP) used to connect virtual stands to the server. The third and fourth sections describe algorithms of the virtual laboratories and rules of writing test sets used for the results assessment. Experimental results are presented in the fifth section. Last section of the paper contains concluding statements.

## 2   Remote Laboratory Control Protocol

RLCP [5, 10] is a protocol developed in ITMO University in order to provide interaction between virtual stand and a server. It defines the basic structure and rules for developing virtual laboratories in order to simplify evaluating process. A structure of the RLCP-compatible virtual laboratory is presented in Fig. 1.

On the client-side virtual laboratory consists of the RLCP Stand (Stand) that is a virtual stand for the laboratory, which is embedded in the Browser. Before the page had been constructed, the assignment for a Lab is generated by RLCP Server (Server) or chosen from the existing task sets. RLCP Client (Client) is responsible to provide interaction between virtual stand and RLCP Server. Browser and Client are connected by HTTP protocol, and RLCP Server is accessed by RLCP protocol. For the assignments that allow intermediate evaluation, a request is sent to the Server and evaluation results are shown to student. As the assignment had been completed, student submits his response to be checked by Server. A request is sent to the RLCP Client that processes it and passes corresponding RLCP request with student's response to the RLCP Server. The response is being evaluated, and assessment results are sent back to the Client that, firstly, stores the results in the Database and then updates Browser page by inserting the results in a special form in order to provide information for a student. A sequence diagram of interaction with RLCP Server is shown in Fig. 2.

Intermediate requests can be classified based on the Lab assignment: requests for intermediate evaluation of student's actions in virtual stand and ones used to receive external information from different resources in order to complete the assignment. Therefore, RCLP contains three main methods: Generate for generating an assignment for a virtual laboratory, Calculate used in Labs that support intermediate evaluation of student's actions, and Check, which is used for the final assessment of student's response. For complete evaluation of the student's response sets of tests are provided for each of the assignments. Type of the RLCP method should be added to the RLCP request header as a state parameter. RLCP Stand API describes one method to be obligatory realized in a virtual stand—getResults—used to send student's response as a string to the Server.

At first, all virtual stands were developed with the use of Java Applet technology. However, during the last few years it had been stated, that Java will be removed from the web browsers, mainly due to the lack in supporting mobile

**Fig. 1** Structure of the RLCP-compatible virtual laboratory

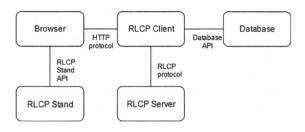

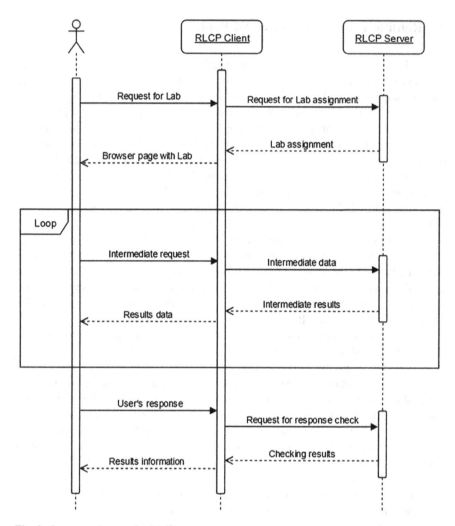

**Fig. 2** Sequence diagram for RLCP

platforms that are widely used nowadays. Only few modern browsers left that still support this technology. Moreover, user interface should have been simplified in order to make it more comprehensible and user-friendly and reduce time and number of attempts spent on the Lab assignments. Therefore, it was decided to modernize all the existing virtual stands to become JavaScript-based. All the modified and newly developed stands support RLCP.

RLCP is successfully applied in AcademicNT LMS (http://de.ifmo.ru) [5, 9–11], developed and used in ITMO University and several other Russian universities. As well as in AcademicNT, RLCP is applied on the Russian national platform "Open Education" (https://openedu.ru/) that was developed based on the Open edX platform. It is applied in the course "Methods and algorithms of the graph theory", where virtual laboratories are used as training and assessment techniques.

## 3 Virtual Laboratories in Computer Science

Virtual laboratories are frequently applied in different disciplines and various education fields. This section describes three main Labs that are used for the online course of Computer Science discipline in AcademicNT LMS. They include interpreters of Post and Turing machines and Multistyle code editor, that are client-server based Lab realizations with a mathematical model for each Lab.

### 3.1 Post and Turing Machines

Post and Turing machines [12, 13] are abstract mechanisms that were described by Emil Post and Alan Turing. They were intended to specify the definition of an "algorithm". These mechanisms are often included in learning programs of the Computer Science in different universities in order to develop practical skills of solving algorithmic tasks.

Benefits of using Post and Turing machines for evaluating learning outcomes include absence of the requirements in knowledge of programming languages. Application of virtual laboratories provides clearness of learning material presentation and reduces time for checking answers as client-server based realizations support automatic student's response evaluation. Therefore, it leads to a complete control of complex learning outcomes in Computer Science, e.g. an ability to develop and analyze algorithms.

AcademicNT LMS provides interpreters for Post and Turing machines as virtual stands for corresponding virtual laboratories, which are used to evaluate specific learning outcomes in Computer Science for the first-year students of ITMO University. Labs have client-server realization and all the students' responses are checked by RLCP-server automatically.

Let us firstly consider the Post machine virtual laboratory. Though, Post machine provides larger commands dictionary than Turing machine, it is less complicated due to the binary alphabet. User interface of the virtual stand is presented in Fig. 3.

Post machine interpreter consists of an endless in both directions tape with cells for marks, a code editor, an information space, where the current command line number and messages are displayed, and an area with control buttons. Number of lines in code editor is endless and access to the necessary code line is provided by scrolling arrows on the right. The command line that is selected by user or currently being executed is highlighted as shown in Fig. 3. User is allowed to fulfil the following actions:

**Fig. 3** Virtual stand of Post machine Lab

- Set a mark in a tape cell, if the cell is empty, or delete mark otherwise;
- Move the tape in both directions;
- Type code in the code editor:
  - 'n'-field states for the code line number;
  - 'K'-field is a command from the Post machine's predefined dictionary. It is chosen from the drop-down list;
  - 'm'-field states for the number of command line to jump to; number of the editable cells depends on the 'K'-field; if the condition operator had been chosen, two cells are available to be filled, otherwise one cell is provided;
- Perform actions provided by the control buttons: insert or delete a code line, clear the tape, run step execution of the code, run the whole code and stop stepping mode if started.

The interpreter supports displaying messages, if an error had been found or program was successfully finished (as shown in Fig. 3). Errors include information about incorrectly filled code, e.g. when 'K' or 'm'-field is empty etc., loops found in the code, e.g. code line $1 > 1$ etc., or errors found when the written program had been executed.

Another virtual laboratory is the interpreter for Turing machine. Turing machine's mechanism is more complicated, as programmer defines the code settings, and number of symbols and states affects algorithm's complexity. User interface of the Turing machine's virtual stand is presented in Fig. 4.

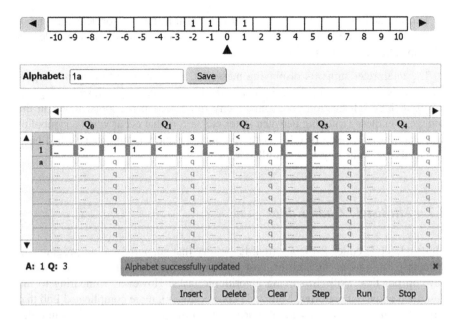

**Fig. 4** Virtual stand of Turing machine Lab

Turing machine interpreter consists of an endless in both directions tape with cells for alphabet symbols, alphabet editor, a code editor part, an information space, where the chosen code part details, i.e. alphabet symbol and state, and messages are displayed, and an area with control buttons. Number of states in code editor is endless in one direction and number of code lines is limited to the number of symbols specified by user. Access to the necessary code part is provided by scrolling arrows on the top and left sides. Each code part consists of three fields, i.e. new alphabet symbol to be inserted in the observed cell, command from the dictionary, and new state that machine should be switched to. User is allowed to fulfil the following actions:

- Insert a symbol on the tape, input is allowed after clicking on the cell;
- Move the tape in both directions;
- Move states and code lines by using scrolling arrows on the top and left side;
- Type code in the code editor. Each code part is filled separately:

  - the first field states for the new symbol to be inserted in the tape cell, it is realized as a drop-down list containing alphabet symbols;
  - the second field is a command from the Turing machine's predefined dictionary, which is chosen from the drop-down list;
  - the third field states for the number of state to switch Turing machine to;

- Perform actions provided by the control buttons: insert state or delete a symbol or state, clear the tape, run step execution of the code, run the whole code and stop stepping mode if started.

The interpreter supports displaying messages, if an error had been found, program was successfully finished or alphabet had been updated (as shown in Fig. 4). Errors include information about incorrectly filled code parts, loops found in the code, e.g. endless switching between states etc., errors that occur when a symbol or state can not be deleted or other errors found when the written program had been executed, e.g. switching to an unfilled state.

## 3.2   Multistyle Code Editor

Multistyle code editor [5] is a virtual laboratory that supports writing programs in different code styles. As well as Post and Turing machines, this Lab is used to develop practical skills in solving algorithmic tasks. However, comparing to the described above laboratories, Multistyle code editor is a more complicated Lab that requires from users to be aware of at least one of the provided code styles in order to complete the assignment successfully. This requirement extends the list of learning outcomes, as it evaluates not only the skill of writing and understanding basic algorithms, but also a competence to apply a programming language to solve an assignment. The list of assignments is also extended as complex data structures and operations are supported by the programming languages and more complicated code can be written by a student. Hence, a virtual laboratory that realizes Multistyle code editor leads to the full assessment of all associated learning outcomes.

The developed Multistyle code editor Lab supports writing programs in three available code styles: Pascal, Basic and C. User interface of the virtual stand is presented in Fig. 5.

The interface of Multistyle code editor consists of five parts:

- Variables panel with the lists of input, output and inner variables, which can be edited by opening the dialog clicking on the 'edit'-icon in the top-right corner;
- The main code editor panel;
- The functions library on the right that provides list of functions for the chosen code style, double click on the nodes insert a particular construction in the code editor
- The bottom panel shows console messages, e.g. errors, info-messages or warnings;
- Code style can be changed by the drop-down list 'Code style'. The list of actions is provided by the control buttons: running code in a debug mode, step execution, run the whole code and stop stepping mode.

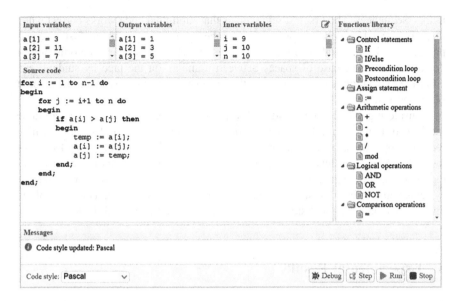

| Input variables | Output variables | Inner variables | | Functions library |
|---|---|---|---|---|

**Fig. 5** Virtual stand of Multistyle code editor

## 4 Evaluation of Student's Response

Student's response can be differently evaluated, e.g. it can be automatically checked by Server with a specified API, based on automata models that can also be used to generate Lab assignment [9], or Hoare-similar logic. Another way is evaluating the answer applying predefined test sets. After a student had finished dealing with the assignment and submitted his response to be checked, request with the submitted answer is sent to the RLCP-server and student's response is being evaluated. Each of the assignments has a particular set of tests used to check the response. The main requirement to the test sets states that they should contain tests with different complexity in order to have person's response been fully assessed. A test set consists of two parts: input parameters that are passed as an input to the code submitted by a person and reference output parameters that describe an output that student's response should produce as a result.

Amount of the test sets depends on the assignment and type of the results differentiation. If the result of response evaluation should be presented in a binary format (correct/incorrect), one test set should be provided. Otherwise, an amount of test sets should be increased considering possible omissions a person may commit. It should be mentioned, that the order of tests should be set in the ascending way based on the omission's weight. Hence, all the responses with various types of omissions will be correctly evaluated.

As an example, let us consider test sets formed for a Post machine virtual laboratory with an assignment of adding number when the carriage is set between marks' sequences. The first test set assumes that submitted code works correct when there is one white space between the numbers on the tape. If the first test was passed, the code is evaluated on the next set, which assumes possible omission of not considering increase of spaces between numbers on the tape by student, etc. After the evaluation had been finished the results were formed based on the amount of passed test sets.

Considering evaluation of students' responses in virtual laboratories it should be mentioned that the multitude of all possible responses is undecidable, which requires application of artificial intelligence methods, e.g. neural networks, fuzzy logic, etc. Specially developed algorithms for the automatic response assessment, e.g. based on automata models [9], let virtual laboratories and learning management systems being referred to smart systems [11] as they allow making a decision of the level of person's practical skills in a particular course, where Lab is applied.

## 5 Experimental Results

All the presented virtual laboratories are used in the online course of Computer Science for the first-year students in ITMO University since 2005. However, JavaScript realizations of the Labs instead of Applets had been firstly approved in 2015. The experiment had been conducted in ITMO University evaluating students' performance for two different types of Lab realizations. First-year students had to pass described above virtual laboratories during the curriculum. They were randomly divided into two groups in a ratio of 2:3: the control group was provided with Applet and active group with JavaScript realization of virtual laboratories. The amount of participants in groups equaled 51 and 78 respectively. Table 1 presents the results for both groups. We considered the following parameters for each student during the evaluation: grade point average (GPA), time spent on handling the assignment and number of attempts used to complete the task.

Application of the method based on confidence intervals revealed that the null-hypothesis for equality of expected values for two groups was rejected at 5 % significance level. Hence, the presented results show positive trend in changes of all the observed parameters during the application of JavaScript realizations of virtual laboratories as the GPA had increased and time spent on dealing with assignment as well as number of attempts was reduced.

**Table 1** Comparison of two different realizations of virtual laboratories

|                                       | Control group | Active group |
|---------------------------------------|---------------|--------------|
| Grade point average                   | 68.90         | 93.12        |
| Average time spent on the assignment  | 25.67         | 11.03        |
| Average number of attempts            | 2.12          | 1.29         |

# 6 Conclusion

Virtual laboratories are perfect instruments that help to develop and assess practical skills in different learning courses. More learning management and MOOCs systems are embedding virtual laboratories and applying them in the learning process. As a result, the main benefits of virtual laboratories are reduction of labor expenditure for preparing assignments for Labs, as they can be generated by the server, and checking students' responses, as well as improving accuracy of learning outcomes measurement by automatic response evaluation. Experience of applying user interfaces of Applet Lab realizations had shown their inefficiency and perception difficulty for students. However, considering the results of implementing JavaScript Labs shows that modified interfaces had become more user-friendly as the amount of time and attempts spent to fulfil the assignment correctly was reduced, as well as significant increase in students' GPA was revealed. Our further work will be related to increasing smartness of virtual laboratories and LMS by using methods of artificial intelligence and particularly smart algorithms for automatic verification of students' response.

**Acknowledgments** This paper is supported by Government of Russian Federation (grant 074-U01).

# References

1. Kay, J., Reimann, P., Diebold, E., Kummerfeld, B.: MOOCs: so many learners, so much potential … . IEEE Intell. Syst. **28**(3), 70–77 (2013)
2. Venugopal, G., Jain, R.: Influence of learning management system on student engagement. In: 2015 IEEE 3rd International Conference on MOOCs, Innovation and Technology in Education (MITE), pp. 427–432 (2015)
3. Djeghloud, H., Larakeb, M., Bentounsi, A.: Virtual labs of conventional electric machines. In: 2012 International Conference on Interactive Mobile and Computer Aided Learning (IMCL), pp. 52–57 (2012)
4. Magyar, Z., Zakova, K.: Using SciLab for building of virtual lab. In: 2010 9th International Conference on Information Technology Based Higher Education and Training (ITHET), pp. 280–283 (2010)
5. Lyamin, A.V., Vashenkov, O.E.: Virtual laboratory: multi-style code editor. In: 23rd ICDE World Conference on Open Learning and Distance Education Including the 2009 EADTU Annual Conference, Maastricht, The Netherlands, 7–10 June 2009. http://www.ou.nl/icde2009
6. Saenz, J., Chacon, J., De La Torre, L., Visioli, A., Dormido, S.: Open and low-cost virtual and remote labs on control engineering. IEEE Access 805–814 (2015)
7. Hristov, G., Zahariev, P., Bencheva, N., Ivanov, I.: Designing the next generation of virtual learning environments—virtual laboratory with remote access to real telecommunication devices. In: 2013 Proceedings of the 24th EAEEIE Annual Conference (EAEEIE), pp. 139–144 (2013)
8. Moritz, D., Willems, C., Goderbauer, M., Moeller, P., Meinel, C.: Enhancing a virtual security lab with a private cloud framework. In: 2013 IEEE International Conference on Teaching, Assessment and Learning for Engineering (TALE), pp. 314–320 (2013)

9. Chezhin, M.S., Efimchik, E.A., Lyamin, A.V.: Automation of variant preparation and solving estimation of algorithmic tasks for virtual laboratories based on automata model. In: Vincenti, G., Bucciero, A., Vaz de Carvalho, C. (eds.) LNICST, vol. 160, pp. 35–43. Springer (2016)
10. Efimchik, E., Lyamin, A.: RLCP-compatible virtual laboratories. In: The International Conference on E-Learning and E-Technologies in Education (ICEEE 2012), pp. 59–64 (2012)
11. Uskov, V., Lyamin, A., Lisitsyna, L., Sekar, B.: Smart e-Learning as a student-centered biotechnical system. In: Vincenti, G., Bucciero, A., Vaz de Carvalho, C. (eds.) LNICST, vol. 138, pp. 167–175. Springer (2014)
12. Post, E.L.: Finite combinatory processes-formulation 1. J. Symbolic Logic **1**(3), 103–105 (1936)
13. Turing, A.M.: On computable numbers, with an application to the Entscheidungs problem. In: Proceedings of the London Mathematical Society, vol. 2, no. 42, pp. 230–265 (1937)

# Combining Science with Art to Educate and Motivate Patients Prior to Colorectal Cancer Screening

Piet C. de Groen, Shreepali Patel, Mariana Lopez, Michael Szewczynski and Rob Toulson

**Abstract** Colorectal cancer (CRC) is the second leading cause of cancer deaths in the US despite wide use of colonoscopy to prevent CRC and CRC-related mortality. Colonoscopy is used to identify and remove lesions that will lead to cancer, however, most deaths occur because lesions are not detected or completely removed during the procedure. Patients play a crucial role in the detection component of colonoscopy: the better the colon is prepared, the higher the chance of detection of all polyps and cancers. In general, patients are instructed to clean the colon by way of a paper or web-based form that lists the objective (scientific) steps involved; unfortunately this too often does not result in a well-prepared colon. Behavior is known to be heavily influenced by emotion. As the first phase of a smart education research project we created an artistic and instructional documentary in which patients engage with the educational content through emotional responses; i.e., we motivate patients to follow instructions by combining scientific with emotional aspects of CRC prevention including preparation of the colon prior to colonoscopy. In the second research phase we will test whether use of the documentary results in improved colon preparation.

**Keywords** Colonoscopy · Colorectal cancer · Emotion · Film · Audiovisual · Education

P.C. de Groen (✉) · M. Szewczynski
Mayo Clinic, Rochester, MN, USA
e-mail: pcdegroen@hotmail.com

S. Patel
Cambridge School of Art, Anglia Ruskin University, Cambridge, UK
e-mail: shreepali.patel@anglia.ac.uk

M. Lopez · R. Toulson
CoDE Research Institute, Anglia Ruskin University, Cambridge, UK
e-mail: mariana.lopez@anglia.ac.uk

R. Toulson
e-mail: rob.toulson@anglia.ac.uk

© Springer International Publishing Switzerland 2016
V.L. Uskov et al. (eds.), *Smart Education and e-Learning 2016*,
Smart Innovation, Systems and Technologies 59,
DOI 10.1007/978-3-319-39690-3_28

# 1 Introduction

Colorectal cancer (CRC) is the second leading cause of cancer-related death in the US [1]. Yet, CRC and CRC-related mortality are mostly preventable if patients participate in a screening and surveillance program that is based on colonoscopy [2]. Indeed, the latest data show that CRC after a colonoscopy is for 80–90 % the result of either a poor colon preparation by the patient or insufficient technique of the endoscopist; in other words, the vast majority of CRC, despite colonoscopy, is not the result of aggressive, uncontrollable tumor biology but the result of sub-optimal human activities [3].

For the last decade the research program of de Groen has focused on improving the human aspects of colonoscopy [4, 5]. His program has developed completely automated software that can measure in real-time how well the patient has cleaned the colon [6]. It can also measure how well the endoscopist inspects the colon mucosa. Several studies have shown that endoscopists can reach the beginning of the colon, the cecum, in a few minutes in a clean colon, and then have 10–20 min to inspect the colon mucosa during withdrawal [7] To the contrary, if the colon is not well cleaned, it takes longer to reach the cecum, and a significant amount of time then is needed to clean the colon to allow inspection. Given a relatively fixed time slot of 30 min per procedure, this means that in a poorly prepared colon there is significantly less time for colon mucosa inspection [8]. Thus, it is crucial that the colon is well cleaned prior to colonoscopy. Unfortunately, this is too often not the case [9].

Currently, patients at Mayo Clinic Rochester are instructed how to cleanse the colon via a paper leaflet that in chronological order lists the objective steps involved in the colon cleansing process. The instruction leaflet holds purely scientific information and objective guidance notes; it has had no specialist design or layout consideration. The colon preparation advice may be included within a larger set of information leaflets and, for example, may be inserted between an explanation of where to go for an ECG and a set of explanations of what to expect during a CT of the abdomen. Yet, whereas all other healthcare instructions, in essence, provide directions and explanations (which really have little or no bearing on the quality of the exam or consultation), colonic preparation prior to colonoscopy is an essential part of the procedure and greatly determines the eventual outcome. Black and white paper instructions do not visually, cognitively and emotionally evoke within the patient an understanding of the importance of the task ahead or the mental state required to get the colon well cleansed. Indeed, it is well-known, that a paper-based method is not effective when compared to instruction by a health professional [10, 11]. Yet personal instruction is not financially feasible given the great volume of patients and the continuously declining reimbursements for procedures.

Research shows that people are much better motivated when a message reaches them both at a cognitive and an emotional level [12]. The present paper explores the creation of a smart educational documentary that allows the Mayo Clinic team to test in a randomized controlled trial whether the concept of smart learning, that combines science, emotion and digital technology in an artistic documentary, better conveys the importance of colon cleansing prior to the colonoscopy than current paper-based instructions.

## 2 Related Work

### 2.1 Enhancing Compassionate and Emotional Responses to Healthcare Issues

Research by Patel and Toulson evaluates the concept of *creative communication*, which describes the sharing and exploration of non-arts knowledge and data in a creative manner—utilizing, for example, art, storytelling, gameplay, filmmaking, sound and music, animation and curated exhibitions. By engaging in creative communication, it has been observed that detailed scientific, technical and even political material can be presented to a wide, non-specialist audience, enhancing impact, engagement, interaction and education. In particular, Patel and Toulson describe the use of creative documentary filmmaking in enhancing compassionate and emotional responses to healthcare issues [13]. Their research has shown that by capturing footage from an artistic viewpoint, it is possible to educate healthcare practitioners and the general public in more compassionate ways, resulting in a deeper engagement with the scientific facts and medical conditions that they need to understand. Patel's film *The Golden Window* follows a neonate who undergoes whole-body cooling during the first 72 h after a traumatic birth [14]. The film captures contradictory and unanticipated emotions that the Neonatal Intensive Care Unit generates, interweaving the candid and immediate thoughts of parents and staff as they experience this world they call 'the bubble'. *The Golden Window* differs from traditional reality and fly-on-the-wall documentaries by focusing particularly on creative aspects of narrative, characters, color, soundscapes and artistic cinematography, and utilizing the latest digital technology (35 mm sensor cameras, macro lenses, audio and visual post-production software such as Da Vinci and Resolve) to enhance and heighten the experience of the viewer.

### 2.2 Endoscopic Multimedia Information System (EMIS)

Since 2003 the Mayo Clinic has worked on creating an automated system to capture, analyze and summarize video files representing an entire endoscopic procedure [4]. The system is called the EMIS for Endoscopic Multimedia Information System and design effort has specifically focused on resolving issues around colonoscopy. Three things need to occur simultaneously in order for a colonoscopy to be of high quality. First the colon needs to be well prepared (*C*lean). Second, most if not all of the mucosa needs to be inspected (*Look Everywhere*). Thirdly, all neoplastic lesions, where possible, need to be completely removed (*Abnormality Removal*) [5, 15]. Collectively, these three features combine to form the CLEAR acronym. EMIS uses computer-based algorithms to analyze the image stream generated during colonoscopy for specific metrics based on the CLEAR principle. EMIS can detect whether the colon is clean, whether the endoscopist removes remaining debris, whether the endoscopist tries to inspect the entire colon and whether polyps are removed. It has been shown that the manual EMIS annotation technique is reproducible among annotators with fair to good inter-operator agreement; inter-operator agreement is best for very low and very high quality procedures, but varies when quality is average [16]. The automated EMIS technology results correlate with manual annotation results, and both manual and automated annotations correlate with Adenoma Detection Rate—the most widely accepted main determinant of colonoscopy quality—for a set of video files representing the work of a single endoscopist or an endoscopy group [7].

## 2.3   Colon Preparation Approaches and Protocols

In general there are three approaches to patient instruction prior to colonoscopy. The first one is using a paper-based or web-based set of instructions. This is the most commonly used method and also used at Mayo Clinic. A second approach includes personal instructions, either in a group session, in person or via telephone. The latter method, personal communication via telephone call with the patient right before the start of the actual preparation, "just-in-time education", has been reported to result in good to excellent colonic preparation [10, 11, 17]. The third approach involves the use of digital technology for conveying instructions, either through online videos or mobile applications. Most video files are a combination of audiovisual instructions that follow the same outline as the paper- and web-based lists. Mobile applications have the advantage of allowing the users to set up reminders throughout the stages of colon preparation, which help keep the process on track. However, while "smart" technology the existing mobile applications do not engage the user through emotional responses [18, 19].

## 3   Research Methods

### 3.1   Research Design

At KES-STET 2014 members of the research teams at Mayo Clinic and Anglia Ruskin University initiated discussions on the creation of a smart educational documentary that combined science and emotional storytelling with the aim of improving screening for CRC. In the months following the meeting the idea of creating a short film focused on CRC and colonoscopy was formed; unlike prior educational movies, this project aimed to expand on the usual scientific information by including an emotionally charged message with the intent to more deeply motivate patients to clean their colon as good as possible: a smarter form of education compared to existing video files. The project included two research phases. In phase one the artistic documentary is created. In phase two, the artistic documentary (the "intervention") is tested by randomly assigning patients to either the usual paper or web-based instructions ("control" group) or the same instructions combined with an expectation of viewing the artistic documentary ("intervention" group). For each group, video files of the colonoscopy procedures will be obtained to allow automated analysis of the colon preparation using EMIS.

This article describes and reflects on phase one of the research, with phase two currently in progress. Phase one was funded by the Mayo Clinic Slaggie Cancer Patient Education Fund.

## 3.2   Collaboration

The project collaboration relied on Anglia Ruskin University (UK) bringing their experience and expertise of artistic documentary filmmaking to deliver Mayo Clinic's (US) specific requirements for communicating better with patients. Collaboration in the design stage of the project utilized a number of networking and sharing tools such as Skype and Dropbox. A key aspect of the early discussions was focused on establishing common language and understanding of each other's professional disciplines. For Anglia Ruskin it was essential to get an insight into the physician's workings, environment and team at Mayo Clinic, which was facilitated by Skype meetings and virtual tours. This was particularly important because a pre-filming recce was not possible due to funding limitations and time constraints.

Anglia Ruskin needed to convey the breakdown of a documentary film production team (producer, director, cinematographer, sound, editor, and graphics) and the production requirements for making a professional film, including potential issues with contributors, equipment, finances, location, sound and data archiving. It was also important to make sure that there was agreement on the film not being a 'movie' (which implies a dramatic approach) with actors but a 'documentary film' with 'contributors' with real life stories and experiences. The ethical requirements and procedures were discussed and agreed upon.

## 3.3   Information Gathering

Director Patel had made a number of films previously (for the BBC and Discovery channels) within the medical environment, and was therefore familiar with ethical protocols and implications; restrictions on movement and equipment; and patient flow. The teams discussed the logistical details of the journey of patients coming in for screening, i.e. the waiting room, registration, pre-screening check-up, changing clothes, moving to the screening room and then eventually to recovery. This was essential to identify the timings and to evaluate how this process could be filmed observationally—considering where to place cameras, what equipment to use, and how the audio was to be recorded. For sound recording it was important to identify when radio microphones would be worn, and by which staff and patients, in order to minimize any counter-productive disruption. The detailed background information and visual understanding of the physical environment allowed potential creative approaches to be identified and evaluated by the director and production team; particularly enabling the uses of associative visual imagery, mnemonic artifacts, graphics and visual landscapes to be clarified and to create a shot list to support the narratives of the film.

During pre-production research, an analysis of the historical and current landscape of public information films and data was conducted, particularly identifying local (i.e., Mayo Clinic specific), national (US) and international approaches [20]. Dry facts dominated 'doctor to patient' web-based instructional videos as well as more 'commercial' celebrity led videos [21, 22]. These examples were evaluated in order to understand the effectiveness of previous approaches [23]. The findings supported the proposal that a purely fact-based instructional video had less impact than a film that presented an emotively led narrative. Further research into healthcare communication campaigns running in the Mid-West area (particularly Minnesota) illustrated that CRC campaigns featured a more abstract approach and used anonymous faces, predominantly actors, rather than 'real' people. Second, a more humorous rather than sympathy charged (i.e., a focus on loss) approach appeared to be more effective [20]. The analysis also highlighted the potential benefits of an international collaboration in such an area, with a UK team applying an approach not ingrained in the institutional culture of the US healthcare system.

## 3.4   Target Audience and Key Messages

The teams decided that the film should be less than 20 min and in an NTSC video format. The target audience would be diverse in all aspects of age, gender, socio-economic and cultural background (including patients with Midwest, Afro-Caribbean and Native American origins) [24]. Additionally, key healthcare stats to be included in the film were identified; the existing method employed by the Mayo Clinic was a dense, informational heavy print literature, which needed to be simplified into consumable 'bites' of information. The key messages to be communicated were identified, particularly the aims and 'call-to-action' of the film, which was to encourage patients to conduct proper preparation prior to colonoscopy and attend regular screenings. In particular the context and statistical data of CRC, needed to be communicated, as well as an understanding of the causes (particularly lifestyle related), the impact, treatment, prevention and procedure of the screening.

In order to achieve these aims, the following issues or 'myths' had to be addressed: perceptions of the preparation prior to the procedure and of the actual colonoscopy process. What would motivate a patient to attend and properly prepare for a CRC screening emotionally? The following messages and 'enablers' were identified for the film:

- Patients could be encouraged to attend with a friend or relative
- Encourage patients to approach colonoscopy as a test for their best interest rather than something that 'must' be done
- If patients attend a colonoscopy they may be around for their children
- If patients catch issues early, then they don't have to deal with the need for more invasive and costly treatment, or the length (and pain) of such treatment

Dr. Topin Austin, a Consultant Neonatologist (Cambridge University Hospitals, UK), shared the story of his father's experience as a patient, who passed away from CRC after a very late diagnosis:

> As a doctor we see patients come and go... [What we] never really saw was the context the disease played in their lives. Like prison - everyone in hospital is in the same uniform - the awful NHS gown. Yet everyone has a backstory - family, work, friends....

This insight, combined with an article written by the Pulitzer prize winning columnist for the Miami Herald, Dave Berry, was particularly eye-opening [25]. Berry reflecting upon his annual colonoscopy, stated "This horrible disease takes so much from us, both caregiver and patient alike, but never let it take away your ability to laugh". The article continues to describe the colonoscopy process from the viewpoint of a patient, from the consumption of the MoviPrep™ ("a nuclear laxative" comparable to the "launch of the space shuttle"), through to the wearing of the hospital garments that make you "feel more naked than when you are actually naked" and the colonoscopy itself; "the moment I had been dreading for more than a decade". Finally Berry describes the results:

> I felt even more excellent when Andy (the consultant) told me that it was all over and my colon had passed with flying colors. I have never been prouder of an internal organ.

The narrative structure applied by Berry provided the design for the 'script' and spine of the film developed.

## 4 Film Production

### 4.1 Patient Selection

The patients were selected after a number of interviews the director conducted via Skype, through which both parties built up a trusting relationship which was essential for the actual interview in the USA to proceed smoothly. What emerged from these conversations was that for the patients their families were their 'reason for living'. As such, the initial patient list, which was predominantly female, was extended to include their partners. Furthermore, it was decided that we also needed to include a patient who was undergoing the actual process—whom the target audience could accompany experientially, through a sense of anticipation, worries and questions. The final selection of contributors were able to deliver the following information in the film: emotion (fear, trepidation, embarrassment); myth busting (preparation and process), humor (the macho approach to colonoscopies) and enablers (guilt, responsibility, companions and understanding).

## 4.2   Filming in Minnesota

A tight and focused schedule was essential to this international collaboration, particularly as the contributors would be travelling long distances to attend the filming at the Mayo Clinic film studio. The visual design was established between the cinematographer and the director. An intimate film style was agreed upon with interview eye line as close to the camera as possible with an anonymous backdrop (depth of field emphasized by intricate, unobtrusive lighting; Fig. 1). Wide angle lenses and a camera 'slider' were utilized to ensure a smooth style and approach.

The technical specifications were finalized according to affordability and avail-ability, and are summarized in Table 1.

**Fig. 1**  Intimate filming style with an abstract backdrop

**Table 1** Technical specifications for film and audio

| Technical specifications—visuals | |
| --- | --- |
| Camera | Sony C500 Digital Camera with Odyssey 7Q external recorder to convert footage to Apple ProRes |
| Format | Apple ProRes 422, 1080i, NTSC 29 fps |
| Technical specifications—audio | |
| Recording System | Double system recording: audio recorded in sync with the picture + on a separate recorder (Sound Devices 702) to be synched to the picture in postproduction. |
| Format | 24-bit 48 kHz wav files |
| Microphones | Radio mics (Sennheiser ew 100 G2) and boom mic (Sennheiser ME 66) |

## 4.3 Creating the Film

The filming took place from 18–22 August 2015 in Rochester, Minnesota. An additional day's preparation was essential for testing the equipment at the facilities house and setting up lighting and audio in the Mayo Clinic studio used for interviews. Ethical issues and consent forms were discussed and completed with the contributors prior to filming.

The intimate set up in studio was created by 'blacking out' the extra members of the crew (audio, producers) behind a cloth or out of eyesight, in order for the interviewee to retain an intimate conversation with the director who was seated with the cinematographer behind the camera. As the interviews were conducted, the partners would sit out of eyesight as well, but could hear the interview. Interviewees were either nervous or curious, but the trust in the relationship between the Mayo Clinic, Anglia Ruskin and themselves, was evident in their level of engagement with the interview and discussion topics.

Questions were asked to provoke a response that was more immediate than pre-planned from the contributors. The spontaneity of response, recollection, pauses, drift—were all essential to building a back-story, a framework that the audience could understand—and to retain that sense of authenticity from their interviews. During the interviews, male contributors provided a very honest male perspective on the procedure, with both humor and emotion. This type of intimate interview can trigger a cathartic reflection as they discuss the emotional journey they have been through with an objective observer. Other insights emerged from these interviews that were built into the evolving post-production script: putting off the process with excuses that there was not time to do it; how to nag a partner into screening and an understanding of the physician's motivation in addressing these issues.

At short notice, the Mayo Clinic team identified a patient who was happy to be filmed through the colonoscopy process. This "live" patient was extremely important, as *The Golden Window* had demonstrated that there is extreme power and emotional connection when talking to someone 'in the moment'. The patient was nervous but also extremely articulate and gave detailed insight into his

OVERVIEW                                    DOCUMENTARY SCRIPT

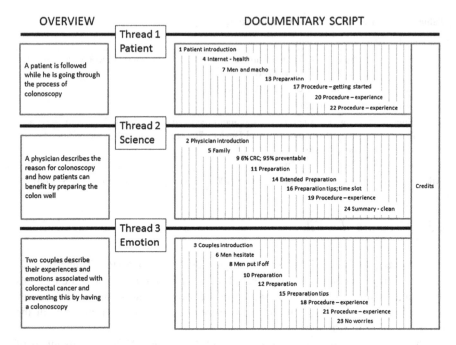

**Fig. 2** Final documentary script

motivation for the screening (he had a wife and four children—he needed to be around for them). He was also approaching the process with humor and patience. An audio only interview was recorded. A radio microphone was then placed on him for the process, as well as on the staff in order for the crew to film observationally without intruding. The patient was confident enough to ask questions before, during and after screening; questions that the target audience would want to ask and have answered. His journey drove the narrative structure, similar to Berry's recollection of his colonoscopy. The story took on a sense of dramatic anticipation; i.e., will he or won't he be fine after the screening. The pre-production script was adapted and revised through the filming process by the director as and when the content was gathered (Fig. 2).

By employing a dramatic narrative arc within the film (i.e. focusing on how the patient would deal with the process and what the outcome would be), the viewer actively engages the viewer with the film and information embedded within the emotional storytelling (Figs. 3 and 4). The 'will he or won't he be ok' strand asks the audience to question what they would do in the same scenario. This pro-active engagement through a sense of anticipation and self-reflection is an essential component of this smart audiovisual application.

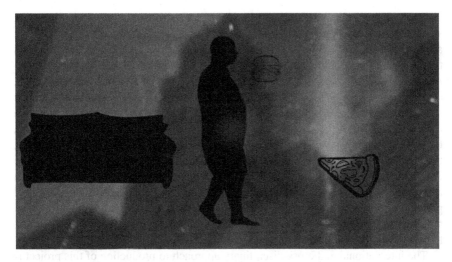

**Fig. 3** Graphics highlight the impact of sedentary lifestyle and poor diet on our health. Whooshes and warning sounds are used to draw attention to the different risk factors

**Fig. 4** Educational graphics that combined with a humorous voiceover track and jazz music, introduce audiences to a patient's approach to colon preparation

## 5 Discussion and Conclusions

Recent scholarly research has established that 'empathy' is the product of both cognitive and emotional processes and that the line between the two is blurred. Decety and Jackson define empathy as "the naturally occurring subjective experience of similarity between the feelings expressed by self and others without losing sight of whose feelings belong to whom" [26]. The construction of the film using smart digital technology based itself on creating this empathy between the viewer and the contributors within the film, focusing on universal themes and questions that the audience could identify with, in order to maximize the impact of information. Audiences are invited to engage with the story and the characters, encouraging the process of self-identification and active engagement with the thought processes of the preparation for colon screening as well as the reflection on the consequences of their actions or 'in-actions'. As such, the contributor selection, the use of humor, easily relatable graphics, quality visual and audio production, together with a designed soundscape of music and sound effects are combined to not just appeal to the 'target audience' but to set a stage and seed of thought which asks them to question their approach and rationale to health screenings and preparation. Indeed, the overall intent was to create a smarter learning environment that emotionally engages the patient to closely follow the evidence-based instructions in order to optimally cleanse the colon prior to colonoscopy.

The international and cross-disciplinary approach to production of this project is unique and constructive. Phase two of the project will provide qualitative data through which to analyze this approach. Previous research demonstrates that focused, high-profile audiovisual campaigns have led to surges in attendance of screenings; this project will hopefully yield long-term sustainable changes in patient behavior to such screenings [27]. We envisage that this project is the first step towards breaking existing barriers to international collaborations between the health and creative sectors. We also expect it to pave the way for the development and understanding of innovative methods utilizing smart technology and creative communication to empower and invigorate patients to take greater responsibility for their health. Lastly, we hope that our work will contribute to the ultimate goal of colonoscopy: to save lives.

**Acknowledgments** This project was supported by Mayo Clinic, Anglia Ruskin University and the Mayo Clinic Slaggie Cancer Patient Education Fund.

# References

1. Siegel, R.L., Miller, K.D., Jemal, A.: Cancer statistics, 2016. CA Cancer J. Clin. **66**, 7–30 (2016)
2. Xirasagar, S., Li, Y.J., Hurley, T.G., Tsai, M.H., Hardin, J.W., Hurley, D.M., Hebert, J.R., de Groen, P.C.: Colorectal cancer prevention by an optimized colonoscopy protocol in routine practice. Int. J. Cancer. J. I. du Cancer (2014)
3. Corley, D.A., Jensen, C.D., Marks, A.R., Zhao, W.K., Lee, J.K., Doubeni, C.A., Zauber, A.G., de Boer, J., Fireman, B.H., Schottinger, J.E., Quinn, V.P., Ghai, N.R., Levin, T.R., Quesenberry, C.P.: Adenoma detection rate and risk of colorectal cancer and death. N. Engl. J. Med. **370**, 1298–1306 (2014)
4. de Groen, P.C.: Advanced systems to assess colonoscopy. Gastrointest. Endosc. Clin. N. Am. **20**, 699–716 (2010)

5. Oh, J., Hwang, S., Cao, Y., Tavanapong, W., Liu, D., Wong, J., de Groen, P.C.: Measuring objective quality of colonoscopy. IEEE Trans. Bio-Med. Eng. **56**, 2190–2196 (2009)
6. Muthukudage, J., Oh, J., Tavanapong, W., Wong, J., De Groen, P.C.: Color based stool region detection in colonoscopy videos for quality measurements. In: Lecture Notes in Computer Science (including subseries Lecture Notes in Artificial Intelligence and Lecture Notes in Bioinformatics), LNCS, vol. 7087, pp. 61–72 (2011)
7. Rex, D.K., Petrini, J.L., Baron, T.H., Chak, A., Cohen, J., Deal, S.E., Hoffman, B., Jacobson, B.C., Mergener, K., Petersen, B.T., Safdi, M.A., Faigel, D.O., Pike, I.M., ASGE/ACG Taskforce on Quality in Endoscopy: Quality indicators for colonoscopy. Am. J. Gastroenterol. **101**, 873–85 (2006)
8. Boroff, E., Crowell, M., Leighton, J., Faigel, D., Gurudu, S., Ramirez, F.: The relationship between withdrawal time and intubation time in colonoscopy: correlation with adenoma detection rate (ADR). Am. J. Gastroenterol. **107**, s806–s807 (2012)
9. Johnson, D.A., Barkun, A.N., Cohen, L.B., Dominitz, J.A., Kaltenbach, T., Martel, M., Robertson, D.J., Richard Boland, C., Giardello, F.M., Lieberman, D.A., Levin, T.R., Rex, D.K.: Optimizing adequacy of bowel cleansing for colonoscopy: recommendations from the US Multi-Society Task Force on Colorectal Cancer. Am. J. Gastroenterol. **109**, 1528–1545 (2014)
10. Abuksis, G., Mor, M., Segal, N., Shemesh, I., Morad, I., Plaut, S., Weiss, E., Sulkes, J., Fraser, G., Niv, Y.: A patient education program is cost-effective for preventing failure of endoscopic procedures in a gastroenterology department. Am. J. Gastroenterol. **96**, 1786–1790 (2001)
11. Liu, X., Luo, H., Zhang, L., Leung, F.W., Liu, Z., Wang, X., Huang, R., Hui, N., Wu, K., Fan, D., Pan, Y., Guo, X.: Telephone-based re-education on the day before colonoscopy improves the quality of bowel preparation and the polyp detection rate: a prospective, colonoscopist-blinded, randomised, controlled study. Gut (2013)
12. Taylor, J.S.: Learning with emotion: a powerful and effective pedagogical technique. Acad. Med.: J. Assoc. Am. Med. Coll. **85**, 1110 (2010)
13. Patel, S., Toulson, R.: Educating and enhancing compassion, emotion and reflective professional practice through contemporary digital filmmaking. In: Smart Digital Futures, vol. 262, pp. 582–591. IOS Press (2014)
14. Patel, S.: The Golden Window. Eyeline Films (2013)
15. Gupta, R., Brownlow, B., Domnick, R., Harewood, G., Steinbach, M., Kumar, V., de Groen, P.: Colon cancer not prevented by colonoscopy. Am. J. Gastroenterol. **103**, S551–S552 (2008)
16. Bakken, J., van Leerdam, M., Enders, F., Tavanapong, W., Oh, J., Wong, J., de Groen, P.: Colonoscopy peer review utilizing automated video capture. Am. J. Gastroenterol. **104**, 1391 (2009)
17. Rosenfeld, G., Krygier, D., Enns, R.A., Singham, J., Wiesinger, H., Bressler, B.: The impact of patient education on the quality of inpatient bowel preparation for colonoscopy. Can. J. Gastroenterol. = Journal canadien de gastroenterologie **24**, 543–546 (2010)
18. Kakkar, A., Jacobson, B.C.: Failure of an internet-based health care intervention for colonoscopy preparation: a caveat for investigators. JAMA Intern. Med. Online first (2013)
19. Kang, X., Zhao, L., Leung, F., Luo, H., Wang, L., Wu, J., Guo, X., Wang, X., Zhang, L., Hui, N., Tao, Q., Jia, H., Liu, Z., Chen, Z., Liu, J., Wu, K., Fan, D., Pan, Y.: Delivery of instructions via mobile social media app increases quality of bowel preparation. Clin. Gastroenterol. Hepatol.: Official Clin. Pract. J. Am. Gastroenterol. Assoc. (2015)
20. Kuruvilla, C.: Minnesota takes aim at colon cancer with cheeky new ad campaign. http://www.nydailynews.com/life-style/health/minnesota-takes-aim-colon-cancer-cheeky-new-ad-campaign-article-1.1296786 (2013). Accessed 18 Jan 2016
21. Butterly, L.: Preparing for a Colonoscopy. https://www.youtube.com/watch?v=xd1N0WOcd5A (2010). Accessed 18 Jan 2016
22. Couric, K.: Katie Couric's Colonoscopy Prep. https://www.youtube.com/watch?v=CbUesuxT1IE (2011). Accessed 18 Jan 2016
23. Galati, J.: Colonoscopy Bowel Prep: what a bad one looks like. https://www.youtube.com/watch?v=VCupasJ2Tbs (2012). Accessed 18 Jan 2016

24. Staff, I.C.: Colorectal cancer rates higher for Minnesota's Natives. http://indiancountrytoday medianetwork.com/2007/10/03/colorectal-cancer-rates-higher-minnesotas-natives-91504 (2007). Accessed 18 Jan 2016
25. Barry, D.: Dave Barry: a journey into my colon—and yours. http://www.miamiherald.com/living/liv-columns-blogs/dave-barry/article1928847.html (2009). Accessed 18 Jan 2016
26. Decety, J., Jackson, P.L.: The functional architecture of human empathy. Behav. Cogn. Neurosci. Rev. **3**, 71–100 (2004)
27. Cram, P., Fendrick, A.M., Inadomi, J., Cowen, M.E., Carpenter, D., Vijan, S.: The impact of a celebrity promotional campaign on the use of colon cancer screening: the Katie Couric effect. Arch. Intern. Med. **163**, 1601–1605 (2003)

# Tangible Interfaces for Cognitive Assessment and Training in Children: LogicART

Fabrizio Ferrara, Michela Ponticorvo, Andrea Di Ferdinando
and Orazio Miglino

**Abstract** This paper describes how to use tangible interfaces for cognitive assessment and training. Assessment and training are two fundamental phases in the learning process: assessment allows to gather information in order to identify the learner starting level and monitor progresses that training can lead to. Assessment uses tests, some of which are addressed to cognitive abilities. Many tests for cognitive abilities are based on verbal materials that are not suitable for population such as young children or people with special needs, and people with hard cognitive disabilities or sensory impairment. In these cases it would be more fruitable to use physical objects. This paper illustrates a hardware-software system, LogicART, that can be used in assessment and training of cognitive abilities, such as reasoning, memory, categorization, etc. exploiting physical materials augmented by RFID/NFC Technology. These materials, employed to study cognitive abilities, offer the advantage to stimulate multisensoriality and manipulation.

**Keywords** Assessment · Testing · RFID/NFC Technology · Deductive reasoning · Tangible interfaces · Technology enhanced learning

## 1 Introduction

In nowadays society, learning is perhaps the most compelling challenge, as it requires to embrace innovation in education, to stimulate new skills and to build and share knowledge in completely new contexts.

In this framework, the starting point for every learning process should be a rigorous cognitive assessment, that is to say an assessment of abilities including memory, reasoning, categorization which are a prerequisite for learning. Once the assessment

F. Ferrara (✉)
Department of Psychology, Second University of Naples, Caserta, Italy
e-mail: fabrizio.ferrara@gmail.com

M. Ponticorvo · A. Di Ferdinando · O. Miglino
Department of Humanistic Studies, University of Naples "Federico II", Naples, Italy

© Springer International Publishing Switzerland 2016        329
V.L. Uskov et al. (eds.), *Smart Education and e-Learning 2016*,
Smart Innovation, Systems and Technologies 59,
DOI 10.1007/978-3-319-39690-3_29

is over and relevant information about the level and the specific difficulties displayed is recorded, it is possible to start a personalized and effective training to improve a specific skill or ability. Assessment and training are two inseparable twins that support every learning enterprise devoted to every cognitive skill.

The present paper describes an integrated hardware and software system specifically addressed to assess and train any cognitive ability; currently it has been developed an activity for reasoning.

Reasoning is the ability to draw conclusions from given premises through inferences. It leads people to acquire new information from the already possessed ones [17] and it is basic for knowledge process and cognitive abilities, as problem solving and decision making. Since Cattell's work [1] this ability has been connected to fluid intelligence, because this capacity to reason and solve novel problems is independent of any knowledge from the past. There are different types of reasoning whose main are deductive, inductive, abductive, and analogical that unroll differenlty along ontogenetic development.

According to some studies [9, 10], for example, analogical reasoning is present since birth. It is extremely important as it can help people, and other animals as well [6], to rapidly acquire new information and knowledge to better adapt. Deductive reasoning is the ability to start with stated rules, premises, or condition and to engage in one or more steps to reach a solution to a problem [19].

Deductive reasoning is therefore the fundamental core in intelligence assessment in educational and clinical contexts, in human resources managing and college admissions. It often assumes the form of sequence completion [8] and it is a prerequisite for many valued abilities in school and business such as coding, that permits to learn basic programming concepts.

Raven's Progressive Matrices [18], one of the most used intelligence test for children, is a nonverbal intelligence test typically used in educational settings. It is composed by items that allow to measure reasoning. In each test item, the subject is asked to identify the missing element that completes a pattern, that are often resented in the form of a matrix. In the coloured form, specifically addressed to children with both typical and atypical development and adults with cognitive disabilities such as mental retardation or decay, Raven's matrices are composed by 36 items that imply analogical and sequential reasoning.

It is worth underlining that all the tests, at our knowledge, include numerical, verbal or visual materials.

After the assessment phase, cognitive abilities, including reasoning can be increased with an appropriate and personalized training. Both assessment and training can be run with different materials. Typical cognitive assessment includes pen and paper-based assessment that offers the advantage to be easily usable by a great amount of people, cheap and allows to record subject replies.

Over the past decade, the cognitive assessment and training market has evolved from traditional pen and paper-based tests to digital assessment techniques. Employing digital materials offers additional advantage, for example the chance to trace the whole process. One drawback of digital material is that they require the use of specific interfaces, for example the mouse, that, even if can be very easily acquired, nonetheless need a preliminary familiarization.

Moreover, if reasoning assessment can be led in very early phases of child development or in presence of cognitive and sensory disabilities, it would be useful to foresee different kind of materials that allow a very direct user-interface interaction, namely physical objects. The use of physical object opens the chance to assess reasoning and other cognitive abilities since the first life months and in groups of people for whom digital or abstract manipulation can be difficult. If physical materials are chosen, it is possible to include every material which is interesting in some respect and include all its features, such as odour, sound, surface; in other word it is possible to use them as multisensory materials. This allows to have a multisensory assessment which takes place in the subject life environment in a transparent way. Another feature conveyed by physical objects, that is particularly useful for children and people with special educational needs, is manipulation: the physical manipulation of objects represents a basic and powerful way to get in touch with the external environment that appears very early in children's life and re-activates in stressful conditions.

Using physical materials allows to overcome the limits of traditional materials used in this kind of tests. For example, numerical materials that require a mathematical expertise together with reasoning; verbal materials that require a linguistic knowledge or visual materials that required sighted people.

What it is proposed in this paper, it is to exploit the advantages to use physical materials in cognitive assessment and training, without giving up to the chances offered by digital materials. This is achieved by using tangible interfaces, augmented and technology enhanced learning (TEL) materials.

Tangible interfaces have been widely used in education, starting from Montessori seminal work [16] and the digital materials inspired by her work [22]. Indeed there is a wide consensus that tangible interfaces can enhance learning [11]. In what follows we propose to use this TEL material for cognitive assessment in particular for reasoning. This approach has been proposed for the assessment of spatial ability [20], a practical ability that is crucial for human adaptation [15], but we do believe that tangible interfaces can be fit for abstract cognitive abilities as well.

## 2  STELT: Smart Technologies to Enhanced Learning and Teaching

The STELT platform, Smart Technologies to Enhance Learning and Teaching [12, 13], links physical and digital applications [14], whose architecture is reported in Fig. 1. It is precisely devoted to build active environments for learning. STELT consists in a software platform that combines the hardware components management (sensors and actuators) and software components (libraries for the storyboard and provision of feedback, authoring systems to be used by non-programmers), supporting learning environments based on physical object handling, crucial for tangible

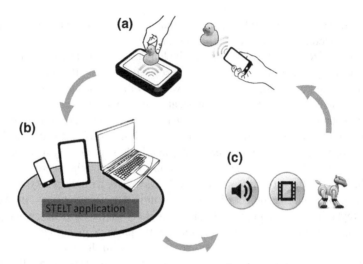

**Fig. 1** STELT platform functional representation

interfaces. STELT implements augmented reality systems based on RFID (radio-frequency identification) and NFC (near field communication) technology, introduced below [21]. The labels RFID/NFC (tags) are very thin transponders that can be applied to any type of object and are detected by small readers. The reader can be connected to a computer with either a wired or wireless connection or integrated into standard equipment on smartphones and tablets (NFC sensor). STELT combines communication protocols with the various hardware devices (readers and output devices), a storyboarding environment for creating various interaction scenarios, a database for tracking user behaviour and an adapting tutoring system [7] that can build a user profile providing customised feedback. This platform works as follows:

a. the learner places a tagged object onto a tablet reader or moves a reader or smartphone close to the tagged object;
b. the signal containing the objects code is sent to a computer (desktop, notebook, tablet or smartphone) containing the STELT platform;
c. once it has entered the STELT system, the signal generates a number of actions from the output devices (audio system, monitors) depending on the current scenario created with the storyboarding module.

Furthermore, the same signal is memorised and analysed by the adapting tutoring module, so that a customised profile is created for the user that guides the subsequent system responses. Human machine interaction takes places solely by the handling/identification of physical objects and by the activation of audio or visual feedback. From a technical point of view, STELT is an SDK (software development kit) containing software libraries for sensor, storyboarding, monitoring and adapting tutoring management and for the creation of applications in Windows (Vista, 7 and 8), Android and iOS environment. STELT platform has been used to implement different products, such as Block Magic, a hybrid physical/software tool that enhances traditional blocks and methods for teaching in kindergarten and primary schools [4, 5]. In this STELT application many tasks for children are present, addressing memory, creativity, logic. For example the *logic train* exercise asks the child involved in the game to complete a sequence with the appropriate missing block. The software shows on the monitor a brief sequence and the child can put the selected block on the table. The software recognizes the block and rewards the child with an approval sentence if the block is correct or encourages her to try again if the block is incorrect.

STELT system has been tested during the cited project, but it is still under development. Up to now it has been tested with many users, especially children, who confirmed the system attractiveness and teachers, who underlined the system efficacy. More details can be retrieved in reports about BM [4, 5].

## 3   LogicART Integrated System

LogicART, Logic Abstract Reasoning Test, is an integrated system consisting of a set of blocks (the augmented Logic Blocks), a tablet device, (the Logic Table) and a specific software (see Fig. 2). It is based on STELT platform introduced above and links together smart technologies and physical materials, uniting the manipulative approach and touch-screen technologies. The physical materials reproduce the Logic Blocks widely used in education [3], made up of a set of blocks (48 pieces) that differentiate for four attributes: geometric shape (triangular, squared, rectangular and circular), thickness (thick and thin), colour (red, yellow and blue) and dimension (big and small).

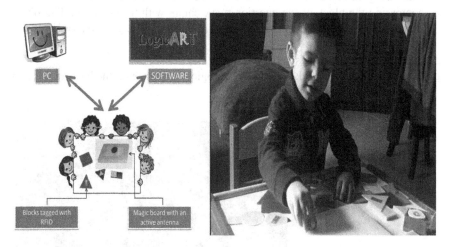

**Fig. 2** LogicART functional representation (on the *left*) and a child playing with the blocks and the Table (on the *right*)

The traditional logic blocks are equipped with RFID tags. This configuration permits to a PC, with LogicART software installed on, to connect with the Logic Table which hosts an hidden antenna that recognizes each block, sends a signal to the PC, and produces a feedback. The LogicART software engine is devoted to receive input from the Logic Table and generate an "action" (aural and visual). These actions implement the direct feedbacks the user can receive interacting with the system. These feedback are regulated by an Adaptive Tutor System [7] embedded that ensures autonomous interaction between the user and the system, receiving active support, corrective indications, feedback and positive reinforcement from the digital assistant on the outcome of the actions performed. The LogicART software is specifically meant to assess and train deductive reasoning.

## 3.1   LogicART Activity

As hinted before, LogicART is a module dedicated to deductive reasoning. In this module, the activity consists in completing a logical sequence made up of four blocks, one of which is unknown and represented by "?". The question mark can be placed in any of the four sequence positions (see Fig. 3). We will refer to type sequence A, B, C or D, depending on the position of "?". The A type has the question mark in the first position, B in the second, etc.

**How to solve logical sequences** Logical sequences are devised so as that they can all be solved using the same simple reasoning strategy (see Fig. 3):

a. grouping the four blocks in two chunks, one with the question mark and one without it;
b. comparing the same chunk blocks, detecting which feature/s vary;
c. comparing the block of the question mark chunk with the corresponding block of the other chunk (i.e. comparing blocks in first and third position, or in second and forth position);
d. detecting which feature/s vary;
e. determining which block completes the logical sequence on the basis of the two previous comparisons.

**Fig. 3** Logical sequences

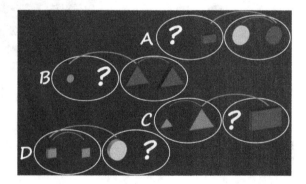

**Fig. 4** Percentage of mistakes for sequence type. Labels indicate how many features vary in chunks (IN) and between chunks (INTER)

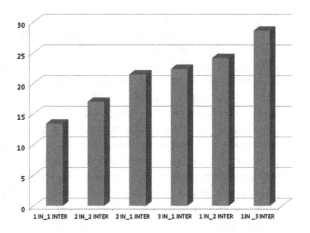

## 3.2 Pretest

A preliminary study was conducted to establish the logical sequences difficulty level in order to calibrate/adapt LogicART for users (adults or children). Sources of variation in logical sequences are the position of the "?" (sequences of type A, B, C, or D); the number of features that varies between blocks of the same chunk (in-chunk variance, from now IN-V) and between correspondent blocks of different chunks (inter-chunk variance, from now INTER-V).

In the pretest, four exercises for each of the 6 IN-V/ INTER-V combination were presented, for a total amount of 24 logical sequences.

Twenty-eight undergraduates students completed the sequences as unpaid and anonymous participants through a Google Drive questionnaire. Since question mark position is a fundamental variation source for logical sequences presented in LogicART, regardless of the users' cognitive level, we considered for analysis only the IN-V/ INTER-V variance. Results (Fig. 4) showed that the number of mistakes increases when three features vary between the corresponding blocks of the chuncks (28,6 %).

Since the software needs a large number of game sessions to be tested, we setted it on a hard level mode (i.e. only with the most difficult type of logical sequence) for undergraduates'usage.

## 3.3 Assessment and Training in LogicART

Starting from the information coming from the pretest, assessment and training in LogicART are designed. LogicART is composed of five exercises: an assessment test and four specific training tasks. According to Dell'Aquila et al. [2], educational games are structured in three layers: hidden layer (the formal rules and game architecture), external layer (what the user will see during the game) and evaluation layer (that analyses users performance, and notifies it). In the description of LogicART we will refer to this definition. LogicART exercises are turn-based games between two players, an artificial agent and the user. Artificial agent presents stimuli (instructions, logical sequences, feedbacks) through images; user can answer and continue the game by putting a block on the table.

In the assessment test user has to solve 24 logical sequences, as hinted before, which differ only for the question mark position in the sequence. The test is made up of numerous scenes and it takes about 20 min to complete it. In the first scene (considering the external layer), a teacher asks, through balloons and speech synthesizer, to help her little pupil to solve some simple logical sequences; after a brief introduction of the task and of the features of the blocks (see Fig. 3), 24 logical sequences to solve are randomly presented. The teacher asks the user to find the block that completes the sequence and to put it on the board.

Considering the hidden layer, every time the user completes a sequence, the system evaluates the answer correctness, counts the overall mistakes number and the specific number of errors given the type of logical sequence. It then marks the sequence as done (so it will be not presented anymore), and calculates how many exercises are left to solve. Therefore, in the external layer users don't have any immediate feedback about the accuracy of the answer in this phase: After completing a logical sequence, she is only informed about how many exercises remain to solve.

When all sequences are completed, the teacher provides a graphic feedback on the number of correctly solved exercises (see Fig. 5), and an evaluation of users performance compared to other users who already completed the assessment test. If correct answers have been less than 23, further information is provided about the correct resolution strategy, and about the type of mistake user committed more frequently (error profile), recommending a specific training tasks.

**Fig. 5** Graphic feedback on correctly solved exercises

Error profiles are calculated matching and combining the mistakes number, given the position question mark. They are:

**AB profile**: Mistakes are more frequent when the question mark is in the first chunk. User shows difficulties in the in-chunk comparison between the third and the forth blocks.

**CD profile**: User made most of errors in sequences of type C and D. Difficulties are linked to the comparison between blocks of the first chunk.

**AC profile**: Mistakes are more frequent when question mark is placed in the first position of the chunks and user has to compare the second and the forth sequence blocks.

**BD profile**: User made most of errors in sequences of type B and D, where "?" is placed in the second position of the chunks and it is required to compare the first and the third blocks of the sequence.

The users are addressed to a training that is specific for their profile. During training the teacher asks to find the block that completes a logical sequence. In the hidden layer, the software selects randomly a sequence among a database of 50 exercises. Nevertheless, this time an immediate feedback is provided to the user about answer accuracy. If the answer is correct, user is invited to solve another sequence, otherwise, in case of wrong response, she is invited again to find the right block.

## 4　Future Directions

In this paper we have described LogicART, an integrated hardware and software system for cognitive abilities assessment and training. Even if the logical sequences are yet to be tested and validated for children, it is possible to affirm that the system offers the appealing chance to design completely new assessing and training enviroments where manipulation and multisensoriality are central. These environments would be fit for precocious life stages and for special educational needs, for example with children with cognitive disabilities or sensory impairments. The chance to go beyond verbal materials opens completely new pathways for assessment: including sense of smell, hearing, touching, thanks to the describe technologies, will allow to assess and train cognitive abilities in everyday life, in a transparent way, also reducing the reactive effects of structured testing settings.

The next steps for this research are twofold: on one side the goal is to propose new activities for different cognitive abilities, such as memory, categorization, decision making, etc. on the other is to validate this platform comparing the results obtained with traditional methods. Moreover in next years the STELT platform will be integrated in other LMS, learning management systems such as Moodles and MOOCs.

**Acknowledgments** This project has been funded in the framework INF@NZIA DIGI.Tales 3.6 project under PON-Smart Cities for Social Inclusion programme.

# References

1. Cattell, R.B.: Theory of fluid and crystallized intelligence: a critical experiment. J. Educ. Psychol. **54**(1), 1 (1963)
2. DellAquila, E., Di Ferdinando, A., Marocco, D., Miglino, O., Ponticorvo, M., Schembri, M.: New Perspective in Educational Games for Soft-Skill Training. Springer (in press)
3. Dienes, Z.P.: Large Plastic Set, and Learning Logic, Logical Games. Herder and Herder, New York (1972)
4. Di Ferdinando, A., Di Fuccio, R., Ponticorvo, M., Miglino, O.: Block magic: a prototype bridging digital and physical educational materials to support children learning processes. In: Smart Education and Smart e-Learning, pp. 171–180. Springer International Publishing (2015)
5. Di Fuccio, R., Ponticorvo, M., Di Ferdinando, A., Miglino, O.: Towards hyper activity books for children. Connecting activity books and Montessori-like educational materials. In: Design for Teaching and Learning in a Networked World, pp. 401–406. Springer International Publishing (2015)
6. Flemming, T.M., Washburn, D.A.: Analogical reasoning in animals. In: Encyclopedia of the Sciences of Learning, pp. 228–230. Springer US (2012)
7. Freedman, R.: What is an intelligent tutoring system? Intelligence **11**(3), 1516 (2000)
8. Korbin, L.: SAT Program Handbook. A Comprehensive Guide to the SAT Program for School Counselors and Admissions Officers, vol. 1, p. 33 (2006)
9. Goswami, U.: Analogical reasoning in children. Perspectives from cognitive science. In: The Analogical Mind, pp. 437–470 (2001)
10. Goswami, U.: The development of reasoning by analogy. Dev. Think. Reason. **49** (2013)
11. Marshall, P.: Do tangible interfaces enhance learning?. In: Proceedings of the 1st international Conference on Tangible and Embedded Interaction, pp. 163–170. ACM (2007)
12. Miglino, O., Di Ferdinando, A., Di Fuccio, R., Rega, A., Ricci, C.: Bridging digital and physical educational games using RFID/NFC Technologies. J. e-Learn. Knowl. Soc. **10**(3) (2014)
13. Miglino, O., Di Ferdinando, A., Schembri, M., Caretti, M., Rega, A., Ricci, C.: STELT (Smart Technologies to Enhance Learning and Teaching): una piattaforma per realizzare ambienti di realtà aumentata per apprendere, insegnare e giocare. Sistemi intelligenti **25**(2), 397–404 (2013)
14. Miglino, O., Gigliotta, O., Ponticorvo, M., Nolfi, S.: Breedbot: an evolutionary robotics application in digital content. Electron. Libr. **26**(3), 363–373 (2008)
15. Miglino, O., Ponticorvo, M., Bartolomeo, P.: Place cognition and active perception: a study with evolved robots. Connect. Sci. **21**(1), 3–14 (2009)
16. Montessori, M.: The Montessori Method. Transaction Publishers (2013)
17. Shaffer, D., Kipp, K.: Developmental psychology: childhood and adolescence. Cengage Learn. (2013)
18. Raven, J.: The Raven's progressive matrices: change and stability over culture and time. Cogn. Psychol. **41**(1), 1–48 (2000)
19. Rips, L.J.: The Psychology of Proof: Deductive Reasoning in Human Thinking. MIT Press (1994)
20. Sharlin, E., Itoh, Y., Watson, B., Kitamura, Y., Sutphen, S., Liu, L.: Cognitive cubes: a tangible user interface for cognitive assessment. In: Proceedings of the SIGCHI Conference on Human Factors in Computing Systems, pp. 347–354. ACM (2002)
21. Shepard, S.: RFID: Radio Frequency Identification. McGraw Hill Professional (2005)
22. Zuckerman, O., Arida, S., Resnick, M.: Extending tangible interfaces for education: digital montessori-inspired manipulatives. In: Proceedings of the SIGCHI conference on Human factors in computing systems, pp. 859–868. ACM (2005)

# On the Process of Mobile-Assisted Teaching and Learning at FIM UHK—Analysis and Reflection

Ivana Simonova and Petra Poulová

**Abstract** The purpose of this article is to introduce results of monitoring and analysis the process of mobile devices exploitation in the process of instruction at the Faculty of Informatics and Management (FIM), University of Hradec Kralove (UHK), Czech Republic. After a two-decade long experience in using the LMS WebCT (Blackboard), the Blackboard Mobile application has been exploited since 2013/14 academic year. Reflecting the users' feedback each online course was tested separately. The results show some problems which could limit the efficient exploitation of LMS on mobile devices. Finally, some recommendations against them are provided.

**Keywords** Mobile-assisted learning · Smart learning environment · Feedback · Mobile devices

## 1 Introduction

Consistently with computers and initial services provided through the Internet two decades ago, latest development in mobile devices brought changes into the 'traditional' process of ICT implementation into education [1]. A shift has been detected from original, i.e. 'immobile', devices and relating technologies to the mobile ones.

The current mobile-assisted process of instruction is characterized by specific features. The crucial ones are recognized as the potential for learning process to be [2]

- personalized,
- spontaneous,

I. Simonova (✉) · P. Poulová
University of Hradec Kralove, Rokitanskeho 62, Hradec Kralove, Czech Republic
e-mail: ivana.simonova@uhk.cz

P. Poulová
e-mail: petra.poulova@uhk.cz

© Springer International Publishing Switzerland 2016
V.L. Uskov et al. (eds.), *Smart Education and e-Learning 2016*,
Smart Innovation, Systems and Technologies 59,
DOI 10.1007/978-3-319-39690-3_30

339

- informal and
- ubiquitous.

As Koole [3] emphasizes, there are some factors having key roles in the use of mobile devices in teaching and learning, e.g.

- physical characteristics of a mobile device such as its size and weight,
- input and output capabilities such as keypad versus touchpad,
- screen size,
- audio-functions and other ones.

In other words, all the factors should be considered within the planning and conducting the process of mobile-assisted teaching and learning. Moreover,

- users's skills,
- prior knowledge and experience with mobile devices for educational purposes,
- learner's attitude towards the learning through mobile devices etc. definitely play the significant role in the efficiency of the mobile-assisted teaching/learning.

Therefore, identically with e-learning and e-teaching, the didactic preparation and training of the users, i.e. both the teacher, who frequently are the designers of online courses, and the students, has been required so that they reach competences for the efficient exploitation for educational purposes [4].

With the respect to the above mentioned, the *main objective of this paper is to define recommendations on the exploitation of mobile technologies reflecting the data collected and analyzed within the process of mobile-assisted learning at the Faculty of Informatics and Management, University of Hradec Kralove, Czech Republic; so that the potential of mobile devices and technologies could be exploited to maximum extent.*

## 2   Theoretical Background

Information and communication technologies have penetrated all spheres of human lives, including education. Reflecting the latest development, mobile devices and technologies are currently under our focus.

## 2.1   Mobile Devices in Education

Mobile devices have been considered the most powerful communication media among modern devices. One of the reasons is they are popular with users of all age groups, starting from very young age groups. This feature directly predetermines them for educational purposes, starting from the pre-primary to lifelong education, despite there definitely are some didactic constrains and technical limitations. With a mobile learning device the learning process can be rather easily conducted any-time and anywhere. Learning through the computer enables the learners to learn in a non-classroom environment, e.g. at home; however, learning through the mobile device provides them with the opportunity to learn beyond these classroom and non-classroom environments, i.e. they can genuinely learn every time and every-where they are. Moreover, the widespread influence of the market increased the popularity of the mobile phone, and this fulfills the need of teachers/designers to provide tools and educational software for the learners in teaching/learning contexts [5].

Two main characteristics of mobile devices are portability, i.e. the possibility to move the device, and connectivity. It means, when been designed, the mobile system must have capability of being connected and communicate with the learning website using the wireless network of the device to access learning material ubiquitously, including short message and e-mail service [6].

Klopfer et al. [7] define the following properties of mobile devices:

- portability—the mobile devices can be taken to different places due to small size and weight;
- social interactivity—exchanging data and collaboration with other learners must be possible;
- context sensitivity the data on the mobile devices can be gathered and respond to the current location and time;
- connectivity—mobile devices can be connected to other devices, data collection devices, or a common network by creating a shared network;
- individuality—the platform could be customized for individual learner.

Moreover, comparing with other wireless devices such as laptop computers, mobile phones are rather inexpensive having functions as Internet browsers avail-able in most devices. With such inexpensive devices accessible to even the poorest areas and having the functionalities of e-mail or SMS, it is now possible to transfer information to and from mobile phones between instructors and learners without any difficulty.

Although learning services through mobile devices have some advantages, they also have some constraints as small screen, reading difficulty on such a screen, data storage and multimedia limitations etc., particularly mobile phones of small size are not designed for educational purposes. That is why, they may be difficult for the learners to be used for learning. This is partly due to the initial design of such devices, and the purpose of their use—none of mobile devices were primarily developed for the purpose of education. However, those which appropriate for specific learning tasks are too expensive for most of the learners to buy [8].

To illustrate the above mentioned problem, Stockwell [9] demonstrated that the learners found some learning activities take too long to complete on mobile devices, and consequently, some of them preferred to use their PCs to do the assigned task. Some of the learners also indicated that they did not intend to use the mobile phones for doing their tasks because of the screen size, the keypad, and last butnot least, because of the cost of Internet access.

In agreement with Kukulska-Hulme and Traxler [8] and Stockwell [9], Sharples et al. [10], when dealing with the theory of mobile learning, focused on the mobility and context. He examined how learning flowed across locations, timed, topics and technologies rather than assumed that learning occured within a fixed location (e.g. classroom), over a bounded period of time (i.e. school lessons). He emphasized that strategies and opinions we formed in childhood provided impact on the way we would learn in the future. He considered the context to be, "… a central construct of mobile learning, not as a container through which we pass like a train in a tunnel." [10: 236].

The reason is the exploration is essentially mobile, either it means physical movement or movement through conceptual space to form new knowledge. Then, the conversation is the bridge connecting learning across contexts, whether through a discussion, or a phone call between people in different locations, or by making a written comments which can be read at a different time or place. The technology in these explorations and conversations place the role of a mediator of learning. To sum up, mobile learning can be characterized as processes (both personal and public) of coming to know through exploration and conversation across multiple contexts amongst people and interactive technologies [10]. Thus when designing mobile learning, the main task is to promote enriching conversations between learners and teachers within and across contexts. This objective is then reflected in understanding how to design technologies, media and interactions to support a seamless flow of learning across contexts, and how to integrate mobile technologies within education to enable innovative practices. In other words, general principles for human-computer interaction defined in the interaction design research by e.g. Jones and Marsden [11], can be applied on mobile devices. These characteristics have been supplemented by more specific findings from mobile learning projects [12]:

- enable quick and simple interactions;
- design flexible materials accessed across contexts;
- consider special affordances of mobile devices;
- exploit the mobile technology not only to deliver but also facilitate learning.

Moreover, Naismith and Corlett (2006) emphasized the design of mobile learning activities should be, like the design of any learning activity, driven by specific learning objectives, as the use of (mobile) technology is not the target the means enabling to conduct activities the learners can benefit from and which are not available in any other ways. They defined five critical success factors for mobile learning thus adding details to the above listed characteristics:

- Access to technology, i.e. to make mobile technology available where and when needed.
- Ownership, i.e. learners either own the technology, or can treat it as if they own it, when exploiting the technology for entertainment and socializing does not

decrease its value as a tool for learning, but rather supports bridging the gap between institutional and personal learning.

- Connectivity, enabling access to learning resources, sharing them, linking people across contexts etc.
- Integration the mobile learning into the curriculum.
- Institutional support, particularly in three areas: institutional management agreement with the process of mobile devices implementation (which includes the mobile learning strategy, providing HW aSW equipment etc.), staff training and Technical and technological support.

To sum up, the process of mobile e-learning implementation should follow the above mentioned findings so that to avoid coincidental activities as at the starting period of e-learning where immobile (non-portable) devices were exploited for educational purposes.

## 2.2 Process of ICT Implementation at FIM

Within the higher education mobile learning is a way how to extend the campus and offer students the opportunity to learn in whatever place or context they prefer. According to a large survey by Benson and Morgan [13], mobile devices are also perceived as a significant contributor to maintaining quality of life; for university students smart devices represent a key social connector and a learning tool. Higher Education institutions consider providing mobile services to students to be an indicator of better performance and higher quality of education.

The process of ICT implementation into education started in 1997 at FIM and widely spread after 2000, when the LMS WebCT (Blackboard) started to be used. Approximately 250 online courses (called e-subjects) have been designed for the FIM students, either to be used in the distance form of education, or to assist the face-to-face teaching/learning process; currently, approximately 50 e-subjects are provided to students for each semester. Traditionally, all e-subjects run within the LMS, and currently they are also available through mobile devices. In other words, the blended model of learning is applied combining three approaches: the face-to-face instruction, work in online courses and individualized approach to them through mobile devices, which satisfies learners' time/place preferences [14] and bridges formal and informal learning [15].

Since 2012/13 academic year the 'virtual desktops' have been available to students and teachers, mainly for work with software not providing free/open access (e.g. MS SQL Server, Enterprise Architect). Moreover, since 2013/14 the Blackboard Mobile LearnTM version 4.0 for Apple and Android devices has been piloted (Blackboard Mobile LearnTM version 4.0 supports iOS6+, i.e. iPhone 3GS, iPad 2 +, IPad mini, iPod Touch 4+ and Android OS 2.3+). Through this application students gain mobile access to their study materials, they can create and/or participate in discussion forums, use e-mail service, blogs, follow their progress and/or comment on the learning process [16].

Having researched the mobile ownership and exploitation of mobile education of FIM students in previous research activities (e.g. [17, 18]) and reflecting the above listed pre-conditions of successful and efficient mobile learning defined by Naii- smithand Corlett [12], we targeted our attention at the question whether the e-subjects are available through mobile devices to the users—FIM students, i.e. whether single tools of each e-subject (Tests, E-mail, Discussions, Announcements, My Grades, Student's Manual, Syllabus, Help??) can be fully exploited through a mobile device and whether the learning content (texts, audio- and video-recordings, presentations, animations, simulations, figure, tables etc.) is clearly visible.

Therefore, following questions were under our focus:

1. Are there any formats which are not supported by any operating system in the learning contents, tests, or other tools?
2. What recommendations should be provided so that to solve these problems?

## 3   Materials and Methods

Data for this study were collected within the project 'Didactic aspects of the Blackboard Mobile Learn implementation into the process of instruction at FIM'. For the purpose of this research a special group of users (called testing team, having eight members) was formed. The testing team members were intentionally trained to systematically monitor the availability of the learning content and tools in single e-subjects in LMS Blackboard. The process of monitoring followed the pre-defined criteria which were approved by the expert team consisting of three academics, three IT experts and the LMS Blackboard administrator responsible for correct operation of the LMS within the faculty. The criteria focused on the availability of online courses in LMS through mobile devices, i.e. (1) whether students have approach to the learning content (study materials) and (2) whether the tools in each course work properly providing their services to the user. In other words, the testing team monitored whether the learning content is presented properly in single files/chapters of the online course and the tools work.

Each online course was monitored and analysed in four areas:

- Learning content. The learning content of each e-subject is defined in Syllabus. The file of Syllabus and other files providing the learning content are the plain texts in pdf format, as required by the Blackboard user manual. Moreover, other formats are included in study materials, such as video-recordings, animations, simulations, presentations, figures etc., which might cause problems.
- Tests. Reflecting the learning content, i.e. the subject the online course sup- ported, various types of test were exploited by the course designers. Totally 17 types of tests are provided within the LMS Blackboard, e.g. Multiple choice

with one correct answer, Multiple choice with two and more correct answers, Either/Or, True/False, Yes/No, Filling in the blank, Fill in the multiple blanks, Matching, Jumbled sentence, Ordering, Calculating formula, Calculating Numeric, Essay, Short answer, Opinion scale/Likert etc. Naturally, different types of questions are used in e.g. Maths, foreign languages, cultural studies etc.

- Communication. Three communication tools are provided to the users in each course: E-mail for private teacher/student or student/student communication and group teacher/students communication; Discussion for group communication; Announcements for teacher's displaying information to students on the board.
- Feedback. Evaluation of students' performance in the cource is provided through the tool My Grades, where test scores and assignments evaluations were displayed, including teacher's comments; and Help tool is also available in need.

In the 2014/15 academic year, totally, 64 e-subjects were available to students covering four areas:

- IT (e.g. Database systems I, II; Programming I, II; Theory of systems I, II; Computer principles, Applied information technologies, Computer networks;
- Economics (e.g. Microeconomics I, II; Macroeconomics I, II, Czech tax system);
- foreign languages (English for specific purposes I–IV, German for specific purposes I, II; Cultural studies of Great Britain; of Africa and India; of New Zealand; of Australia, Czech language for international students);
- general subjects (e.g. Maths I, II, Psychology I, II, Methodological seminar, Law; Academic writing);
- projects (e.g. Geography for tourism and management; Distance online courses for tourism and management);

All courses were monitored by three types of smartphones (Nokia Lumia G20, Asus Zenfone 4, HT Desire 500) and two types of tablets (Lenovo Yoga, iPad) using three different operating systems (Windows, iOS, Android), following the above mentioned criteria.

## 4 Results and Discussions

The collected data are presented in the form of tables, described and discussed.

As mentioned above, 64 e-subjects (online courses) were monitored by the eight-member testing team by a smartphone, tablet and iPad with Windows, Android and iOS operating systems. In Table 1 the frequency analysis of problems detected on mobile devices and operations systems i.e. displayed. The data show that the highest frequency of problems was detected with smartphones (20.3 %), half occurrence was with tablets (9.4 %) and the lowest share of problems was caused by iPads (6.25 %). From the view of operating systems, the highest

**Table 1** Problems with the availability of the online course content on mobile devices and operating systems

| Mobile device | n | % | Operating system | n | % |
|---|---|---|---|---|---|
| Smartphones | 13 | 20.3 | Windows | 8 | 12.5 |
| Tablet | 6 | 9.4 | Android | 6 | 9.4 |
| iPad | 4 | 6.25 | iOS | 10 | 15.6 |

**Table 2** Problems with the availability of the online course content in single tools

| Tool in online course | n | % |
|---|---|---|
| Study materials | 19 | 29.7 |
| Tests | 12 | 18.75 |
| Syllabus | 8 | 12.5 |
| My grades | 8 | 12.5 |

frequency of problems was detected with iOS (15.6 %), followed by Windows (12.5 %) and Android (9.4 %).

In each course the single tools were monitored. As displayed in Table 2, the highest frequency of problems was detected with Study materials (29.7 %) followed by Tests (18.75 %); with Syllabus and My grades (both 12.5 %). Other tools, i.e. Announcements, Calendar, E-mail service and Discussions were available (i.e. worked correctly) on all types of mobile devices and in all types of operating systems.

To answer the Question 1 whether there are any formats not supported by any operating system in the learning contents, tests, or other tools, a deeper analysis was made. We discovered that:

Problems with availability of single files in *Study materials and Syllabus* were not caused by the pdf format in which they (texts, chapters etc.) were uploaded in the online course—these files worked correctly, they could be accessed through all types of mobile devices and operating systems. The problem was if *links to other sources* were included in the text files—these links did not work and additional sources could not be displayed; e.g. if a video file, animation, simulation were added directly to the text, the they could be played, if the link was inserted only, it did not work. Moreover, in 10 % of courses another problem was detected, when page one of the text was displayed only and following ones did not appear. In this case, the problem was in the size of screen of mobile devices. Therefore, the courses with some problems detected were also tested on the desk computers in the computer laboratory at FIM (having the operational system Microsoft Windows 8 installed). In this case no problems appeared, all files were available to the users and the links to other sources worked properly. With some types of mobile devices the screen is of identical or similar small size but with none of them the size was large enough to display a small arrow in the right top corner; by clicking on the arrow the user moves to the next page, which was not visible on these mobile devices. Students have this problems rather frequently, they do not notice the arrow even if using the desk computer or notebook with large screen because the arrow is not highlighted. This fact is pointed out by the tutor during the starting tutorial.

In the *Tests* tool different frequency in using single types was detected, e.g. multiple choice tests with one correct answer were the most frequently exploited (in 96 % of e-subjects), Multiple choice with two and more correct answers (11 %), Either/Or (7 %), True/False (21 %), Yes/No (26 %), Filling in the blank (6 %), Matching (39 %), Jumbled sentence (8 %), Ordering (5 %), Calculating Numeric (4 %), Essay 13 %), Short answer (43 %), Opinion scale/Likert (9 %); Calculating formula and Fill in the multiple blanks were not used in any e-subject. Within the 18.75 % of tests which did not work properly of mobile devices (i.e. 12 tests), in seven ones the test settings were not set properly—these tests did not work on desk computers and notebooks either. In the other five tests (all of them were in Maths I and Maths II and were designed by one tutor) *links to other sources* were included in the tasks. It means in the case of tests the purpose of improper functionality was identical with Study materials and syllabus.

The *My grades* tool did not work properly in eight online courses, one member of the testing team found out. Having analysed this result in detail, it was discovered that it was him who had not worked with this tool in an appropriate manner. After that discussion, no problems were detected with this tool.

To sum up, problems with the exploitation of mobile devices for running online courses in Blackboard Mobile Learn application were caused neither by the type of mobile device, nor the operating system. The cause was that the *links to additional information sources* did not function properly. Moreover, the size of screen brought some problems because small icons in the corners of the screen were not seen and thus they could not be exploited by the users.

Our answer to the Question 2 on what recommendations should be provided so that to solve the above mentioned problems, the fact must be reflected that mobile devices have become an unseparable part of the young generation lifestyle, the exploitation of the Blackboard Mobile Learn application was considered meaningful for the process of education on mobile devices, particularly within autonomous learning, at FIM UHK. Identically to e-learning implementation into higher education, the mobile-assisted process was running randomly, non-intentionally.

Reflecting the studies on mobile-assisted learning, the focus is paid to foreign language learning (e.g. to vocabulary by Alzahrani [19], Wang and Shih [20], Wu [21]; listening comprehension by Kyung-Mi [22], de la Fuente [23]; reading comprehension by Lin [24] etc.). Moreover, the role of intrinsic and extrinsic motivation (e.g. Liu and Chen [25]) is researched, as well as daily family life (e.g. Ronka et al. [26]). The meta-analysis of effects on learning languages with mobile devices were summarized by Lee et al. [27]. They analysed 44 studies published within 1993–2013 and detected a shift from computer-assisted to mobile-assisted learning and statistically significant mean effect size. Sanchez et al. [28] reflect the fact there is a sharp increase in the use of mobile devices. This leads many users to exploiting them to access LMS from these devices for which neither these systems, nor their contents were not designed, as mobile devices contain various features in their interaction media and there are no standards defined. This variety of features changes from the type of display, keyboard and sensors, to operating systems. A natural task to mobile device users would be to access any LMS in a simple, effective and efficient way from any mobile device. In their paper they present an analysis of the functionality of LMS focused on student role users using different mobile devices. They propose the developers should write a native application for each mobile platform, or at least to apply a middleware-based architecture. Our results head towards identical conclusions.

## 5   Recommendations and Conclusion

Above all, rules for mobile devices exploitation should be defined, identically to the situation two decades ago, when the process of computer-assisted education started. In the case of mobile devices, the size of screen should be a feature strongly limiting the process of their implementation into education. Duman et al. [29] analyzed studies published in 2000–2012 to examine their characteristics and trends in research and problem solving in this field. The increased pace of the process from 2008 was indicated, reaching the peak in 2012. As Kukulska-Hulme states [30], the use of mobile phones and other portable devices have shown an impact on how learning takes place in many disciplines and contexts, thus providing the potential for significant change in teaching and learning practices. She emphasizes the mobile-assisted is not a stable concept; therefore its current interpretations need to be made explicit so that to highlight what is distinctive and worthwhile about mobile learning.

The system of mobile learning-recommendation services supporting small-screen display interfaces, which was designed by Chen [31], could be a solution. Moreover, challenges and opportunities of mobile-assisted learning described by Elias [32] result in principles towards building flexibility of instructional design of learning content and operating system so that they could be used by a wide range of students:

- equitable use,
- flexible use,
- be simple and intuitive,
- provide perceptible information,
- feel tolerance for mistakes,
- require low physical and technical effort,
- feel support as being part of community of learners, and
- live in the positive instructional climate.

From the long-term view, didactic principles having been defined and applied in e-learning, it is high time technical constraints of mobile devices were seriously considered and reflected. Then the process will become efficient and motivating for learners [33].

**Acknowledgments** The paper is supported by the SPEV Project 2016.

# References

1. Šimonová, I.: Mobile-assisted language learning in technical and engineering education. In: Proceedings of the International Conference on Interactive Collaborative Learning (ICL), 20–24 Sept 2015, Florence, pp. 169–176. IEEE (2015). doi:10.1109/ICL.2015.7318020
2. Mosavi Miangah, T., Nezarat, A.: Mobile-assisted language learning. Int. J. Distrib. Parallel Syst. (IJDPS) **3**(1) (2012)
3. Koole, M.: A model for framing mobile learning. In: Ally, M. (ed.) Mobile Learning: Transforming the Delivery of Education and Training, pp. 25–47. AU Press, Athabasca (2009)
4. Šimonová, I., et al.: Klíčové competence a jejich reflexe v terciárním e-vzdělávání [Key Competences and Their Reflection within Higher e-Education]. Hradec Kralove, WAMAK (2011)
5. Alonso, F., Manrique, D., Martinez, L., Vines, J.M.: Study of the influence of social relationship among students on knowledge building using moderately constructivist learning model. J. Educ. Comput. Res. **51**(4), 417–439 (2015)
6. Huang, C., Sun, P.: Using mobile technologies to support mobile multimedia English listening exercises in daily life. In: The International Conference on Computer and Network Technologies in Education (CNTE 2010). http://cnte2010.cs.nhcue.edu.tw/. Accessed 20 Jan 2015
7. Klopfer, E., Squire, K., Jenkins, H.: Environmental detectives: PDAs as a window into a virtual simulated world. In: Proceedings of IEEE International Workshop on Wireless and Mobile Technologies in Education, pp. 95–98. IEEE Computer Society, Vaxjo, Sweden (2002)
8. Kukulska-Hulme, A., Traxler, J. (eds.): Mobile Learning: A Handbook for Educators and Trainers. Routledge, London (2005)
9. Stockwell. G.: Investigating learner preparedness for and usage patterns of mobile learning. ReCALL **20**(3), 253–270 (2008)
10. Sharples, M., et al.: Mobile learning: small devices, big issues. In: Technology Enhanced Learning: Principles and Products (2007)
11. Jones, M., Marsden, G.: Mobile Interaction Design. Wiley, Chichester (2006)
12. Naismith, L. Corlett, D.: Reflections on success: a retrospective of the mLearn conference series 2002–2005. In: Proceedings of MLEARN 2006 Conference, 22–25 Oct 2006, Banff, Canada (2006)
13. Benson, V., Morgan, S.: Student experience and ubiquitous learning in higher education: impact of wireless and cloud applications. Creat. Educ. **4**(8A), 1–5 (2013)
14. Pieri, M., Diamantini, D.: From e-learning to mobile learning: new opportunities. In: Mobile Learning. Transforming the Delivery of Education and Training, pp. 183–194. Athabasca University Press, Athabasca (2009)
15. Abdullah, M.R.T.L., et al.: M-learning scaffolding model for undergraduate English language learning: bridging formal and informal learning. TOJET: Turk. Online J. Educ. Technol. **12**(2), 217–233 (2013)
16. BlackBoard.com.: Transforming the Experience with Blackboard Mobile (2012)
17. Simonova, I., Poulova, P.: Learning ESP in engineering education through mobile devices. In: Jefferies, A., Cubric, M. (eds.) Proceedings of the 14th European Conference on e-Learning, pp. 568–575. Hatfield, University of Hertfordshire (2015)
18. Simonova, I., Poulova, P.: Social networks and mobile devices in higher education: pilot project. In: Ahamed S.I., et al. (eds.) Proceedings of 2015 IEEE 39th Annual International Conference Computers, Software and Applications Conference, vol. 2, pp. 851–856. The Institute of Electrical and Electronics Engineers, Inc., Taichung, Taiwan (2015)

19. Alzahrani, H.: Examining the effectiveness of utilizing mobile technology in vocabulary development for language learners. Arab World Engl. J. **6**(3), 108–119 (2015)
20. Wang, Y.H., Shih, S.K.H.: Mobile-assisted language learning: effects on EFL vocabulary learning. Int. J. Mobile Commun. **13**(4), 358–375 (2015)
21. Wu, Q.: Pulling mobile assisted language learning (MALL) into mainstream: MALL in broad practice. PLoS ONE **10**(5) (2015)
22. Kung-Mi, O.: The effectiveness of mobile assisted language learning on L2 listening comprehension. Multimedia-Assist. Lang. Learn. **18**(2), 135–158 (2015)
23. de la Fuente: Learners attention to input during focus on form of listening tasks: the role of mobile technology in the second language classroom. Comput. Assist. Lang. Learn. **27**(3), 261–276 (2014)
24. Lin, C.C.: Learning English reading in a mobile-assisted extensive reading program. Comput. Educ. **78**, 48–59 (2014)
25. Liu, P.L., Chen, C.J.: Learning English through actions: a study of mobile-assisted language learning. Interact. Environ. **23**(2), 158–171 (2015)
26. Ronka, A., Malinen, K., Jokinen, K., Hakkinen, S.: A mobile-assisted working model for supporting daily family life: a pilot study. Family J. **23**(2), 180–189 (2015)
27. Lee, Y.S., Sung, Y.T., Chang, K.E., Liu, T.C., Chen, W.C.: A meta-analysis of the effects of learning languages with mobile devices. In: New Horizons in Web Based Learning (ICWL 2014, Tallin, Estonia), LNCS, vol. 8699, pp. 106–113 (2013)
28. Sanchez, D.V., Rubio, E.H., Ledesma, E.F.R., Viveros, A.M.: Student role functionalities towards learning management systems as open platforms through mobile devices. In: 2014 International Conference on Electronics, Communications and Computers (CONIELECOMP), pp. 41–46 (2014)
29. Duman, G., Orhon, G., Gedik, N.: Research trends in mobile assisted language learning from 2000 to 2012. ReCALL **27**, 197–216 (2015)
30. Kukulska-Hulme, A.: Will mobile learning change language learning? ReCALL **21**(2), 157–165 (2009)
31. Chen, H.-R.: Mobile learning-recommendation services supporting small-screen display interfaces. In: Xie, A., Huang, X. (eds.) Advances in Computer Science and Education, AISC 140, pp. 177–182. Springer, Berlin, Heidelberg (2012)
32. Elias, T.: Universal instructional design principles for mobile learning. Int. Rev. Res. Open Distrib. Learn. **12**(2) (2011)
33. Burston, J.: MALL: pedagogical challenges. Comput.-Assist. Lang. Learn. **27**(4), 344–357 (2014)

# Assessment of Outcomes in Collaborative Project-Based Learning in Online Courses

Dmitrii A. Ivaniushin, Andrey V. Lyamin and Dmitrii S. Kopylov

**Abstract** In this paper the problem of assessment of interdisciplinary and social and personal learning outcomes is covered. The teaching and assessment tools are reviewed. The self, peer and expert assessment methods are described. The approach of assessment learning outcomes in collaborative project-based learning is proposed. This approach uses students' behavior analysis to increase grade reliability and validity. The method allows determine irresponsible students and assign low final grade to them. The tool for carrying out collaborative project-based learning is developed. It is based on the XBlock SDK of Open edX platform.

**Keywords** MOOC · Project-based learning · Cooperative learning · Self assessment · Peer assessment · Expert assessment

## 1 Introduction

Massive open online courses (MOOCs) are the one of the most actual trend of e-learning development nowadays, and an increasing number of higher education institutions are engaged in the development of this courses. According to The New York Times, 2012 became "the year of the MOOC" because of the starting of the one of the most known MOOC-providers, such as Udacity (founded by Sebastian Thrun, David Stavens and Mike Sokolsky in 2012), Coursera (founded by computer science professors Andrew Ng and Daphne Koller from Stanford University) and edX (created by Massachusetts Institute of Technology and Harvard University) [1].

D.A. Ivaniushin · A.V. Lyamin · D.S. Kopylov (✉)
ITMO University, Saint Petersburg, Russia
e-mail: dima@cde.ifmo.ru

D.A. Ivaniushin
e-mail: d.ivanyushin@cde.ifmo.ru

A.V. Lyamin
e-mail: lyamin@mail.ifmo.ru

© Springer International Publishing Switzerland 2016                                         351
V.L. Uskov et al. (eds.), *Smart Education and e-Learning 2016*,
Smart Innovation, Systems and Technologies 59,
DOI 10.1007/978-3-319-39690-3_31

According to the e-learning experts, two general instructional designs of MOOCs are [2]:

1. cMOOC are online courses where the connectivism theory is used [3]. In this courses learner obtain knowledge by making connection between domain objects and by making connections between other learners.
2. xMOOC are online courses where the behaviorism theory is used. This theory is based on content of the course. The course includes instruments of automatically learning and assessment of learning outcomes.

Upon success of MOOCs other types of online courses started to appear. Small Private Online Courses is one of them. One of describing features of this type of course is its usage among in-campus students. These courses allow to maintain blended learning and flipped classroom learning. In [4] it is shown that usage of SPOCS increases students' earned points by 10 % in comparison of usage of traditional materials.

Typical MOOC consists of video lectures with textbooks, tests and forums [5]. The test items can be represented by various types, for example, multiple choice, text input, drag-n-drop, programming assignments and virtual laboratories with automatic check [6]. Some of the platforms use the gamification methods for increasing student's engagement and emotional response [7, 8]. Also some several of MOOC-providers provide peer assessment problems, where student should give an answer by uploading an artifact, such as essay, picture, etc. [9–12]. Student should complete assignment following to course author instructions and upload the results to the system. After that he needs to participate in training assessment and assess artifact made by course author. The system checks whether student gives the same grade as course author did. Student can participate in peer review only after his given grade is the close enough to course author grade. Student should review assess works of other students. After that he should assess his own work—self assessment stage [13]. The system calculates final grade of every student after they made peer review. Grades that are given as a part of self assessment are not taken into account.

There is an assignment that cannot be solved by using existing assessment tools. It is the assessment of the interdisciplinary and social and personal learning outcomes. These learning outcomes can be enhanced by using project-based and cooperative learning only. These learning outcomes are especially important in engineering education; because of engineers often work in a team in the project. The individual success impacts on the success of the entire project. In this paper we propose an approach for learning outcomes assessment in the technological-based collaboration courses and a tool that automates assessment.

Our approach differs from the ones shown above (self and peer assessments) by usage of smart algorithms that take into account particular student's background of studying the course and social and personal qualities.

## 2 Background

### 2.1 Teaching Methods

**Project-based learning**. Behind project-based learning lies development of cognitive skills, abilities to construct own knowledge, abilities to orient in information field, development of critical and imaginative thinking. Working on specific project requires ability to integrate data from different scientific fields in order to reach the goal. Effectiveness of this tool is defined with development of student's creative skills, initiative, self discipline and increase of interest in the wake of work. Applying project-based learning allows developing following competences:

1. Reflecting skills;
2. Research abilities;
3. Teamwork skills;
4. Management skills;
5. Communicative skills;
6. Presentational skills.

**Cooperative learning**. Forms of cooperative and collaborative work are applied for intensification of learning process and increase effectiveness of students' work. Usage of such education forms involves into working process the least communicable students, because being among their team members where each of them solve some problems in coordination to others, they have no choice to refuse do their own job, otherwise they are to be blamed, and they set high value of their team members' opinion. Besides working in micro-society, competitiveness may take place, which increase intensity and gives emotional coloring. Collaborative work should be especially important for the student; he should understand and admit it. Student fills need only in this case, so he can expect on the result of the effort.

Term team stands for cooperation of people who work together, share specific interest and are able to achieve common aims each of autonomous and consistently with minimal input control. Core factor for the whole team can be attaining the goals as an output for collaborative work. Independence and consistency of actions each of team member, which is required for achieving the goal, is the balanced behavior that others expect of.

Characteristics of formed team are unity of purpose, interest consistency, collaboration, overall responsibility for outcome, self sufficiency, individual specialization and complementarity. Team members achieve better results working together that one at a time. Thus, synergy between team members takes place.

## 2.2   Assessment Methods

While choosing and analyzing assessment methods special aspects of learning process and specific problems related to collaborative work should be taken into account. Major problem of assessing team members working on same project or problem is reliability and validity of individual grade. Grading system should be as objective as possible and should not give students any chance to manipulate particular result grade for themselves or anyone else. Assessment individualization is difficult because major work is held out of classroom.

Self assessment is a process when students self reflect and assess their own learning outcome, judge of degree of achievement and compliance to specified criteria, define strengths and weaknesses of own achievements. Making judgments of learning process is essential part of overall learning process, as self judgment is based on natural temptation to check own learning process via defining increment of own learning outcome. If student cat define his current achievements and get what else he should learn, his further learning can be motivated. Moreover, self judgment encourages thoughts of own learning outcome and it may increase responsibility and independency of student.

Most simple and most effective form of self assessment is complex index calculated on own grades of knowledge, experience and abilities with following levels: "high", "normal", and "low". Level "high" stands for 1 point, "normal" for 0.5 and "low" for 0. This total index (coefficient of learning outcome) is calculated as:

$$k = \frac{k_1 + k_2 + k_3}{3} \tag{1}$$

where $k_1$—points for self assessed level of knowledge, $k_2$—points for self assessed level of experience, $k_3$—points for self assessed level of ability to apply knowledge and experience.

**Peer assessment** besides of getting the grade is aimed on retaining student's knowledge and analyzing their progress. This method can be widely used in e-learning, especially is humanitarian subjects. While using peer assessment course author sets number of peers required and peer selection rules.

Peer assessment has following advantages:

- It allows usage of test items that are beyond automated or machine grading
- It provides individual feedback to course students, even for courses having thousands of students
- It provides ability to learn roleplaying "teacher" and "student" that is a pretty effective tool.

Being a peer in peer assessment is allowed for students who solved the item themselves and passed specific conditions like earned required grade in the course in total. Grading uses several criteria defined by course author; each of them has a scale with detailed description. In accordance with the course rules minimal number of reviews may be set, students may be forced to give short text review on each graded item. Reviewing more items than required is encouraged but optional. Peer assessment is held anonymously for same course students. Grades and reviews given may be edited until deadline of peer assessment stage. After deadline system calculate final grade based on peer grades and reviews given.

**Expert assessment**. Expert assessment is a grading procedure based on expert review. Expert is a competent specialist who examines, provides a review on particular subject using his own knowledge, past experience and intuition in applying generalities and particular patterns of creating specific solutions and provides optimality (objectiveness of conclusions, opinions, recommendations, and grades). While using expert assessment, external experts (assessors) assess work outcome. Number of experts and grading rules, in case of conflict between experts, must be stated in grading rules. Expert assessment is based on ranking, pairwise comparison, direct grading, mathematical statistic. There are two groups of expert grades: individual, based on independent anonymous expert reviews, and collective, based on collective expert opinion.

**Measuring the personal qualities**—psychometric constructs—became particularly spread with usage of point rating systems. Measuring the personal qualities like discipline, responsibility, initiative etc. requires portion of points up to 10 % of total grade. While in classroom education this points may be charged by particular subject instructor. While assessing learning outcome head of department can use such possibility. In e-learning points for personal qualities may be charged for keeping schedule of training, discussion in forums etc.

## 2.3 Assessment Tools

**Survey**. Written survey can be used for collective phenomenon, e.g. in education. Survey is a massive data collection tool, which uses specially prepared questionnaire. Respondent in-puts data in survey by answering prepared in advance standardized questions with specified answer types, usually with no direct contact with researcher. Quality of survey and observance of methods concerning its usage define sustainability and fidelity of research outcome.

Usage of surveys in e-learning can be fully automatized. Survey can be uses as entry surveys to discover student expectation of the course. It also can be used as mid- and end of course polls. Exit survey is necessary feedback to course developers.

**Behavior analysis**. LMS provides several fold possibilities to analyze user behavior by monitoring numerous user events like opening course page, pausing or unpausing video lectures, time spent on answering test items, time spent on self or peer assessing. Date and time, location determined by IP-address, type of user device accessing the course can be registered for each of such event. Heart rate variability [14] and eye-tracking [15] are ones of promising methods of user behavior analysis.

Too little time spend on assessing teammates is one of factors which can decrease final student grade. This tells about little care paid to assessing his colleagues. On the other hand, student may achieve lower grade if he spent little time for self assessing.

## 3 Approach of Assessment Learning Outcomes in Collaborative Project-Based Learning

As the result of review above, it may be stated that self, peer and expert assessment should be involved in assessing outcome of project-based learning. Future employers are more viable to be encouraged than regular teachers. Employers participation into project assessment process, each of whole project and its elements, will help to estimate student skills that modern job market needs.

Written survey is viable in methods concerned above. Course author create questions to particular item. If item consists of several stages, different surveys should be created for each stage.

Combination of project-based and collaborative learning develop teamwork and management skills, and skills to work individually on problem solution. Usage of these tools provides ability to model future real work, because not all the employers are ready to teach their novice-employees but want to see experienced ones in students just out of university.

Self assessment, peer assessment and expert assessment are characterized by great labour intensity and may lead to decreasing grade reliability and validity. That's why automated behavior analysis should be applied to increase effectiveness and accuracy of pedagogical measurement.

Let's cover rules of calculating total grade for completing team project that consists of several items. Total grade that student may earn can be calculated as:

$$G = W_s G_s + W_p G_p + W_e G_e \tag{2}$$

where $G$—total student grade for completing the project, $G_s$—student grade earned by self assessment, $G_p$—student grade earned by peer assessment, $G_e$—student grade earned by expert assessment; $W_s$, $W_p$ and $W_e$—grade weight set by course author for different types of assessments: self, peer and expert assessments respectively. At that $G$, $G_s$, $G_p$, $G_e$, $W_s$, $W_p$ and $W_e$ belong to [0; 1] and

$$W_s + W_p + W_e = 1 \tag{3}$$

Weights above ($W_s$, $W_p$ and $W_e$) can be calculated and adjusted with usage of neural network with supervised learning [16, 17]. Training of this network must be performed within current project assignment under supervision of course author or staff.

Let $I$ items be assigned to student of interest in some project. Course author may assign weights to item applying to different assessment methods while creating it: $C_{si}$—item $i$ weight while self assessing, $C_{pi}$—item $I$ weight while peer assessing, $C_{ei}$—item $i$ weight while expert assessing. At that $C_{si}$, $C_{pi}$ and $C_{ei}$ belong to $(0; +\text{inf})$.

Grade earned by student while self assessing can be calculated as:

$$G_s = \frac{\sum_{i=1}^{I} \frac{G_{si}}{C_{si}}}{I} \cdot W_i \tag{4}$$

where $G_{si}$—grade for item $i$, $W_i$—student's weight while assessing response item $i$,

$$W_i = f(p) \tag{5}$$

where $f$—function that determines student's weight while assessing response item $i$, $p$—data structure, containing information about student behavior during assessment and previous rating, calculated by smart e-learning system. For example, this structure may contain time spent on assessing and minimal time that should be spent on assessing (set by course author). In that case, function returns value less than 1, when peer spent less time on assessing. Moreover, the function may analyze other factors that impact on final grade, like keystroke pattern, fulfilling required conditions while assessing etc. There are following constraint:

$$0 \leq G_{si} \leq C_{si} \tag{6}$$

and $W_i$ lie in $[0; 1]$.

Grade that can be earned by student being assessed by his teammates:

$$G_p = \sum_{j=1}^{J} \frac{\sum_{i=1}^{I} (G_{pij}/C_{pi}) \cdot W_{ij}}{I} \tag{7}$$

where $J$—number of teammates who assess current student, $G_{pij}$—grade, which student $j$ gave for response item $i$ of current student, $W_{ji}$—peer weight, which is calculated as (5). Besides:

$$0 \leq G_{pij} \leq C_{pi} \tag{8}$$

and total peer weight should satisfy condition:

$$\frac{\sum_{j=1}^{J} \sum_{i=1}^{I} W_{ij}}{J} = 1 \tag{9}$$

Grade earned by student through expert assessment is calculated as:

$$G_e = \sum_{k=1}^{K} \frac{\sum_{i=1}^{I} (G_{eik}/C_{pi})}{I} \cdot W_k \tag{10}$$

where $K$—number of experts assigned to grade student, $G_{eik}$—grade given by expert $k$ to solution to item $i$ of current student, $W_k$—individual expert $k$ weight, given by course author. Additionally:

$$0 \le G_{eik} \le C_{ei} \tag{11}$$

and total weight of all experts should satisfy following condition:

$$\frac{\sum_{k=1}^{K} W_k}{K} = 1 \tag{12}$$

The approach shown proposed above is smart due to automatic data processing which is collected during students' work on project assessments, grading system adoption for particular project and automatization of processes for self and peer assessments. E-learning system built in accordance of such approach will have smart abilities to adapt, to sense and to infer.

## 4 Tool for Carry Out of Collaborative Project-Based Learning

At the present day only edX of largest online courses-platform providers have published source code of its platform for all comers. This platform named Open edX and it is free open-sourced software. Thus everyone may deploy his own instance of the platform to organize his courses.

Backend of this platform is written in the Python programming language with usage of Django framework. Frontend is written in HTML markup language and CSS, with usage of JavaScript and Coffee-Script programming languages for dynamic user interface.

Open edX platform allows to integrate new user modules like new item response types with help of application programming interface (https://open.edx.org/xblocks). It claims to reach the goal of software development community to take part in the development of the edX educational platform and the next generation of online courses. Implementing collaborative project-based learning requires development of new response type using XBlock SDK.

While applying collaborative project-based learning, it is useful to take advantage of project management systems like Redmine to increase productivity (http://www.redmine.org/). This system as Open edX is free open-sourced software. Also

Redmine allows to integrate version control systems like Subversion (http://www.subversion.apache.org/) or Git (http://git-scm.com/) and visualize history of source code changes.

We shall consider usage of assessing cross-disciplinary, social- and personal learning outcomes approach within collaborative project-based learning. Let the students be to learn to work in team, which specialize in web-site development. Online course is necessary for their education. This course may contain following elements:

- Video lectures, where theory materials are given.
- Quizzes for self assessment during video lectures.
- Assignments concerning web-site development to develop skills.

Assignments "Web-site development" is a complex one, which is shared between students attending this course. Such item is mapped as project in project management system Redmine, where assignments defined by course creator are imported in as so-called tickets. Students are to develop web-site while they should split into teams (by themselves or with aid of tutor) and assign tickets.

There are two supposed ways of work process in our approach: students share roles between each other (e.g. like manager, designer, layout designer and programmer etc.), or each student develops his own section of web-site completing all the related tasks (e.g. forum, user profile, polls etc.). Students should create source code as the work output, send it to version control system and associate it to particular tickets.

Team members should assess their own work, their team members' work assigning grade according to criteria given by course author. In virtue of these grades, final grade of particular student is calculated. Further external experts invited by course team should assess students' responses on course items. As external experts higher-level students or employees can be taken. Each expert has his own weight, which depends on personal experience, knowledge etc. Course author or course team assigns expert weight. Final student's grade for project work is calculated as in (2). Second-year students of ITMO University were offered to create team project with usage of the tool within discipline "Introduction to information systems development". The analysis showed that student being a peer with low weight (as in formula 3) earned lower final grade while peer assessing. In addition, he had low grade while self assessing though assessed himself high. All this tells that he completed his assignments badly and did not take care of assessing others.

## 5 Conclusion

The approach of assessment of cross-disciplinary, social- and personal learning outcomes with automated analysis of students' behavior and tool for organization and carrying automated assessment of collaborative project-based learning outcome were developed. The approach allows to automatically calculate students' rating with formula proposed. Originality of the approach described lies within a combination of self, peer and expert assessing and usage of students' behavior analysis (e.g. time spent on particular actions) with aid of smart system for grading learning outcome. While using expert assessment it is not viable to involve tutors but future employees. Using described approach may be mapped on any type of online courses: massive open online courses, small private online courses etc.

**Acknowledgments** This paper is supported by Government of Russian Federation (grant 074-U01).

## References

1. Papano, L.: Massive Open Online Courses Are Multiplying at a Rapid Pace—The New York Times. http://www.nytimes.com/2012/11/04/education/edlife/massive-open-online-courses-are-multiplying-at-a-rapid-pace.html
2. Rodriguez, C.: MOOCs and the AI-Stanford like courses: two successful and distinct course formats for massive open online courses. Eur. J. Open Distance E-Learn. (2012)
3. Downes, S.: 'Connectivism' and Connective Knowledge http://www.huffingtonpost.com/stephen-downes/connectivism-and-connecti_b_804653.html
4. Fox, A.: From MOOCs to SPOCs. In: Communications of the ACM—2013, vol. 56, Issue 12, pp. 38–40 (2013)
5. Lisitsyna, L.S., Lyamin, A.V.: Approach to development of effective e-Learning courses. In: Frontiers in Artificial Intelligence and Applications, IET—2014, vol. 262, pp. 732–738 (2014)
6. Efimchik, E.A., Lyamin, A.V., Chezhin, M.S.: Automation of variant preparation and solving estimation of algorithmic tasks for virtual laboratories based on automata model. In: E-Learning, E-Education, and Online Training, IET—2015 (2015)
7. Lisitsyna, L.S., Pershin, A.A., Kazakov, M.A.: Game mechanics used for achieving better results of massive online. In: Smart Innovation, Systems and Technologies, IET—2015, vol. 41, pp. 183–193 (2015)
8. Kopylov, D.S., Fedoreeva, M.K., Lyamin, A.V.: Game-based approach for retaining student's knowledge and learning outcomes. In: 9th IEEE International Conference on Application of Information and Communication Technologies, AICT 2015—Conference Proceedings, IET—2015, pp. 609–613 (2015)
9. Creating Peer Assessments—Creating a Peer Assessment documentation. http://edx.readthedocs.org/projects/edx-open-response-assessments/en/latest/
10. Cruz Matos, G.: Online peer assessment: an exploratory case study in a higher education civil engineering course. In: 15th International Conference on Interactive Collaborative Learning (2012)
11. Sterbini, A., Temperini, M.: Peer-assessment and grading of open answers in a web-based e-learning setting. In: Information Technology Based Higher Education and Training (ITHET), 2013 International Conference, Antalya, pp. 1–7 (2015)

12. De Marsico, M., Sterbini, A., Temperini, M.: Experimental evaluation of OpenAnswer, a Bayesian framework modeling peer assessment. In: Advanced Learning Technologies (ICALT), 2014 IEEE 14th International Conference, Athens, pp. 324–326 (2014)

13. Mortazavi, B.: Self assessment surveillance using e-Portfolio. In: E-Learning and E-Teaching (ICELET), 2010 Second International Conference, Tehran, pp. 72–77 (2010)

14. Lisitsyna, L.S., Lyamin, A.V., Skshidlevsky, A.S.: Estimation of student functional state in learning management system by heart rate variability method. In: Smart Digital Futures 2014, pp. 726–731. IOS Press (2014)

15. Lyamin, A.V., Cherepovskaya, E.N.: Biometric student identification using low-frequency eye tracker. In: 9th IEEE International Conference on Application of Information and Communication Technologies, AICT 2015—Conference Proceedings, IET—2015, pp. 191–195 (2015)

16. Mustafa, H.M.H., Al-Hamadi, A., Hassan, M.M., Al-Ghamdi, S.A., Khedr, A.A.: On assessment of students' academic achievement considering categorized individual differences at engineering education (Neural Networks Approach). In: 2013 International Conference on Information Technology Based Higher Education and Training (ITHET), pp. 1–10 (2013)

17. Liu, K.F.-R., Chen, J.-S.: Prediction and assessment of student learning outcomes in structural mechanics a decision support of integrating data mining and fuzzy logic. In: 2010 2nd International Conference on, Education Technology and Computer (ICETC), vol, 3, pp. 499–503 (2010)

# Vagrant Virtual Machines for Hands-On Exercises in Massive Open Online Courses

Thomas Staubitz, Maximilian Brehm, Johannes Jasper, Thomas Werkmeister, Ralf Teusner, Christian Willems, Jan Renz and Christoph Meinel

**Abstract** In many MOOCs hands-on exercises are a key component. Their format must be deliberately planned to satisfy the needs of a more and more heterogeneous student body. At the same time, costs have to be kept low for maintenance and support on the course provider's side. The paper at hand reports about our experiments with a tool called Vagrant in this context. It has been successfully employed for use cases similar to ours and thus promises to be an option for achieving our goals.

**Keywords** MOOC · Online learning · Virtual machines · Virtualization

T. Staubitz (✉) · M. Brehm · J. Jasper · T. Werkmeister · R. Teusner · C. Willems · J. Renz · C. Meinel
Hasso Plattner Institute, Potsdam, Germany
e-mail: thomas.staubitz@hpi.de

M. Brehm
e-mail: maximilian.brehm@student.hpi.de

J. Jasper
e-mail: johannes.jasper@student.hpi.de

T. Werkmeister
e-mail: thomas.werkmeister@student.hpi.de

R. Teusner
e-mail: ralf.teusner@hpi.de

C. Willems
e-mail: christian.willems@hpi.de

J. Renz
e-mail: jan.renz@hpi.de

C. Meinel
e-mail: christoph.meinel@hpi.de

© Springer International Publishing Switzerland 2016                    363
V.L. Uskov et al. (eds.), *Smart Education and e-Learning 2016*,
Smart Innovation, Systems and Technologies 59,
DOI 10.1007/978-3-319-39690-3_32

# 1  Introduction

MOOCs are often criticized for being mere distributors of instructional videos and theoretical quizzes. Real learning, however, requires more than mere instruction. According to constructivist theory, learning is not providing "'true' representations of an objective environment", but enabling the construction of a "relative fit with the world as it is experienced by the learner"–a process called adaptation [1]. Practical exercises and assignments are an essential element in this process, providing the possibility for trial and error and thus to construct knowledge in an active way. More and more courses, particularly in the area of computer science, are aiming to provide hands-on experience and practice online. However, distributing and maintaining practical exercises in a scalable environment poses technical challenges. In this paper we show how existing approaches satisfy the needs of users and providers, and evaluate a novel technique based on virtual machines for providing hands-on experience in IT focused online courses.

## 1.1  Metrics for Hands-On Exercises

To carry out hands-on exercises in MOOCs, the teaching team needs to provide an exercise environment to the students. As presented in Sect. 1.2, such environments can be provided in a variety of ways. To evaluate the different approaches we defined the following metrics.

**Setup costs**  The amount of effort and time the user has to invest in order to get started with the exercise.

**Support costs**  The amount of effort and time the course provider has to invest to deliver the exercise or to help users with the setup.

**Hosting costs**  The amount of money and resources the course provider has to spend on providing the required infrastructure.

**Control**  The degree to which the course provider can control the behavior of provided tools or to which it has the ability to help users.

**Real world application**  The degree to which the exercise prepares students for the use of tools as used by professionals and confronts them with real world workflows.

**Responsiveness**  The time needed by a tool to react to user input.

**Exploration**  The degree to which the environment allows students to learn and explore on their own initiative.

**Integration**  How well can the working environment be integrated with the course platform.

**Target audience**  The user group for which an environment is suited.

## 1.2   Current Solutions for Hands-On Exercises

Existing approaches usually differ in where applications are installed and how students interact with them. We evaluated several of the most common scenarios using the metrics named above. Table 1 gives an overview of this evaluation by estimating values on a scale from 0 to 10 where 10 indicates the best solution for the challenge.

**Local setup**   Setting up applications such as programming environments, IDEs or other required tools on the students' machines seems to be the most naive approach as it is closest to the setup of professional users. It teaches the use of real tools and guarantees best possible responsiveness. However, the installation and configuration of such environments goes beyond the capabilities of novice users and results in high demand of support. Furthermore, the course provider has no control over the tools and thus cannot handle unexpected behavior.

**Browser based environments**   cover the other extreme of the spectrum. The user is provided with browser-based tools running the students' work and test it on correctness. The provider can tailor the tools to the needs and abilities of the course participants and remain in control over their behavior. The user has no setup costs at all and can switch seamlessly between the MOOC environment and practical exercises. Users do not, however, train with real world tools and never undergo the associated learning process. For a detailed overview of existing solutions see [2].

**VMs on user machines**   Virtual machines provide the opportunity to combine benefits of the approaches named above. Course providers can prepare the machine and install software tailored to their needs, thus preventing setup costs and other issues with the students. At the same time computation is performed on the participants' machines, therefore, lowering costs for the provider. Students use a system as it would be used by a professional, thus, training them for real world applications. The issue with this approach is that such VM images can easily take several

**Table 1**   Comparison of different ways to bring hands-on exercises to the students

|                 | Local setup | Browser based | VM at user | VM at prov. |
|-----------------|-------------|---------------|------------|-------------|
| Setup           | 1           | 10            | 7          | 10          |
| Support costs   | 1           | 9             | 7          | 10          |
| Hosting costs   | 10          | 3             | 9          | 1           |
| Control         | 1           | 10            | 6          | 8           |
| Real world app. | 9           | 2             | 9          | 8           |
| Responsiveness  | 10          | 2             | 9          | 9           |
| Exploration     | 10          | 2             | 10         | 9           |
| Integration     | 1           | 10            | 5          | 8           |
| Target audience | adv.        | beg.          | beg./adv.  | adv.        |

Gigabytes of memory. Distributing such VMs–in a way that is both cost-efficient for the provider and comfortable for the user–is the primary focus of this work.

**VMs hosted by the course provider**  can provide fully functional systems whose setup would go beyond the scope of a course. An example for courses on network security was introduced in [3]. These server-side VMs are beyond the scope of this paper.

## 2  Related Work

The rising demand for practical assignments in MOOCs is well documented. Willems et al. [4] describe the introduction of hands-on assignments to openHPI.[1] Here, the focus lies on applications installed on the students machine. The inherent heterogeneity was used to diversify the assignment and underline the complexity of distributed applications. The paper does, however, report massive support efforts and problems and suggests moving to more homogeneous environments. Staubitz et al. [5] report about a different approach on the same platform, which provided the students with hands-on coding assignments. Here the focus lies on cost efficient and easy to set up practice environments by integrating browser-based third party coding tools into existing MOOC infrastructures including automated assessment.

Staubitz et al. [2] provide an overview on the wide range of possible approaches to implement hands-on exercises similar to Sect. 1.2. While they focus on coding assignments and the use of web-based editors, they do provide deep insights into various testing and assessment methods.

There have been few courses and research projects exploring the use of virtualization software on the user side. E.g. a course on software defined networking was conducted by Princeton University on Coursera,[2] which was based on the Mininet VM [6]. In another course on Coursera, *Audio Signal Processing for Music Applications*[3] a VM was provided with the required open source software pre-installed. Berger et al. [7] conducted a course on database systems using VMs as an optional means for the assignments. They suggest that VMs could not only be helpful for reducing setup and support costs for hands-on exercises, but also for collecting data on the learning process.

## 3  Virtualization Software and Vagrant

As suggested in Sect. 2, virtual machines can, and have been, successfully employed to form the basis for realistic hands-on exercises in MOOCs. The question that remains is how to deploy such VMs to the participants. Traditionally, VMs are dis-

---

[1] https://open.hpi.de.

[2] https://www.coursera.org/course/sdn1.

[3] https://class.coursera.org/audio-002.

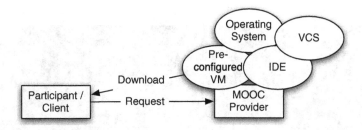

**Fig. 1** Downloading a preconfigured VM–including operating system, integrated development environment, and version control system–from the MOOC provider's server

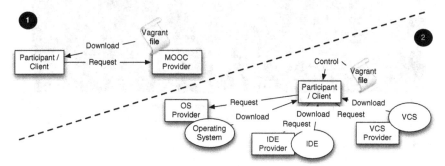

**Fig. 2** (1) Downloading the Vagrantfile from the MOOC provider's server. (2) Downloading the required tools from their original location

tributed through proprietary file formats that contain a description of the virtual machine and its state including its hard drive.

While this meant that every machine is virtually identical, such files rapidly grow to multiple Gigabytes, which makes them hard (respectively expensive) to distribute (see Fig. 1). With Vagrant[4] a relatively new technique was established to deploy virtual machines. In contrast to traditional VMs–where the VM is created, preconfigured, stored, and then distributed–a Vagrant box is defined only by a text file called *Vagrantfile* (see Fig. 2). Using this file, Vagrant creates a virtual machine on the host system using existing VM providers such as VirtualBox.

## 3.1 Vagrantfile

An exemplary Vagrantfile is displayed in Fig. 3. The VM on the users machine is created and configured according to the description given in this file (see also Fig. 2 (2)). The first commands define which base box should be used and some network settings. The commands in the block `config.vm.provider` include settings specif-

---

[4]https://www.vagrantup.com/.

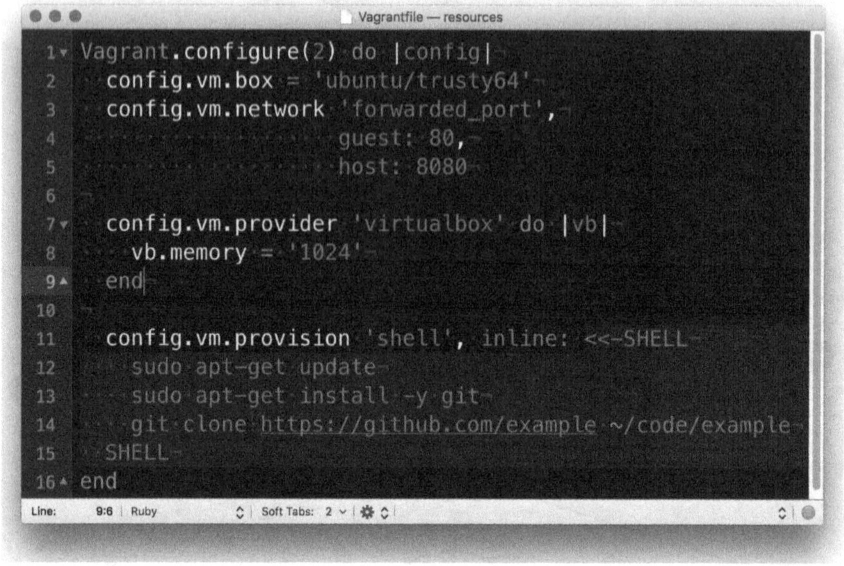

```
● ● ●                        Vagrantfile — resources
1▾  Vagrant.configure(2) do |config|
2      config.vm.box = 'ubuntu/trusty64'
3      config.vm.network 'forwarded_port',
4                        guest: 80,
5                        host: 8080
6
7▾     config.vm.provider 'virtualbox' do |vb|
8          vb.memory = '1024'
9▲     end
10
11     config.vm.provision 'shell', inline: <<-SHELL
12         sudo apt-get update
13         sudo apt-get install -y git
14         git clone https://github.com/example ~/code/example
15     SHELL
16▲ end

Line:    9:6 | Ruby          ⬍ | Soft Tabs: 2 ⌄ | ⚙ ⬍                          ⬍ | ●
```

**Fig. 3**  Basic Vagrantfile

ically directed at VirtualBox. With `config.vm.provision` the provisioning steps
are defined. These steps are executed after the setup of the machine and are used to
install software or configure the system. It is this provisioning step where the VM is
tailored to the needs of the MOOC it is used in. To deploy a basic setup only a small
text file has to be provided on the MOOC platform. All other resources are pulled
from their original locations on the internet. If additional custom content has to be
deployed, Vagrant can pull this from any location on the internet, e.g. the MOOC
platform, or cloud storage. Thus hosting costs, for storage and traffic, on the MOOC
provider's side are kept at a minimum.

### 3.2  Customization of Vagrantfiles

As Vagrant builds on nothing but the Vagrantfile, customizing a machine requires
a mere substitution. Willems et al. [4] suggest using heterogeneity to diversify the
learning process or tailor it to specific users. Vagrant promises to enable teachers
to achieve the goal of per user customization more easily. Rather than providing
a single Vagrantfile, individual Vagrantfiles could be generated on demand from a
template by substituting single system variables. Customization on a per user level
also has the distinct advantage that user specific information such as access tokens
can be embedded in the VM. This could be put to use for the hand-in procedure
as discussed in Sect. 4. As a proof of concept we implemented a tool that writes

arbitrary content into a Vagrantfile template depending on the user it is supposed to be delivered to.

## 3.3 Using Graphical Applications

Vagrant virtual machines are usually accessed via command line *ssh* sessions, more specifically with the Vagrant specific command *vagrant ssh*. Standard Vagrant boxes in their default configuration do not come with a graphical user interface (GUI). Access via *ssh* would pose a great challenge to the novices, who are our main target group. As one of our intended use cases involves graphical applications such as Integrated Development Environments (IDEs)we examined several methods for GUI access to a Vagrant box. The traditional approach for coding within Vagrant is to use a locally installed editor while the VM serves only as the execution environment. The simplest approach would be to use a shared folder that is accessible both by the host and the guest system. However, since the idea of our approach is to deliver a fully preconfigured environment, this option misses the point.

Alternatively, the GUI of an application running in the VM can be forwarded to the host system using X11. After connecting to the Vagrant box with `vagrant ssh` one may simply launch GUI applications and display them at the local X Server. However, as Windows does not support X Server natively, additional setup is required. This way we only replace one complexity with another.

Finally, Virtual Box can be configured to deactivate the headless mode and launch a window showing the whole desktop. This method works on all operating systems with low configuration overhead. We therefore favor this option when targeting a wide audience. To improve the performance of the VM, we selected the LUbuntu operating system as it comes with the lightweight LXDE desktop instead of a more standard Unity, Gnome, or KDE desktop.

## 4   Implementation

In this paper we consider two different scenarios: A Java programming course at an intermediate level and a security lab in an Internet security course. Those two courses are very different in nature and each has its own requirements and challenges. Both courses have already been taught in a similar form on openHPI. The problems that have been reported by some participants with setting up the required environments on their own have been a main inspiration for the paper at hand. Next to the bare setup, we will consider the tasks of downloading new exercises, updating faulty exercises and submitting solutions.

## 4.1 Java Programming Course

The envisioned Java course is a follow-up to an introductory hands-on Java programming course. After basics of the language have been taught using a browser-based learning environment, this course teaches the use of real world tools, particularly the Eclipse IDE, by having the participants apply their knowledge in a Java programming project. It requires a JDK, the Eclipse IDE, and git to be installed and configured correctly. For novices this can be a serious challenge. Using Vagrant we can reduce the frustrating setup procedures to installing VirtualBox and Vagrant and deliver a perfectly configured course environment to the student. Throughout the course students download weekly assignments onto their virtual machines, solve them and submit the results back to the course provider. All of the following automated solutions might benefit from templated Vagrant files as described in Sect. 3.2. Those templated Vagrant files could introduce user IDs and authentication tokens as environment variables into the VM.

**Task Retrieval and Update** Assignments should be delivered to the users' machines on a regular basis. Also, a method for updating buggy or incomplete assignments needs to be in place as mistakes tend to occur even with an optimal preparation. Often, students download assignment files into a folder of their choice. As the manual transport of files into the VM can be tedious, we recommend having a clickable script that downloads the newest versions inside the VM. These scripts might be customized with the user's ID or a token to authenticate with the course provider. In the simplest form, the scripts download the files as a zip and unpack them into the user's Eclipse workspace. Hash sums can guarantee that the scripts download new or updated files only. Somewhat more complex, the scripts can retrieve the assignment files using VCS software, such as git. These tools might save resources using incremental updates over the zipped solution if the assignment files are very large.

**Solution Upload** After working on an assignment, students need to upload solutions to the course provider. It is important that the upload happens in an automated fashion, because file transfer between the host and the VM can be error prone. Again, the simplest solution might be zipping the assignment solution and sending it to the course provider for grading via a simple HTTP upload. If course providers and students are interested in keeping a history of the uploaded solutions it makes sense to use an VCS tool for uploading. It comes at cost of hosting a repository for every student, however.

## 4.2 Security Labs

The requirements of a security lab are very different from those of a programming course. Usually, a security lab involves multiple machines with a specific network configuration and configured services. Oftentimes, instead of producing a piece of software the assignment is to obtain information by intercepting communication or breaking into systems (penetration test). Security labs can involve software that is difficult to obtain on users machines and that can cause harm if used improperly. Usually, these two challenges have been tackled by giving the user access to remotely hosted virtual machines. Using virtual machines at the users side reduces hosting cost for the organizers, but introduces the concern of users obtaining the information or cheating by introspection of the local machines. In the following, if there are multiple machines involved, the term attacker machine refers to the VM that is equipped with the required penetration testing tools for the student. Accordingly, the term defender machine refers to a VM that exposes exploitable vulnerabilities.

Multiple machines can be configured in a single Vagrant file. Afterwards, network configuration is a matter of assigning IP addresses and naming the virtual networks. Multi-host environments include scenarios, such as remote exploitation with one attacker and one defender machine or man in the middle attacks with one attacker and two defenders.

**Separation from the Outside World**    To provide a safe learning environment for the student we want to prevent accidental attacks of uninvolved targets. Those uninvolved targets can be machines on the learners local network or hosts on the the the internet. For a proof of concept, we use Linux' standard firewall *iptables* and the extension *Conntrack*. First, we disallow any traffic leaving the machine to the outer network by using *iptables* to prevent any new outgoing connections on Vagrant's default NAT adapter. *Conntrack* now helps to allow an exception: Only outgoing traffic of connections that were initiated from the outer world to our machine is allowed to leave. Hence, users can establish an *ssh* connection to the machine and have a two way communication, but they cannot open a new connection from inside of the machine to the outside. The shown provisioning steps have to be executed for every VM of the scenario.

**Task Retrieval and Update**    Assignments might require exploitation of various services and operating systems. Hence, security labs need to be carried out using assignment specific VMs. This reduces the reusability of the provisioned VMs. For each assignment that requires a new set of VMs a new Vagrant file must be downloaded and in turn the VMs must be provisioned. After that, users can attempt to solve the assignment.

**Task Submission**    Often, users have to obtain information or so called *flags* in security labs. Those flags might be a password, a user name or a random string. They can be obtained from various activities, such as password cracking, reverse engineering or exploiting services. Given that they are only strings, submitting flags can be as simple as pasting them into a browser form. The feasibility of this approach has been demonstrated by Staubitz, et al. [5] for a different use case but would work here exactly the same way.

**Cheating**    Many concerns about cheating arise if the defending machines are run by the students themselves. There are several ways to obtain the information in unrighteous ways. Students might try to get access to the machines by guessing login credentials, using hypervisor functionality or mounting the virtual machines hard drives to another machine. There are technical remedies for some of these back doors, but it is ultimately very difficult to protect an application from the user running it. Also, these remedies introduce technical complexity. While course providers can prevent students from mounting the hard drives by encrypting them and hiding the key on the boot partition, this introduces a new source of errors and lowers reusability. A more feasible way of preventing cheating might be embedding the submission of flags into a quiz of the respective learning section. Students can prove that they rightfully obtained the flag by demonstrating knowledge of the involved techniques, tools, and steps.

## 5   Evaluation

Assuming that VMs can reduce local setup issues for practical tasks in a MOOC environment, we prototypically have set up two basic scenarios: a penetration testing setting in an Internet Security course and a programming environment in a Java course. With this experiment we have shown that the heterogeneity of different environments, maintenance, and support efforts for the teaching teams can be reduced by employing VMs. We have also shown that the benefits of employing VMs are differing from scenario to scenario. For the Java programming environment described in Sect. 4.1 we recorded a number of metrics.[5] The Ubuntu base image accounts for 378 MB which have to be downloaded and stored locally. The required packages add another 498 MB to download. The fully built VirtualBox VM has a size of 3598 MB. Provisioning the box, not counting the time required for downloading the base box, took 11:37 min. In comparison, an Ubuntu box without a GUI and having some lightweight tools only takes up roughly 1500 MB. This shows that providing Vagrantfiles instead of fully pre-configured VMs can reduce the costs of a MOOC provider significantly.

## 6   Future Work

We intend to employ this approach in one of our upcoming courses and evaluate its acceptance among the participants and its perceived usefulness in terms of the defined learning outcomes.

---

[5]The machine that has been employed was a standard PC with 8 GB RAM, i5 CPU 2.67 GHz, SSD running Windows 7.

# 7 Conclusion

In this paper, we examined the use of Vagrant as a tool to deploy preconfigured, hands-on learning environments to our online students. Vagrant provides the opportunity to deliver customizable and cheap to host VMs that create realistic and responsive hands-on environments. Furthermore, we introduced eight relevant metrics to evaluate the different solutions for hands-on tasks in MOOCs and assess their benefits and shortcomings. We have shown that Vagrant reduces the friction for creation, distribution, setup and update of Virtual Machines. We demonstrated the use of Vagrant in the context of different scenarios and discovered and discussed several challenges in carrying out hands on assignments using virtual machines. In summary, we can say that the use of Vagrant, in complex server scenarios, can reduce hosting costs, improve the user experience for learners compared to traditional virtualization software and thus reduce support efforts on the course provider's side.

# References

1. von Glasersfeld, E.: Learning and adaptation in the theory of constructivism. Commun. Cogn. **26**(3/4), 393–402 (1993)
2. Staubitz, T., Klement, H., Renz, J., Teusner, R., Meinel, C.: Towards practical programming exercises and automated assessments in massive open online courses. In: International Conference on Teaching, Assessment, and Learning for Engineering (TALE), pp. 23–30. IEEE (2015)
3. Willems, C., Meinel, C.: Tele-lab it security: an architecture for an online virtual it security lab. Int. J. Online Eng. (iJOE) **4**, 31–37 (2008)
4. Willems, C., Jasper, J., Meinel, C.: Introducing hands-on experience to a massive open online course on openHPI. In: International Conference on Teaching, Assessment and Learning for Engineering (TALE), pp. 307–313. IEEE (2013)
5. Staubitz, T., Renz, J., Willems, C., Jasper, J., Meinel, C.: Lightweight ad hoc assessment of practical programming skills at scale. In: Proceedings of IEEE Global Engineering Education Conference (EDUCON), pp. 475–483. IEEE (2014)
6. Lantz, B., Heller, B., McKeown, N.: A network in a laptop: rapid prototyping for software-defined networks. In: Proceedings of the 9th ACM SIGCOMM Workshop on Hot Topics in Networks, p. 19. ACM (2010)
7. Berger, O., Gibson, J.P., Lecocq, C., Bac, C.: Designing a virtual laboratory for a relational database MOOC. In: Proceedings of the 7th International Conference on Computer Supported Education, pp. 260–268 (2015)

# Sequencing Educational Contents Using Clustering and Ant Colony Algorithms

María José Franco Lugo, Christian von Lücken
and Esteban Ramírez Espinoza

**Abstract** This work presents a model to optimize the presentation order of educational contents in the Moodle e-Learning platform. The objective here is to infer those learning paths for which it is expected the students may achieve the best performance. The foundations of the proposed model are (i) a clustering of similar students according to a student model, and (ii) a metaheuristic to obtain an improved educational content sequence for each group. The clustering of similar students is achieved by a modified k-prototypes algorithm. Then, for each group of students, an Ant Colony Optimization algorithm is used for self-organize the sequencing of the educational content that better adapts to its learning characteristics. Finally, to evaluate the proposal, synthetic and real data were used for testing purposes. Experimental results show the viability of the proposed approach.

**Keywords** Sequence of educational content · e-Learning · Clustering algorithms · Ant Colony Optimization

## 1 Introduction

Since the late nineties the use of so-called Course Management Systems (CMS) has increased continuously. Students in an online course may have different prior knowledge, preferences, goals, and learning styles. Developing a course that considers these differences may become a difficult task. One of the difficulties is to

M.J. Franco Lugo (✉) · C. von Lücken · E.R. Espinoza
Facultad Politécnica, Universidad Nacional de Asunción,
San Lorenzo, Paraguay
e-mail: mjfranco@pol.una.py

C. von Lücken
e-mail: clucken@pol.una.py

E.R. Espinoza
e-mail: eramirez@pol.una.py

© Springer International Publishing Switzerland 2016       375
V.L. Uskov et al. (eds.), *Smart Education and e-Learning 2016*,
Smart Innovation, Systems and Technologies 59,
DOI 10.1007/978-3-319-39690-3_33

determine sequence of content or teaching units most likely enable a given student to achieve his goals [1].

The Ant Colony Optimization (ACO) is an optimization technique inspired by the use of a trail pheromone by real ants to find the shortest path from their nest to their food source [2]. The ACO approach has proven to be efficient for solving various problems of finding optimal paths in different domains. Then, several authors have developed proposals that demonstrate their viability as an alternative for the construction of learning paths in e-Learning environments [3–7].

Optimization of the content sequence using ACO was first proposed in [3] to develop a commercial learning platform that uses both the paths followed by students and their performance to determine the best way of learning. ACO based on using only the path of students who have participated in the course to build the optimal path systems are presented in [4] and [5], the difference between these systems is that [4] considers only those have managed to complete the course successfully, while [5] considers all students. Given that students may have different learning styles, [6] presents an approach where they are considered different categories of students according to their learning style, only students who belong to the same group are considered for the construction of their learning path. In this case, it is possible that a different learning path exist for each category. Finally, [7] presents a prototype that combines a rule-based planning with ACO to optimize learning paths.

Previously cited works [3–5] used prototype or proprietary code for their implementations, also, with the exception [6], they provide a single learning path for all of the students. This paper proposes to optimize the sequence content considering groups of similar students using Moodle [15], an open source CMS widely used. To this aim, it first uses the concept of learning styles [9–12] and modified clustering algorithm based on the k-prototypes [13] to group similar students as they enter the system [14]. Moreover, to optimize the learning path, is proposed an ACO that obtains an optimal sequence of learning content, for each group, based on the average of grades and a questionnaire on the usefulness of the content provided for each group of students. The various codes and scripts required to incorporate the proposal made in this paper on this platform are available in [16].

This article is organized as follows: Sect. 2 describes the architecture of the proposed model and how the concepts are articulated. Section 3 presents the clustering algorithm used; Sect. 4 shows the use of ACO for optimal training of content sequences; Sect. 5 the tests model is presented; in Sect. 6 the results of the tests are analyzed. Finally, Sect. 7 presents the conclusions.

## 2 Proposed Adaptive Model

This section presents the adaptation model proposed in this work, which is based on the use of a clustering technique to form groups of similar students and the use of an ACO to optimize the content sequence for each of these groups. Figure 1 presents a

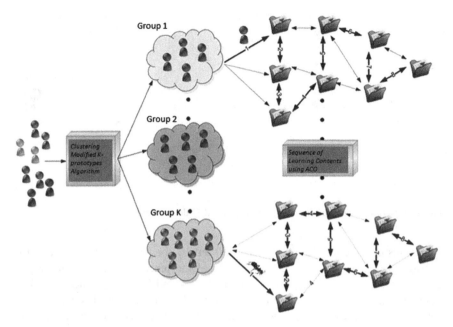

**Fig. 1** Adaptive model for contents sequence

sketch of the adaptive model, where each group has a learning path tailored to its corresponding students.

A clustering procedure serves to divide a set of objects into subsets of similar items called clusters [14]. The goal of clustering is to find groups minimizing the distance between the objects of each cluster and, maximizing the distance between clusters, in order to have groups defined and representatives. To obtain the groups of similar students in this paper the k-prototypes algorithm [13] was modified for the sequential arrive of students, to the e-Learning platform.

Grouping of students is done taking into account their learning styles for which an initial test is taken. The concept of learning style is basically about the characteristics that indicate modes and ways to learn from a student [12]. The model of student used in this work is based on three models of learning styles, which serve to provide a high degree of student information. Learning styles are complimented with the basic data stored students in the e-Learning platform Moodle. The models of learning styles used here are: Felder-Silverman [11], cerebral quadrants [10] and the cerebral hemispheres [9]. The student model used in this work can be extended to save more information about the student, such as prior knowledge and usage of other models of learning styles.

For optimization of the sequence of educational content using ACO the objective to maximize is the student achievement throughout the sequence of contents. The student performance is represented by the average of the grades obtained in each educational content. Each student navigates contents on a graph with precedence

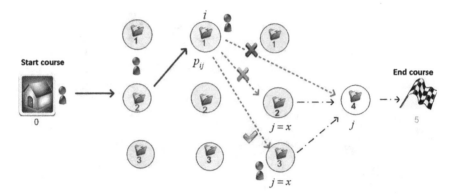

**Fig. 2** Selection of the learning content using the ACO decision rule

constraints considering the pheromone trail left by the previous students and it is considered as a virtual ant. As an example of the precedence constraints, Fig. 2 represents a course with six contents 0, 1, 2, 3, 4 and 5. Lines connecting nodes represent the possible sequences to take starting from node 0. In this example, all contents can be presented in any order to the students but considering the restrictions node 1, 2 or 3 must be preceded by node 0, and node 4 must be visited only after nodes 1, 2 and 3.

The pheromone trail used in the ACO here is proportional to the grade average obtained by the students in all educational content. The pheromone trails of each student is deposited at the end of travel in the route suggested by the algorithm and that was presented to the student. In addition to the pheromone trail, the ACO model contemplates the use of visibility, which is used to get more information about the preferences of previous students and thus accelerate the process of adaptation. This visibility is considered in each transition edge between contents and represents the value that the destination content have for students who have previously taken the edge and finished the corresponding content, this visibility is measured using a score that is requested students at the end of each unit or content. In order to avoid the premature convergence, the implemented ACO implements an evaporation mechanism based on the use of the transitions between contents.

## 3  Modified k-Prototypes Algorithm

The k-prototypes algorithm [13] provides an efficient grouping of objects with numerical and categorical attributes when the number of objects and groups to be formed is known. In this algorithm, initially the set of objects to be partitioned, the number of desired groups and a centroid for each group is provided. Then, for each object, the nearest centroid is determined and the object is added to the group associated with that centroid. Next, for each group generated in the previous step its centroid is recalculated. This process is repeated until a stop condition, for example, achieve a given number of iterations, or that the difference between the centroids of two consecutive iterations is smaller than a given threshold.

A drawback of the use of the k-prototype algorithm in e-Learning systems is that the final number of students is not known and hence the number of groups to be formed, therefore, a mechanism to adapt the algorithm for sequentially entering data over time is necessary.

To solve this problem, in this paper, the students are analyzed sequentially as they are recorded in the system [17]. For every student $e$ that enters the system, the distance $d$ of the student to the nearest centroid is calculated. If this distance is less than or equal to a threshold $\omega$, the standard k-prototypes algorithm is executed with all students without changing the already calculated centroids, otherwise, if the distance $d$ is greater than the threshold is creates a new group and the current student is the centroid.

Most clustering techniques are based on a function that defines the degree of similarity between two objects [17]. This function must be defined according to the type of objects to group, strongly depends on the nature of the object and the characteristics that differentiate them.

An array of 10 elements is used to represent each student. The first 4 elements are numerical values for the most representative dimensions of the Felder-Silverman model. Then, 6 array elements are for categorical data: 4 elements represent the Cerebral Quadrants of Hermann model and 2 elements represent the left and right hemispheres of the Cerebral Hemispheres model, respectively.

To define the distance between objects with numeric and categorical attributes the metric presented in [18] is used. This metric defines the distance between two objects $X_i$ and $Q_l$, $r$ numerical attributes and $c$ categorical attributes as follows:

$$d(X_i, Q_l) = \varpi d_{num}(x_i^r, q_l^r) + \delta d_{cat}(x_i^c, q_l^c) \tag{1}$$

The values $\varpi$ and $\delta$ in (1) correspond to the weights to be given to numerical and categorical attributes respectively, such that $\varpi, \delta \geq 0$, and $\varpi + \delta = 1$. The function $d_{num}(x_i^r, q_l^r)$ calculates the distance among numeric attributes and it is based on the Euclidean distance, the function $d_{cat}(x_i^c, q_l^c)$ provides the distance among categorical attributes and it is based on the Hamming distance.

# 4 Optimizing the Learning Path Using ACO

Optimizing the sequence of educational content using ACO is achieved analyzing the students behavior through the course navigation graph, students are considered as virtual ants [4–8]. The system will present a new topic to students when culminating a previous one. When a student a given educational content, he evaluates it, which in turns serves to determine the content to be proposed by the ACO heuristic.

Artificial ants move through the graph of contents by applying a probabilistic decision policy based on two parameters called pheromone trails $\tau$ and visibility $\eta$ [2]. The pheromone trails reflect that, the most visited edges by similar students are highly desirable. While, visibility aims to measure importance to the rating by students to educational content were already completed. This assessment was obtained through a survey that students must complete to finish each content. This survey will provide a content rating assigned by the student and that is reflected in the arch of transition following the student to get to this content. Formally, the probability of choosing the edge from node i to node j is given by:

$$p_{ij} = \frac{\tau_{ij}^{\alpha} \times \eta_{ij}^{\beta}}{\sum_{x \in K_i} \tau_{ix}^{\alpha} \times \eta_{ix}^{\beta}} \qquad (2)$$

where:
$K_i$   is the set of feasible nodes located one hop from node $i$.
$\tau_{ij}$   pheromone, for the edge between node $i$ and node $j$.
$\eta_{ij}$   visibility, corresponding to the edge between node $i$ and node $j$.
$\tau_{ix}$   pheromone, for the edge between node $i$ and node $x$, where $x \in K_i$.
$\eta_{ix}$   visibility, corresponding to the edge between node $i$ and node $x$, where $x \in K_i$.
$\alpha$   defined a priori, is the relative importance of the pheromone.
$\beta$   defined a priori, is the relative importance of visibility.

The decision rule defined in (2) is applied to select the next node (learning content) to be assigned to the student through the learning graph as shown in Fig. 2. In this case the node 3 has a probability of 70 % being selected, while the node 2 has a 20 % chance of being selected and the node 4 carries a 10 %. The final choice of the node that will be presented to the student is made by the roulette algorithm [2] which makes this choice based on the prior obtained probability.

Each ant deposits an amount of pheromone $\Delta\tau$ that is in direct ratio to the grade average of the contents taken by the student after they have completed the course. In Fig. 3, the pheromones update on the path followed by the student is displayed,

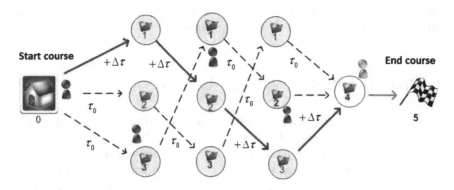

**Fig. 3** Pheromones update

initially all the edges have a pheromones value of $\tau_{ij} = \tau_0$ when the student completes the course following the path $\{0-1-2-3-4-5\}$, represented by the path indicated by the solid lines, the values of $\tau_{ij}$ are updated by adding $\Delta\tau$ to each edges of the path.

Visibility is updated taking account of the expression (3).

$$\eta_{ij} = \eta_{ij} + \eta_{ij1} \tag{3}$$

where $\eta_{ij1}$ is the value of the perception the student that assigns to the current content, this assessment is represented by qualitative value the student provides.

In case of the visibility update, when a student completes the educational content 3, the edge immediate above is updated according to the assessment that the student has given to the content of the node and this it considered a visibility for the ACO model.

To avoid convergence to local optima, an evaporation mechanism is added, as in the actual context of ant colonies [2]. The intensities or values of pheromone trails progressively decrease, favoring the exploration of different edges during the search process on the content graph. $\rho$ is a value between 0 and 1 the coefficient of pheromone evaporation, evaporation is performed by decreasing the value of each $\tau_{ij}$ according to:

$$\tau_{ij} = (1-\rho)\tau_{ij} \tag{4}$$

Instead of reducing all values of pheromones over time, the edge values are only updated when a student goes through that edge or a neighbor [8]. In this model of pheromone evaporation, a neighbor edge is one edge that points to at the same node learning.

## 5 Test Cases

As noted above the model presented was implemented for the Moodle platform. The code need to incorporate these features to the platform in version 1.9.5 is available in [16]. The tests performed in this study are divided into two stages, first clustering model is verified, and, in the second stage the sequencing of content using the ACO model for each group of similar students is verified.

### 5.1 Test for the Clustering Method

The tests performed to the clustering are based on external validation and internal validation of the considered modified k-prototypes algorithm. These validations are performed using synthetically generated data and with data provided by real students. The external validation is performed to measure the quality of the groups without considering the nature of the data. To perform this validation, synthetic data representing objects with numerical and categorical attributes were generated with a predefined cluster structure where each object belongs to a previously established group. Then, the modified k-prototypes is run with the synthetic data and the Rand index [19] is calculated at each execution. The Rand index measures the similarity between the predefined grouping of synthetic data and the clustering result of the algorithm, in order to quantity the effectiveness of the method.

The internal validation provides the necessary parameters for configuring the model taking into account the nature of the objects to group, in this case students. The main parameter required by the modified k-prototypes algorithm is the threshold $\omega$ for creation of new groups. To obtain this parameter we used real data obtained through questionnaires completed by 487 students of various ages and genders in order to get their learning styles and to establish the student model for each. With these real data the best possible grouping is obtained, taking into account the sum of squared error [19] as a measure of the quality of clusters. Once formed this best grouping, the threshold creating new groups $\omega$, is obtained as the average of the distances between the centroids of this best grouping.

## 5.2 Test for the Sequencing of Contents Using ACO

The tests about the sequencing model contents ACO were performed using a simulation with a fully connected graph of 10 nodes representing educational content where all these can be taken in any order by students. The rate of convergence in this model is based on the approximate number of students who must complete the content for a sub-optimal initial path that has been optimized into a more promising route for students in each group.

As the aim is to obtain the optimal learning path, then, the goal is to find the path with the better overall average grade in all content of courses taken by students. In this simulation the marks obtained by students in each node are in a range between 0 and 100 with a normal distribution. In a special route is assigned the higher score to students with the aim of that sequencing model converges to this route once a certain number of students completed the content. In this best route the notes have an average of 90 and a dispersion of 10 which simulates that students get good results. To simulate lower notes, in all other routes an average of 60 and a dispersion of 20 is used.

To set a trend of convergence speed the ACO sequencing model, the simulation starts with different sub-optimal learning path provided by the teaching team. Each of these initial routes have a pre-established percentage difference with respect of the expected optimal route, is best route in the simulation it is arbitrarily chosen because in reality can be different for each group of students. These synthetic tests are developed with initial learning paths that differ by 20 % between them and the expected optimal route.

# 6 Results

The results of the external validation of the clustering model shows that in all cases more than 90 % of the objects were assigned to the group it belongs and in some cases a 100 % effective. As shown in Fig. 4, the modified algorithm k-prototypes is executed with different sizes of synthetic objects and number of groups, for example, the first bar from the left of the figure shows that the Rand index for a quantity of 100 objects with 2 predefined groups (100-2) has had an efficiency of 100 %, for a total of 100 objects with 4 groups (100-4) has had an efficiency of 95 %, etc. Results show that the implemented clustering algorithm performs adequately.

Measurements of execution runtime the modified k-prototypes it was verified that the developed model has a quasi linear execution time relative to the number of objects to be grouped within the system.

In the internal validation of the clustering model using the actual data provided by the students, the optimal value for k is obtained. The assigned value was k = 3

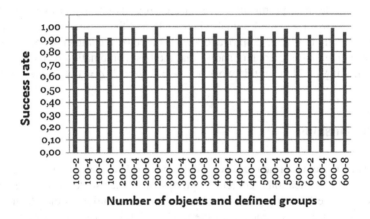

**Number of objects and defined groups**

**Fig. 4** Percentage of correct group formation for the modified k-prototypes algorithm

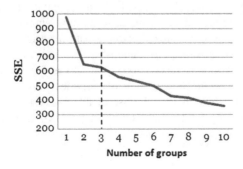

**Fig. 5** Results for the k optimal value

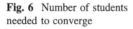

**Fig. 6** Number of students needed to converge

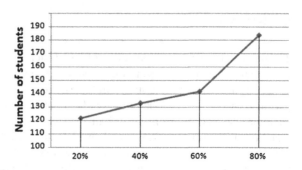

since that point is the most obvious point of inflection and a balance between the number of groups and the squared sum error or SSE (*Sum Square Error*) as shown in Fig. 5.

Having k groups to form, the best partitioning with real data was obtained by repeated execution of the k-prototypes and the application of SSE. Using this partitioning the threshold to group formation was set to $\omega = 0.8875$.

The results of the conducted reveal that the number of students needed to converge to the most promising route for the group is linearly proportional to the difference between the sub-optimal initial route and the optimal route, this can be seen graphically in Fig. 6. Empirical evidence based on the implemented simulation for sequencing learning content shows that the best combination of parameters for the sequencing problem in this work are: $\alpha = 1$, $\beta = 2$, $q = 0.4$ and $\rho = 0.6$.

## 7   Conclusions

This paper presents an adaptation model developed to optimize the sequence of presentation of educational content to groups of similar students and was implemented in the e-Learning platform Moodle as a contribution to the Open Source community. During construction of the model its different parts where evaluated. The conclusions drawn from the tests results are: (i) by modifying the clustering algorithm k-prototypes to support sequential entry to group objects was achieved an effective clustering method in the selected e-Learning environment; (ii) by using the ACO metaheuristic independently with each group of similar student's rapid convergence is achieved at an optimized learning path for each student group.

Some future work includes: (i) establish a mechanism to account for prior knowledge of students; (ii) consider other objectives for optimal sequencing of content, such as the estimated duration of activities (a-priori) and the effective duration of activities; (iii) analyze the use of alternative visibilities for the ACO such as ratings given by the teaching staff by type of content presented.

# References

1. Ros, M.Z.: Secuenciación de contenidos y objetos de aprendizaje. Revista de Educación a Distancia (2005)
2. Bianchi, L., Dorigo, M., Gambardella, L.M., Gutjahr, W.J.: A survey on metaheuristics for stochastic combinatorial optimization. Nat. Comput. **8**(2), 239–287 (2009)
3. Semet, Y., Lutton, E., Collet, P.: Ant colony optimization for e-learning: Observing the emergence of pedagogic suggestions. In: Swarm Intelligence Symposium, 2003. SIS'03. Proceedings of the 2003 IEEE, pp. 46–52. IEEE (2003)
4. Van den Berg, B., Van Es, R., Tattersall, C., Janssen, J., Manderveld, J., Brouns, F., Koper, R.: Swarm-based sequencing recommendations in E-learning. In: 5th International Conference on Intelligent Systems Design and Applications, 2005. ISDA'05. Proceedings, pp. 488–493. IEEE (2005)
5. Gutiérrez, S., Pardo, A., Kloos, C.D.: Finding a learning path: toward a swarm intelligence approach. In: Proceedings of the 5th IASTED International Conference on Web-Based Education, pp. 94–99. ACTA Press (2006)
6. Wang, T.I., Wang, K.T., Huang, Y.M.: Using a style-based ant colony system for adaptive learning. Expert Syst. Appl. **34**(4), 2449–2464 (2008)
7. Wong, L.H., Looi, C.K.: Adaptable learning pathway generation with ant colony optimization. Educ. Technol. Soc. **12**(3), 309–326 (2009)
8. Gutiérrez, S., Valigiani, G., Collet, P., Kloos, C.D.: Adaptation of the ACO heuristic for sequencing learning activities. In: Proceedings of the EC-TEL, pp. 17–20 (2007)
9. Despins, J.P.: Connaitre les styles d'apprendissage pour mieux respecter les façons d'apprendre les enfants. Vie Pedagogique **39**, 10–16 (1985)
10. Herrmann, N.: The creative brain*. J. Creat. Behav. **25**(4), 275–295 (1991)
11. Felder, R.M.: Matters of style. ASEE Prism **6**(4), 18–23 (1996)
12. Keefe, J.W.: Profiling and Utilizing Learning Style. NASSP, 1904 Association Drive, Reston, VA 22091-1578 (1988)
13. Huynh, O.M.S.V.N., Nakamori, Y.: k-Prototypes Algorithm for Clustering Mixed Numeric and Categorical Data (2003)
14. Berkhin, P.: A survey of clustering data mining techniques. In: Grouping Multidimensional Data, pp. 25–71. Springer, Berlin, Heidelberg (2006)
15. Moodle.org.: Open-source community-based tools for learning. Enero 2016. http://moodle.org/ (2016)
16. Franco, M.J.: Aprendizaje Web. Enero 2016. http://www.aprendizajeweb.com/ (2016)
17. Hong, C.: A Survey of Model-Based Clustering Algorithms for Sequential Data (2002)
18. Malo Herrera, E.: Modelo de clustering para el análisis de los datos poblacionales del cáncer. Universidad de Valparaíso, Chile (2007)
19. Hubert, L., Arabie, P.: Comparing partitions. J. Classif. **2**(1), 193–218 (1985)

# Students' Assessment Preferences in ESP in the Smart Learning Environment

**Ivana Simonova**

**Abstract** The paper deals with the assessment of learners' knowledge, particularly how the learners can show the teacher what they know. The main objective of this article is to detect and analyze students' assessment preferences when exploiting the smart learning environment where an individualized offer is provided to them. The sample group consisted of 203 students of the Faculty of Informatics and Management, University of Hradec Kralove, Czech Republic. The method of questionnaire was applied, when students evaluated various assessment methods on the five-level scale. Then, the method of multiple regression was exploited to prove whether there exists any correlation between learners' assessment preferences and assessment formats. The results show the most preferred way of assessment is the semester project work. Unfortunately, no statistically significant preference was discovered between the assessment preferences in the sample group and those reflecting individual learning styles. Finally, a didactic recommendations defined by Comenius centuries ago were emphasized.

**Keywords** Smart learning environment · Learning management system · Blackboard Mobile · Smart application · Assessment preferences

## 1 Introduction

Latest technologies have been providing strong impact on the process of teaching/learning for decades. Currently, living at the junction of education and technology [1], a shift has been detected from digital technology-enhanced processes to activities running in smart learning environments, supported by smart devices, exploiting smart solutions. According to Mikulecky [2], smart environments, as providing new facilities, can be considered a new level of computer-enhanced learning.

I. Simonova (✉)
University of Hradec Kralove, Rokitanskeho 62,
Hradec Kralove, Czech Republic
e-mail: ivana.simonova@uhk.cz

© Springer International Publishing Switzerland 2016
V.L. Uskov et al. (eds.), *Smart Education and e-Learning 2016*,
Smart Innovation, Systems and Technologies 59,
DOI 10.1007/978-3-319-39690-3_34

Reflecting this fact, the field of Ambient Intelligence can be studied not only from the view of technological, social or ethical perspective but also the educational perspective focusing on the field of proper education should be considered, particularly the context-aware ubiquitous learning. Yang et al. [3] define it as a computer-supported learning paradigm providing integrated, interoperable, pervasive and seamless learning experience, when identifying learner's surrounding context and social situation. In other words, it means the anytime/anywhere learning is shifted to learning at the right time and the right place with right sources and right collaborators. Moreover, other authors, e.g. Hwang et al. [4], consider the context-aware ubiquitous learning to be an innovative approach integrating wireless, mobile and context-aware technologies that, reflecting the learner's situation in the real world, are able to provide appropriate support and guidance. However, the same authors [4] admit, there exist various views of the term of the context-aware ubiquitous learning, from a rather wide approach of anytime/anywhere learning mentioned above, which does not consider whether the wireless communication and/or smart devices are implemented in the process of learning, to special cases, e.g. the ISTAG scenario 4: Annette and Solomon in the Ambient for social learning, which is a vision of sophisticated smart learning environment (SLE) based on learning as a social process. Yang et al. [3] defined following eight aspects of the context-aware ubiquitous learning: mobility, location awareness, interoperability, seamlessness, situation awareness, social awareness, adaptability and pervasiveness. Moreover, Hwang et al. [5] set the potential criteria of context-aware ubiquitous learning as the adaptability to learner's behaviour and context of the real and cyber world, active personalized support reflecting learner's individual and surrounding context, seamless learning independent on the place, and the ability to adapt the learning content to meet technical requirements of different mobile devices. Then, information can be provided not only anytime/anywhere but the right information can be said at the right time to the right person [6]. Defining the permanency, accessibility, immediacy, interactivity and situation to be the main characteristics of ubiquitous learning, Yang introduced an identical approach, when sharing three dimensions of learning resources (learning collaborators, contents and services) to provide "intuitive ways for identifying right learning collaborators, right learning contents and right learning services in the right place at the right time." [2: 217].

In this paper, the smart learning environment is perceived as technology-supported learning environment that can make adaptations and provide appropriate support (e.g. guidance, feedback, hints or tools) at the right place and at the right time based on individual learners' needs. These are determined through analyzing their learning behaviours, performance and the online and real-world contexts they are situated in [7]. Smart devices, particularly smartphones, tablets, pads etc., are generally understood as those connected to other devices or to networks via various wireless protocols (such as Bluetooth, WiFi, 3G etc.) and operating interactively and on the principle of ubiquitous computing [8]. While technology has changed what is possible to learn and how students can be supported in their learning, the principles of effective instruction have not changed. The technology does not drive learning, it is simply what mediates and supports the process, Means states [9].

Therefore, smart devices, been widely exploited for private purposes by learners of all age groups, are naturally used for learning. Students shifted from listening to music, watching films, video-recordings etc. to exploiting the smart devices for educational purposes, particularly firstly for foreign language learning. Within the higher education, English for Specific Purposes (ESP) is widely under the focus.

In our previous research activities, when reflecting the structure of the learning process including four phases (motivation, explanation, practising and assessment [10]), we focused on the explanatory phase and dealt with the role of smart study materials available in individualized online courses in the learning management system (LMS). The smart qualities of LMS were assured by the smart application added to the standard learning management system. The smart application reflected (1) the pattern of individual student's learning styles detected before and (2) the evaluation of each type of study materials as follows: minus one $(-1)$ means this type of study material, activity, assignment, communication etc. is refused, i.e. does not match the given learning style; zero (0) is the middle value, i.e. the student neither prefers, nor refuses, but accepts this type; one (1) means this type is preferred, it matches the given learning style.

Considering these two criteria we expected the smart application could adjust the learning content to individual student's learning preferences and result in learners' better knowledge [11].

As this hypothesis did not prove, in this paper our attention was paid to the process of assessment, particularly to learner's assessment preferences. As described below, various assessment formats are available within the LMS. Identically to the exploitation of smart study materials provided to the learners in preferred formats generated by the smart application, the process of detecting knowledge was tailored to learner's assessment preferences, i.e. the most appropriate formats were provided to the learners in the individualized order within online courses. Additionally to this smart approach, we monitored learner's real wishes. Therefore, the main objective of this paper was to detect students' real assessment preferences, consider them in relation to the smart solution and propose didactic recommendations towards efficient assessment their knowledge in the future.

## 2 Theoretical Background

Despite education has been transformed to the smart learning environments, the process of instruction still keeps the 'traditional' rules defined centuries ago by the Czech scholar Jan Amos Komensky (Johann Amos Comenius) [10]. Briefly summarized, he emphasized the learning content should be mediated to learners through as many as possible ways so that each learner could choose the one which corresponded to their learning preferences. This approach was applied in our previous research [11]. To improve the learning management system (LMS) towards having smart qualities, the smart application was added to each online course which correlated individual learning preferences of learners to the appropriateness of single didactic means used in the online course and generated the individualized learning process for each of them [11].

Our approach was based on the theory of Unlocking the will to learn by Johnston [12]. To describe the whole process of learning, she uses the metaphor of a combination lock saying that cognition (processing), conation (performing) and affectation (developing) work as interlocking tumblers; if aligned, they unlock an individual's understanding of his/her learning combination. She compares human learning behaviour to a patterned fabric, where the cognition, conation and affectation are the threads of various colours and quality. It depends on individual weaver (learner) how s/he combines threads and what the final pattern is.

Johnston designed the Learning Combination Inventory (LCI) consisting of 28 statements, responses to which are defined on the five-level Likert scale, and three open-answer questions. It emphasizes not the product of learning, but the process of learning; it focuses on how to unlock and what unlocks the learner's motivation and ability to learn, i.e. a way how to achieve student's optimum intellectual development. Learner's responses to LCI describe the schema (pattern) that drives their will to learn. Respondents are categorized into four groups where sequential, precise, technical and confluent ways of processing information are combined. The sequential processors are defined as the seekers of clear directions, practiced planners, thoroughly neat workers; the precise processors are identified as the information specialists, info-details researches, answer specialists and report writers; the technical processors are specified as the hands-on builders, independent private thinkers and reality seekers; and the confluent processors are described as those who march to a different drum beat, creative imaginers and unique presenters. Reflecting the individual pattern of each learner, six types of study materials were provided to them to tailor the online courses to their individual learning preferences: full-texts, short texts structured for the distance learning, presentations displaying the learning content in bullets, animations and video-recordings, images and figures, links to other sources including dictionaries, encyclopaedia etc. Unfortunately, no statistically significant differences were detected in the increase of knowledge with students who had learned from the appropriate types of study materials generated by the smart application compared to those not having this possibility available [11].

Therefore, another approach was applied which followed the design of Leither's methodology [13]. She emphasized the testing, i.e. assessment-related problems should be taken into account from the view of assessment formats. Most educators have been well aware of the individualization factor and have adjusted their teaching to learners' preferences; they have included approaches and strategies preferred by single types of learners and excluded those not appreciated. But what about the field of assessment of learners' knowledge and skills? Are the 'testing styles', i.e. assessment preferences of single students, taken into account? In this context, Johnston within the LCI also asked the question 'How would you show your teacher what you have learned?'. Unfortunately, despite the assessment is recognized a crucial part of the process of instruction, teachers often tend to use tests (either standardized or non-standardized ones) of the same types for all learners, i.e. learners' individual preferences in assessment are not reflected at all, as Leither mentions [13]. On one side, teachers are pushed to make assessment more systematic, transparent, objective, so that to provide all students with the same conditions. However, on the other side, this "fair" treatment is the cause of the "unfair" conditions from the point of individual preferences in assessment. Leither started experimenting with giving students choices on their exams when offering the option to taking exams in the multiple-choice or open-answer form. As expected, the group where assessment preferences were reflected reached significantly higher test scores (the difference was 5.51 %, $p < 0.005$ level) [13: 417]. The data produced by the whole Leither's pedagogical experiment were also analysed according to several other criteria, including e.g. whether learners' choice of a type of test (i.e. the assessment format) is relevant to their learning styles, if the choice matches the learning style, what students' opinions on assessment and exam formats are and others.

These activities led us to conducting this research to detect (1) what our learners' assessment preferences are at the Faculty of Informatics and Management compared to assessment formats generated by the smart application added to online courses in LMS and (2) how the results could be exploited in the didactic context.

## 3 Materials and Methods

### 3.1 Research Sample

Data were collected within the group of 203 bachelor study programme students of the Faculty of Informatics and Management, University of Hradec Kralove. The sample group is described by four criteria as follows:

- gender—male (60 %), female (40 %);
- study programme—Applied Informatics (AI3, 41 %) and Information Management (IM3, 22 %) bachelor study programmes, bachelor study programmes Financial Management (FM, 10 %) and Tourism & Management (TM, 27 %;
- form of study—full-time (60 %), part-time (40 %);
- age—below 20 years (2 %), 20–24 years (70 %), 25–29 years (13 %), 30–39 years (11 %), 40+ years (4 %).

### 3.2 Teaching/Learning ESP at FIM

Starting from 2001, the LMS WebCT was exploited for running online courses; since 2008 the SLE Blackboard has been used and since 2013/14 the Blackboard Mobile Learn$^{TM}$ version 4.0 for Apple and Android devices has been piloted (Blackboard Mobile Learn$^{TM}$ version 4.0 supports iOS6+, i.e. iPhone 3GS, iPad 2+, IPad mini, iPod Touch 4+ and Android OS 2.3+). Currently, approximately 250 online courses supporting single subjects are available to FIM students, either to assist the teaching/learning process, or to be used in the distance form of education. In total, 21 of them have applied mobile-assisted language learning (MALL) principles, namely in four courses of ESP for IT students, Business English, ESP for Tourism & Management, History and Culture: UK, History and Culture: Australia, History and Culture: New Zealand). All online courses run traditionally within the LMS, and currently they are available on smart mobile devices as well. In other words, the blended learning model is applied which combines the face-to-face instruction, work in online courses and individualized approach to them through smart mobile devices which satisfies learners' time and place preferences [14] and bridges formal and informal learning [15]. Within this research formats of assessment in six ESP courses for IT students were under our focus.

## 3.3   Methods

The research methodology followed the above mentioned Leither's pedagogical experiment [13] focused on learners' preferences in assessment formats, i.e. how the learners would show their teacher what they have learned. It was structured into two phases.

First of all, the questionnaire 1 was administered to students to express their preferences in various assessment formats. As smart learning and assessment is discussed, totally eight written (W) formats were listed and described in Table 1 to be selected from. Students articulated their preferences on the five-level scale (strongly preferred, rather preferred, accepted, rather rejected, strongly rejected) where the same evaluation could be applied repeatedly if their preferences in the format were identical. Both the individual (I) and group (G) assessment formats were included in the list.

Second, the individual pattern of learning preferences defined by Johnston's LCI was correlated to these eight assessment formats. The data were processed by SPSS

**Table 1**  Written assessment formats

| AF | F | Assessment format description |
|---|---|---|
| I | I | A question (problem, topic) from pre-defined list is set for essay writing |
| II | I | A question (problem, topic) from unknown list is set for essay writing |
| III | I | Multiple-choice test with 1 correct answer |
| IV | I | Multiple-choice test with 2+ correct answers |
| V | I | Yes/No format |
| VI | I | True/False format |
| VII | G | Students introduce results of the project they worked on during the semester; topic was set at the beginning of the semester |
| VIII | G | Students introduce results of the project they worked on during the exam day; topic is selected from the unknown list |

*AF* assessment format; *F* form; *I* individual; *G* group; *W* written format statistic software by the method of multiple regression where four variables of the pattern of individual learning preferences were tested against single assessment formats.

## 4  Research Results

The data collected in the questionnaire 1 were processed by the method of frequency analysis by NCSS2007 statistic software. The results are displayed in Fig. 1.

As expected,

- the strongest preference (black colour) and partial preference (dark grey) were given to the group assessment format of project work running through the whole semester (assessment format VII);
- the essay writing on the topic from the pre-defined list of question (assessment format I) also belonged to preferred formats—this type of assessment is widely used from lower secondary school level;
- rather strong rejection was detected with both multiple-choice formats (assessment formats III and IV);
- multiple-choice format with 2+ correct answers was the most frequently rejected one.

On the other side, we were rather surprised that

- the project work running at the place and within the time of exam (assessment format VIII) belonged neither to strongly preferred and rather preferred, nor to strongly rejected and rather rejected formats;
- the essay writing on the topic from the unknown list was widely accepted from the same reason of frequent use of this format in the Czech education system;
- there were no strong rejections to the Yes/No and True/False (assessment formats V and VI);

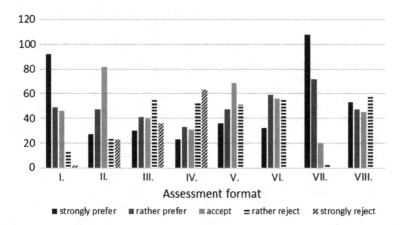

**Fig. 1** Written assessment format preferences (Author's own source)

**Table 2** Written assessment formats results of multiple regression analysis

| Assessment format | I | II | III | IV | V | VI | VII | VIII |
|---|---|---|---|---|---|---|---|---|
| Significance | 0.184 | 0.904 | 0.459 | 0.809 | 0.457 | 0.396 | 0.176 | 0.761 |

Significance < 0.05

- multiple-choice format with 1 correct answer did not belong to preferred formats (assessment format II).

The statistically significant correlation between the individual pattern of learning preferences defined by LCI and single assessment formats was not discovered—the multiple regression coefficients were too high with all eight assessment formats (statistical significance was below 0.05). The values closest to the significance coefficient were detected with assessment formats VII (0.176; Students introduce results of the project they worked on during the semester; topic was set at the beginning of the semester.) and assessment format I (0.184; A question, problem, topic from pre-defined list is set for essay writing.). These results correlate with those in Table 1. The data of multiple regression are displayed in Table 2.

## 5 Conclusion

In conclusion, the comparison of the data collected in the research showed on one side large differences in preferences in single assessment formats, e.g. assessment format I (individual essay writing on the topic from the pre-defined list) and format VII (group semester project solving); but on the other side learners' preferences in assessment formats detected by the questionnaire and displayed in Table 1 were identical if the preferences detected by LCI were considered. Multiple-choice (format III and IV) and dichotomous formats (IV, V) were not preferred to the expected extent. The reason might be that students are the LMS users experienced enough to know that if tests are reliable and valid, it is difficult to 'guess' the answer without appropriate knowledge, despite it seems to at first sight. Both these findings *do not prove the importance of an assessment format for showing the teacher what the learner really knows, despite some preferences were detected.*

Considering the strengths of group semester project work compared to the project work running within the exam, the independent work at the right time at the right place and with the right person/s is appreciated by students, including the possibility to use all necessary materials—if supervised within the exam, students do not feel comfortable and might have feelings of been limited to some extent.

Irwin and Hepplestone [16] followed the call for changes in assessment practices in the higher education towards increasing its flexibility and giving learners more control over the assessment process. Instead of others, they also focused on the role of ICT in this process, particularly on new opportunities to select the assessment format to show the teacher what they learned. Undergraduate students' expectations and preferences in assessment were examined by Sander et al. [17]. They found out students expressed the preference for writing essays, running projects and showing how to solve real problems. Moreover, the assessment format should consider and facilitate the acquired learning content and reflect the level of understanding, thus supporting learners in developing higher order thinking skills, Buyukozturk and Gulbahar stated [18]. Reflecting the results of Wegelaar-Jansen et al. [19], the discovery and participation were detected as preferred types of activities within the process of learning. Both of them are included in the project work, which is in accord with our results.

From the didactic view, various recommendations could be provided, both for the teachers and students. However, each of the recommendations should follow the Comenius' didactic principles defined centuries ago, saying every teacher should present the learning content to the learners through as many as possible ways, as well as ask the learner various types of questions to let them show what they know. Such an approach seems to sound very easy but the opposite is true. It depends on teacher's professional qualification and 'student's will to learn', as both Comenius [10] and Johnston state [12].

At the Faculty of Informatics and Management, the project work is included in each of four ESP courses. Students work either individually, when searching and processing professional texts for the e-book called IT Reader, which is the set of texts with their recordings where unknown professional vocabulary is translated. After every academic year a current issue of the IT Reader is edited. Students use the IT Reader for independent practising the listening comprehension skills during the year; finally, this activity is part of the final exam in summer semesters [20]. The group project work focuses on designing ICT-enhanced didactic aids, usually e-applications which can support students learning of ESP, e.g. for practising and revising vocabulary, tailoring the course content to student's preference (as mentioned above) etc. Thus the students' project work is exploited to make the ESP learning easier for them. Moreover, the highest added value of such activities is in using students' experience, improving their professional image among the peer students and supporting their motivation to learn, both the ESP and IT subjects.

To conclude, teachers should be aware each student is different, exploiting different ways both to learning and showing the knowledge. Different approaches do not mean they are not appropriate [10]. Compared to his and later times, modern technologies brought changes to all spheres of our lives, including education; mobile technologies performed a paradigm shift from the traditional learning processes to those in smart learning environment which are distinguished by mobility, interoperability, seamlessness, social awareness, adaptability, and pervasiveness [3].

**Acknowledgments** This paper is supported by SPEV Project 2016.

# References

1. Punya Mishra's Web. http://punya.educ.msu.edu/2008/01/12/mishra-koehler-2006/. Accessed 12 Jan 2016
2. Mikulecky, P.: Smart environments for smart learning. In: 9th International Scientific Conference on Distance Learning in Applied Informatics DIVAI 2012, pp. 213–222, Nitra, UKF (2012)
3. Yang, S.J.H., Okamoto, T., Tseng, S.S.: Context-aware and ubiquitous learning (Guest editorial). Educ. Technol. Soc. **11**(2), 1–2 (2008)
4. Hwang, G.J., Yang, T.C., Tsai, C.C., Yang, S.J.H.: A context-aware ubiquitous learning environment for conducting complex science experiments. Comput. Educ. **53**, 402–413 (2009)
5. Hwang, G.J., Tsai, C.C., Yang, S.J.H.: Criteria, strategies and research issues of context-aware ubiquitous learning. Educ. Technol. Soc. **11**(2), 81–91 (2008)
6. El-Bishouty, M.M., Ogata, H., Rahman, S., Yano, Y.: Social knowledge awareness map for computer-supported ubiquitous learning environment. Educ. Technol. Soc. **13**(4), 27–37 (2010)
7. Hwang, G.J.: Definition, framework and research issues of smart learning environments—a context-aware ubiquitous learning perspective. Smart Learn. Environ. **1**(4), 1–14 (2014)
8. Weiser, M.: The computer for the twenty-first century. Sci. Am. **265**(3), 94–104 (1991)
9. Means, B., et al.: Evaluation of Evidence-Based Practices in Online Learning: A Meta-Analysis and Review of Online Learning Studies. U.S. Department of Education, Washington (2010) https://www2.ed.gov/rschstat/eval/tech/evidence-based-practices/finalreport.pdf. Accessed 14 Jan 2016
10. Comenius, J.A.: The Gate of Languages Unlocked, or, A Seed-Plot of All Arts and Tongues: Containing a Ready Way to Learn the Latin and English Tongue. T.R. and N.T. for the Company of Stationers, London (1946)
11. Šimonová, I., Poulová, P.: Learning Style Reflection in Tertiary e-Education. Hradec Kralove, WAMAK (2012)
12. Johnston, C.A.: Unlocking the Will to Learn. Corwin Press Inc., Thousand Oaks (1996)
13. Leither, A.: Do student learning styles translate to different "testing styles? J. Polit. Sci. Educ. **7**(4), 416–433 (2011)
14. Pieri, M., Diamantini, D.: From e-learning to mobile learning: new opportunities. In: Ally, M. (ed.) Mobile Learning. Transforming the Delivery of Education and Training, pp. 183–194. AU Press, Athabasca (2009)
15. Abdullah, M.R.T.L., et al.: M-learning scaffolding model for undergraduate English language learning: bridging formal and informal learning. TOJET: Turk. Online J. Educ. Technol. **12**(2), 217–233 (2013)
16. Irwin, B., Hepplestone, S.: Examining increased flexibility in assessment formats. Assess. Eval. **37**, 773–785 (2012)
17. Sander, P., Stevenson, K., King, M., Coates, D.: University students' expectations of teaching. Stud. High. Educ. **25**(3), 309–323 (2000)
18. Buyukozturk, S., Gulbahar, Y.: Assessment preference of higher education students. Egitim arastirmalari-Eurasian J. Educ. Res. **10**(41), 55–72 (2010)
19. Wegelaar-Jansen, A.M., van Wijngaarden, J., Slaghuis, S.S.: Do quality improvement collaboratives' components match the dominant learning style preference of the participants? BMC Health Serv. Res. **15**, Article number 239 (2015)
20. Šimonová, I.: Multimedia in the teaching of foreign languages. JoLaCE—J. Lang. Cult. Educ. **1**(1), 112–121 (2013)

# Smart Technologies in Psycho-Oriented Strategies of Adaptive Education

**Alexander Gein, Dmitriy Istomin and Andrey Sheka**

**Abstract** This document is a survey of research on user modeling for Adaptive Learning. We suggest a model for automated detection of deviations or statistically stable behavior using smart technologies. Proposed model allows to explain test results using documented deviations. We have analyzed the main psychophysiological parameters which can be used in presented model for student's behavioral analysis. The proposed research can help to improve the adaptivity of computer-based education systems.

**Keywords** Behavioral analysis · User modeling · Adaptive learning · EEG · ECG · Eye tracking

## 1 Introduction

The main purpose of the usage of smart technology in the learning process is the creation of a learning environment where students, getting an access to a huge amount of educational resources, will be able to build an individual trajectory of their education. One of the slogans of smart technologies in education is the idea of the active students' participation in self-development of the curriculum and even academic disciplines, teachers at this stage are mainly tutors helping students to understand their educational goals and formulate them in terms of school subjects. The process of

Supported under the Agreement 02.A03.21.0006 of 27.08.2013 between the Ministry of Education and Science of the Russian Federation and Ural Federal University.

A. Gein (✉) · D. Istomin · A. Sheka
Ural Federal University, Yekaterinburg, Russian Federation
e-mail: A.G.Geyn@urfu.ru; A.G.Gein@urfu.ru

D. Istomin
e-mail: D.V.Istomin@urfu.ru

A. Sheka
e-mail: AndreySheka@urfu.ru

implementing of these opportunities is based on the principle of an informal educational institutions association carrying out joint educational activities. Top position here is the ability to create The United European University with an extremely wide variety of educational resources and variations of educational strategies. However, a wide variety and variability sets a number of problems. In this case, students have to choose from a much larger list of courses and tutors are required to be competent to provide appropriate assistance to students. To deal with these problems, the use of adaptive automated systems is required which can respond to as the students' specific needs, as to students abilities, for example, psychological and cognitive features. We should note that they must be taken into account not only at the design stage, but directly in the learning process with the use of electronic information resources. It is important, for example, to monitor the level of student involvement in the learning process, or to adapt the course dynamically, or to be sure that students have listened carefully to the proposed course. And if "offline" teachers can see the external behaviour of the students, then in the Internet services it is necessary to provide such a tracking without human participation with a high accuracy, which can be carried out, for example, via video analysis, oculography sensors, etc. The developed patterns of human behavior that are based on the obtained data, allow to perform behavioral analysis and build adaptive educational programs.

## 2  The Development of the Basic Model

The main role in any system of adaptive learning plays the model of how a student obtains knowledge and skills. It is according to this model that the dynamics of the learning process can be viewed. Adaptive learning technology takes into account the characteristics of each stage of the transition to the new productive level and the accomplishment of training activities on each stage. The student will be offered a portion of new learning material, if the previous material has been learnt well or, if not, there will be restudy. Students will be asked to do the task of the next level if he passed the previous one, or he will be sent to revise the material that did not allow him to pass the test. It means that the key part of adaptive education is to monitor how a student gains knowledge and skills, as well as landmark and final control of educational achievements. It is essential that this important component of education was paid much attention by specialists dealing with the problem of introduction of electronic tools and smart technologies. The main teaching tool here is testing, which allows highly formalized examination, ensuring uniformity of assessment and its objectivity. Control tasks, even shown not in the test form, are presented in the computer form, their results are also registered in the computer, then checked and handled either directly or transmitted to a single specialized center via the network technologies.

However, all technologies are based on the pretext that all students differ only in the level of knowledge, sometimes there can be an attempt to take into account differences in cognitive styles (first of all, in preferred types of making decisions:

declare-logical vs algorithmic or abstract and specific, etc.). The main role in the right interpretation of test results is really played by subjective factors such as motives orientation, mental stability, reaction speed, etc. Even in the traditional conditions, monitoring and control of educational achievements are conducted by a teacher, who seems to take into consideration specific characteristics of a student, but there were cases when a student had failed such type of testing and further he appeared to be a rather successful student. These facts are frequently presented as historical extraordinary cases, though it is rather possible that the students, who suffered from such assessment mistakes, paid a high price for them.

In terms of individual e-learning, it is required to create the environment for the student, close to the comfortable full-time education with a teacher. This implies the necessity to register non-verbal reactions to the process of obtaining educational material as well. Here, in our opinion, such factors as the orientation of the motives, the stability of nervous system, quick response and some others are rather essential. They can be viewed in behavioral reactions during the control tests.

To develop the use of smart technologies in which you could take into account all these aspects, you should have an appropriate model of a student. The fact of such a model itself is being discussed now due to the development of adaptive systems of education [10]. But up to today there is no single opinion on which factors determine this model and what its components are [5–8, 10, 12]. In the narrow meaning, a student model is formal requirements for personal and professional features of this student, for his skills, knowledge and abilities in various subjects, characteristics of physical and mental condition, etc. In fact, the main parameters of the student model are determined by the standards of education. This model is usually named normative. further control system is based on this model: the types of questions and the choice of tasks to test students' knowledge; the selection of criteria for assessment of each task and final test in general, etc. In the broad meaning, the student model, in addition to the normative one, includes starting level of a student's knowledge of the course, students' knowledge of other subject, which are necessary for the study of this course, the current knowledge of the course, the individual characteristics of a student (learning ability, skills level, etc.).

The viewed models are mainly oriented to build an adaptive process of knowledge, skills and abilities obtaining but not monitoring and control. We would like to pay attention to its part, that is the stage of control. Firstly, we mean tracking of deviations in the behavior of the student during the control test, for example, attempts to get information from various external sources which are not allowed by the rules of the test. If we speak about monitoring, there can be statistically stable student behavior in such situations. Nowadays, this function, as a rule, is fulfilled by a person-supervisor, who presents at the lesson or test or supervises distantly. We suggest a system of automated detection of deviations or statistically stable behavior with the use of smart technologies. At the same time the use of our proposed model gives the opportunity to explain the results of tests basing on the documented deviations.

Speaking about the motivational component of our proposed model, we should note that the main reason for deviant behavior here is the desire to pass a test on a formal level, but not to demonstrate really acquired knowledge and skills. The motives for this can be different—belief in the fact that these knowledge and skills will not not be in great demand in the future career, difficulty in mastering these skills and knowledge which objectively or subjectively led to their missing, lack of self-confidence, etc. The causes of this are beyond recording and interpreting of deviant behavior themselves.

The other part is special cognitive features of a student, when it appears that the student have not acquired material enough or this material have not been structured in an appropriate way.

The third component is non-confidence in the right choice of how to do the test.

Teaching experience of the authors shows that these are the key factors of deviant behavior when passing tests.

At the same time, these three components can be revealed when there is character tracking of deviant behavior in tests. We can say that there are three possible cases of deviant behavior. The first is when deviant behavior is recorded since the very start of an exercise. Secondly, such behavior is seen after making some part of the exercise and a student got any problems in the middle. Finally, the third case is at the end of the exercise when the student has difficulty with typing the right answer to the checking system. Our interpretation is just an example and is based on the only parameter—approximate time of a student to shift to the deviant behavior. In fact, it is based on a complex of psychophysiological reactions of the student, which can be recorded by electronic means and detected using smart technologies. However, it should be emphasized that each of these interpretations is just probable in any of three cases.

The use of our proposed model in monitoring allows to get rather detailed psychological portrait of a student. It enables individualizing of adaptive education not just according standard parameters of flexible passing from one knowledge level to another, but also forming the complex of knowledge itself.

In the following, we are going to view technical part of implementation of our proposed model.

## 3 Possible Technical Means to Use

It is a student's feedback in the basis of adaptive education that is used to correct educational program. The most frequent mechanism in adaptive courses is behavior analysis of a student at the educational site. In particular, the analysis of the user activity when watching some video. For example, it is known that students can review some part of a video lecture for several times. This demonstrates the difficulty of this part.

Beside recording statistics, one can use additional devices which register psychophysiological reactions to the material. Let us show the main psychophysiological parameters that can be used in student's behavioral analysis.

The electrical skin resistance is a good parameter for the determination of emotions. Many emotions cause a response from sweat glands by increasing secretion. Though this increase is usually small, but the sweat contains water and electrolytes which increase the electrical conductivity, thus reducing the electrical skin resistance. The electrical resistance of skin can respond to the reactions such as anger, fear, changes in spa orientation [1]. Another common display is the dilation of facial blood vessels when a person is confused.

The cardiovascular system is also significantly responds to changes in the psychological state of a person. The simplest indicators of the cardiovascular system is pulsation and blood pressure. In particular, heart rate and blood pressure can increase in some stressful situations because of the adrenaline release. These changes can be recorded to monitor person's condition [11].

The analysis of cardiovascular activity and the electrical skin resistance are widely used to monitor the condition of the body, for example, in fitness trackers.

Breathing is primarily regulated by metabolic processes. However, respiration may be adjusted in response to emotional changes, such as feeling of good luck, or, on the contrary, feeling alarm or fear. In most cases, this results in increasing the frequency and amplitude of breathing.

Electroencephalography (EEG) is the method of brain investigation, based on the tracking electrical impulses, coming from the particular zones and areas. This is a sensitive research method. It reflects the slightest changes in the function of the cerebral cortex and deep brain structures, providing a millisecond time resolution. EEG enables the qualitative and quantitative analysis of the functional brain state and its reactions to the stimulus. Depending on the frequency range, amplitude, wave shape, topography and the type of reaction EEG rhythms can be distinguished. Human activities can be characterized by the rhythm. In spite of medicine, EEG is widely used in producing neurocomputer interface. In particular, the research of robot controlling, cars and other devices are carried out [2, 4].

Registration of eye movement is researched in Eye tracking. When reading eye movement is characterized by two conditions: short stops, called fixations, and sharp movements, called saccades. Fixations and saccades characterize the material that a person is studying. In particular, studying the change of a sight, you can determine the information which the person mostly pays attention to. In addition, it is possible to measure the loss of a person's attention, if he starts to get distracted by some other sources. Today Eye tracking is widely used in the analysis of the usability of web sites. For this purpose, there are specialized devices that track the eye direction on the monitor. In combination with specialized software these devices allow you to make heat maps for websites, showing the user's attention.

Modern technologies through a real-time image from a webcam can identify key points on the face. The key points are, for example, the corners of the lips, eyebrows, eyes, chin. Basically, algorithms detect 68 key points [9]. The relative positions of the points can characterize human emotions. In particular, there are algorithms detecting the following emotions: anger, contempt, disgust, fear, sadness, happiness and surprise [3]. Besides, key points allow to define the orientation of the head according to the camera, which can also characterize a person's condition. Key points of the face are used in multimedia applications with the aim to transmit emotions into the virtual game space. Key points also make it possible to build a man-machine dialogue at the brand new level.

The emotional reaction of a person can be tracked by voice as well. The mere fact of having voice in this or that situation characterizes the person's condition. Moreover, the pronounced words and the way they are pronounced can bring additional information. Thus, the voice response can provide essential information for the analysis of complex human behaviour. Voice is widely used in mobile devices, when it is easier for a person simply pronounce the search query than to type it on the keyboard.

The foregoing demonstrates a very wide range of developments in the use of various electronic means for the analysis of psychophysiological state of a person. Many of these tools are quite mobile, and their use is either remote or possible with minimal restrictions on human activity.

## 4   The Implication of Smart Technologies for Developing Psycho-Oriented Strategies in Adaptive Education

Depending on the diagnostic tools described above, we will distinguish three levels of research. The first level: audio-video monitoring. It is a non-contact, some means of visual and audible monitoring of behavior of the student are used there. Video observation has three main areas: workplace of a student (surface of the table and part of his environment), the student himself (his face, hands) and a student's working area environment as a whole.

Monitoring of a workplace includes the control of any use of information devices that are not allowed by the rules of an outgoing training. As it was noted in paragraph 1, it is important to record not only the fact of using unauthorized means, but also the time when the student began to use them, as well as the duration of its use relative to the total run-time of the test. The type of a source can give an additional information—the general reference book or an information object specially made by the student (such as a crib). For automated diagnostics, we apply here existing recognition algorithms. The task is greatly simplified if the student is trying to use electronic devices. Audiomonitoring allows to track attempts to obtain information from other students.

The eye tracking and the use of facial key points allow to see the orientation of students' attention. Here, we can add algorithms for recognising their emotional state. It is together with the analysis of the working place environment of the student, that these methods help to increase the value of situation diagnostics, and psychological state as well. If these procedures are carried out in the framework of a single test, than the diagnostics is based on the average behaviors of students in these situations. If the procedures are carried out in the context of systematic monitoring, than it is possible to obtain a psycho-pedagogical portrait of a particular student in terms of his learning style. To automate the process of such portraiting it is natural to use self-learning recognition programs.

The simple example is monitoring of how the student acquires the part of the material with some new information. The important parameters here are the speed of acquaintance ( relevant to the speed of acquaintance with other parts) in general, underlining the most important facts, showing the attention accentuation, the number of returns to this part in further study of examples with the given new notions and so on. It is clear that effectiveness of the use of smart technologies will be higher, the more structured learning material is. In case of system monitoring, it can lead to the possibility to produce an adequate cognitive portrait of a student.

The loss of attention while studying is possible, too. It can be caused by outside factors, or by individual features of a student. The loss of attention can be monitored by means of Eye tracking and EEG. When the loss appears, you can stop demonstration of the material and ask a student some questions to return his attention back.

These means can also give an additional information about the difficulty of the material. If the student watches the same part of the video for several times, focuses for a long time, then the teaching material should be adjusted.

The reaction of a student is also important when doing tests. Technical means can measure the stress rate, as it is the indicator if a student cheats or not, in complex with all other factors. Monitoring of the workplace environment plays an additional role of control. If the student can not be monitored by all other means.

The second level includes audio-video devices and compact means of electronic control of pulse, blood pressure, electronical skin resistance, etc. They can be in the form of bracelet that is worn on the wrist without influencing the study efficiency. Of course, the student should be accustomed to this procedure, without causing an extra psychic pressure on the student, distorting typical picture of a student's psychological condition. What is for final data interpreting, it is very similar to what we outlined in the above, so we are not going to describe them in particular.

The third level—complex monitoring of a student during education including EEG. The use of such means can be restricted not only by necessity of special conditions, but by considerable psychic influence on a student as well. In our opinion, the motive for a student to use these tools can be his understanding that they allow to make a detailed psychological portrait, consequently, to determine individual strategy of adaptive education in a more appropriate way.

Brief information on the levels is presented in the Table 1.

These means can be used in extracurricular activities to view general state of a student. It is very naive to think that a student will assimilate the material if he is hungry, did not sleep well or has just run the marathon. One should be sure that the student is able to acquire the new material. All of these can be provided by physiological parameters measured by fitness trackers.

We should note, that in some cases a student has no possibility to be in perfect physiological condition for the perception of information. That is why it is necessary to consider the current state of a student. For example, you can limit the amount of video that he will watch or change the tempo of the material layout.

**Table 1** Brief description of levels

| Level | Device | Measuring parameters | Monitored activity | Student's discomfort |
|---|---|---|---|---|
| 1. | Videocamera | Facial keypoints and emotions, gaze tracking, head position | Facial, head and eye movements | External monitoring |
| 2. | Wrist bracelet | Pulse, ECG, blood pressure, skin resistance | Cardiovascular system reaction | The device does not restrict body movements |
| 3. | Head device | EEG | Brains activity | Student should sit |

## 5 Conclusion

We can make a conclusion that the use of such smart technologies can provide high productivity in model recognition, self-learning algorithms and others in the sphere of education. Their use together with student behavior monitoring means can help to improve the adaptivity of computer-based systems of education.

## References

1. Bechara, A., Damasio, H., Damasio, A.R., Lee, G.P.: Different contributions of the human amygdala and ventromedial prefrontal cortex to decision-making. J. Neurosci. **19**(13), 5473–5481 (1999)
2. Bell, C.J., Shenoy, P., Chalodhorn, R., Rao, R.P.N.: Control of a humanoid robot by a noninvasive brain-computer interface in humans. J. Neural Eng. **5**(2), 214–220 (2008)
3. De la Torre, F., Vicente, F., Cohn, J.: IntraFace. In: 2015 11th IEEE International Conference and Workshops on Automatic Face and Gesture Recognition (FG), pp. 1–8. IEEE, May 2015
4. del Millan, J.R, Renkens, F., Mourino, J., Gerstner, W.: Noninvasive brain-actuated control of a mobile robot by human EEG. IEEE Trans. Biomed. Eng. **51**(6), 1026–1033 (2004)
5. Dorça, F.A., Lima, L.V., Fernandes, M.A., Lopes, C.R.: A stochastic approach for automatic and dynamic modeling of students learning styles in adaptive educational systems. Inf. Educ. **11**(2), 191–212 (2012)
6. Esichaikul, V., Lamnoi, S., Bechter, C.: Student modelling in adaptive e-learning systems. Knowl. Manage. E-Learn. **3**(3), 342–355 (2011)
7. Hohl. H., Böcker, H.D., Gunzenhäuser, R.: Hypadapter: an adaptive hypertext system for exploratory learning and programming. User Model. User-Adap. Inter. **6**(2–3), 131–156 (1996)
8. Jing, C., Quan, L.: An adaptive personalized e-learning model. In: 2008 IEEE International Symposium on IT in Medicine and Education, pp. 806–810. IEEE, Dec 2008
9. Kazemi, V., Sullivan, J.: One millisecond face alignment with an ensemble of regression trees. In: 2014 IEEE Conference on Computer Vision and Pattern Recognition, pp. 1867–1874. IEEE, June 2014
10. Martins, A.C., Faria, L., De Carvalho, C.V., Carrapatoso, E.: User modeling in adaptive hypermedia educational systems. Educ. Technol. Soc. **11**(1), 194–207 (2008)
11. Schandry, R.: Heart beat perception and emotional experience. Psychophysiology **18**(4), 483–488 (1981)
12. Wu, D., Bieber, M., Hiltz, S.R.: Asynchronous participatory exams: internet innovation for engaging students. IEEE Internet Comput. **13**(2), 44–50 (2009)

# Elderly People and Their Use of Smart Technologies: Benefits and Limitations

Blanka Klimova

**Abstract** Currently, there is a rising number of elderly people worldwide. By 2020 the percentage of elderly people aged 60+ years should reach 30 % out of the total number of population living in the developed European countries This aging process, however, results in new economic and social issues Therefore there is constant effort to prolong and maintain the active age of these elderly people, who want to lead active, fulfilling and quality life in terms of inclusion, socialization and independence. This can be achieved not only by a continuous support from their family members, but also with the help of modern information and communication technologies (ICT), which can assist elderly people in this process. The purpose of this article is to explore benefits and limitations of using ICT, particularly the smart ones, by the elderly people. In addition, the author of this study emphasizes the importance of appropriate training about the benefits of smart technologies.

**Keywords** Elderly · Smart devices · Benefits · Limitations

## 1 Introduction

Currently, there is a rising number of elderly people worldwide. By 2020 the percentage of elderly people aged 60+ years should reach 30 % out of the total number of population living in the developed European countries [1].

The aging process, however, results in new economic and social issues [3, 4]. Therefore there is constant effort to prolong and maintain the active age of these elderly people, who want to lead active, fulfilling and quality life in terms of inclusion, socialization and independence. This can be achieved not only by a continuous support from their family members, but also with the help of modern information and communication technologies (ICT), which can assist elderly people

B. Klimova (✉)
University of Hradec Kralove, Rokitanskeho 62, Hradec Kralove,
Czech Republic
e-mail: blanka.klimova@uhk.cz

© Springer International Publishing Switzerland 2016                     405
V.L. Uskov et al. (eds.), *Smart Education and e-Learning 2016*,
Smart Innovation, Systems and Technologies 59,
DOI 10.1007/978-3-319-39690-3_36

**Fig. 1** Proportion of the population aged 60+ : world and development regions in the period of, 1950–2050 (based on the data from the United Nations [2]—author's own processing

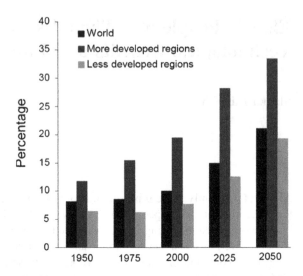

in this process. The technological devices, which are primarily aimed at the elderly people, are also known under the common name gerontechnology because they try to meet the needs of aging society [5].

In fact, present seniors are more and more involved in their use and these devices are becoming part and parcel of their everyday life [6]. Haukka [7] claims that people aged 55–74 years are adapting to use of the Internet, mobile telephones, tablets, and gaming technologies nowadays. Some studies (e.g., [8]) already talk about cyberseniors who are defined as active elderly people on the Internet with ease using services offered online such as information search, communication with family and friends, use of social networks, or conducting online payments. In addition, they are eager to participate in life-long education. Nevertheless, the exploitation of ICT by the majority of older people still seems to be too basic and more attention should be paid to the training of the elderly people in the use of ICT and the benefits these technological devices can bring them [6] (See Fig. 1).

The purpose of this article is to explore benefits and limitations of using ICT, particularly the smart ones, by the elderly people. In addition, the author of this study emphasizes the importance of appropriate training about the benefits of smart technologies.

## 2   Methods

The author used a method of literature review of available sources exploring research studies focused on the use of ICT for elderly people in the acknowledged databases and a method of comparison and evaluation of their findings. This review was done by searching databases of Web of Science, Scopus, ScienceDirect, and

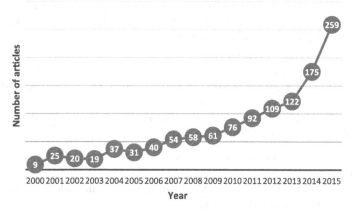

**Fig. 2** A growth of research studies on the topic elderly people and ICT (author's own processing based on the data from ScienceDirect [9])

Springer in the period of 2000–2015, using combinations of the following key words: *elderly people* and *ICT; elderly people* and *smart technologies*. In addition, other relevant studies were reviewed on the basis of the reference lists of the research articles from the searched databases. The search showed that there was an increasing trend dealing with the elderly people and their use of ICT. Figure 2 illustrates this trend in the growth of research studies on this topic in the period of 2000-2015 in the database ScienceDirect.

Most recently, a number of studies have examined an attitude of the elderly people towards mobile technologies such as mobile phones or tablets.

## 3 Smart ICT Technologies for the Elderly People

As it has been already mentioned above, research studies [10, 11] confirm that older generation of people, aged 58–77, are nowadays much more digitally aware than they used to be ten years ago. More than 70 % of elderly people are able to work on a computer and use a mobile phone. More than 50 % also use tablets (cf. [12]). For example, Campbell [13] claims that more people are now beginning to use the tablets because of the bigger size screen. In addition, more than one third of these elderly people started to use smartphones [14, 15]. And in future this number of the users of smart technologies is expected to grow since the producers of mobile technologies began to adjust them to elderly people's needs.

Nevertheless, based on the definition of smart learning environment [16–19], this smart technology for the elderly people must be fully supportive and adaptive to their behaviour, performance and the situation they are in.

Campbell [13] lists the main design principles which should meet these needs and focus on the following aspects:

- vision and hearing of the elderly people (e.g., appropriate font size, usually bigger than 16 pixels, contrast ratios with text, or provision of subtitles when video or audio content is fundamental to the user experience);
- motor control (e.g., bigger buttons—at least 9.6 mm or bigger screen device);
- device use (user-friendly);
- relationships (e.g., enable connection with a smaller, but a more important group of people such as their family members and friends);
- cognitive capacities of the elderly (e.g., provision of services such as reminders and alerts as cues for habitual actions).

Figure 3 provides an example of such a technological device designed for the elderly people.

Chen and Chan [21–23] summarize the main services the elderly people usually demand with respect to smart technologies:

- *Healthcare and monitoring the state of health*: At present, there is an increasing number of elderly people who exploit the so-called mobile healthcare, i.e. remote care service due to the limitation of aged care resources. This remote care service might include, for example, obtaining information on their health, receiving reminders for scheduled visits, medication instructions, or consulting a doctor at a distance [24].
- *Leisure and sales service*: Personal leisure (entertainment or self-education) is important for senior citizens in what constitutes their free time occupation. They can also do shopping via mobile phones, order meals or play computer games in order to maintain quality of their life in case they are not able to go out regularly.

**Fig. 3** An example of a smart technology for the elderly people [20]

**Fig. 4** An example of smart home designed for the elderly people [26]

Pull cord

Fall detector

Smoke alarm

Carbon monoxide detector

Flood detector

Sounder beacon

Temperature extremes sensor

Movement detector

Bogus caller button

Natural gas detector

- *Safety and privacy service*: This is considered as the most critical aspect for senior citizens. The user's activity must be monitored by using presence sensors and be analysed with consideration to different scenarios.

All the aspects described above are also reflected in their homes which start to resemble smart homes that represent places equipped with technologies such as cooking hob and oven safety control, sleeping pattern monitoring, emergency alarm or automatic lighting system that enable people to maintain living independently while in charge of their own healthcare and its costs [25]. A detailed picture of more safety devices is provided in Fig. 4.

The devices illustrated in Fig. 3 are often in connection with the local setting and emergency call centres.

Table 1 below then summarizes the key benefits and limitations of smart ICT devices for elderly people.

**Table 1** Outline of the key benefits and limitations of smart ICT devices for elderly people (author's own processing)

| Benefits | Limitations |
|---|---|
| • Providing healthcare, socializing, independence, entertainment, safety and security<br>• Overall improvement of quality of life of elderly people<br>• Reduced potential costs on care and treatment of elderly people | • A limited offer of these devices for this target group on the market<br>• Sometimes inappropriate design to meet the needs of elderly<br>• High costs of some smart devices for seniors<br>• Low awareness of the benefits of using smart technologies |

## 4   Importance of Training the Elderly People in Using Smart Technologies

However, research studies also indicate that the elderly people need more training in order to be able to use new technologies and thus avoid their anxiety about using a computer and difficulties in use [27]. Rain, Svarcova [28], for example, proposed four key methodological principles on motivating older people to use the Internet. These include: motivation and elimination of barriers; general identification with terminology, basic principles and links; practical use of the particular Internet service; and regular use of the acquired computer skills. In addition, the involvement of older people in the use of ICT is supported both at the international and national level by different projects which are specifically aimed at the exploitation of ICT by the older people.

For example, Benacova and Valenta [1] in their study present four projects which proved to be successful in this respect. They were conducted within the framework of EU Socrates Programme and Grundvig Projects and they were as follows: Self-organised Learning in Later Life (SoLiLL); European Computer Network (EuCoNet); Older people in Network (SEN-NET); and e-Learning in Later Life (eLiLL).

Furthermore, in the Czech Republic seniors are involved in studying at the University of the Third Age which enables them on a regular basis throughout the whole academic year (September-June) to attend tailored-made classes to their needs and interests. This contributes not only to the maintenance and expansion of their cognitive functions, but also to social inclusion since they meet people of the same age. As [1] present in their study, there are about 400 different educational programs of the Universities of the Third Age, in which already about 16, 000 older people have already participated all over the Czech Republic, both in face-to-face classes and marginally in virtual classes, particularly suitable for those who are not from the cities where the university is located. The topics of classes range from the field of sociology, art, history up to science. For instance, the syllabus of ICT class based on elderly people's needs is as follows [12]:

- Introduction to ICT
- Current modern technologies (types and their use)
- Internet and multimedia (Internet search, security basics, use of multimedia, web sites and blogs, safe on-line shopping)
- E-mail (its use, SPAM, spoofing, phishing, other threats and security)
- Socializing on the Internet (social networks, tools for the Internet communication—Skype, safety of social networks)
- Digital photos (devices for taking digital photos and software for their editing, print settings, possibilities for online presentation of digital photos, open source graphics processors)
- Office software (licensed and free and their differences, downloads and installation of open source office software)
- E-learning and distance education (their definitions, courses)

# 5 Conclusion

At present people are experiencing a decline in the growth of active population groups. On the contrary, there is a dramatic rise of aging population, which will even expand in future. Therefore societies all over the world try to deal with this new phenomenon and find the ways how to maintain quality of life of these older citizens. Smart technologies seem to be promising devices in this process because they appear to meet the requirements and needs of elderly people [5].

Future research in the field of the exploitation of smart technologies by the elderly people should focus on the followings:

• extensive trials examining the attitude of the elderly at the age of 55 and above towards the use of different smart technological devices;
• extensive trials exploring the effect of the use of smart technologies on the improvement of quality of life of the elderly people;
• studies examining the critical factors (e.g., age, education, or economic factors) in the use of smart technological devices by the elderly people.

**Acknowledgments** This work is supported by SPEV project titled Impact of Mobile Technologies and Social Networks on the Development and Maintenance of Cognitive Processes run at the Faculty of Informatics and Management in Hradec Kralove.

# References

1. Benacova, H., Valenta, M.: Moznosti informaticke vyuky senioru v ČR a EU, [Possibilities of computer science teaching of elderly people in the Czech Republic and EU]. Systemova Integrace **4**, 77–86 (2009)
2. United Nations, World Population Ageing: 1950–2050 (2001). http://www.un.org/esa/population/publications/worldageing19502050/
3. Klimova, B., Maresova, P., Valis, M., Hort, J., Kuca, K.: Alzheimer's disease and language impairments: Social intervention and medical treatment. Clin. Interv. Aging **10**, 1401–1408 (2015)
4. Maresova, P., Klimova, B., Kuca, K.: Alzheimer's disease: Cost cuts call for novel drugs development and national strategy. Ceska Slov. Farm. **64**(1–2), 25–30 (2015)
5. Plaza, I., Martin, L., Martin, S., Medrano, C.: Mobile applications in an aging society: status and trends. J. Syst. Softw. **84**, 1977–1988 (2011)
6. Klimova, B., Simonova, I., Poulova, P., Truhlarova, Z., Kuca, K.: Older people and their attitude to the use of information and communication technologies—a review study with special focus on the Czech Republic (Older people and their attitude to ICT), Educational Gerontology (2015). doi:10.1080/03601277.2015.1122447
7. Haukka, S.: Older Australians and the Internet. Brisbane (2011)
8. Machado, L.R., Behar, P.A., Doll, J.: Pedagogical practices to teacher education for gerontology education. In: Uskov, V.L. et al. (eds.), Smart Education and Smart e-Learning, Smart Innovation, Systems and Technologies, vol. 41, pp. 403–413 (2015)
9. A growth of research studies on the topic elderly people and ICT (2015). http://www.sciencedirect.com/science?_ob=ArticleListURL&_method=list&_ArticleListID=-914280094&_sort=r&_st=13&view=c&md5=e4d3bfeb75f82f70523892c7523a585d&searchtype=a

10. Hernandez-Encuentra, E., Pousada, M., Gomez-Zuniga, B.: ICT and older people: beyond usability. Educ. Gerontol. **35**(3), 226–245 (2009)
11. Sayago, S., Sloan, D., Blat, J.: Everyday use of computer-mediated communication tools and its evolution over time: an ethnographical study with older people. Interact. Comput. **23**, 543–554 (2011)
12. Vacek, P., Rybenska, K.: Research of interest in ICT education among seniors. procedia—social and behavioral sciences, no. 171, pp. 1038–1045. (2015)
13. Campbell, O.: Designing for the elderly: Ways older people use digital technology differently (2015). http://www.smashingmagazine.com/2015/02/designing-digital-technology-for-the-elderly/
14. Herma, J.: Tretina senioru vlastni chytry telefon, mobilni internet je vsak zatim neoslovil, [One third of seniors own a smartphone, however, the mobile Internet has not addressed them yet] (2014). http://smartmania.cz/bleskovky/tretina-senioru-vlastni-chytry-telefon-mobilni-internet-je-vsak-zatim-nenadchl-8710
15. Kubec, P.: Cesti seniori ovladli mobilni technologie, Telefonuji vice nez mladi, [Czech seniors have dominated mobile technologies, They phone more than young people] (2014). http://zpravy.tiscali.cz/cesti-seniori-ovladli-mobilni-technologie-telefonuji-vice-nez-mladi-243395
16. Mikulecky, P.: Smart Environments for smart learning. In: Proceedings of the 9th International Scientific Conference on Distance learning in Applied Informatics, Nitra, UKF, pp. 213–222 (2012)
17. Hwang, G.J., Tsai, C.C., Yang, S.J.H.: Criteria, strategies and research issues of context-aware ubiquitous learning. Educ. Technol. Soc. **11**(2), 81–91 (2008)
18. Hwang, G.J.: Definition, framework and research issues of smart learning environments—a context-aware ubiquitous learning perspective. Smart Learn. Environ. **1**(4), 1–14 (2014)
19. Spector, J.M.: Conceptualizing the emerging field of smart learning environments. Smart Learn. Environ. **1**, 2 (2014)
20. Miller, J.T.: Simplified smartphone options for tech-shy seniors (2015). http://www.huffingtonpost.com/jim-t-miller/simplified-smartphone-opt_b_6791776.html
21. Chen, K., Chan, A.H.S.: Cell phone features preferences among older adults: a paired comparison study. Gerontechnology **13**(2), 184 (2014)
22. Gao, J., Koronios, A.: Mobile application development for senior citizens (2010). http://www.pacis-net.org/file/2010/S05-03.pdf
23. Lapinsky, S.E.: Mobile computing in critical care. J. Crit. Care **22**, 41–44 (2007)
24. Bujnowska-Fedak, M.M., Pirogowicz, I.: Support for e-health services among elderly primary care patients. Telemedicine J. E-Health **20**(8), 696–704 (2014)
25. Chernbumroong, S., Atkins, A.S., Yu, H.: Perception of smart home technologies to assist elderly people. In: Proceedings of the 4th International Conference on Software, Knowledge, Information Management and Applications (SKIMA 2010), pp. 1–7. Bhutan (2015)
26. An example of smart home designed for the elderly people (2015). http://yesgroup.eu/telecare-information-2
27. Formosa, M.: Digital exclusion in later life: A Maltese case study. Humanit. Soc. Sci. **1**(1), 21–27 (2013)
28. Rain, T., Svarcova, I.: Internet and seniors. J. Effi. Responsib. Educ. Sci. **3**(2), 79–85 (2010)

# Part IV
# Smart Education: Software and Hardware Systems

# Smart Classroom

Jean-Pierre Gerval and Yann Le Ru

**Abstract** This paper sets out the methods and the technologies used to design a captive portal to redirect users to the URL (Uniform Resource Locator) of a course taking place in a given room. The captive portal is designed on a Raspberry Pi 2 carrying an Apache HTTP (HyperText Transfer Protocol) server and using iptables for redirections. It has a web configuration interface, developed with AngularJS, which communicates through HTTP request to the server side, developed in PHP, following the principle of a REST (REpresentational State Transfer) architecture. In addition to redirect users to the URL of a course, the interface is configurable in two modes: (1) fixed URL that sets an URL to which the user is redirected automatically, (2) hosting a local website which is used to load a web site in zip format on the device and then redirect users to this web site even when the device is not connected to any Ethernet network.

**Keywords** Smart objects · Internet of things · Ubiquitous learning · Wifi network · Captive portal

## 1 Introduction

Connected objects (also called smart objects, or Internet of Things) appeared recently. They are connected to the Internet so they can communicate with other systems to obtain or to provide information. This is made possible by the super miniaturization of electronic components.

J.-P. Gerval (✉) · Y. Le Ru
Institut Supérieur de l'Electronique et du Numérique – Brest,
20 rue Cuirassé Bretagne – CS 42807, 29228 Brest Cedex 2, France
e-mail: jean-pierre.gerval@isen-bretagne.fr

Y. Le Ru
e-mail: yann.le-ru@isen-bretagne.fr

© Springer International Publishing Switzerland 2016
V.L. Uskov et al. (eds.), *Smart Education and e-Learning 2016*,
Smart Innovation, Systems and Technologies 59,
DOI 10.1007/978-3-319-39690-3_37

They can for example:

- Collect and store information according to their environment: heart rate, cellar hygrometry, etc.
- Trigger actions based on the information gathered on the web, such as watering a lawn on the eve of a severe drought day.

To transform an everyday object into a connected object, simply connect it to the Internet and enable the object to react according to available data: weather, stock quotes, and user's action onto a smartphone…

The real "intelligence" of the object lies in the interface, and not in the object itself [1]. For example, a basic lamp, if it is connected to the internet via an interface (i.e. at the power socket level) it can be controlled from a smartphone, and thus becomes a "connected lamp". To make yourself a connected object, it is now possible to buy at very reasonable price "kits", or using nano-computers such as Raspberry Pi, coupled to relays to operate devices connected to the electrical network.

The main idea of the work presented in this paper concerns the implementation of a system which will automatically download, in a given room, all the data linked to the course taking place in this room.

Main targets of such a work are as follows:

- To save time at the beginning of practical activities, courses…
- To facilitate access to resources, to simplify access to resources: more and more services are available and it becomes increasingly slow to arrive directly on the right page.
- To force access to resources (quiz, homework, etc…) through a specific network that does not give access to the Internet (i.e. to prohibit networks access from 3G or 4G smartphone).
- To provide useful information in the physical location where and when it is needed.

## 2   System Overview

### 2.1   On the Hardware Side

An electronic device with wireless technology enables sending and receiving data by an automated manner. It must be efficient enough to run a Linux distribution and to support an Apache server type. It must be scalable, with at least one Ethernet port and two USB ports (the first for the Bluetooth key and the second for the wireless key).

The device we have chosen is a Raspberry Pi 2 [2]. There are many other devices that are similar in their characteristics either at interfaces, dimensions or performance, often for a price of around € 50.

A criterion which is justifying the choice of such a device is its popularity, which implies a great number of resources that facilitates developments. Raspberry Pi 2 device seems to be the most suitable for the application according to its characteristics (performance, peripherals and consumption) and its popularity (large documentation and community).

In view of the aforementioned criteria, it has four USB ports allowing them to easily connect a Bluetooth dongle, a wireless dongle, a keyboard and a mouse to program easily. There is also the possibility of using a RJ45 port/Ethernet network connection and Bluetooth/Wi-Fi key to share or transmit the data contents.

## 2.2 On the Software Side

It is a Client/Server application that will automatically download in a given room all course materials available on Moodle [3], corresponding to the course that takes place there according to the timetable.

### Front-End Technologies

On the client side there is the possibility of using different technologies JavaScript, jQuery, frameworks, or generated PHP. To meet standards and new server architectures, the choice is to let PHP on server side and create a REST (REpresentational State Transfer) architecture explained below. It remains the choice between JavaScript and frameworks. Frameworks offer ease of maintenance and quite handy code. New frameworks are quite powerful and structured [4]. Our choice is AngularJS [5] for its ease of learning, maintainability and its large community. The community around a framework is extremely important for the development: slightest problems can be solved according to problems that have already been solved by the community. There are opportunities to interact easily with other people. If the community is active and large then it gets faster to learn. AngularJS well respects the MVC concept (Model View Controller) that separates software components to enable programmers to structure their code and to facilitate the maintenance of software.

### Back-End Technologies

JAVA and PHP are two different programming languages that would fit us for this application. PHP seems to be suitable. This scripting language is most often used server side. It is linked with an Apache server which is the software we have selected server side. This couple enables easy data retrieval from databases. We do not have chosen technology like Node.js on server side because speed and flexibility with respect to the number of users is not an important criterion for this application. The system architecture follows the diagram below (Fig. 1). The client simply communicates via HTTP (HyperText Transfer Protocol) requests which are redirected to the web server that provides information depending on the type of

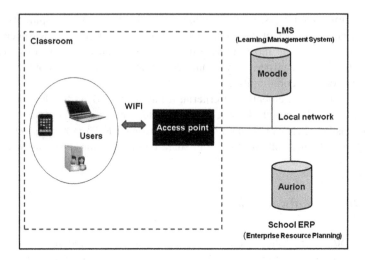

**Fig. 1** System architecture

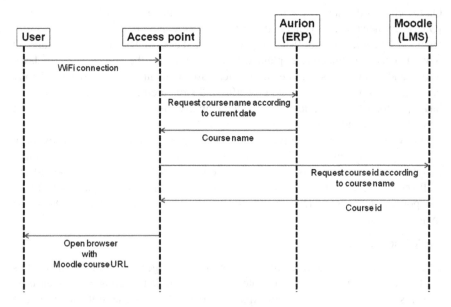

**Fig. 2** Sequence diagram

method which is used. The server communicates with the local network and sends requests (Fig. 2) to the Aurion [6] database (School Enterprise Resource Planning) and the Moodle database (Learning Management System).

## The Kernel

The system installed on the device (Fig. 3) is a recent version of Raspbian which is an adapted version of Linux Debian to the ARM (Advanced Risc Machine) architecture of Raspberry.

The web application consists of two parts. The first one is the main page showing the smart content for the room. This is the visible part for users. The second one is the administration of the device allowing administrators to choose the mode, to upload local site and to define the Service Set Identifier (SSID) of the wireless network. The administration access appears in the bottom right of the page with a small icon (Fig. 4).

The device communicates with the client by means of a WiFi wireless Internet connection and with our databases on the local network by means of a RJ45 Ethernet link. A DHCP (Dynamic Host Configuration Protocol) server configures clients with IP (Internet Protocol) addresses and the device as a gateway. The administration application creates iptables rules redirecting HTTP flow to the smart content delivered by the server. Nevertheless according to the mode selected by the administrator for the device, contents available are not the same. We will detail these modes in the following chapter.

Two iptables rules allow us to filter and redirect the client only to the local web server's home page. This one is just a framework including an "iframe" where the content depends on the configuration.

### iptable rule #1

- A PREROUTING ! -d $IP_origine/32 -p tcp –dport 80
- j DNAT –to-destination 192.168.8.1:80

### iptable rule #2

- A PREROUTING ! -d $IP_origine/32 -p tcp –dport 443
- j DNAT –to-destination 192.168.8.1:443

192.168.8.1 is the address of the interface pointing to the server to which the user is redirected when attempting to access content other than the address defined by the IP address located after "! –d". This address can change dynamically

**Fig. 3** Raspberry Pi 2

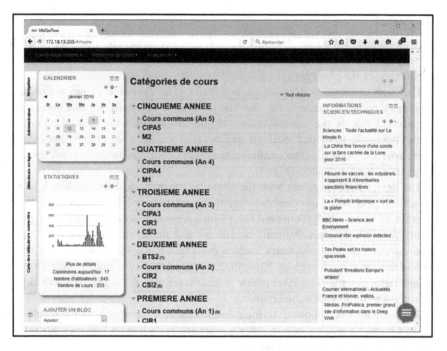

**Fig. 4** Captive portal with Moodle inside

according to the server configuration: IP address or hostname provided by the client URL (Uniform Resource Locator).

We use the "hostapd" software to create a WiFi access point available to the client. When the system gets a connection, a DHCP server "isc-dhcp-server" assigns an IP address to the client. Then as described above iptables rules redirect the client to the interface and the server. The authorized content which is included in the iframe is requested through the Ethernet interface.

## 3   Experimentation

Two test rooms have been equipped. Users who go into a room where an access point is available can connect to it. Following this connection, a page is automatically opened in the browser of the client showing the course supposed to take place in the room (Fig. 4). The system retrieves and sends pages to clients, taking into account schedules and rooms.

In order to increase the flexibility of this system, two additional operating modes have been added:

- Fixed URL mode—The system redirects clients to a website, previously configured.

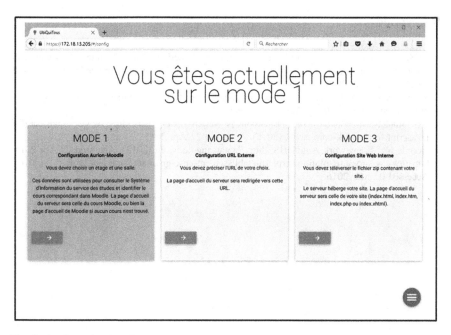

**Fig. 5** Configuration interface

- Hosting a website locally—The system redirects users to a website hosted locally on the cart.

A specific interface (Fig. 5) enables to configure the device for each operating mode:

- To define the room number and then to set up automatically the WiFi SSID of the device.
- To choose an URL to redirect the client to.
- To upload the web site that will be stored locally on the device.

## 4 Conclusion

By the end of the experiments we noted that in order to avoid interference we had to select channels that did not overlap between various rooms and we also had to pay attention to limit the power of devices.

Finally this system could obviously be improved. For instance, a Bluetooth connection or another type of connection could be added to the device. In an era of increasingly digital and interactive, this project should certainly fit with other contexts. Such as museums or exhibitions, where various contents should be delivered taking into account the location of the end user.

# References

1. CESER Auvergne (Janvier 2015), Les Usages du Numérique pour la Santé, l'Enseignement Supérieur et la Nouvelle Production Industrielle. http://www.cesdefrance.fr/pdf/13713.pdf (2015). Accessed 18 June 2015
2. Raspberry Pi Foundation.: https://www.raspberrypi.org (2015). Accessed 06 Apr 2015
3. https://moodle.com/ (2016). Accessed 03 Feb 2016
4. Uri Shake, AngularJS versus Backbone.js versus Ember.js.: https://www.airpair.com/js/javascript-framework-comparison(2015). Accessed 27 Apr 2015
5. Google Inc. Framework AngularJS.: http://www.angularjs.org (2015). Accessed 27 Apr 2015
6. https://www.auriga.fr/solutions-erp/erp-etablissement-enseignement-superieur.html (2016). Accessed 02 Mar 2016

# Creating a Smart Virtual Reality Simulator for Sports Training and Education

Emil Moltu Staurset and Ekaterina Prasolova-Førland

**Abstract** Virtual reality (VR) has been increasingly applied in a training context, including sports. VR provides an opportunity to train in an immersive, safe and controlled environment, with support for accurate performance measurement and user feedback. In this paper we present a prototype of a smart VR simulator for ski jumping training. It is modelled after the ski jumping hill of Granåsen in Trondheim, Norway that has been venue for several international skiing competitions and is now competing for hosting World Ski Championship in 2021. The prototype supports Oculus Rift and has been evaluated with both athletes and representatives for the general public. The paper presents the results of the implementation and evaluation, outlining directions for future work.

**Keywords** Smart sports training and education · Virtual reality · Ski jumping

## 1 Introduction

Virtual reality (VR) has been successfully applied in a wide range of educational situations such as military training [1], sports training [2] and medical education [3]. Jonathan Steuer [4] defines VR as "a real or simulated environment in which a perceiver experiences telepresence" where telepresence is defined as "a mediated perception of an environment". Today there exist a wide variety of VR technologies with varying degree of immersion, spanning from regular computer screens, to head mounted displays such as Oculus Rift [5], and the more advanced CAVE (cave automatic virtual environment) systems. Interface technologies such as haptic and gesture recognition are contributing to increase immersiveness in VR applications. VR technology has several traits that makes it useful for education and training such

E.M. Staurset · E. Prasolova-Førland (✉)
Norwegian University of Science and Technology, 7491 Trondheim, Norway
e-mail: ekaterip@ntnu.no

E.M. Staurset
e-mail: emil.staurset@gmail.com

© Springer International Publishing Switzerland 2016
V.L. Uskov et al. (eds.), *Smart Education and e-Learning 2016*,
Smart Innovation, Systems and Technologies 59,
DOI 10.1007/978-3-319-39690-3_38

as the ability to provide meaningful interaction in simulated contexts with high immersion, safely and at a relatively low cost [6]. In particular, there are several potential advantages for sports training [7]:

- Allowing standardized scenarios to be created by coaches
- Additional information can be given to players to enrich the environment and guide their performance
- It is possible to quickly and randomly change the environment to match any competitive situations

In sports, athletes need to react quickly to the cues and information provided by their environment. This is especially relevant for duels between two players where the opponents need to anticipate each other's actions. Traditionally, these situations were analyzed by showing a video on a wide-screen, stopping at strategic time events, so that athletes could then guess the remainder of the sequence. Bideau et al. [8] suggest that VR could be an alternative for this, presenting a simulation for handball goalkeepers where a controlled virtual environment enables fine-tuning of parameters, and measuring the parameters' effect on the athlete's behavior. VR has also been successfully applied in other sports, for instance Komura et al. [2] introduce a prototype for a virtual batting cage that enables batters to face a variety of baseball pitchers. Park and Moon [9] present how VR can be used for teaching beginners the basic of snowboarding. Stinson and Bowman [10] show that VR has a great potential in psychological training in sports.

In this paper, we focus on solutions for ski jumping training that could be adapted to the needs of both athletes and amateurs. Ski jumping is a form of Nordic skiing where athletes descend a take-off ramp, called the in-run, jump and fly as far as possible [11]. During the in-run the athlete tries to gather as much speed as possible by adopting an aerodynamically favorable body position. While in-flight the athlete attempts to hold a favorable body position, avoiding to land for as long as possible. In competitions, points are awarded both for the style and the length of the jump. The skis are normally 11 cm wide and can be up to 146 % of the total body height of the skier. The ski jumping hill consists of two main parts: the in-run and the landing slope. The in-run typically starts with an angle of 38–36° and then curves into a transition often called the take-off radius. The transition is then followed by the take-off ramp, which typically has an angle of 7–12° downhill. The landing slope has a smooth curve that follows the profile of the jump. Because of this, the skier is never more than 6 m above the ground [11].

Several ski jumping games have been developed over the years. Deluxe Ski Jump 4 [12] was by many considered to be the best ski jumping game on the market in 2015. Deluxe Ski Jump has 24 accurately modeled jumping hills, and the game is based on real physics. Wii Fit ski jumping [13] is another example. The game uses Wii Balance Board, so the player is using his body to control the jump. In 2014 the Alf Engen Ski Museum in Park City, Utah opened a new virtual ski jump exhibition called "Take Flight". The game is displayed on a TV-monitor, and it uses some kind of a force scale for the jumping, similar to the Wii Fit jumping game [14]. In

March 2015 the Slovenian insurance company Triglav released the "Planica 2015 Virtual ski flying". According to Sporto Magazine [15] this is a VR ski jumping application based the real ski jumping hill Planica, one of the largest ski jumping hills in the world with a hill record of 248.5 m. The application is using Oculus Rift Development Kit 2 [5], and it presumably only uses the head tracker of the Oculus for controlling the jump.

In this paper we present a prototype of a smart VR application for training ski jumping for different user groups. It is modelled after the ski jumping hill of Granåsen in Trondheim, Norway that has been venue for several international skiing competitions and is now competing for hosting World Ski Championship in 2021. The application is adaptable to users with different needs and abilities, from tourists and children to professional athletes. The simulator may have several applications: teaching the basics of ski jumping to the general public, being a tool for recruiting aspiring young ski jumpers, promoting Trondheim and Granåsen as a tourist destination, and aiding ski jumpers in a training situation. The simulator provides the users with a responsive environment, enabling instant feedback on their performance as well as supporting an immersive and risk-free ski jumping experience.

## 2 Implementation

The prototype of the simulator was implemented in two iterations. In this section, we will explain the techniques used in the process, with the focus on two major aspects: modelling the hill and the physics of the jump and implementation in a game engine.

**Model of the hill and physical model.** A model in a 3D space consists of vertices, edges and faces. In the first iteration of the development process the official Granåsen hill certificate [16] was used as a reference for modelling. The virtual jumping hill was created by performing operations on basic cubes and curves. A detailed description of the process is beyond the scope of this paper and can be found in [17]. As follows from the evaluation with athletes and experts (as described in subsequent sections), it was clear that this model had several faults compared to the actual hill. This was not something that the general public would recognize, but an experienced jumper would notice that at once. The model had to be improved, if the application was going to be valuable for professional ski jumpers. Therefore, in the second iteration we used a LIDAR system [18] to create an accurate model of the hill. LIDAR is a remote sensing technology that accurately measures distances to a target by illuminating it with a laser and analyzing the reflected light [18]. The LIDAR system was attached to a drone that flew over a large area around Granåsen. The output from the LIDAR system is a point cloud in the form of a LAS-file, i.e. a set of data points in 3D space. The point cloud had a granularity of 10 cm. The surface of the hill was reconstructed using the Poisson surface reconstruction-algorithm [19]. However, the in-run was not of high enough quality to be used in the application. The solution was to recreate it in Blender [20] using the reconstructed point cloud as a reference.

A ski jump is a mathematically complex and computationally expensive process. The state of the art aerodynamic computations using computational fluid dynamics is too complex to be performed in real time. Hence it was necessary to devise a simplified physical model describing a ski jump. It was decided to implement the model proposed in [21], which describes the forces acting on the ski jumper while in flight. It takes a simplified Newtonian approach to the problem where only two forces act on the skier: the gravitational force $F_g$ and the dynamic fluid force F. The magnitude of the dynamic fluid force is proportional to the air density $\rho$, the surface area A (an intersection of the plane perpendicular to the relative jumper's motion and his body) and the square of the jumpers velocity $v$ relative to the air. The equation for the fluid force can be expressed as follows (1):

$$F \propto \rho A v^2 \tag{1}$$

After testing the first prototype with a professional ski jumping coach it was found that the physical model was sufficient, and gave a good impression of how it felt to perform an actual jump.

**Implementation in Unity**. The application was developed in Unity 3D [22] (Fig. 1). Unity is a cross-platform game creation system, including a game engine and an IDE (integrated development environment). Unity also has its own integration package for Oculus Rift, which makes it relatively easy to create VR-compatible applications.

When creating an application in Unity the game is organized in scenes. Each unique scene can be thought of as a unique game level. The scenes may contain numerous GameObjects that are containers for components that can be attached to

**Fig. 1** Granåsen ski jumping simulator developed in Unity with Oculus Rift

the object. Through a well-documented API (application program interface) it is possible to access and manipulate the GameObjects and other components via scripting. The player behavior was implemented using the Unity First Person Controller as a starting point. The controller was extended by adding the physical model while the skier is in the air, and changing the behavior while the skier is in the in-run and out-run of the jump. During the second iteration, several improvements in the implementation of the player behavior have been added, for example in the rotation of the skier's body when transitioning to flight and landing, head rotation, skier's body rotation while in air and so on. Another improvement is an algorithm for jump-height, where the force generated by the jump is dependent of how well-timed the jump is.

The first implementation was using a Playstation controller to control the player (jump, landing, and angle of attack). In the second iteration the goal was to let the player use his/her body as input for jump detection, landing and changing the angle of attack. For jump detection we tried using the Wii Balance Board, having the player jump off the board to make the jump. Unfortunately, there was a delay in the board, which made it unsuitable for the purpose. For measuring the angle of the body we ended up using the Wiimote as an accelerometer that lets us read the orientation of the controller. The Wiimote is attached to the player's torso using a belt. The accelerometer is not very accurate but the main point is to encourage the player to emulate the movements of a ski jumper to improve immersion.

Apart from the visual feedback before, during and after the jump, the player is provided with an assessment of his/her performance (timing of the jump, angle of attack and landing) in the form of a calculated jump distance. In addition, the player can review the jump by accessing a JumpEvaluation scene in the simulator.

## 3 Evaluation

The prototype has been evaluated with two major user groups: ski jumpers and general public with no or very limited ski-jumping experience. In the following subchapters we will explain how the tests were conducted, and the results from the evaluation.

### 3.1 Athlete Evaluation

The evaluation with ski jumpers was conducted with athletes and coaches from the Norwegian national ski jumping team. The evaluation was in the form of a semi-structured interview. The interview gave valuable insight into the strengths and weaknesses of the prototype. The visual impression the participants got was that the in-run was too short, mainly that the straight part of the in-run should be 20 m

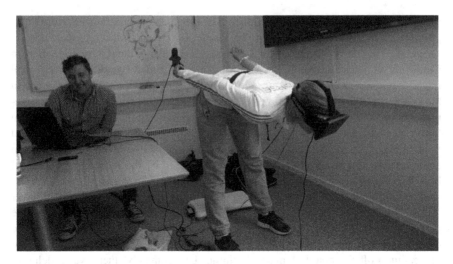

**Fig. 2** A female athlete from the Norwegian national ski jumping team is testing the simulator

longer. They also thought that the radius seemed a little flat. However the out-run looked very similar to the actual out-run in Granåsen.

The evaluation was conducted with the Oculus Rift Development Kit 1, with the Balance Board for jumping, and the Wiimote for body angle measurement (Fig. 2). The athletes really struggled with hitting the timing of the jump, always being too late on the take-off ramp. We concluded that this must be due to a delay with the Balance Board. While the delay may be no more than 100 ms, it has a great impact on the jumper on the take-off ramp as a normal jump takes about 0.3 s from the skier starts the jump until he/she leaves the ground. Hence we had to improvise and find a method that could satisfy the athletes. We ended up discarding the Balance Board, and letting the athletes use the Playstation 4 Controller for jumping. Even though it was not necessary, the testers still jumped physically as they pushed the button, presumably to connect the physical and visual experience of the ski jump.

The skiers found that there was a problem with the head rotation, as they were not able to see the take-off ramp while sitting in a normal in-run position. This is very important because the skier uses a combination of visual and physical cues when he/she jumps. Typically, the skier starts the jumping movement as soon as the acceleration from the take-off radius is released. Fortunately, this was rather easy to fix, making the system feel much better according to the athletes. As for the flying phase, the skiers said that it was highly realistic, and feels very similar to how it is to jump in real life. "The feeling of "take-off" in a long jump feels very good, and reminds me much about how it is in real life", one of the athletes said. Another jumper said that he felt that the profile of the jump felt a little off, i.e. he was too high above ground after take-off, and a little low above ground at the end of a jump. Still, this is rather individual for every ski jumper, and is dependent on many factors such as jump timing, how much force the jumpers generate, and flying technique of the jumper among other things. However, the athletes pointed out that there was a problem with the physical model. The way it is implemented now the lift increases as the angle of attack decreases, i.e. the lower the angle of attack the longer the jump. In real life the skier's body and the skis act like a wing to minimize the drag and maximize the lift forces acting on the skier. Ideally, the skier should have an angle of attack of about 11°. Another aspect that the model does not catch is the importance of the rotation. A ski jumper should have a small rotation forward during a jump. However if he/she adjusts the rotation forward too fast he/she may get an over-rotation, and if the rotation is too small the skier will be pushed back, thus generating more drag resulting in a shorter jump. The athletes suggested that there should be implemented a kind of a breaking point in the physical model where the lift is reduced if the angle is below 11°. Apart from this, they were pleased with the physical model.

All the jumpers agreed on that the in-run seemed too short, and the major problem was that the straight part of the in-run should be at least 15 m longer. They also said that the straight part felt a little too flat. As we found out, the reason for that is that the LIDAR-system has failed to recreate the top part of the in-run. The consequences of the short in-run is not very severe, as it has successfully recreated the radius and the take-off ramp. Fixing the in-run is rather straightforward, and when this is done, the in-run should be satisfactory according to the skiers. One of the jumpers suggested that the in-run might feel flatter because they are standing on the flat ground while they are used to feel the steepness of the in-run in their feet. Another factor is that they do not feel the acceleration forces act on the body as they do in a real ski jump.

The jump is now implemented so that the player jumps immediately when the jump button is pushed. Other possible solutions were discussed, including using infrared lights attached to the shoulder, hip, knee and ankle of the jumper. These IR-lights could be recaptured by the Wiimote to reconstruct a dynamic 2D-skeleton of the skier. This skeleton could be possibly used for jump detection, also allowing to create a replay system where one could watch the skier's 2D skeleton as he/she jumps. It was also suggested to use force scale to detect the jump, using the force generated by the skier as an input. With these improvements, the system could also be used to evaluate the skiers' jumping technique.

The athletes of the Norwegian jumping team seemed to be immersed into the virtual reality, and one of the coaches noted that the skiers were adjusting their bodies in the in-run as they would do in a real jump. When asked whether the prototype could be used for training purposes, the skiers stated that if the in-run is extended and the jump button gets a 0.3 s delay, the prototype could be used as a tool to practice the timing of a jump. The athletes also said that the simulator could be used as a tool for visualization, giving a feeling in how it is to jump in the modeled hill. One of the coaches suggested that if all the hills the skiers compete in the World Cup have been modelled, the prototype could be used to prepare the skiers before competitions as the time allocated to physical jumps is rather limited. As for the flying phase the technique is too simplified so it does not offer any training value. However, the feeling of flying is very similar to how it feels in reality, and can give amateurs an authentic experience of performing a real ski jump. Finally, the athletes believed that the application would be useful to trigger interest for ski jumping among non-professionals.

## 3.2  Evaluation with General Public

The prototype was also tested with the general public. The tests were conducted at the Trondheim Science Centre, a scientific "hands-on" experience center and Open Day at the Norwegian University of Science of Technology (NTNU), attended by high school students. The plan was to have the participants answer a questionnaire afterwards, but the participants at the Science Centre were mainly children younger than 10 years old, so the questionnaire was not suitable. A simplified questionnaire was developed with smileys where the testers could give a rating to the system from 1 (lowest) to 5 (highest), and answer if they wanted the simulator to be part of the exhibition at the Science Centre. The system was tested by 62 children and 58 of them gave it a rating of 5 out of 5. The remaining 4 gave it a rating of 4 out of 5. Furthermore, all of the testers wanted the system to be a part of the exhibition at the Science Centre. Initially the test was conducted with the Wii Balance Board for jump detection, and the Wiimote for measuring the angle of attack. However, this setup proved to be difficult for the children, and 3 out of 5 of the first testers fell on the ground after jumping off the Balance Board. So it was decided to let the rest of the children test the system using the Playstation controller for jumping and adjusting the angle of attack.

On the Open Day at NTNU the test was conducted with Oculus Rift DK1, the Balance Board for jumping and Wiimote for measuring the angle of attack. After testing the prototype, the testers answered a questionnaire. The questionnaire was designed to focus on user experience, realism and the sense of presence. In total 32 persons answered the questionnaire, and of these 28 were male and 4 female. Of these participants, 24 had no experience with ski jumping, 7 had little experience, and 1 had much experience. Most of the participants were experienced with computer games. Some of their responses are summarized below:

The questionnaire results show that the participants mostly agreed that Oculus Rift improved the realism of the prototype. The same applies to Wii Balance Board and Wiimote. The participants did not seem to feel like they really visited Granåsen. We believe that this is because the testers still felt like they were playing a game, and not physically visiting a place. Adding audio to the application would improve the feeling of visiting the physical place, according to the testers. It would also be interesting to have a fan blow "wind" on the player adding to the impression of physically performing a ski jump. While the majority of the participants seemed to feel that it was indeed them who jumped, the participants did not seem to be completely convinced of the realism of the prototype. This finding was rather surprising as the athletes who tested the prototype seemed to rate the realism rather high. The participants seemed to agree that the prototype would be useful to trigger interest for ski jumping. Also, the majority seemed to agree that the prototype could be used as a training tool for ski jumpers, and to promote Trondheim to tourists.

# 4   Conclusions and Future Work

In this paper we have described the process of developing and evaluating a smart virtual reality simulator for training ski jumping. On the overall, the major goals of this project have been achieved, as the simulator could be adapted to and used by different user groups for training purposes with different interface configurations. By involving a simple set of sensors (Playstation, Wiimote and in-built head tracking in Oculus Rift), the application provides the user with a responsive environment with constant feedbacks and performance measurements. With some minor improvements, professional ski jumpers can use the application for practicing the timing of the jump. In addition, the simulator provides a safe and immersive experience of performing a real ski jump and enables non-specialists to learn some basic techniques. This and other similar examples, such as [23], show how VR technologies can be used in 'smart physical training' [23] and other training contexts.

**Limitations of the study**. The study has certain limitations. Most of the participants in the general public group we had access to, have been children or young adults. As these participants did not have much time for testing the system, we had to limit the number of questions in the questionnaire. In addition, due to time limitations, most of the participants only tried 2–5 jumps. The tests were conducted using the Balance Board for jump detection, which was not accurate enough according to the athletes. Still, this study allowed us to explore how an immersive VR environment can be used for sports training for various user groups.

**Contribution in the context of related work**. The training simulator presented in this paper exhibits a number of innovative features compared to other similar applications. Most of the applications such as Take Flight, Wiimote and Deluxe Ski Jump 4 [12–14] only use a regular TV-monitor for displaying the graphics, which provides a less immersive experience than by using VR interface. The two former applications use Balance Board or a variant of it for input, but do not require the player to mimic the technique of a real ski jump. Deluxe Ski Jump 4 (DSJ) has a very sophisticated physical model that is able to capture the rotation of the skier's body, and the ski's behavior while in the air, but it does not allow assessing how the skier's body is positioned. One of the coaches that tested our game is an experienced DSJ-player, and he said that the ideal would be to have an application with DSJ's physics and our user interface. Since the introduction of Oculus Rift, we have seen examples of various simple VR ski jumping applications such as VR Ski Jump [24]. However, we are aware of only one currently existing VR ski jumping application in addition to our simulator that recreates an existing jumping hill, which is "Planica 2015 Virtual ski flying" application (Planica is also competitor to

Trondheim in World Championship 2021 application). Based on information in [15], it only uses the head tracker of Oculus Rift as input to determine the timing of the jump. Therefore, we believe that even without the Balance Board for jump detection, our application has a user interface that mimics the ski jumping technique more accurately than the applications mentioned above and therefore should be more useful for training purposes. The empirical evaluations with different user groups constitute another contribution of this work.

**Future work**. In the current version of the prototype, there are still some functions that rely on traditional user interface (buttons/keys), such as landing and jump. For a fully 'smart physical training', these functions should have been implemented with gesture recognition, e.g. with Kinect, to make the interface more natural and responsive to the user's movements. As mentioned in Sect. 3.1, alternative mechanisms for jump detection should be explored, such as attaching IR-lights to the jumpers body and using the Wiimote as an IR-camera. Still, this method might be more costly and involve extensive set-up between each user, making it problematic for public demonstrations when the turnover of users is rather high. Another solution could be using a force scale to detect the jump. In addition, the graphical user interface and the visual experience should be improved, adding more visual, sound and other effects (such as air flow), making the iconic Granåsen jumping hill easily recognizable to locals and tourists. Other additions might include 'smart coaching', where the simulator provides in-depth analysis of the jump and concrete feedbacks on how the technique could be improved.

In general, more research is needed in the field of 'smart physical training' and VR, to develop adaptive, immersive and responsive environments with support for intelligent coaching, not only in ski jumping, but in different sports, including team sports. Apart from training and education, smart sports entertainment and smart physical rehabilitation are related and highly relevant research areas.

**Acknowledgments** The authors would like to thank the employees at the Olympiatoppen, Visit Trondheim and Trondheim Science Centre, the athletes from the Norwegian national jumping teams, as well as all the evaluation participants.

# References

1. Bowman, D.A., McMahan, R.P.: Virtual reality: how much immersion is enough? Computer **40**(7), 36–43 (2007)
2. Komura, T. Atsushi, K., Yoshihisa, S., NiceMeet, V.R.: Facing professional baseball pitchers in the virtual batting cage. In: the 2002 ACM symposium on Applied computing (SAC '02), pp. 1060–1065. ACM, New York, NY, USA (2002)
3. Seymour, N.E., Gallagher, A.G., Roman, S.A., O'Brien, M.K., Bansal, V.K., Andersen, D.K., Satava, R.M.: Virtual reality training improves operating room performance: results of a randomized, double-blinded study. Ann. Surg. **236**(4), 458–464 (2002)
4. Steuer, J.: Defining virtual reality: dimensions determining telepresence. J. Commun. **42**(4), 73–93 (1992)
5. Oculus (2016). https://www.oculus.com/

6. Cram, A., Hedberg, J.G., Gosper, M.: Beyond immersion–meaningful involvement in virtual worlds. In: 2nd Global Conference on Learning and Technology, pp. 1548–1557 (2011)
7. Miles, H.C., Pop, S.R., Watt, S.J., Lawrence, G.P., John, N.W.: A review of virtual environments for training in ball sports. Comput. Graph. **36**(6), 714–726 (2012)
8. Bideau, B., Multon, F., Kulpa, R., Fradet, L., Arnaldi, B.: Virtual reality applied to sports: do handball goalkeepers react realistically to simulated synthetic opponents? In: 2004 ACM SIGGRAPH VRCAI Conference, pp. 210–216. ACM (2004)
9. Park, C., Moon, J.: Using game technology to develop snowboard training simulator. In: HCI International 2013-Posters' Extended Abstracts, pp. 723–726. Springer, Berlin (2013)
10. Stinson, C., Bowman, D.A.: Feasibility of training athletes for high-pressure situations using virtual reality. IEEE Trans. Vis. Comput. Graph. **20**(4), 606–615 (2014)
11. Ski Jumping. Wikipedia (2016). https://en.wikipedia.org/wiki/Ski_jumping
12. Deluxe Ski Jump 4 (2016). http://www.mediamond.fi/dsj4/
13. Wii fit balance games (2016). http://wiifit.com/training/balance-games.html
14. Engen-Museum. New virtual ski jumping exhibit (2014). https://www.engenmuseum.org/news/new-virtual-ski-jumping-exhibit
15. Zavarovalnica triglav offering virtual reality in planica, Sporto (2015). http://sporto.si/en-us/novica/87/zavarovalnica-triglav-offering-virtual-reality-in-planica
16. Granåsen official hill certificate (2016). http://www.skisprungschanzen.com/photos/nor/trondheim_granaasen/HS140_2011.pdf
17. Staurset, E.M.: Creating an immersive virtual reality application for ski jumping. Master thesis. http://brage.bibsys.no/xmlui/bitstream/handle/11250/2353626/13193_FULLTEXT.pdf (2015). Accessed June 2015
18. LIDAR, Wikipedia (2016). https://en.wikipedia.org/wiki/Lidar
19. Kazhdan, M., Bolitho, M., Hoppe, H.: Poisson surface reconstruction. In: the 4th Eurographics symposium on Geometry processing (SGP '06). Eurographics Association, Aire-la-Ville, Switzerland, pp. 61–70 (2006)
20. Blender (2016). https://www.blender.org/
21. Marasovic, K.: Visualised interactive computer simulation of ski-jumping. In: the 25th International Conference on Information Technology Interfaces, (ITI 2003), pp. 613–618 (2003)
22. Unity (2016). https://unity3d.com/
23. Smart physical training in Virtual Reality. CITEC/Bielefeld University (2016). http://www.sciencenewsline.com/news/2016021817340003.html
24. VR Ski Jump (2016). https://share.oculus.com/app/vr-ski-jump

# Using Embedded Robotic Platform and Problem-Based Learning for Engineering Education

Fredy Martínez, Holman Montiel and Henry Valderrama

**Abstract** This paper shows the use of an embedded robotic platform of low cost and high performance, hand in hand with problem-based learning strategies, to professional training in Electrical Engineering at the District University Francisco José de Caldas (Colombia). These technical training and research tools involves several innovations, among which stand out the robot itself, which is inexpensive, robust and with high performance (suitable for both training and research), the study of real problems and the support with software tools that complement a smart learning environment. The robot has a mechanical differential platform that is easy to build and modify, a processing hardware supported in a 900 MHz quad-core ARM Cortex-A7 CPU able to run a graphical OS, and ROS as communication and control software. As advantages of its implementation has documented a better appropriation of theoretical concepts, increased student enthusiasm, improved ease of communication and teamwork, and greater interest in participation in research activities.

**Keywords** Electric engineering · Problem-based learning · Robotic platform · ROS

## 1 Introduction

Embedded systems, especially mobile platforms, today play a fundamental role in human activity. This is especially true in the processes of education and training. If these systems are used correctly, they become a very important complementary

F. Martínez (✉) · H. Montiel · H. Valderrama
District University Francisco José de Caldas, Bogotá D.C., Colombia
e-mail: fhmartinezs@udistrital.edu.co
URL: http://www.udistrital.edu.co

H. Montiel
e-mail: hmontiela@udistrital.edu.co

H. Valderrama
e-mail: henrygyovas@hotmail.com

© Springer International Publishing Switzerland 2016
V.L. Uskov et al. (eds.), *Smart Education and e-Learning 2016*,
Smart Innovation, Systems and Technologies 59,
DOI 10.1007/978-3-319-39690-3_39

tool in education [6]. These embedded systems consist of hardware and software that allows autonomous interaction anytime, anywhere. Typically possess specialized hardware with a central processor and various communication systems, sensors and actuators, and an OS that allows interaction and communication. In many processes of formation, training institutions take advantage of the existence of a hardware in the hands of the student (smartphone) [6, 15, 20], while in other cases it is designed a custom hardware [4, 18]. In either of the two cases, access to the system or the low cost of acquisition, enables each student to have their own platform, allowing access to specialized equipment that students can use in activities outside the classroom.

The software is a key element in both embedded systems and other training process. The software tools can be introduced relatively easily in laboratory activities in order to show specialized workplaces (design and simulation tools for example [2, 9, 16]), to facilitate the visualization and analysis of information, and facilitate the programming and development of solutions to problems [5]. Today many curriculum make use of online platforms where the students can design and verify their designs with an individual feedback [3]. These platforms accompany the training process, allow to keep a detailed record of progress, and responds to individual student needs. In addition, the impact on the process is much higher if they include specialized design tools [11]. The power of these software tools is greatly multiplied when they are incorporated in an embedded system that allows portability and direct interaction [19].

Robotics has been used many times in training processes. It involves strong motivating elements (both working with robots as the large number of contests and competitions that are encouraged in the field), and induction to research [7, 22]. Its potential is seen in many cases where different platforms are programmed by different students to find their own solutions to real problems [2, 13].

Teamwork and the communication skills are an important aspect of undergraduate engineering which ensures an adequate social and professional development of graduates [1, 17]. To develop and strengthen these skills, many teachers use inductive training strategies, such as problem-based learning, together with demands for socialization of progress. These strategies can always strengthen by technological tools to form a smart pedagogy [8].

Finally, a technological element that must be present in a process of smart pedagogy is the interaction and support on line [12, 21]. These are general characteristics taken into account in the design of our robotic platform and its introduction into academic courses as part of the curriculum project policies to encourage and motivate the academic performance.

The paper is organized as follows. Section 2 details the structure and configuration of the robotics embedded platform. Section 3 describes the learning strategies used with the groups of students. Section 5 shows some issues to overcome and future work. Finally, conclusion and discussion are presented in Sect. 6.

## 2  Embedded Robotic Platform

The embedded robotic platform was designed to allow programming, operation and communication versatile and with high performance. It consists of a mobile mechanical structure, Raspberry PI as processing and control unit, and ROS OS as programming software platform.

### 2.1  Raspberry PI

The first step is the initial configuration of the Raspberry Pi card, the student must install an operating system compatible with Raspberry Pi. To download the operating system image, the student must use a Micro SD memory, which will make the task of ROM. The recommended image (NOOBS) is on the page http://www.raspberrypi.org. The reason for its choice is the affinity with Arch Linux, OpenELEC, Pidora, RISC OS, Raspbian and RaspBMC.

#### 2.1.1  Raspbian Image

The next step is to insert the memory into Raspberry Pi as well as peripherals (mouse, keyboard, screen). Once the installation script is run, the student accesses the initial Raspbian menu. The last question of entry is to type the username and password.

#### 2.1.2  Wireless Connectivity

For applications it is essential to have Internet connection. Because it is a mobile robot, we suggest students use a WiFi antenna.

#### 2.1.3  PuTTY

The student must install an SSH client, on Windows we suggest PuTTY. With it you may access the Terminal remotely. Once downloaded and installed, the student access PuTTY to connect with the Raspberry Pi card. Then he enters the hostname where the IP address to which is connected the Raspberry Pi board is specified. Must be specified port 22.

#### 2.1.4  Remote Connection Using VNC

The procedure is as follows:

- The student installs by Terminal the VNC server on the Raspberry Pi card. This takes care of loading the remote desktop. This command can be executed using PuTTY.

```
$ mkdir -p ~/catkin_ws/src
$ cd ~/catkin_ws/src
$ catkin_init_workspace
$ cd ~/catkin_ws/
$ catkin_make
```

**Fig. 1** Creating the workspace

- When the installation concludes, the card must be restarted. Then, the following command is run:
- The command specifies the remote desktop will use. Additionally, in the first use of Raspberry Pi via VNC the system prompts to create a password.
- The next step is to install VNC through Chrome, with VNC Viewer for Google Chrome option. In this window, the IP address used by the Raspberry Pi board and desk number used to connect to the Internet is entered.

## 2.2 ROS OS Groovy Galapagos

ROS Groovy Galapagos is the sixth ROS distribution release for the Raspberry Pi board (December 31st 2012), and it is the proven and recommended distribution for work with the tool. The installation is done from binary packages.

### 2.2.1 Creating the Workspace

For the creation of the workspace the student must use the list of Fig. 1.

When the process of creating workspace ends, the catkin ws folder get two subfolders: build and devel. In devel the .sh files are stored. To conclude the installation, the student should address to Raspbian with the ROS system and new packages or workspaces.

### 2.2.2 Arduino IDE and ROS OS Integration

We recommend using an Arduino board as a bridge between Raspberry Pi and sensors/actuators of the robot. This provides a layer of security against hardware problems. The integration of the two systems is performed by means of Rosserial package.

After installation of Rosserial, the ros lib folder containing the libraries for communication between Arduino and ROS is created. The package must be moved to be recognized both ROS and Arduino IDE.

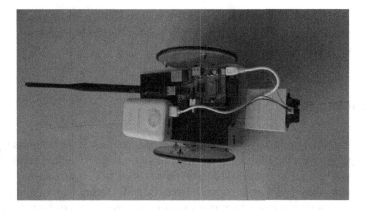

**Fig. 2** Embedded robotic platform: differential wheeled robot and Raspberry Pi

## 2.3  Robotic Platform

We encourage students to use the Serb robot to implement navigation tasks [14]. This is a differential wheeled open source robot moved by two servomotors. The original design was modified to fit the Raspberry Pi board, the distance sensors and rechargeable batteries for power supply. Given the characteristics of movement of the robot, the tasks of navigation and obstacle avoidance are designed. The final structure of the robotic platform is complemented with the Arduino Uno board, the infrared sensor IR GP2Y0A41SK0F, two portable chargers, two servomotors, one breadboard, the Raspberry Pi board and the wireless antenna (Fig. 2).

### 2.3.1  Robotics Task

Execution of robotics application is as follows:

- When the Raspberry Pi card is turned on, which is connected by wireless, and previously synchronized with the computer that performs the task of monitoring, the user accesses the Terminal to start ROS.
- The next step is to synchronize ROS OS with Arduino through a new Terminal. This is so that the system locates the workspace.
- Initiate communication between ROS OS and Arduino.
- The servomotors control is done in a new Terminal, where it is possible to can change the speed and direction of the task.
- For control of movement, the student indicates the servomotor which is assigned speed and direction, this feature is modified through the angle. For this case, the servomotor 1, the maximum speed and advancing direction is assigned by an angle of 180° .

- Rqt_plot: The rqt_plot tool works with graphics in two dimensions, where a sampling of the signal takes when the infrared sensor detects a nearby object within five centimeters is displayed.

## 3 Problem-Based Strategy

Problem-based learning was included in electronic courses of the program of Electrical Engineering at the District University Francisco José de Caldas (Technological Faculty) as a strategy to increase student motivation, as well as processes of research training, autonomy and teamwork were strengthened. During its implementation it has also seen a strong impact on appropriation of theoretical concepts, and better integration with other courses in the curriculum.

In particular, courses of semiconductor devices, microcontrollers and control have been strengthened with the robotic approach. This approach goes from the design of robotic platforms to implementing complex control schemes. The embedded robotic platform described in Sect. 2 was defined as a fundamental tool in this process.

At the beginning of the semester, the platform structure is described, and students are encouraged to its construction. The research group has laser cutting equipment and 3D printer that are available to students for prototyping. While the platform is fully functional and documented in detail, students are encouraged to modify and even to re-design it completely (Fig. 3). This process is used to encourage ingenuity, teamwork and communication skills.

Student groups have weekly show at the course details of its design and progress on it (Fig. 4). This allows feedback to the process of formation of each group, and fed back to their classmates on similar issues, promoting co-evaluation in a critical environment. Thanks to this collaborative work, students share their prior and acquired knowledge, discussing the course contents.

**Fig. 3** Custom designed robotic platform. Non-holonomic four wheel robot implemented by a group of students

**Fig. 4** Public presentation of designs by students. During this, feedback it is performed both the group and all students of the course

The platform becomes a real example of a complex system that addresses from engineering. Students use it to build knowledge in electrical, mechanical and control design throughout the semester. In the latter case, students work in both traditional control schemes and motion planning strategies, primarily on observable static reactive environments.

Each course moves around a loosely structured problem that the group of students must solve. At the end of the semester, almost 100 % of the groups achieved successfully complete this task. In addition, students have openly reported increased motivation, in particular face the problem of programming robots, an area that has traditionally been very weak in the academic program.

## 4 Smart Systems in the Training Process

There are many documented work about the use of many smart systems developed for education. Some of them focus in the classroom, and others in the development of smart environments. In the latter case takes great importance the use of prototypes of smart systems, systems in which specific processes are strengthened. In electronics, these processes are related to sensing activities, information processing, communication, management of actuators, and especially with smart answers and adaptability. The work with our electricity students has benefited from the development of some smart prototypes, in particular, small autonomous robots.

Robotic platforms are for us one of the most important tools [10]. It is particularly important the use of high capacity embedded processing systems, as in the case of Raspberry Pi, together with a software system of coordination and communication (ROS) that enables the implementation of complex control schemes. Students use these tools to solve real planing path problems, and implement classical navigation schemes.

The smart education system is designed and developed as a smart student-centred autonomous robotic system with certain features of smart systems (sensing, data processing, transmission, activation of actuators) and some degree of intelligence and autonomy (modelling, inferring, learning, adaptation, and self-organization).

## 4.1 Impact on Students

One of the most important reasons for the use of specialized tools in the training process is the motivational enhancement of students in theoretical courses with high conceptual content. To evaluate this aspect, we apply some surveys to students at the end of the courses. The surveys were implemented during the last year (2015) to two different groups of students, one of 27 students during the first half of 2015, and another one of 25 students in the second half of 2015.

The questionnaire was applied to students just before they made their final course evaluation. The questionnaire had five questions, and the answers were limited to five incremental options: Extremely, Very, Moderately, Slightly and Not. In general, our analysis considers the first two grades of responses as positive responses, and the last two as negative responses.

Figures 5 and 6 shows some results of survey conducted with students. Figure 5 shows the opinion of students at the question: How helpful is the tool? It shows that 63 % of students found the tool extremely or very helpful, while 9 % found it slightly or not helpful. Figure 6 shows the opinion of students at the question: How well the use of the tool motivates you to work on the course? It shows that 44 % of the students agree that the tool motivated their course work, while 18 % think that this motivation was slight or nonexistent.

The survey results show that students consider very useful the tool for the development of their training processes. They are also motivated by its use in courses.

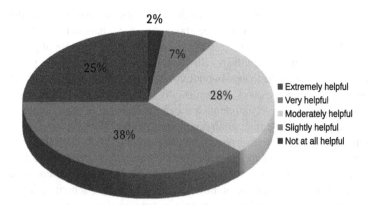

**Fig. 5** Student opinion face the question: how helpful is the tool?

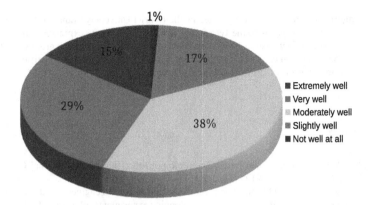

**Fig. 6** Student opinion face the question: how well the use of the tool motivates you to work on the course?

However, the motivational level found was less than expected. Subsequent consultations with students indicated that there is some level of complexity in configuring and tuning of the tool, aspects that need improvement.

## 5 Outreach and Future Research

We must be mindful of the methodological limitations when considering the results. The most critical variable is that the different methodological techniques were applied to different groups of students in different academic periods. Although the student community is more or less homogeneous, and widely characterized by the District University, learning styles and the dynamics that exist between students are unique. To increase the validity of the results, we propose to apply the same tool and strategies to many more groups of students to analyze statistically the results.

## 6 Conclusions

This paper presents a tool (hardware/software) linked with a problem-based learning strategy to strengthen the specific training of students of electrical engineering at the District University Francisco José de Caldas. This tool is a robotic platform consisting of a movable mechanical structure, a processing unit supported on an ARM Cortex-A7 processor and a programming environment for high performance (ROS OS). The use of this tool in several courses has proved to be an excellent motivator, as well as forming a solid conceptual foundation in students. The communication and teamwork skills also have benefited from the introduction of the tool and the methodological strategy.

**Acknowledgments** This work was supported by the District University Francisco José de Caldas, in part through CIDC, and partly by the Technological Faculty. The views expressed in this paper are not necessarily endorsed by District University. The authors thank the research groups DIGITI and ARMOS for the evaluation carried out on prototypes of ideas and strategies, in particular Andres Moreno and Daniel Paez for their work with the platform.

# References

1. Abidin, A.Z., Saleh, F.: Team-based electronic portfolio. In: 3rd International Congress on Engineering Education (ICEED 2011), pp. 48–53. IEEE (2011)
2. Aloulou, A., Boubaker, O.: Enhancing technical skills of control engineering students in robotics by using common software tools and developping experimental platforms. In: International Conference on Education and e-Learning Innovations (ICEELI 2012), pp. 1–5. IEEE (2012)
3. Baneres, D., Clariso, R., Jorba, J., Serra, M.: Experiences in digital circuit design courses: a self-study platform for learning support. IEEE Trans. Learn. Technol. **7**(4), 360–374 (2014)
4. Bayindir, R., Kabalci, E., Kaplan, O., Oz, Y.E.: Microcontroller based electrical machines training set. In: 15th International Power Electronics and Motion Control Conference (EPE/PEMC 2012), pp. DS3e–12. IEEE (2012)
5. Fares, D., Khaddaj, S., Joujou, M.K., Kabalan, K.Y., et al.: A learning approach to circuitry problems using MATLAB and PSPICE. In: IEEE Global Engineering Education Conference (EDUCON 2012), pp. 1–5. IEEE (2012)
6. Guerra, M.A., Francisco, C.M., Madeira, R.N.: PortableLab: implementation of a mobile remote laboratory for the android platform. In: IEEE Global Engineering Education Conference (EDUCON 2011), pp. 983–989. IEEE (2011)
7. Krasnansky, P., Toth, F., Huertas, V.V., Rohal'-Ilkiv, B.: Basic laboratory experiments with an educational robotic arm. In: International Conference on Process Control (PC 2013), pp. 510–515. IEEE (2013)
8. Leo, T., Falsetti, C., Manganello, F., Pistoia, A.: Team teaching for web enhanced control systems education of undergraduate students. In: IEEE Education Engineering (EDUCON 2010), pp. 1525–1530. IEEE (2010)
9. López, C.: Evaluación de desempeño de dos técnicas de optimización bio-inspiradas: algoritmos genéticos y enjambre de partículas. Tekhnê **11**(1), 49–58 (2014). ISSN: 1692-8407
10. Martínez, F.: Robótica autónoma: Acercamientos a algunos problemas centrales, vol. 1. Universidad Distrital Francisco José de Caldas (2015). ISBN: 9789588897561
11. Music, G., Hafner, I., Breitenecker, F., Korner, A.: Learning petri net dynamics through a matlab web interface. In: 8th EUROSIM Congress on Modelling and Simulation (EUROSIM 2013), pp. 324–329. IEEE (2013)
12. Navrapescu, V., Chirila, A.I., Deaconu, A.S., Deaconu, I.D.: Educational platform for working with programmable logic controllers. In: 2015 9th International Symposium on Advanced Topics in Electrical Engineering (ATEE), pp. 215–220. IEEE (2015)
13. Nitschke, C., Minami, Y., Hiromoto, M., Ohshima, H., Sato, T.: A quadrocopter automatic control contest as an example of interdisciplinary design education. In: 14th International Conference on Control, Automation and Systems (ICCAS 2014), pp. 678–685. IEEE (2014)
14. Oomlout: Arduino controlled servo robot. http://oomlout.com/a/products/serb/ (2013). Accessed 1 Mar 2013
15. Pal, S., Choudhury, P., Mukherjee, S., Nandi, S.: Feasibility of mobile based learning in higher education institutions in India. In: 2nd International Conference on Business and Information Management (ICBIM 2014), pp. 17–22. IEEE (2014)
16. Perez, A., Reyes, M., Ortiz, E.: Work in progress-homer: an educational tool to learn about the design of renewable energy systems at the undergraduate level. In: Frontiers in Education Conference (FIE 2012), pp. 1–6 (2012)

17. Rohrbach, S., Ishizaki, S., Werner, N., Miller, J., Dzombak, D.: Improving students' professional communication skills through an integrated learning system. In: IEEE International Professional Communication Conference (IPCC 2013), pp. 1–6. IEEE (2013)
18. Sanchez-Azqueta, C., Gimeno, C., Celma, S., Aldea, C.: Using the wiimote to learn MEMS in a physics degree program. IEEE Trans. Educ. (2015)
19. Shoubaki, E., Amarin, R., Batarseh, I., et al.: Java based symbolic circuit solver for electrical engineering curriculum. In: IEEE Global Engineering Education Conference (EDUCON 2012), pp. 1–4. IEEE (2012)
20. Simonova, I.: Mobile-assisted language learning in technical and engineering education: tools and learners' feedback. In: International Conference on Interactive Collaborative Learning (ICL 2015), pp. 169–176. IEEE (2015)
21. Toth, P.: New possibilities for adaptive online learning in engineering education. In: International Conference on Industrial Engineering and Operations Management, pp. 1–5 (2015)
22. Weinert, H., Pensky, D.: Mobile robotics in education—South African and international competitions. In: 5th Robotics and Mechatronics Conference of South Africa (ROBOMECH 2012), pp. 1–6. IEEE (2012)

# Learning Object Assembly Based on Learning Styles

**Aldo Ramirez-Arellano, Juan Bory-Reyes
and Luis Manuel Hernández-Simón**

**Abstract** The goal of this paper is to develop a system, referred to as the Management System for Merging Learning Objects (msMLO), which offers an approach that retrieves learning objects (LOs) based on students' learning styles and term-based queries and produces a new outcome. The first step ranks LOs using a unified learning style model and creates better LOs by merging the top-ranked LOs. The second step maps LOs onto a hierarchy of concepts to avoid including duplicated topics in the merged LO. Fifty-six students were randomly split into experimental and control groups. The experimental group browsed the LOs retrieved by the msMLO based on the students' learning styles, term-based queries and merge functionality, whereas the control group browsed the LOs retrieved based on the students' learning styles and term-based queries. The results demonstrated that the experimental group improves their learning performance, thus msMLO is a promising approach.

**Keywords** Learning objects aggregation · Learning styles · Learning technologies

## 1  Introduction

The manual review of Learning Objects (LOs) is a complex task because of the growth of many educational resources. Learning Management Systems (LMSs) and Learning Object Repositories (LORs) are designed to retrieve learning resources in

A. Ramirez-Arellano (✉) · J. Bory-Reyes · L.M. Hernández-Simón
Department of Systems Engineering, National Polytechnic Institute,
SEPI-ESIME Zacatenco Edificio 5, Col., C.P. 07738 Linadavista, Mexico
e-mail: aramirezar@ipn.mx

J. Bory-Reyes
e-mail: juanboryreyes@yahoo.com

L.M. Hernández-Simón
e-mail: lmhernan10@hotmail.com

© Springer International Publishing Switzerland 2016      447
V.L. Uskov et al. (eds.), *Smart Education and e-Learning 2016*,
Smart Innovation, Systems and Technologies 59,
DOI 10.1007/978-3-319-39690-3_40

an easy manner; in most cases, the user provides a query, and the LOR will then search and rank learning resources according to this query.

The main difficulties in recommending LOs are *the appropriate selection of LOs according to the user's profile, the complexity of LO resources and the top-ranked LOs fulfill the user's needs* [1, 2]. Most retrieved hits by a recovery system can make the selection process of the best LOs tedious because the user must browse them individually. For example, a user interested in a linear regression topic (the theoretical background) and examples (problems solved using a spreadsheet) may identify two different resources that cover these topics; however, it might be more appropriate to merge both LOs instead of thumbing through them separately.

The main goal of this paper is to develop a system, referred to as the Management System for Merging Learning Objects (msMLO), to offer an approach that retrieves LOs based on student learning styles and term-based queries, and to create new LOs with better scores.

The advantages of filtering the LO based on the student's learning style are as follows: *avoiding the cold-start problem, mapping user needs onto items and immediately reflecting the change of student's preferences* [3]. Furthermore, limiting the top-ranked items reduces the number of LOs to be merged; thus, the scalability problem is addressed [3, 4]. The remaining sections describe the process of retrieval LOs and the experimental results. Finally, the discussion, conclusions and further topics for investigation are considered.

## 2  Related Work

To place this work in context, the next subsection provides a brief review of learning style theories and their use in an adaptive and educational personalized recommender system (ERS). Next, recent works focused on an adaptive systems, ERS and LO aggregation are described.

Considerable efforts have been made to integrate the learning styles into recommender and adaptive systems; for example, the interconnection among Honey and Mumford's [5] learning style model, activities, teaching, learning methods and types of LOs are depicted in [6]. The relationship between the learning profile is *split into categories, including learning style*, and LOs are addressed by linking the metadata of LOs with a student's learning style [1]. Resources, such as text, graphics, video and web pages, are categorized according to Kolb's [7] learning style model [8, 9]. Felder-Silverman's learning style [10] and LOM standard [11] are the cornerstones of the personalized LO recommendations in [12].

Anaya, Luque and García-Saiz designed a recommender system that is able to warn students and tutors about potentially problematic circumstances; thus, this system can propose a recommendation that can solve the problem [13]. Students' browsing patterns and the learning material's attributes are modeled as a tree structure to improve the quality of the recommended LOs [14]. The recommender system developed in [15] recommends courses to first-year students based on other students' choices among a particular set of courses collected from a LMS. Web-mining techniques form the basis for recommending the next link to visit within an adaptable educational hypermedia system [16]. The previous approaches are based on data mining techniques; thus, they are susceptible to the cold-start problem.

To overcome the cold-start problem, other research has computed the similarity between student profiles and learning material by means of a function, as in [17], where the quality of the materials is taken into account as a factor in the recommendation process. The adaptation based on the mastery of competencies is addressed by considering the characteristics of *the competencies of each subject, the designed activities and the student's individual profile* [18]. This approach requires that additional efforts be made by faculty members who assign an importance degree weight to each competence. A personalized english reading sequencing approach [19] used a ubiquitous learning, location-aware technology in conjunction with a similarity function that compares unfamiliar vocabulary by determining how often such words occur in a portfolio of reading materials.

The adaptive system in [20] selects the appropriate components (learning materials) based on learning style; thus, it presents the materials using a specific layout. This approach is inconvenient because different learning materials for the same topic have to be designed according to each learning style. An ontology represents the interconnections among student learning styles, preferred learning activities, and relevant teaching/learning methods that help students identify suitable LOs [6]. Similarly, an enriched domain ontology is used to index and retrieve educational resources based on query concepts, pedagogical knowledge and user contexts [21]. A term-based query, language, educational level and repository are the personal settings that recommend LOs in a multi-agent-based systems [22].

Our approach is based on a similarity measure that retrieves and creates new LOs instead of refactoring the query [23, 24, 25] or adapting the result [26, 27, 28, 29]. The proposed function measures the similarity between students' learning styles and the LOs.

A framework that helps instructors locate learning design documents and reuse them in other contexts is presented in [30]. Similarly, a system that assists in the aggregation of html pages is presented in [31]. These approaches are inconvenient because the aggregation of documents is semi-automatic and is performed by the instructor. On the other hand, an automatic approach for merging LOs, represented by ontologies, is depicted in [32]. These issues are overcome by the msMLO using an automatic approach that builds a hierarchy of concepts based on the content of each LO, thus a domain ontology is not required.

Koper stated that a learning unit's reusable parts are physical resources, such as assessments and lectures [33]. Furthermore, the multigranularity reuse of learning resources [34] means that an LO's fragments are available for aggregation as modules of a large learning resource at multiple levels of granularity. The *merge* function, presented later, is based on a multigranularity property that allows us to consider a leaf item of a packaged LO, using the SCORM [35] standard, to be a learning resource of granularity and an inner item to be a composite learning resource [36].

## 3   Retrieval Process for LOs

The two-step retrieval process is illustrated in Fig. 1. First, the LOs are scored using the student's learning style and term-based query. The term-based query also restricts the LOs to the student's topic of interest [22]. Second, the msMLO will attempt to identify new top-ranked LOs by combining n LOs obtained in the first step, which are known as the source LOs. We chose n = 10 because the most suitable LOs are typically within the first 10 positions [37]. The msMLO's capabilities are limited compared with those of a LMS; thus, the approach uses some dimensions of the Unified Learning Style Model (ULSM) [38], as shown in Table 1.

### 3.1   Labeling LO Items

The LOs, packaged using a SCORM [35] standard, contain several items, which include resources such as html pages, images, and documents. The items often have LOM metadata [11] and can be labeled independently from one another. The LOM metadata [11] describe the content of an LO through descriptor categories. The categories adopted in this work are: *General, Technical and Educational*.

The text in the fields of the general category, html pages and documents is used to extract the frequency with which a term t appears in an LO, the number of LOs that contains the term t, and the number of LOs stored in the msMLO (see [37] for further details). These parameters are necessary to measure the angle between vectors q and d in Eq. (1), as explained in the next subsection. Technical and educational fields are useful for mapping the value of the LOM field onto preferences (see Table 2); for example, a LOM field value = *figure, text* matches the *visual, verbal and reflective* preferences.

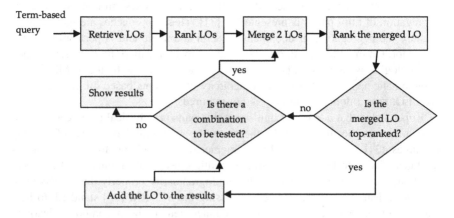

**Fig. 1** The retrieval process based on learning styles and LO aggregation

**Table 1** Links among learning styles, resource types, metadata and review sequences

| Dimension | Preferences | Description | Resources | LOM field values | Sequence (SCORM) |
|---|---|---|---|---|---|
| Perception | Visual | Students best remember what they see | Pictures, diagrams, schemas, videos | Figure, video, film | Simple, sequenced |
| | Verbal | Students learn best from spoken or written words | Documents, notes, podcast | Text, sound, lecture | Simple, sequenced |
| Processing information | Abstract | Students rely on conceptual interpretation | Documents, html pages | Expositive | Simple, sequenced |
| | Concrete | Students rely on immediate experiences | Simulations, experiments | Active | Simple, sequenced |
| | Sequential | Students prefer linear progression (step-by-step approach) | – | – | Sequenced |
| | Holistic/Global | Students prefer the overall picture and patterns | – | – | Simple |
| | Active | Students prefer experiments and exercises | Simulation, experiments | Exercise, simulation and experiment | Simple, sequenced |
| | Reflective | Students prefer to think and draw parallels | Documents | Questionnaire and text | Simple, sequenced |
| Field dependence | Dependence | Students recognize theme and main ideas without paying ample attention to details | – | – | Simple |
| | Independence | Students pay substantial attention to the details but less to the context | – | – | Sequenced |
| Reasoning | Deductive | Students prefer to reason from general to specific | Presents documents then simulation, experiments | Lecture, problems, simulation, experiments | Sequenced (from fundamentals to applications) |
| | Inductive | Students prefer to reason from particular facts to a general conclusion | Presents problems, experiments, simulation and then documents | Problems, simulation, experiments, lecture | Sequenced (from problems to a general conclusion) |

**Table 2** Matrix for mapping LOM values to preferences

| | Perception | | Processing information | | | | Reasoning | |
|---|---|---|---|---|---|---|---|---|
| | Visual | Verbal | Abstract | Concrete | Active | Reflective | Deductive | Inductive |
| Figure | X | | | | | | | |
| Video | X | | | | | | | |
| Film | X | | | | | | | |
| Text | | X | | | | X | | |
| Lecture | | X | | | | | X | X |
| Sound | | X | | | | | | |
| Expositive | | | X | | | | | |
| Active | | | | X | | | | |
| Simulation | | | | | X | | X | X |
| Experiment | | | | | X | | X | X |
| Exercise | | | | | X | | | |
| Questionnaire | | | | | | X | | |
| Problems | | | | | | | X | X |

The *sequential, holistic/global, dependence* and *independence* preferences are not related to any resource; however, they reflect a way of browsing content. The LO sequences are modified to adapt the LOs instead of being used to label the LOs. For example, if the student's preference is *sequential*, the LO's browsing will be adapted using the SCORM standard [39]. A sequential review indicates that the student will not be allowed to open the next item if the previous item has not yet been reviewed. By contrast, in a simple review, the student browses the LO in a free manner. To solve the issue of missing LOM [11] fields, the proposed approach involves mapping the file extension and MIME type [40] of the LO resources onto a particular preference; for example, an "avi" file will be mapped to the visual preference.

## 3.2 Computation of the LO Score

The approach uses the vector space model [37] and similarity measure to compute the score of the LOs using the following formula:

$$Score = [abs\left(\frac{c}{lp - sp - c}\right)][\frac{1}{sp}\sum_{i=1}^{sp}\frac{spi}{lp}][\frac{\vec{d}\cdot\vec{q}}{|\vec{d}||\vec{q}|}] \tag{1}$$

The first term of Eq. (1), $abs(c/(lp - sp - c))$, measures the similarity between the student's learning style and preferences of the given LO, where $c$ is the number of common preferences between the student's learning style and current LO. $sp$ is the number of preferences for the student's learning style whereas $lp$ is the number of preferences for the LO. Because the LOs comprise different items [35], a preference

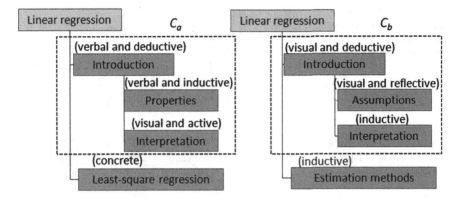

**Fig. 2** Example of an LO labeled according to preferences

may appear more than once. The second term, $[1/sp \sum sp_i/lp]$, aids the repeated preferences within the LO, where $sp_i$ is the number of times that a student's preference appears in the LO. A simple example is shown to illustrate these effects.

Let $S = \{visual, reflective, inductive\}$ represent the student's preference and $L = \{verbal, verbal, visual, concrete, deductive, inductive, active\}$ represent the preferences of the LO on the left-hand side of Fig. 2; thus, $c = 2$, $sp = 3$ and $lp = 7$. Counting the appearance of the items of $S$ in $L$, we obtain $sp_{visual} = 1$, $sp_{reflective} = 0$ and $sp_{inductive} = 1$. The partial result is $2/(7 - 3 - 2)$ $(1/3)$ $(1/7 + 1/7) = 2/21$. The last term of Eq. (1), $(\vec{d} \cdot \vec{q})/(|\vec{d}||\vec{q}|)$, in which $q$ is a vector of the query terms and $d$ is a vector of the terms in the title, metadata and content of the LO, determines the size of the angle between the two defined vectors using an inner product "$\cdot$". After a pre-processing task, the title, metadata and content of the LOs fit together in a Lucene document [41]. The preprocessing task includes deleting stop words. Furthermore, a stemmer is used to derive roots from words, for example "runs", "running", and "run" all map to "run". For a more detailed discussion regarding how to obtain the $q$ and $d$ vectors, we refer the reader to [37].

## 3.3 Selection of LOs for the Merging Process

To reduce the number of LOs browsed by students, we propose a merging process to combine two LOs and produce a new LO with a higher score, provided that a new pivot was found. To solve the issue of deciding which LOs will be combined, we use an *encoded genealogical tree* (EGT). A genealogical tree depicts the ancestors and descendants of a given LO. Figure 3 shows the source LOs (solid squares), which are stored in the msMLO and the ancestor of any merged LOs (dashed square). For example, the ancestors of F are encoded as a sequence of

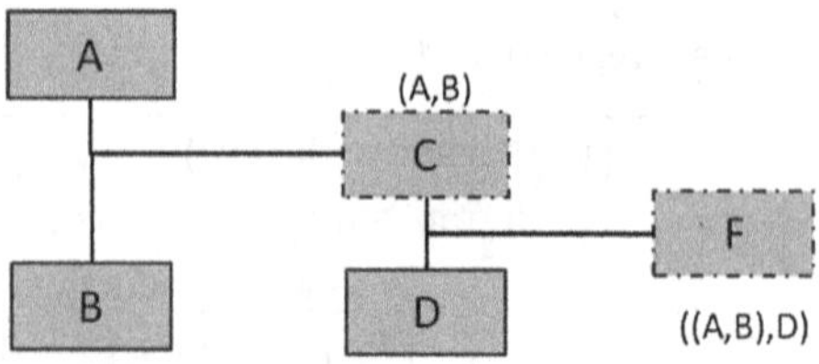

**Fig. 3** EGT showing the ancestors of F, which are encoded as a sequence of characters ((A, B), D)

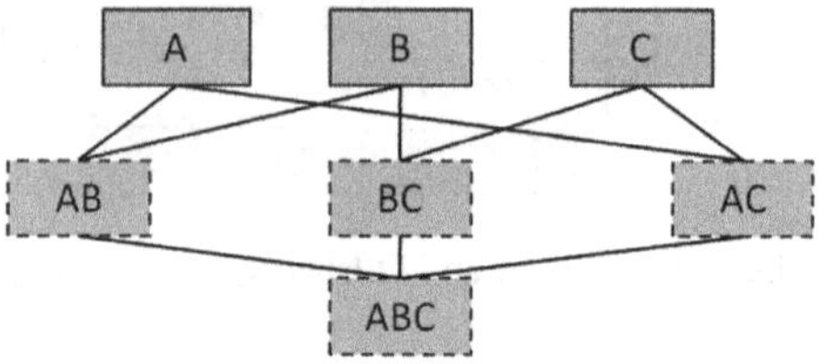

**Fig. 4** Lattice structure with n = 3. Each node represents an LO

characters ((A, B), D) referred to as an EGT. Analyzing the EGT, the msMLO is able to detect whether two LOs have a common ancestor.

Given a set of $n$ LOs ranked by Eq. (1), the msMLO must perform $2^n - n - 1$ combinations of LOs (see Fig. 4). In the first stage, the lattice structure has only source LOs (solid square) at the top; therefore, the algorithm must perform four merging processes to obtain the remaining LOs. Testing all the combinations makes the process of merging LOs complex and impractical. The rationale of our algorithm is to search for better LOs by performing the minimum number of merging processes with the EGT.

We describe the relationship-based algorithm (RA), shown later, as follows. Let $P_0$, $LO_i$ and $LO_c$ be LOs, where $P_0$ is the pivot, $LO_i$ is the candidate to be merged and $LO_c$ is the outcome of *merge* function. The set $S$ has LOs ranked according to their scores. In the first stage, the pivot is the top-ranked $LO_i$, where $LO_i \in S$. $|LO|$ indicates the score of the LOs assigned by Eq. (1). The *relationshipdetection* function tests whether the LOs have a common ancestor. The *merge* function produces a combination of two given LOs, and it will be explained in the next section. Furthermore, the *merge* function can produces a new pivot (see RA, line 6) for each iteration; the *relationshipdetection* function will discard most combinations of the current level of the lattice (see Fig. 4). For example, let A be the pivot in the first level of the lattice; then, the first merging process can produce the pivot (A, B) in the second level of the lattice. Now, $S = \{AB, A, B, C\}$. Because the *relationshipdetection* function avoids merging the pivot with A or B, the remaining combination is ABC. The RA only calls the *merge* function once for each level of the lattice, regardless of the combination serving as the current pivot; thus, the RA performs at most $n - 1$ merging processes, where $n$ is the number of the source LOs at the top of the lattice.

1. $P_0$=top-ranked LO
2.     do
3.         for each $LO_i \in S$
4.             if(not relationshipdetection($P_0$, $LO_i$) &$P_0 \neq LO_i$)
5.                 $LO_c$=merge($P_0$, $LO_i$)
6.                 if($|LO_c|>|P_0|$)
7.                     $P_0$= $LO_c$
8.                 end if
9.                 S=S $\cup$ $LO_c$
10.        end if
11.        end for
12.     while(S has been changed)
13. return $P_0$

Relationship-based Algorithm (RA).

## 3.4   Merging Process

The Merge Algorithm (MA) uses a hierarchy of concepts that represents a given LO to identify similar items between two different LOs; then, the most suitable item will be selected and added to the merged LO.

1. $LO_c$=the merged LO
2.     for each $C_a \in H_a$
3.         Cb= sim($C_a$, $H_b$)
4.         If $C_b$ is null
5.             add $Ic_a$ to $LO_c$
6.         else
7.             if score(S, $Ic_{a,}$)> score(S, $Ic_b$)
8.                 merge($C_a$, $C_b$, $LO_c$), merge($C_b$, $C_a$, $LO_c$)
9.                 add $C_a$ to $H_c$
10.            else
11.                merge($C_b$, $C_a$, $LO_c$), merge($C_a$, $C_b$, $LO_c$)
12.                add $C_b$ to $H_c$
13.            end if
14.        end If
15.     end for
16. return $LO_c$

Merge Algorithm (MA).

A pre-processing task is conducted before using the approach described in [42] to obtain the concepts for each LO. To extract the concepts, each LO is treated as a document, and the top-five LOs ranked by Eq. (1) are considered the minimum support (see [42] for further details). The result of this step is a hierarchy of concepts, which are mapped onto each LO item; thus, the relationships among the items are identical among the concepts in the hierarchy (see Fig. 2).

The MA combines two LOs, and it produces a new LO without redundant items. It is based on the $sim(C_a, H)$ function [43] to identify the most similar concept. The *sim* function compares the ancestors and descendants of two given concepts. For example, the "Introduction" concept for the LO on the left-hand side of Fig. 2 is similar to the "Introduction" concept for the LO on the right-hand side because they and their parents match. For further details on the *sim* function, we refer the reader to [43].

Let $C_b \epsilon H_b$ to $C_a \epsilon H_a$ represent concepts from two hierarchical representations, $H_a$ and $H_b$, of two LOs; let $Ic$ represent the mapped item from which concept $C$ was extracted; and let $S$ represent the student's learning preferences and scores computed by Eq. (1). If *sim* identifies a similar concept $C_b$, it evaluates $C_a$ and $C_b$. For example, $C_a$= "Introduction" for the LO on the left-hand side of Fig. 2 is similar to $C_b$ = "Introduction" for the LO on the right-hand side. Let $S$ = *{verbal, holistic, inductive}* represent the student's preferences $C_a$ = *{verbal, verbal, visual, deductive, inductive, active}* and $C_b$ = *{visual, visual, deductive, reflective, inductive}*. The scores are $Ca = 0.16$ *and* $Cb = 0.12$; thus, $Ca$ will be added to *LOc*. Because the "Introduction" concepts for the LOs on the left- and right-hand sides of

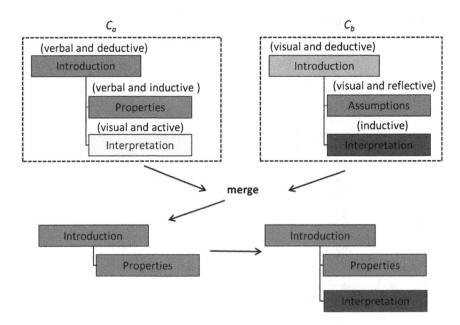

**Fig. 5** Process of merging two similar items

Fig. 2 are sub-hierarchies, they must be merged (lines 8, and 11 of MA). Now, *sim* shows that $C_a$ = "Interpretation" for the LO on the left-hand side of Fig. 5 is similar to $C_b$ = "Interpretation" for the LO on the right-hand side; consequently, $C_a$ = {visual, active} and $C_b$ = {inductive}. The scores are $C_a = 0$ and $C_b = 0.111$; hence, $C_b$ will be added to $LO_c$ (see Fig. 5). The best item is chosen according to its score (see line 7 of MA). The merged LO contains fewer items than the sum of the items of the source LOs used to create it.

# 4 Results of Experiments

This section will present the experimental results of a learning activity for the "Applied Computing in Biological Sciences" course at the Instituto Politécnico Nacional (IPN) in Mexico City. The 127 LOs initially stored in the msMLO and used in the first experiment were developed by the teachers of this course. The 500 LOs used in the second experiment included the 127 LOs from the first experiment, and the remainder were mostly gathered from support learning resources for face-to-face courses at IPN. The LOs covered several subjects. The experiments were conducted using a server with two 2.4 GHz CPUs, 8 GB RAM and a HD of 465 GB.

## 4.1 Experimental Design

In the first experiment, 56 undergraduate students were randomly divided into two groups: *the control group (n = 28) and the experimental group (n = 28)*. The students were trained to use the msMLO, and the Wayne State University Learning Styles Questionnaire (WSULSQ) [44] was subsequently distributed to determine the students' learning styles. The students in the experimental group solved a learning activity by browsing the LOs recommended by the msMLO based on their learning styles, their term-based queries and the merge functionality. The students in the control group browsed the LOs retrieved by their term-based queries.

The learning activity was designed based on Bloom's educational objectives [45] and was scored by four faculty members. It included five multiple-choice items, three open-ended questions and two practical problems (a total score of 10). A diagnostic test was administered, and the results ensured that the participants in the study had the same level or prior knowledge regarding the concepts related to the learning activity. The time spent completing the learning activity started when the student sent the term-based query and ended when the student uploaded a file that contained the solution of the activity. Students had no time limit for completing the learning activity.

A second experiment was conducted to determine the scalability of the msMLO when the number of stored LOs increased. Ten of 28 profiles from the experimental group were selected; thus, the retrievals were based on Eq. (1) and the merge functionality. The number of LOs merged by the users' profiles were 2, 3, 4, 5, and 8, and their frequencies were 1, 3, 3, 2, and 1, respectively. The retrieval process for each profile was tested 10 times, and the number of LOs stored varied from 100 to 500.

## 4.2 Results

Table 3 shows the descriptive statistics for the learning activity grade, the time spent on the activity and the number of LOs reviewed. Independent t-tests were used to analyze these factors. The differences between the grades, completion times and LOs of the experimental and control groups were significant ($p < 0.05$).

The msMLO was able to create a new LO for all the queries of the experimental group. The time spent executing the retrieval process includes the ranking of LOs using Eq. (1) and the process of creating a new LO using the RA and MA.

Figure 6 shows a linear relationship between the average of the difference between the scores for each merged LO and the top-ranked source LO grouped by

**Table 3** Descriptive statistics and t test of the grade, LOs reviewed and time

|            | Group        | N  | Min | Max | Mean  | S.D. | T    | P          |
|------------|--------------|----|-----|-----|-------|------|------|------------|
| Grade      | Experimental | 28 | 8   | 10  | 9.20  | 0.59 | 6.57 | 1.086 E-07 |
|            | Control      | 28 | 5   | 10  | 7.35  | 1.36 |      |            |
| LO         | Experimental | 28 | 1   | 4   | 1.85  | 0.95 | 7.77 | 9.112 E-09 |
|            | Control      | 28 | 4   | 16  | 6.55  | 3.39 |      |            |
| Time (min) | Experimental | 28 | 38  | 51  | 43.17 | 3.17 | 8.21 | 7.376 E-11 |
|            | Control      | 28 | 41  | 67  | 52.79 | 6.03 |      |            |

**Fig. 6** The linear relation between the difference of the score of each merged LO and the top-ranked source LO of experimental group

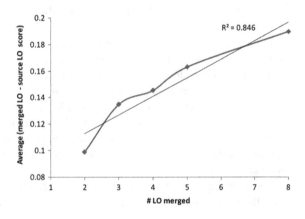

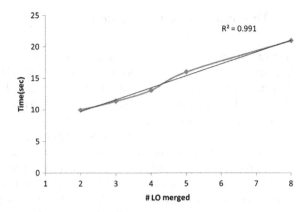

**Fig. 7** The linear relation between the number of LO merged and the time (sec) of retrieval

the number of LOs merged. Similarly Fig. 7 shows a linear relationship between the number of LO merged and the time consumed in the retrieval process. The time includes the ranking of LOs using Eq. (1) and the process of creating a new LO using the RA and MA. Figure 8 shows the results of testing 10 retrieval processes that vary in terms of the number of LOs stored in the msMLO. The time spent on the ranking stage, using Eq. (1), was the shortest part of the entire retrieval process.

This stage represented approximately 7 % of total time. The merging stage consumed approximately 93 % of total time. On the other hand, the time spent on the merging stage increased based on the number of LOs to be merged, not on the number of LOs stored in the msMLO (see Figs. 7 and 8).

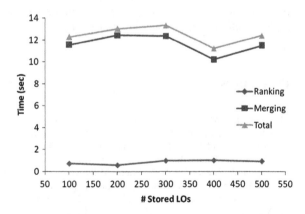

**Fig. 8** The time (sec) consumed by ranking and merging stages varying the stored LOs

## 5   Discussion and Conclusion

The results presented in Table 3 suggest that the overall learning performance of the experimental group improved significantly because of the learning style retrievals and the merging of LOs. The LOs that were created by the msMLO reduced the time that the experimental group spent on the learning activity.

The experiment did not allow us to know the individual impact of the merged LOs and learning style retrievals on the learning performance and the number of LOs reviewed by students. These results pave the way for new research.

The results in Fig. 6 suggest that a linear relationship exists between the number of LOs merged to create the new LO and the difference between the merged LO score and the top-ranked source LO score. Figure 8 suggests that the ranking and merging stages are not significantly affected by the number of stored LOs. The RM computes Eq. (1) using the parameters explained in Sects. 3.1 and 3.2. These parameters are stored and updated in the Lucene [41] index each time that a new LO is added by the SM; thus, the time spent computing Eq. (1) can be considered constant. Based on the results presented in Fig. 8, the number of merged LOs is the only factor that increases the time spent on the retrieval process. The number of LOs stored in the msMLO that are used to test the scalability of the ranking and the merging stages is not large enough to infer a convincing conclusion; however, our findings can elucidate how the PRM can scale and which factor affects the scalability of the retrieval process.

The findings have demonstrated that the msMLO is a promising approach. The merged LOs improve learning performance, reduce the number of LOs reviewed and reduce the time spent solving the learning activity. The msMLO retrieves LOs that are packaged using well-known standards; thus, no additional information, such as ontology or annotation, is needed.

## References

1. Zaina, L.A.M., Jose, F., Rodrigues, J., Bressan, G.: An approach to design the student interaction based on the recommendation of e-learning objects. In: Paper presented at the Proceedings of the 28th ACM International Conference on Design of Communication. Sao Carlos, São Paulo, Brazil (2010)
2. Zapata, A., Menéndez, V.H., Prieto, M.E., Romero, C.: A framework for recommendation in learning object repositories: an example of application in civil engineering. Adv. Eng. Softw. **56**, 1–14 (2013). doi:10.1016/j.advengsoft.2012.10.005
3. Manouselis, N., Drachsler, H., Vuorikari, R., Hummel, H., Koper, R.: Recommender systems in technology enhanced learning. In: Ricci, F., Rokach, L., Shapira, B., Kantor, P.B. (eds.) Recommender Systems Handbook, pp. 387–415. Springer (2011)
4. Verbert, K., Manouselis, N., Ochoa, X., Wolpers, M., Drachsler, H., Bosnic, I., Duval, E.: Context-Aware recommender systems for learning: a survey and future challenges. IEEE Trans. Learn. Technol. **5**(4), 318–335 (2012). doi:10.1109/tlt.2012.11
5. Honey, P., Mumford, A.: The Manual of Learning Styles. Peter Honey Publications, UK (1992)

6. Kurilovas, E., Kubilinskiene, S., Dagiene, V.: Web 3.0—Based personalisation of learning objects in virtual learning environments. Comput. Hum. Behav. **30**, 654–662 (2014)
7. Kolb, D.A.: Experiential Learning: Experience as the Source of Learning And Development. Prentice Hall, Englewood Cliffs, NJ (1984)
8. Graf, S., Liu, T.-C., Kinshuk, Chen N.-S., Yang, S.J.H.: Learning styles and cognitive traits—their relationship and its benefits in web-based educational systems. Comput. Hum. Behav. **25** (6), 1280–1289 (2009)
9. Sahabudin, N.A., Ali, M.B.: Personalized learning and learning style among upper secondary school students. Procedia—Soc. Behav. Sci. **103**, 710–716 (2013)
10. Felder, R.M., Silverman, R.K.: Learning and teaching styles in engineering education. J. Eng. Educ. **78**(7), 674–682 (1988)
11. IEEE: IEEE standard for learning object metadata. Institute of Electrical and Electronic Engineer (2002)
12. Ibrahim, K.M.F.: Semantic retrieval and recommendation in adaptive E-Learning System. In: Paper presented at the International Conference of Computing and Information Technology. Saudi Arabia (2012)
13. Anaya, A.R., Luque, M., García-Saiz, T.: Recommender system in collaborative learning environment using an influence diagram. Expert Syst. Appl. **40**(18), 7193–7202 (2013)
14. Salehi, M., Nakhai Kamalabadi, I.: Hybrid recommendation approach for learning material based on sequential pattern of the accessed material and the learner's preference tree. Knowl.-Based Syst. **48**, 57–69 (2013)
15. Aher, S.B., Lobo, L.M.R.J.: Combination of machine learning algorithms for recommendation of courses in E-Learning System based on historical data. Knowl.-Based Syst. **51**, 1–14 (2013)
16. Romero, C., Ventura, S., Delgado, J., Bra, P.: Personalized links recommendation based on data mining in adaptive educational hypermedia systems. In: Duval, E., Klamma, R., Wolpers, M. (eds) Creating New Learning Experiences on a Global Scale, vol 4753. Lecture Notes in Computer Science, pp. 292–306. Springer, Berlin (2007). doi:10.1007/978-3-540-75195-3_21
17. Tejeda-Lorente, Á., Porcel, C., Peis, E., Sanz, R., Herrera-Viedma, E.: A quality based recommender system to disseminate information in a university digital library. Inf. Sci. **261**, 52–69 (2014)
18. Serrano-Guerrero, J., Romero, F.P., Olivas, J.A.: Hiperion: A fuzzy approach for recommending educational activities based on the acquisition of competences. Inf. Sci. **248**, 114–129 (2013)
19. Wu, T.-T., Huang, Y.-M., Chao, H.-C., Park, J.H.: Personlized English reading sequencing based on learning portfolio analysis. Inf. Sci. **257**, 248–263 (2014)
20. Yang, T.C., Hwang, G.J., Yang, S.J.H.: Development of an adaptive learning system with multiple perspectives based on students' learning styles and cognitive styles. Educ. Technol. Soc. **16**(4), 16 (2013)
21. Ruiz-Iniesta, A., Jimenez-Diaz, G., Gomez-Albarran, M.: A semantically enriched context-aware OER recommendation strategy and its application to a computer science OER repository. IEEE Trans. Educ. **57**(4), 255–260 (2014). doi:10.1109/te.2014.2309554
22. Casali, A., Gerling, V., Deco, C., Bender, C.: A recommender system for learning objects personalized retrieval. In: Santos, O.C., Boticario, J.G. (eds.) Educational Recommender Systems and Technologies: Practices and Challenges. IGI-GLOBAL, p. 362 (2011)
23. Zhou, D., Lawless, S., Wade, V.: Improving search via personalized query expansion using social media. Inf. Retrieval **15**(3–4), 218–242 (2012)
24. Chirita, P.-A., Firan, C.S., Nejdl, W.: Personalized query expansion for the web. In: Paper Presented at the Proceedings of the 30th Annual International ACM SIGIR Conference on Research and Development in Information Retrieval. Amsterdam, The Netherlands (2007)
25. Yin, Z., Shokouhi, M., Craswell, N.: Query Expansion Using External Evidence. In: Boughanem, M., Berrut, C., Mothe, J., Soule-Dupuy, C. (eds.) Advances in Information Retrieval. Lecture Notes in Computer Science, vol. 5478, pp. 362–374. Springer, Berlin Heidelberg (2009)

26. Katakis, I., Tsoumakas, G., Banos, E., Bassiliades, N., Vlahavas, I.: An adaptive personalized news dissemination system. J. Intell. Inf. Syst. **32**(2), 191–212 (2009)
27. Vallet, D., Cantador, I., Jose, J.M.: Personalizing web search with folksonomy-based user and document profiles. In: Gurrin C, He Y, Kazai G et al. (eds) Advances in Information Retrieval, Lecture Notes in Computer Science, vol. 5993, pp. 420–431. Springer, Berlin (2010). doi:10. 1007/978-3-642-12275-0_37
28. Stamou, S., Ntoulas, A.: Search personalization through query and page topical analysis. User Model. User-Adap. Inter. **19**(1–2), 5–33 (2009)
29. Steichen, B., O'Connor, A., Wade, V.: Personalisation in the wild: providing personalisation across semantic, social and open-web resources. In: Paper Presented at the Proceedings of the 22nd ACM conference on Hypertext and Hypermedia. Eindhoven, The Netherlands (2011)
30. Yang, S.J.H., Chang, A.C.N., Chen, I.Y.L: Enhancing learning resources reusability with a new learning design framework. In: Fifth IEEE International Conference on Advanced Learning Technologies, 2005. ICALT 2005, pp. 414–418, 5–8 July 2005. doi:10.1109/icalt. 2005.140
31. Huang, K.H., Huang, Y.S., Ssu, K.F., Tsai, S.R., Huang, Y.C., Ho, S.P., Wen, C.K.: A system for the sharing and reuse of learning objects. In: 2014 International Conference on Teaching, Assessment and Learning (TALE), pp. 8–10, 66–172, Dec 2014. doi:10.1109/tale.2014. 7062612
32. Ching-Chieh, K., Chien-Sing, L.: Learning objects reusability and retrieval through ontological sharing: a hybrid unsupervised data mining approach. In: Seventh IEEE International Conference on Advanced Learning Technologies, 2007. ICALT 2007, pp. 18–20, 548–550, July 2007. doi:10.1109/icalt.2007.177
33. Koper, R.: Combining re-usable learning resources and services to pedagogical purposeful units of learning. In: Littlejohn, A., Shum, S.B. (eds.) Reusing Online Resources: A Sustainable Approach to eLearning, pp. 46–59. Kogan Page, London (2003)
34. Meyer, M., Rensing, C., Steinmetz, R.: Multigranularity reuse of learning resources. ACM Trans. Multimedia Comput. Commun. Appl. **7**(1), 1–23 (2011). doi:10.1145/1870121. 1870122
35. ADL: ADL SCORM 2004 4th Content Aggregation Model, Versión 1.1. ADL (2004)
36. Allena, C.A., Mugisab, E.K.: Improving learning object reuse through OOD: a theory of learning objects. J. Object Technol. **9**(6), 51–75 (2010)
37. Baeza-Yates, R.A., Ribeiro-Neto, B.: Modern Information Retrieval. Addison-Wesley Longman Publishing Co. Inc. (1999)
38. Popescu, E.: A unified learning style model for technology-enhanced learning: what, why and how? Int. J. Distance Educ. Technol. **8**(3), 65–81 (2010)
39. ADL: ADL SCORM 2004 4th Run-Time Environment, Versión 1.1. ADL (2004)
40. IANA Multipurpose Internet Mail Extensions (MIME) and Media Types: The Internet Assigned Numbers Authority (IANA). http://www.iana.org/protocols#index_M (2015). Accessed 03 July 2015
41. Michael, M., Erik, H., Otis, G.: Lucene in Action, Second Edition: Covers Apache Lucene 3.0. Manning Publications Co. (2010)
42. Toledo-Alvarado, J.I., Guzmán-Arenas, A., Martínez-Luna, G.L.: Automatic building of an ontology from a corpus of text documents using data mining tools. J. Appl. Res. Technol. **10** (3), 398–404 (2012)
43. Guzman-Arenas, A., Olivares-Ceja, J.M.: Measuring the understanding between two agents through concept similarity. Expert Syst. Appl. **30**(4), 577–591 (2006)
44. Giuliano, C.A., Moser, L.R., Poremba, V., Jones, J., Martin, E.T., Slaughter, R.L.: Use of a unified learning style model in pharmacy curricula. Currents Pharm. Teach. Learn. **6**(1), 41–57 (2014)
45. Bloom, B.S.: Taxonomy of Educational Objectives Book 1: Cognitive Domain 2nd edn. Addison Wesley Publishing, New York (1984)

# Transmedia Storytelling for Social Integration of Children with Cognitive Disabilities

Miguel Gea, Xavier Alaman and Pilar Rodriguez

**Abstract** This paper proposes the use of transmedia storytelling techniques to stimulate the day living of children with cognitive disabilities in the context of their social integration on future Smart Cities. By using virtual immersive world we focus on authoring stories and contents by these children to stimulate their abilities, and therefore, the exploration of their skills on a transmedia environment for therapeutically and employment purposes.

**Keywords** User experience · Transmedia storytelling · Mixed reality · Inclusive design

## 1 Introduction

Recent research approaches have been explored to identify new models of interaction and aids for everyday life and inclusive design [1]. These proposals can be categorized according the tools applied, the kind of users and purposes, but these scenarios have a common goal: the development of tools and interaction techniques for future citizen living on a more comfortable and suitable environment for all: the smart city. Smart cities represent an emerging trend towards a seamless integration of different and sustainable technologies on different scopes (governance, living, mobility, etc.) to overcome, for instance, architectural and technological barrier for

M. Gea (✉)
Dpt. Lenguajes y Sistemas Informáticos, Computer Science, University of Granada, Granada, Spain
e-mail: mgea@ugr.es

X. Alaman · P. Rodriguez
Dpt. Ciencia de la Computación e Inteligencia Artificial, Universidad Autónoma de Madrid, Madrid, Spain
e-mail: xavier.alaman@uam.es

P. Rodriguez
e-mail: Pilar.Rodriguez@ii.uam.es

© Springer International Publishing Switzerland 2016
V.L. Uskov et al. (eds.), *Smart Education and e-Learning 2016*,
Smart Innovation, Systems and Technologies 59,
DOI 10.1007/978-3-319-39690-3_41

people with disabilities. This is a service oriented architecture offering connectivity and different multilevel information mechanism to transform our surrounding in an augmented reality environment for citizen, and therefore, it is a key issue in the H2020 European strategy [2]. This envisioning model allow us to identify two strategic and interconnected layer:

- The interaction level, using and adopting of different technologies to interact with the surrounding and
- The content creation level, managing different kind of information on different formats and purposes

These layers are specially suited on emerging paradigms based on mixed-reality experiences, a convergence of the digital and physical environments where virtual and real objects complement and interact with each other for a expanded environment [3]. This interactive paradigm must also be inclusive allowing everyone to take part on these layers without barriers. Several projects are related with use an adoption of technologies for different purposes, for instance, using augmented reality inclusive games for children [4] or immersive virtual reality environments to enhance social interaction for children with Autism [5].

These approaches aim to explore different ways of inclusive interactive scenarios where experts design the content layer. Our goals focus on children with cognitive disabilities, facilitating several tools and interactive techniques to explore their creativity by telling digital stories for different media. Digital storytelling is a first-person experience within a fantasy world, in which the user may create, interact, and observe a character whose choices and actions affect the course of events just as they might in a play [6]. Interesting case studies are based on a storytelling approach such as the CHESS Project (Cultural Heritage Experiences through Socio-personal interactions and Storytelling) is the enrichment of museum visits through personalized interactive storytelling, enabling user dependent stories according to a variety of factors such as visitor choices, past actions, visitor features or locations [7].

Creativity is a relevant aspect of research on these interactive scenarios, as shown using virtual reality to increase the stimuli for student [8], but also for people with behavioral disorder or disabilities. These objectives are part of the e-Integra project [9], the Spanish research project in the context of the Information Society, particularly in the area of attention to diversity, and focusing on the development of technology to provide comprehensive training to people with special needs. These people have some limitations that affect their daily activities, making them less independent and leading them to a potential risk of social and labour exclusion. These objectives allow children with cognitive disabilities to manage technologies based on emerging interactive paradigms and create authoring contents for these environments. Section 2 shows some of the recent results of this project exploring user experiences on different interactive scenarios. Section 3 focuses specifically on creativity and exploring different ways of storytelling, while the main task for the development of a 3D virtual museum explained in Sect. 4.

## 2 Engaging User Experiences for Inclusive e-Training

In the context of the e-Integra project, several technologies and outcomes are evaluated to promote the learning engagement of user with cognitive disabilities o difficulties for social inclusion. For such purposes, different strategies are adopted, thus analyzing the feasibility of these interactive environments for such purposes. Some of the most relevant are shown below:

- Virtual Touch (a tangible user interface) is a toolkit to develop educational activities using tangible elements allowing the interconnection of a virtual world with the real world to provide comprehensive training [10]. The "mixed reality" experience was used for training on different school in Spain where invermsive virtual reality is combined with tangible User Interface to interact with games scenario for educational purposes (Fig. 1).
- DEDOS: an educational authoring tool for based on collaborative tasks on multi-touch surface system [11]. Problem solving activities are proposed to Down syndrome users, which they interact with digital objects and the rules to manage them (Fig. 2).
- Detection of emotions in written natural language (from different sources) to enhance motivation and identify possible stressing situations [12].
- Other experimental approaches using mobile devices as assistive technologies for people with cognitive disabilities to orient and perform their daily life activities, helping on orientation and sequencing activities [13].

The approaches of e-Integra project reveal a heterogeneous level of technology for learning and training activities oriented to users with similar difficulties in cognitive skills. These researches and outcomes show the development of smart technologies for an inclusive interaction level. However, a further step will be discussed in next section, the creation of digital content layer by storytelling.

**Fig. 1** Virtual Touch toolkit teaching sorting methods using physical dices

**Fig. 2** DEDOS, a multi touch surface for cooperative learning

## 3 From User Experience to Storytelling

The learning user experience for children with disabilities was the first approach of using smart technologies in the e-Integra project adopting multiuser virtual environment, mobile and multi touch technology for a smart and ubiquitous environment. The further step is to promote creativity by these new media for therapeutically and employment purposes. These media and interaction activities represent motivating tasks in schools. For instance, some researches explore the features of virtual reality environments to enhance stimuli and creativity to students [14]. In this case, we use narratives methods to create digital stories. Digital storytelling is narrative process that reaches its audience via digital technology and media usually with the goal of entertainment, and where interactivity plays a central role. The art of telling stories is also a useful instrument to engage user on immersive experiences [15]. User plays an active role and the interaction is the keyword to follow and continue the story. In practice, approaches for interactive storytelling are based on the following principles [16]:

- Story are malleable, it is not fixed in advance
- Nonlinear neither non chorological stories
- The user co-create the story
- It's an experienced actively
- Different outcomes are possible

Thus, the story experience can be told across multiple platforms and formats using digital technologies in a transmedia approach [17]. Transmedia storytelling is an emerging narrative method to tell stories across multiple media and platforms and across time. Each one to these media acts as a standalone story, but there are some interlaced points. Also, each different media tells the story all the best it can do. So therefore, any media is relevant, interconnected and gives some plot-points to change from one model to another. This is a pretty interesting metaphor to understand stories between multiple artifacts and technologies on day living.

But on the other hand, we achieve a new paradigm shift where new media interact in complex ways with old media and contexts. The convergence goes to the nexus of all media content where individual acts as consumers or producers, playing and taking part of a collective intelligence. Transmedia anatomy is based of the following pillars:

- The story, that is, something to tell (for educational purposes, hobbies, training, etc.) using characters, tales and objectives.
- The real world where things happens.
- The user participation, implication or interaction in the context of the story, and finally,
- Gaming view, defining the roles you may take part and goals to achieve and finally
- Dimension of the experience: individual or shared as social.

Therefore, we have to adapt and adopt the contents to the better strategies, skills and abilities of user. This is an ambitious approach to achieve relevant outcomes, but the process is interesting itself because we can identify motivations, hidden skills and also, social recognition to this citizen usually with low rate of employment of low-skilled jobs. To achieve these goal mush be accomplished in a stepwise approach:

- Focus on a digital technology and interaction paradigm. At first glance, we will focus on virtual reality environments.
- Focus on the user experience, that is, the user perception as they interact with the products or services [18] as a method to analyze the use of technology and its effectiveness.
- Focus on creativity and authoring for content creation, facilitating the use of skilled task to develop and create digital contents and annotations from personal experience.
- Enrich with annotations, collaboration and personal experiences, made with other kind of media or platforms, allowing more complex stories.

The first project using this approach is under development: *musegrades*, a transmedia experience of a virtual immersive museum. The aim is to develop a collaborative and immersive museum using virtual reality environment (Open Simulator [19]). The user can build their rooms for exhibit their art crafts, put their content and include comments and personal comments. Further goals are to promote collaborative activities (guided tours) and exporting to other media (cardboard, and mobile devices using alternative technologies) (Fig. 3).

There are several tasks to accomplish for user on the immersive environment:

- Walk through: discover the 3D immersive world, acquiring skills for 3D movement, orientation and object localization.
- Character building. Identify their avatar, and personalize according to their preferences,

**Fig. 3** *musegrades* project at OpenSim

- Building activities: create walls, upload media content and putting on the 3D environment.
- Communication activities. Manipulating chat and other interactive elements where they can interact with the 3D world.
- Collaborative activities, such as follow a group, meet someone, etc.

Each one of these goals represents different level of complexity. In these situations, we can identify two dimensions in the user experience: level of activity and socialization as shown in Fig. 4.

These levels and dimensions allow expert to evaluate different level of skills on user with cognitive disabilities, and promoting the engagement by means of active participation and social interactive experience. These aspects are key topics for creating artifacts and ICT-driven services for society, and in our case, for training user with cognitive disabilities and special needs for their social integration. Nowadays, we are collecting user and their digital content in a Spanish school for people with cognitive disabilities [20]. This school creates handcrafted material with regular collective exposition (Fig. 5).

This virtual museum is therefore a new model of creating user-generated digital contents (as shown in Fig. 6) an outcome: from traditional printed calendars towards an immersive multiuser experience, allowing other ways of socio-labor activities and funding. For instance, the virtual museum might be created by crowd-funding activities, or by other kind of social support for a sustainable model.

Future work will be oriented to recompilation of successful stories and connecting them and with other user (outside this school) and other media. Thus, we will evaluate possible incomes to create future skill for digital artist.

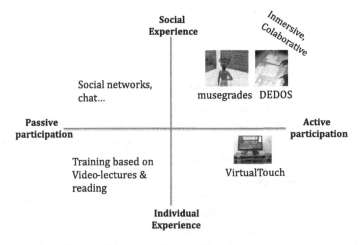

**Fig. 4** Dimensions of transmedia *musegrades* experience

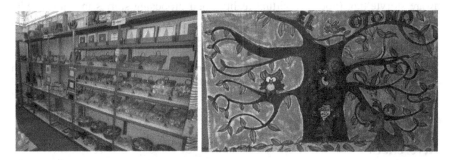

**Fig. 5** Handcrafted exhibitions at Asprogrades School

**Fig. 6** Creating different user-generated content models

# 4 Conclusions

This paper shows the relevance of user experience in the design of technology for smart cities. The heterogeneous interaction paradigms for smart environments gives us the opportunity to analyses a stepwise approach:

- Focusing on UX and ways to map user experience
- Using interactive storytelling to narrate each scenario
- Integrating these narrations on a transmedia approach with connections and relationships between these episodes.

This is an ongoing research, and nowadays we are selecting and training on Virtual reality immersive environments to different users with cognitive disabilities from Spanish school. We have authoring content and we are dealing with tools to create a 3D immersive virtual museum, with the aims to evaluate the progress of these users. Finally, this museum may be a useful tool to generate content and incomes (using virtual stores, ticket visit or any other kind of business).

In the future we plan to evaluate different authoring tools for context and social aware storytelling (not only virtual reality), and proposing stories in a transmedia framework (considering a pool of media technologies, goals and their relationships) for training and integrating people with special needs.

**Acknowledgments** The work described in this paper was partially funded by the Spanish National Plan of I + D+i (TIN2013-44586-R) Project: e-Training and e-Coaching for the social and laboral integration.

# References

1. Ylipulli, J.: A smart and ubiquitous urban future? contrasting large-scale agendas and street-level dreams. Observatorio (OBS*) J. Media City: Spectacular, Ordinary Contested Spaces 85–110 (2015)
2. Manville, C., Cochrane, G., et al.: Mapping Smart Cities in the EU, Directorate General For Internal Policies Policy Department A: Economic And Scientific Policy, IP/A/ITRE/ST/2013-02 (2014)
3. Milgram, P., Kishino, A.F.: Taxonomy of mixed reality visual displays. IEICE Trans. Inf. Syst. **12**, 1321–1329 (1994)
4. Brederode, B., Markopoulos, P., Gielen, M., Vermeeren, A., de Ridder, H.: pOwerball: the design of a novel mixed-reality game for children with mixed abilities. In: Proceedings of the 2005 Conference on Interaction Design and Children, IDC05, pp. 32–39. ACM, NY (2005)
5. Ke, F., Im, F.: Virtual-reality-based social interaction training for children with high-functioning autism. J. Educ. Res. **106**(6), 441–461 (2013)
6. Millner, C.H.: Digital Storytelling: A Creator's Guide to Interactive Storytelling. Focal Press, Elsevier (2008)
7. Vayanou, M., Katifori, A., Karvounis1, M., et al.: Authoring personalized interactive museum stories, interactive storytelling. In: 7th International Conference on Interactive Digital Storytelling, ICIDS, LNCS 8832, pp. 37–48 (2014)

8. Wong Lau, K., Yuen Lee, P.: The use of virtual reality for creating unusual environmental stimulation to motivate students to explore creative ideas. Interact. Learn. Environ. **23**(1), 3–18 (2015)

9. Alaman, X.: e-Training and e-Coaching for the social and laboral integration. Spanish National Plan of I + D+i (TIN2013-44586-R) Project (2013)

10. Mateu, J., Lasala, M.J., Alamán, X.: Developing mixed reality educational applications: the virtual touch toolkit. Sensors 2015 **15,** 21760–21784 (2015)

11. Roldán, D., Martín, E., Haya, P.A., García-Herranz, M.: Adaptive activities for inclusive learning using multitouch tabletops: an approach. Int. Workshop Personalization Approaches Learn. Environ. **732**, 42–47 (2011)

12. Rodriguez, P., Ortigosa, A., Carro, R.M.: Detecting and making use of emotions to enhance student motivation in e-learning environments. Int. J. Continuing Eng. Edu. Life-Long Learn. (IJCEELL), SI Adv. Adapt. e-Learn. via Interact. Collaborative Emotional Syst. **24**(2), 168–183 (2014)

13. Gomez, J., Montoro, G.: User study and integration of assistive technologies for people with cognitive disabilities in their daily life activities. J. Ambient Intell. Smart Environ. **7**(3), 389–390 (2015)

14. Wong Lau, K., Yuen Lee, P.: The use of virtual reality for creating unusual environmental stimulation to motivate students to explore creative ideas. Interact. Learn. Environ. **23**(1), 3–18 (2015)

15. Nakevska, M., van der Sanden, A., Funk, M., Hu, J., Rauterberg, M.: Interactive Storytelling in a Mixed Reality Environment: The Effects of Interactivity on User Experiences. ICEC 2014, LNCS 8770, pp. 52–59 (2014)

16. Millner, C.H.: Digital Storytelling: A Creator's Guide to Interactive Storytelling. Focal Press Elsevier (2008)

17. Jenkins, H.: Convergence Culture: Where Old and New Media Collide. NYU Press (2006)

18. Kuniavsky, M.: Smart Things: Ubiquitous computing User Experience Design. Morgan Kaufmann (2014)

19. OpenSimulator: http://www.opensimulator.org. Accessed 15 Jan 2016

20. Asprogrades: Asociacion a favor de las personas con discapacidad intelectual, http://www.asprogrades.org. Accessed 15 Jan 2016

# Application of Semantic Web Technologies to Facilitate Use of E-Learning System on Mobile Devices

Boban Vesin, Aleksandra Klašnja-Milićević and Mirjana Ivanović

**Abstract** Evaluation of mobile learning components is an important and popular research topic in the field of education. We have developed the modern e-learning system, named Protus 2.1. In order to personalize the learning process and adapt content to each learner, this system uses strategies that have the ability to meet the needs of learners and perform the adaptation of teaching materials. This paper presents the design and implementation of mobile version of Protus 2.1 which will allow its comfortable use for adapting the learning contents to the learners competences, to the learners context and to his/her mobile device.

**Keywords** Mobile learning · Semantic web · Ontology · Tutoring system

## 1 Introduction

The essential goals and functionalities of e-learning systems are to offer learners capability to access courses from various platforms, any time and anywhere [14]. Mobile learning is based on use of mobile devices in an educational process. Mobile technologies can provide a way to engage learners [10] and are recognized as an emerging technology to facilitate educational strategies that exploit individual learners' context [5].

B. Vesin (✉)
Department of Computer Science and Engineering,
University of Gothenburg/Chalmers University of Technology, Gothenburg, Sweden
e-mail: boban.vesin@cse.gu.se

A. Klašnja-Milićević · M. Ivanović
Faculty of Science Department of Mathematics and Informatics,
University of Novi Sad, Novi Sad, Serbia
e-mail: akm@dmi.uns.ac.rs

M. Ivanović
e-mail: mira@dmi.uns.ac.rs

© Springer International Publishing Switzerland 2016
V.L. Uskov et al. (eds.), *Smart Education and e-Learning 2016*,
Smart Innovation, Systems and Technologies 59,
DOI 10.1007/978-3-319-39690-3_42

473

Recent developments and technological achievements in communications and wireless networks have resulted in mobile devices (e.g., mobile phones, tablets) becoming widely available, more convenient, and less expensive [14]. However, mobile learning brings challenges in the adequate processing and delivering learning content [15]. Modern trends dictate the content is accessible and easily adaptable to different kinds of devices.

Semantic web presents a collaborative environment in which the information has a meaning understandable to machines [3]. To achieve this goal, addition of metadata (e.g., XML, RDF, Ontologies, or RSS) and adaptation rules [10] to existing e-learning systems are required.

In our previous research, we implemented tutoring system named Protus 2.1 (PROgramming TUtoring System) that is used and tested for programming topics. Protus is designed to recognize the behaviour patterns of learners and identify their learning styles, form clusters (categories) of similar learners, based on their learning styles and categorize teaching materials based on their rating and present recommended learning materials for learners.

Some parts of system are implemented using ontologies and adaptation rules for knowledge representation [13]. All Protus 2.1 components are implemented in the form of educational ontologies, while personalisation options are presented with SWRL (Semantic Web Rule Language) adaptation rules. The system is completely operative but only on desktop computers. In order to increase the usability of the system on mobile devices with smaller screens, it is necessary to perform dynamic adaptation of user interface to a lower resolution screens. This paper presents the implementation of mobile version of Protus 2.1 in order to provide same e-learning functionalities to mobile users. It is necessary to adapt the *Learner interface ontology* of Protus 2.1 with adding certain SWRL rules that would select the appropriate form of presentation for particular devices. Although this paper shows an application in programming tutoring system considered approach can be applied in a variety of other learning domains.

The rest of the paper is organized as follows. In the Sect. 2 appropriate related work is analysed. Section 3 describes components of our system, including modified ontologies and dynamic selection of presentation. User interface adapted to mobile devices is described in Sect. 4. Results of the Protus 2.1 testing are presented in Sect. 5. Section 6 concludes the paper.

## 2  Related Work

Nowadays learners can use smartphones, portable video consoles, and GPS navigators for learning [10]. Numerous research papers has focused on several broad areas of inquiries such as the effectiveness of mobile learning and the development of systems for mobile learning [4]. Different institutions have been concentrating their research efforts on mobile learning and have proposed various content processing techniques. Some of the techniques focus on the design of adaptive and personalized learning systems. Learners in those systems [2, 5, 8, 15] are provided with adaptive and personalized learning experiences that are tailored to their particular educational needs and mobile devices.

Semantic applications are based on the idea of systems being able to extract meaning from information on the web and to provide personalized services and information according to user needs [10]. Authors in [12] introduced a systematic approach to service personalisation for mobile learners in pervasive environments. They presented service-oriented framework that integrates semantic technologies for learner modelling and personalised reasoning. The use of Semantic web technologies and in particular ontologies for development of mobile application for preschool cognitive skills learning is presented in [1]. CogSkills mobile application is developed to evaluate specific knowledge model.

The research purpose of majority of mobile learning studies focuses on effectiveness and mobile learning system design [14]. Surveys and experiments were used as the primary research methods. The effects of mobile learning on learners achievements and attitude were presented in [10]. The study revealed that mobile learning keeps the learners engaged, and it is possible to deliver learning that is authentic and informal via the mobile learning technologies [10].

Mobile phones and tablets are currently the most widely used devices for mobile learning [14]. Therefore it is important to adapt e-learning systems to variety of technologies. Not many e-learning systems exist that are adapted for wide range of devices with different screen sizes. Mainly, those systems are adapted specifically for mobile devices or only for desktop computers.

Although there are many tools for mobile learning of programming languages, Protus 2.1 stands out by using personalization for adapting content to specific users [7]. Our goal is to enable the use of the Protus 2.1 both through mobile application and via web browser, using a single user account.

In this paper, we will present the possibilities for adaptation of an existing tutoring system to cope with diversity of platforms on which it runs. System will also provide functionalities for testing of acquired knowledge over mobile devices unlike many of those systems [4, 7, 10, 14].

The main contributions of the paper are (1) the learning approach adopted to take advantages of the learners context (mobility and specific needs) (2) the flexible architecture of Intelligent learning system and its modular and fine grained components (3) the mechanism provided for handling the learners context and the model on which it is based and (4) the semantic and ontological descriptions of the learners mobile context, the learning approach as well as the components and services provided by the system.

# 3   Protus 2.1

Protus 2.1 is a tutoring, interactive system that allows learners to use teaching material and test acquired knowledge for introductory Java programming course [13]. Learner's interface of Protus 2.1 is a series of web pages that provide: taking lessons and testing learner's knowledge. All data about a learner and his progress in the course, as well as data about tutorials, tests and examples are stored on the system's server. Learners attend courses through the web interface [9] that:

- review of the offered courses and teaching materials,
- various display of teaching materials adapted to learning styles,
- testing of acquired knowledge,
- communication with the mentor and other learners,
- reports about progress, test results, coursework and their own learning styles.

Due to the highly fragmented mobile technology trends and rapidly evolving standards, there is no single solution to make existing educational content working for every possible mobile or desktop device. Educators are forced to design new learning content or reformat existing learning materials for delivery on different types of mobile devices [15].

Tutorials and tests in Protus 2.1 are structured in html, therefore it is suitable for presenting on different devices. In order to adapt Protus 2.1 itself to different devices, its user interface should be flexible and scalable, based on semantic web and appropriate rules.

## 3.1   System Architecture of Protus 2.1

From a learner's perspective, learning content is always the key element in education delivery, not the mobile technology itself. Implemented architecture of Protus 2.1 improves the ontology utilization, where the representation of each component is made by a specific ontology (Fig. 1): *Domain ontology*, *Learner model ontology*, *Task ontology*, *Teaching strategy ontology* and *Learner interface ontology*. Various adaptation conditions in Protus 2.1 were captured in the body of SWRL rules. As a result of the execution of rules, recommendations in the form of various content presentations are generated [6].

Protus 2.1 has achieved a remarkable impact on learners self-learning [9]: learners have gained more knowledge in less time. The next step in development of the system is its adaptation to mobile devices, i.e. to provide more possibilities to its users, because learners can learn more conveniently if they are not limited by specification of hardware used in learning process.

Current form of the Protus 2.1 components allow easy maintenance of the system. Changes will have to be made in *Learner interface ontology* and to SWRL adaptation rules for selection of presentation templates. User interface must be designed to adapt the application to the screen resolution of mobile devices. Current implementation is undertaken in four stages:

- design of the learner interface layouts for different screen resolutions,
- adaptation of *Learner interface ontology* to precisely designed layouts,
- design of the required SWRL rules that selects presentation layout and
- testing on different devices.

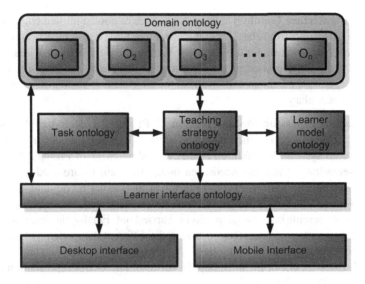

**Fig. 1** Protus 2.1 Semantic web architecture

**Learner Interface Layouts**. In order to provide adapted user interface of the existing tutoring system for mobile devices, appropriate changes must be incorporate
only to the *Learner interface ontology*. Therefore, the advantage of this approach
is that learners will basically use the same system over different devices, but with
appropriately adapted user interface (Fig. 2).

In case of higher screen resolution of device, the first layout (that simultaneously
presents all page segments) will be presented. When requests arrive from devices
with smaller resolution screen, only chosen page segments will be presented.

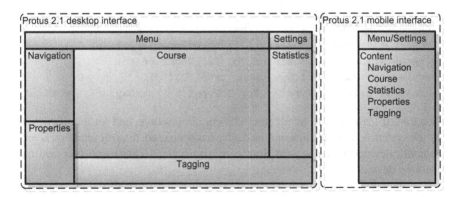

**Fig. 2** Different user interface layouts in Protus 2.1

**Learner Interface Ontology** defines final stage of course implementation. System reads a decision from the *Teaching strategy ontology*, and based on that decision it creates presentation for the particular learner and device. System generates an interface for the learner based on the chosen layout. *Learner interface ontology* will be used to specify the content and layout of the pages and to standardize the content and query vocabulary.

**Adaptation Rules for Mobile Version of Protus 2.1**. Rule-based reasoning enables a more expressive method for inference of processes and interaction in Semantic web system [12]. While *Learner model ontology* of Protus 2.1 provides the representations of learners needs, reasoning mechanisms are used to infer presentation and services of the system in terms of the learner model and application context. These mechanisms are part of the *Task ontology* and the *Teaching strategy ontology* that presents how the adaptation is carried out. Finally, the *Interface ontology* presents result of communication among the different components of the Protus 2.1 architecture.

SWRL rules are one of the most popular forms of knowledge representation [11]. Rules used in Protus 2.1 are:

- *Learner modelling rules* that add knowledge about a learner, inferring new learner features from other already existing features of that learner. They are necessary for the building and updating of the of *Learner model*.
- *Adaptation rules* that define strategies of adaptation, taking into account domain features, system adaptation goals, user features, context and used presentation methods. Rules that will alter presentation layouts of Protus 2.1 fall in this category.

At the presentation level, Protus 2.1 requests information about the current device and current course level to make decision which element will be presented on the screen and in which form. Adaptation rule for selecting appropriate presentation layout is:

```
Learner(?x)^Interaction(?y)^hasInteraction(?x,?y)^
isPartOf(?r,?y)^Resource(?r)^isRecommended(?r, true) ^
Presentation (?p)^consistOf(?p, ?r)^Decision(?d) ^
implements(?d,?p)^ofType(?mob, true)^UI(?u)^
creates(?p,?u)^hasType(?u,?t)^UIType(?t)->hasLayout(?t,mobile)
```

Therefore, if learner x has interaction with certain resource, and for that particular session system determined that mobile device has been used, than presentation layout for mobile devices should be used. Similar rule is triggered when requests come from a desktop device.

Exactly one of these rules is triggered every time learner sends request to server, and based on the response, appropriate user interface will be generated.

# 4  User Interface for Mobile Devices

Proposed modified ontology architecture will provide comfortable use of Protus 2.1 on different devices. The user interface is generated based on the ontology data and SWRL adaptation rules. Protus 2.1 functionalities and pages viewed from a mobile device are: for presenting lectures Fig. 3, an overview of the courses—Fig. 4, page with personal course statistics for a learner Fig. 5.

Protus 2.1 application dynamically selects the layout based on the dimensions of the used screen and selected teaching material. This ensures that a student is clearly presented with the educational material and navigation options in every moment. We tested correctness of display on different platforms and ensured that education material is displayed as on desktop computers.

**Fig. 3** *Course* Page

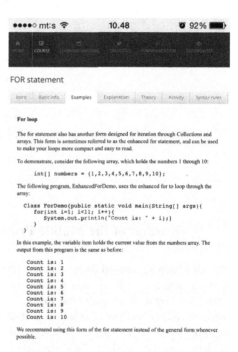

**Fig. 4** *List of Courses* Page

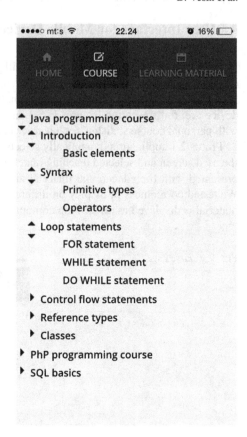

## 5 Application of the Mobile Prototype of Protus 2.1

We have been performed an experiment of using Protus 2.1 for the delivery of adaptive learning activities via mobile devices. We evaluated the responsiveness of the application by its deploying onto various mobile devices with different screen sizes using various networks with a group of 26 students. All students have previously used the same application on desktop computers. The intention was to investigate students opinions on the mobile version of the system.

Students explored Protus 2.1 independently but they followed the learning path proposed by the teacher, in order to test all pages and functionalities. Students used devices they owned, on different platforms (Apple iOS, Google Android and Microsoft's Windows), with different screen sizes (from 3.5 in to 7.9 in) using various connectivity options (3 G, 4 G, EDGE, Wi-Fi). Student were expected to occasionally change the method of access (via different networks).

They were stimulated to test the Protus 2.1 over mobile device and reflect on their impressions related to its use, speed of execution and responsiveness of the pages they visited and discover possible problems. After that, a brief survey has

**Fig. 5** *Statistics* Page

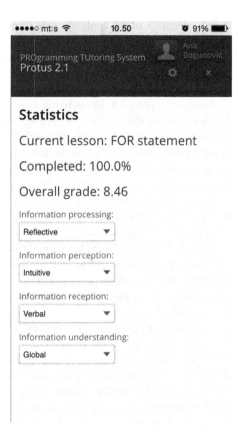

been conducted. Students are expected to provide estimates of the extent to which they agree with the statements in the questionnaire (using 1 to 5 scale, 1—do not agree, 5—fully agree) (Table 1).

The test results showed that mobile devices have not decreased the overall functionality. Students highly assesses the user interface display on mobile devices. Most of the students noticed the slightly reduced visibility of educational material but all user interface components were displayed properly. They also noted significantly

**Table 1** Questionnaire details

| No | Assertion | Overall grade |
|----|-----------|---------------|
| 1 | User interface elements are properly displayed | 4.1 |
| 2 | Educational material is clearly presented | 3.2 |
| 3 | Execution is slower on mob. devices compared to computers | 3.9 |
| 4 | The application is responsive | 4.3 |
| 5 | Overall satisfaction with the use of the app. over mobile device | 4.2 |

slower execution of the application, but found useful ability to access the system on a variety of devices.

High final score of 3.9, for overall impression on the use of the application over mobile devices, testified that despite some problems (reduced visibility and execution speed) students enjoyed the mobile learning environment.

The second phase of the study was conducted with 51 students from two courses. The goal of this phase was to let the students experience of our mobile e-learning environment but instead of just giving their feedback on our application, we wanted them to give us their perceptions on the potential role of mobile learning in programming area. The students used this system for three weeks to access and discuss the class materials. This was followed by a survey using the same 5-point Likert scale used in the first phase. However, the survey questions were modified to emphasize not just on our implementation of the system but on the use of mobile applications, in general, for programming learning. The results of the second survey are shown in Table 2. The results from the second phase show that the students expect mobile learning system as an effective learning tool or service, making flexible access from anywhere and suitable to use application. Students also perceive an important supplementary role for mobile devices in e-learning and are effective in delivering personalized content. The survey results were also grouped by student agreement and disagreement on the ten questions (see Table 2) of our survey. As shown in Fig. 6 there was strong agreement or agreement on all the ten questions. Students in general support the use of mobile devices in learning and foresee a strong role for these devices in improving flexibility and efficiency of the learning environment. The qualitative comments from the students support the quantitative results. Students liked the convenience, ease-of-use, ability to be reminded, and the mobility factor which allowed them to utilize any dead-time for productive learning activity. But they also disliked the small screen-size, tedious process of typing on phone keypads, and slow connection speeds, response times, lack of pictures and visual stimulation.

**Table 2** Results from 2nd survey of 51 students

| No | Assertion | Overall grade |
|----|-----------|---------------|
| 1 | Protus adds value to e-learning | 4.74 |
| 2 | Protus allows instant access regardless of your location | 4.3 |
| 3 | Protus is useful to supplement to an existing course | 3.61 |
| 4 | Protus is an effective learning aid or assistant for students | 4.23 |
| 5 | Protus is an effective method of providing personalisation | 4.75 |
| 6 | Protus allows to convert any wait time into productive | 3.89 |
| 7 | Protus convenient access to discussions anywhere and anytime | 4.44 |
| 8 | Protus that sends the information via messages may be better | 3.50 |
| 9 | Protus also allows access to information from the website | 4.15 |
| 10 | Protus can be used as a supplemental tool for any existing course | 4.75 |

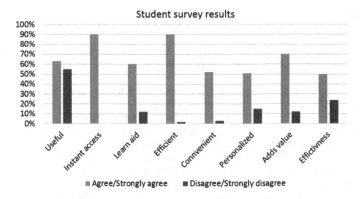

**Fig. 6** Student survey agreement/disagreement analysis

The evaluation study was successful because it gave us some feedback on what the students think about our e-learning system, helped us determine whether the students find the flexibility of mobile devices useful for learning and their opinion on the role of m-learning in education. Our study confirms the findings of previous research discussed earlier. Mobile device usage is bound to increase in the future and they will have a significant impact on the quality of student learning.

# 6 Conclusion

Mobile devices as an emerging technology possess the huge potential to support strategies that exploit individual learners context. The widespread use of mobile technology, supported with the availability of efficient mobile broadband connections, offers a unique opportunity to develop innovative e-learning applications.

Variety of learning platforms, standardisation of content, use of Semantic web technologies, enables semantic search engines to effectively retrieve and present educational material and provide advanced services to learners.

This paper presented the implementation of mobile version of Protus 2.1 system in order to provide its functionalities to mobile users. The framework, in conjunction with the Semantic web technologies, will offers new opportunities for mobile learning, not limited to different hardware specifications, but enhanced with portability.

Given SWRL examples are useful but also complicated, as they require a fairly complex mapping of learning resources to presentational and layout metadata. They place a burden on the authors of materials. For the future work, we plan to simplify whole semantic structure and to direct an attention on measuring the influence of specific classes on performance of the application.

We hope that this research will reveal that computer based instruction can still be effective in situations where the novelty of mobile technologies distracts the learners from the task.

# References

1. Abbas, M.A., Ahmad, W.F.W., Kalid, K.S.: OntoCog: a knowledge based approach for preschool cognitive skills learning application. Procedia—Soc. Behav. Sci. **129**, 460–468 (2014)
2. Coccoli, M., Torre, I.: Interacting with annotated objects in a Semantic Web of Things application. J. Vis. Lang. Comput. **25**(6), 1012–1020 (2014)
3. Devedzic, V.: Semantic Web and Education. Springer, Berlin (2006)
4. Fulantelli, G., Taibi, D., Arrigo, M.A.: framework to support educational decision making in mobile learning. Comput. Hum. Behav. **47**, 50–59 (2015)
5. Gómez, S., Zervas, P., Sampson, D.G., Fabregat, R.: Context-aware adaptive and personalized mobile learning delivery supported by UoLmP. J. King Saud Univ.—Comput. Inf. Sci. **26**, 47–61 (2014)
6. Ivanović, M., Mitrović, D., Budimac, Z., Vesin, B., Jerinić, L.: Different roles of agents in personalized programming learning environment. In: Lecture Notes in Computer Science (including subseries Lecture Notes in Artificial Intelligence and Lecture Notes in Bioinformatics), pp. 161–170. Springer (2014)
7. Jacob, S.M., Issac, B.: The mobile devices and its mobile learning usage analysis. In: Proceedings of the International MultiConference of Engineers and Computer Scientists 1 Hong Kong (2008)
8. Kasaki, N., Kurabayashi, S., Kiyoki, Y.A.: Geo-location context-aware mobile learning system with adaptive correlation computing methods. Procedia Comput. Sci. 593–600 (2012)
9. Klašnja-Milićević, A., Vesin, B., Ivanović, M., Budimac, Z.: E-Learning personalization based on hybrid recommendation strategy and learning style identification. Comput. Educ. **56**(3), 885–899 (2011)
10. Martin, F., Ertzberger, J.: Here and now mobile learning: an experimental study on the use of mobile technology. Comput. Educ. **68**, 76–85 (2013)
11. Romero, C., Ventura, S., Hervas, C., Gonzalez, P.: Rule mining with GBGP to improve web-based adaptive educational systems. In: Data Mining in E-Learning, pp. 171–188 (2006)
12. Skillen, K.L., Chen, L., Nugent, C.D., Donnelly, M.P., Burns, W., Solheim, I.: Ontological user modelling and semantic rule-based reasoning for personalisation of Help-On-Demand services in pervasive environments. Future Gener. Comput. Syst. **34**, 97–109 (2014)
13. Vesin, B., Ivanović, M., Klašnja-Miliević, A., Budimac, Z.: Protus 2.0: ontology-based semantic recommendation in programming tutoring system. Expert Syst. Appl. **39**(15), 12229–12246 (2012)
14. Wu, W.H., Jim Wu, Y.C., Chen, C.Y., Kao, H.Y., Lin, C.H., Huang, S.H.: Review of trends from mobile learning studies: a meta-analysis. Comput. Educ. **59**, 817–827 (2012)
15. Yang, G., Chen, N.-S., Sutinen, E., Anderson, T., Wen, D.: The effectiveness of automatic text summarization in mobile learning contexts. Comput. Educ. **68**, 233–243 (2013)

# Applications of Innovative "Active Learning" Strategy in "Control Systems" Curriculum

**Dmitry Bazylev, Alexander Shchukin, Alexey Margun, Konstantin Zimenko, Artem Kremlev and Anton Titov**

**Abstract** The paper describes the learning course "The integrated design and control systems" of "System analysis and control" master program at ITMO University with using of active learning methods. The key feature of the course is application of active learning methods on lectures and seminars, namely "Problem-based lectures", "Lectures with planned errors" and "Press conference lectures". In order to implement a practical knowledge background of students the laboratory bench ASUD-2 was developed. Results of the course implementation have shown a high level of students training and improvement of their professional skills.

**Keywords** Active learning approach · Master student · Control system · Laboratory bench

## 1 Introduction

Active learning methods are given much attention in last years, for example [1–3]. These methods are such organization and management of the learning process, which are aimed at all-round intensification of teaching and learning activities of students

D. Bazylev (✉) · A. Shchukin · A. Margun · K. Zimenko · A. Kremlev · A. Titov
ITMO University, Kronverkskiy av. 49, Saint Petersburg 197101, Russia
e-mail: bazylevd@mail.ru

A. Shchukin
e-mail: shhukinaleksandr1990@yandex.ru

A. Margun
e-mail: alexeimargun@gmail.com

K. Zimenko
e-mail: kostyazimenko@gmail.com

A. Kremlev
e-mail: kremlev_artem@mail.ru

A. Titov
e-mail: antony.titov@gmail.com

© Springer International Publishing Switzerland 2016
V.L. Uskov et al. (eds.), *Smart Education and e-Learning 2016*,
Smart Innovation, Systems and Technologies 59,
DOI 10.1007/978-3-319-39690-3_43

485

through a wide, preferably complex, using of pedagogical (teaching), organizational and management tools [4]. Practically speaking, active learning is defined as some method of teaching, that involves students to the learning process, provides to make learning activities meaningful and comprehended. A lot of works are devoted to the investigation of different active learning methods (see [5–8]). Papers [8] and [14] describe a comprehensive examination of the cognitive, motivational and emotional processes underlying active learning approaches; their effects on learning and transfer; the core training design elements (exploration, training frame, emotion control) and individual differences (cognitive ability, trait goal orientation, trait anxiety) that shape these processes.

Numerous experiments show that active learning in education improves the understanding of the subjects, improves thinking, practical skills and other cognitive functions [4–12]. In [10] data from over 6000 physics students in 62 introductory physics courses was reviewed. It was founded that students in classes that utilized active learning and interactive engagement techniques achieving an average gain of 48 % on a standard test of physics conceptual knowledge. In [11] a curriculum for introductory mechanics that emphasizes interactive engagement and conceptual understanding using the studio format is described. It is shown that the curriculum much improved the students' conceptual understanding compared to the traditional course without significantly affecting the scores on a traditional final exam.

Recently, a new learning method in studying the course "Integrated Systems Design and Control" for the masters program "System Analysis and Control" was proposed in [13]. See [6, 9], where the relevant topic is discussed. A key feature of the method in [13] is in integrated application of classical learning methods and techniques of active learning, that increases the effectiveness of the education program and strengthens the student's interest (motivation) in training. Designed course is supported by practical activities with actual problems, arising from the current challenges in the field of industrial electronics and telecommunication systems. Therefore, the implementation of the active learning method, introduced in [13], reduces the gap between theoretical knowledge and practical application.

As an object for practical and experimental research, the team of Control Systems and Informatics Department in ITMO University developed a laboratory bench "ASUD - 1". This bench is designed for control home automation and based on KNX/EIB technology with usage of web-interface.

The theoretical material of the course is presented during classical lectures, combined with exercises for every student (problem-based lectures). On the practical lessons students receive a complex task related to designing, configuration and assembling of various home automation control systems. Developed training programs not only effectively distribute the material during the school hours, but also provide a high urgency of taught knowledge.

The performance of the learning method was analysed at the end of the course through the round table discussions and practical tests. It was determined that the students did not have any problems in reading (understanding) and designing of electrical systems and circuit diagrams. However, they experienced difficulties in setting interactions between devices. Namely, the lack of programming skills was identified. For this reason the department staff decided to upgrade the laboratory bench "ASUD - 1" and modernize the training program to face new challenges.

## 2  Laboratory Bench "ASUD - 2"

Laboratory Bench "ASUD - 2" as well as its first version "ASUD - 1" is based on the widespread and open technology KNX/EIB. This standardized network communications protocol is compatible with a large number components produced by different manufacturers, has high functionality and integration with other technologies, and provides the ability to use both wired and wireless technology.

Laboratory bench "ASUD - 2" is a modular system, which consists of two main parts. The first part is shown in Fig. 1. It provides operation of the basic systems of "smart home" and comprises the following units:

- lighting control modules (four channel dimmer of incandescent bulbs, control gateway of lighting system DALI, dimming and light scenes sequencer);
- analog devices (two 8-channel input/output devices for connecting sensors, controllers 0–10 V to control blinds and curtains);
- various sensors installed on the bench panel and connected via plug (temperature and humidity sensors, motion sensor, leak detector, etc.);
- other applications (IP-camera, power consumption counter, ball valve for water supply system and weather station).

A wide range of different devices and sensors provides a variety of designs of smart home system with specific features. Modular structure of the bench increases the mobility of its installation. Therefore, it can be used not only in a special laboratory, but also in any other classroom.

**Fig. 1**  The first part of the laboratory bench "ASUD - 2"

**Fig. 2** The second part of
the laboratory bench
"ASUD - 2"

The interaction between the devices is performed by means of the script programming language LUA. To make constructed technical system friendly for the user or operator, students also develop the user interface, that presents the second module of the laboratory bench. The user interface that comprises interactive room map, KNX switches and touch panels performs a variety of control methods, which depend on the needs and desires of the customer. The second module of the bench illustrated in Fig. 2 includes the scheme of an apartment, KNX switches and touch control panels. The scheme of an apartment (or building), which serves to indicate the events taking place in it, is located in the upper part of the module. This scheme is similar to the one that is mainly used in the dispatch consoles. In the middle part of the plant there are different KNX switches used to change the state of devices (for example, the value of brightness), or switching between predefined scenarios of operation. Particularly, lightning sensors are usually installed at the entrance of the premises.

Touch control panels are mounted at the bottom of the module. Such panels are the most popular control devices of smart home systems. They can be embedded into a wall or be portable providing control of building (apartment) in any part of the planet (if there is an access to the internet). For instance, Apple IPAD mini tablet with preinstalled IRIDIUM MOBILE application can be applied as a turnkey solution. The other common way is to use a standard browser with an ad hoc WEB-interface shown in Fig. 3.

Development of the user interface is an another one fascinating problem, which requires the knowledge of programming basics and customer requirements. According to the training course students design their own unique user interfaces depending on a set of used devices. The main criteria are operability, functionality and efficiency of the entire technical system.

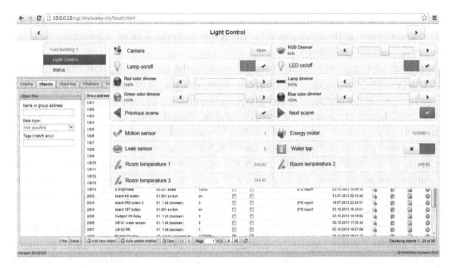

**Fig. 3** User WEB-interface used to control the smart home system

## 3 Training Course Description

Along with modification of the laboratory bench the course material and training program was upgraded. The previous version of the course, presented in [13], was mainly focused on electrical engineering. The new program pays equal attention to the circuit design and computer science. Now, the first phase of training includes three laboratory works devoted to three main areas of the first part of the bench. The first task is in development of different scenarios for the lighting in the room. For example, students can change the color or brightness of the LED strip depending on the temperature of the apartment. The necessary practical knowledge of engineering system control, namely, for water supply and air conditioning, are presented in the second laboratory work. The third practical task imparts skills of working with sensors and other peripheral devices. A typical example is a development of a system that notifies the user if there is a presence of movement in the building. The operation algorithm can be the following: when a motion is detected and it continues for more than 5 seconds, then IP-camera and a light indication turn on. Such problem statement requires the use of the motion sensor and IP-camera, but students can use, for instance, incandescent lamp or color LED indication strip for a light indication - both of the selected decisions are correct.

The laboratory works can be performed by students individually or in small groups. The initial data and problem formulation are provided in the form of a technical specification that is typical for modern engineering industry. As in the previous version of the teaching program students have freedom to choose the final device for project implementation if it is not specifically established in the specification.

Passing the first stage of the training course, students receive the necessary practical knowledge on the interaction between the main devices of automation systems of "smart home" (building).

The second stage is a creative work based on the second supplementary module of the bench: creation of a custom user interface for three laboratory works, executed earlier at the previous stage. Students are also given the choice of components and tools, herein, rationale and justification of selection are estimated as well. For example, if the camera is not used and there is no need to display any type of information, control can be realized on the use of KNX switches jointly with the indication scheme of the apartment, installed in the upper part of the second module. Such solution can be the most optimal, since it meets the requirements of the customer, provides reliability and simplicity of practical realization and significantly reduces production costs.

After completing laboratory and creative tasks students prepare technical reports. These reports contain results concerning to entire project implementation: problem statement, justification of element choice, calculations, functional schemes of control systems, electrical schematic diagrams, spreadsheets of group addresses of program ETS (special program for KNX), as well as a listing of scripts.

Evaluation of student work is carried out on a 100-point scale via distance learning center (DLC). With the help of this system master students can monitor their marks and, therefore, take timely action to correct their subject status. The assessment is based on the correct implementation, deadlines (timely preparation of the work) and accuracy of the report. In Table 1 forms of control and maximum scores for their implementation are presented. The maximum number of points for implementation of the laboratory works is 10 points for each lab. Passing the second stage of the course and implementing creative works students can score maximum 15 points for the programming block and the same number for the user interface design. Activeness of students in the classroom and timely execution are evaluated separately – a student can get up to 10 points for active participation and visiting workshops as well as 10 points. At the end of the course student is allowed to pass the exam if all the laboratory works are executed.

**Table 1** Evaluation of educational achievements of master students

| # | Forms of control | Maximum score |
|---|---|---|
| 1 | 1st task | 10 |
| 2 | 2nd task | 10 |
| 3 | 3rd task | 10 |
| 4 | Activeness during labs | 5 |
| 5 | Programming tasks | 15 |
| 5 | User interface design | 15 |
| 6 | Activeness during creative works | 5 |
| 7 | Timely execution | 10 |
| 8 | Exam | 20 |
| | Total | 100 |

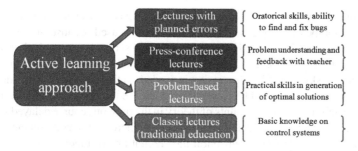

**Fig. 4** Structure of the proposed active learning method

The plan of the training program was upgraded according to the new theoretical and practical material. Several minor shortcomings were uncovered in the previous curriculum, proposed in [13]. In order to receive feedback previous course was also evaluated via students' questionnaires and their marks dynamics. Acquired information was analyzed and detected defects were taken into account during the new course development.

Two tutorials were prepared and issued by the team of Control Systems and Informatics Department. The first one is devoted to the theory foundations of home (building) automation, KNX technology and the basics of working in the specialized program ETS. The second tutorial includes examples of work with the scripting language LUA and laboratory works description.

The key feature of the new course is an application of active learning methods at lectures and seminars. The updated program of training comprises "Problem-based lectures", "Lectures with planned errors" and "Press conference lectures" (see Fig. 4). These new classes are mixed with traditional education (classic seminars and lectures).

## 3.1 Problem-Based Lectures

The problem-based lectures are devoted to control systems and include problem statements, prerequisites and necessary theoretical material. Students try to find optimal solutions to execute assigned tasks. A teacher introduces short theory parts to them, closely monitors and helps students during the execution of task, pushes them to the right direction of thoughts and new ideas. Through these activities the student understands how presented theoretical material is used to solve certain control problems for real control systems. For example, it is necessary to select temperature sensor when performing a system of smart home conditioning control tasks. A variety of types, design performances of sensors, different price categories and, accordingly, the accuracy characteristics constitute the task of multiparametric optimization of sensors selection. Typically, the students offer a simple method of measuring temperature using a resistive sensor. The internal resistance changes its value in the relation of temperature change. This method is often used because of the low price and simple electronic design. However, this type of sensor has a relatively low accuracy and requires a calibration. Therefore, this technology may be not applicable for a number of tasks, where high precision is needed. In this case, the teacher describes other possible solutions and helps students to choose the optimal component base.

Representation of large amounts of theory in the previous course delayed the solution of the problem or made it trivial. In order to increase the comprehensibility of knowledge, it was decided to abandon the "recitation" of theoretical material during problem lectures and make it completely practical. Therefore, students on this lessons fully focus on the task. Necessary theoretical base is given on the classic lectures.

## 3.2 Lectures with Planned Errors

In the beginning of lectures with planned errors teacher declares that there are various mistakes (for example, in some technical solutions) and students search for these errors during the sessions. Such lessons develop the ability of students to analyze information quickly and express their views in the presence of colleagues.

According to the results presented in [13] conducting of lectures with the planned errors is also changed: preparation of a multi-level, complex error requires too much time-consuming for teachers, but finding these errors are not related to the practical work on discipline. It was decided to consider the cases of the actual practice of building automation on lectures with the planned errors. We consider the implementation of a smart house complex and its individual parts, which caused a problem in the system. It can be various errors, which could be approved during the design phase. One of the typical task is in wiring diagram analysis: to find and fix the wrong connections between the smart home components. The other problems contain errors in the software (incorrect program algorithm or designation of variables). Such lessons show the vulnerabilities of control systems and show development aspects that require maximum concentration of the designer.

## 3.3 Press-Conference Lectures

Before the beginning of the press-conference lectures teacher asks students to write their questions. Next, the teacher organizes the questions and conducting a lecture via one of two possible scenarios. The first one is consistently answering on asked questions in value chain on specific topics. The second scenario is a classic press-conference, during which the teacher takes questions from students in turn and answers them. Conducting of such activities causes students to formulate their questions correctly and accurately. Press-conference lectures are carried out at the end of the block or course of lectures and take the form of a round table, students prepare questions in advance. Elements of an interactive method of teaching and conducting business games as workshops are remained unchanged: students are divided into groups and distribute the duties themselves. The group completing task (case) faster than others receives more points.

Press conference lectures are particularly useful when students are faced with technical problems in the process of laboratory works. For example,students successfully select component base and structure of the control system but they have problems with the plotting of humidity and temperature graphs. Control system should save sensor data during the day in the database and demonstrate to the user dynamics of change in a convenient way. Herein, two possible problems arise. The first one is data overflow. For instance, it can occur when the data volume from the temperature and humidity sensors exceeds maximum value of data type. The second problem is the wrong plotting of the graph with data from sensors. For clarity, introduce one of the possible answers of the teacher: "Sensor Readings are displayed in real time on the user interface, and data changes are taken every 30 min for plotting. Thus, the data remain valid at any time, and the graph is constructed with an acceptable margin of error".

## 4  Conclusion

Permanent adaptation of the education program is important not only for ensuring the relevance of knowledge, gaining experience and development of specific competencies of the student, but also to increase the availability and understanding of the material. This paper describes the learning course on the discipline "The integrated design and control systems" of "System analysis and control" masters program with using of active learning methods. In order to implement the program the laboratory bench ASUD-2 for practical works is developed. New education program is formed taking into account questionnaires proposed to the previous students for evaluation of education course and detection of its strengths and weaknesses. New education program and bench pay special attention to the programming of interaction between the various devices and systems of "smart home" and the development of the user interface. In each academic semester developed learning course is evaluated via students' questionnaires and their marks dynamics for future adaptation to the modern technical level and labor market. The results of the new course implementation demonstrate strengthening of the student motivation and improvement of their professional skills in comparison with the previous training program.

The future work is devoted to modification of problem-based lectures. In order to increase the efficiency of training we plan to involve professionals from industry and arrange factory tours for students. The second direction of research is in extension of the developed course to the other related subjects, namely, electromechanical systems and robotics.

**Acknowledgments** This work was partially financially supported by Government of Russian Federation, Grant 074-U01. This work was supported by the Ministry of Education and Science of Russian Federation (Project 14.Z50.31.0031).

# References

1. Feiman-Nemser, Sh: From preparation to practice: designing a continuum to strengthen and sustain teaching. Teach. Coll. Rec. **103**, 1013–1055 (2001)
2. Smith, K.A., Sheppard, Sh.D., Jonson D.W., Jonson, R.T.: Pedagogies of engagement: classroom-based practices. J. Eng. Educ. **94**(1), pp. 87–101 (2005)
3. Prince, M.: Does active learning work? A review of the research. J. Eng. Educ. **93**(3), pp. 223–231 (2004)
4. Bonwell, C., Eison, J.: Active Learning: Creating Excitement in the Classroom. ASHE-ERIC Higher Education Reports, p. 121 (1991)
5. Hoic-Bozic, N., Mornar, V., Boticki, I.: A blended learning approach to course design and implementation. IEEE Trans. Educ. **52**(1), pp. 19–30 (2009)
6. Bazylev, D., Margun, A., Zimenko, K., Kremlev, A., Rukujzha, E.: Participation in robotics competition as motivation for learning. Proc. Soc. Behav. Sci. **152**, 835–840 (2014)
7. Huntzinger, D.N., Hutchins, M.J., Gierke, J.S., Sutherland, J.W.: Enabling sustainable thinking in undergraduate engineering education. Int. J. Eng. Educ. **23**(2), pp. 218–230 (2007)
8. Holbert, K.E., Karady, G.G.: Strategies, challenges and prospects for active learning in the computer-based classroom. IEEE Trans. Educ. **52**(1), pp. 31–38 (2009)
9. Zimenko, K., Bazylev, D., Margun, A., Kremlev, A.: Document Application of innovative mechatronic systems in automation and robotics learning. In: Proceedings of the 16th International Conference on Mechatronics, Mechatronika, pp. 437–441 (2014)
10. Hake, R.R.: Interactive-engagement versus traditional methods: a six-thousand-student survey of mechanics test data for introductory physics courses. Am. J. Phys, **66**(1), pp. 64–74 (1998)
11. Hoellwarth, C., Moelter, M.J.: The implications of a robust curriculum in introductory mechanics. Am. J. Phys. **79**(5), pp. 540–545 (2011)
12. Michael, J.: Where's the evidence that active learning works? Adv. Physiol. Educ. **30**(4), pp. 159–167 (2006)
13. Bazylev D., Margun, A., Zimenko, K., Shchukin, A., Kremlev, A.: Active learning method in System Analysis and Control area. In: Proceedings—Frontiers in Education Conference, FIE, pp. 1–5 (2015)
14. Bell, S.S., Kozlowski, S.W.J.: Active learning: effects of core training design elements on self-regulatory processes, learning, and adaptability. J. Appl. Psychol. **93**(2), pp. 296–316 (2008)

# Learning Strategies for Cryptography Using Embedded Systems

Edwar Jacinto, Fredy Martínez and Fernando Martínez

**Abstract** In the technological formation of students, specially for Technology and Electronic Engineering at the District University Francisco José de Caldas of Colombia, South America, it is required that students in addition of the theorical bases and the use of tools possess skills in implementation of real applications, where the use of mathematics in daily problems of engineering is necessary. Therefore, it is sought to do academic exercises to have theorical components along with a strong practical component, in this case, a secure information application, specifically a lightweight block cipher type, designed in embedded systems and made in both hardware and software. With this exercise, it is want to apply the knowledge acquired in the area of embedded digital devices, whether using digital devices configurable FPGA type or microcontroller devices. In this paper, an exercise of application using the smart pedagogy of lightweight cypher that is used as a tool for teaching the use of embedded hardware and software platforms.

**Keywords** Problem-based learning · Significant learning · Embedded systems · Cypher block

## 1 Introduction

An application in digital electronic area using the know-how concept is wanted to generate [9] and along with skills in the telecommunications area, in this way, to use it as a smart teaching strategy in students of technology in electronic and especially for

E. Jacinto (✉) · F. Martínez · F. Martínez
District University Francisco José de Caldas, Bogotá D.C., Colombia
e-mail: ejacintog@udistrital.edu.co
URL: http://www.udistrital.edu.co

F. Martínez
e-mail: fhmartinezs@udistrital.edu.co

F. Martínez
e-mail: fmartinezs@udistrital.edu.co

© Springer International Publishing Switzerland 2016    495
V.L. Uskov et al. (eds.), *Smart Education and e-Learning 2016*,
Smart Innovation, Systems and Technologies 59,
DOI 10.1007/978-3-319-39690-3_44

students from undergraduate and postgraduate of telecommunication engineering, using the smart pedagogy strategy.

The need for secure communications, for different types of applications is analyzed, such applications can be made in configurable hardware [1, 18, 23], using software embedded systems [5, 20], as an application for a mobile device or personal computer, therefore the know-how applied in a smart pedagogy for the students to teach a developer embedded systems, the main objective is that they acquire certain skills in the know-how of the firmware developer, for this reason, this paper shows the step by step of the implementation of the lightweight block cipher for applications with small data transfer as a educational tool [3], for this task, a low cost embedded hardware was used. In this way, they learn to make real applications in devices with a low computational capacity [2, 8, 27].

When choosing block ciphers [4, 15] as one of the standard encryption schemes [25], students must make a bibliographic revision of different algorithms and their features [10, 14, 15] and the tests over embedded software systems type microcontroller [5, 13, 20]. To acquire the knowledge of data encryption specifically in block ciphers they must understand the basics steps of AES algoritthm (Advanced Encryption Standard) [5, 14, 26] with the necessary mathematical bases.

In order to the students get the mathematics knowledge for the implementation of called algorithms, this academic exercise tries to implement a block cypher to achieve with the philosophy of lightweight by the possibility of making a hardware and software.With this prior knowledge and making the respective bibliographic revision, all possible algorithms to work, it is decided to work with the PRESENT algorithm [6–8, 18, 23], since this is standardized by NIST and NSA [16, 19].

PRESENT is the smallest symmetric encryption algorithm as it uses substitution and permutation blocks such as 4-bit basis. This word size allows that it suits to reduced hardware devices or processors with a small bus also, it has wide keys relatively small compared with other algorithms of the same type and requires a reduced amount of rounds; on the other hand, from the point of view of design and implementation, it presents an interesting challenge, because it requires a strategy taking into account the balance between amount of resource used and communication speed system [3, 21, 24].

The paper is organized as follows. In Sect. 2 the basic concepts of cryptography are discussed, Sect. 3 describes the design methodology of embedded system. Section 4 analyses the performance by simulation. Section 5 shows some results. Finally, conclusion is presented in Sect. 6.

## 2 Basic Concepts of Cryptology

The student before having read on the theoretical foundations of cryptographic topography along with some concepts in finite fields of arithmetic, also to have clear concepts of logical device programming and programmable microcontroller devices, you must know the basic blocks of a block cipher. Now, some basic concepts of

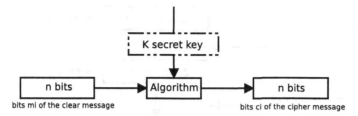

**Fig. 1** Block Cipher

cryptography necessary to understand the academic exercise and with which the student will generate respective data-path or pseudo codes to make both hardware and software.

## 2.1 Block Cipher

In this type of encryption, the plain-text message is divided into blocks of $n$ bits each [7, 18]. The main feature of such ciphers is that each block is encrypted in the same way, regardless of its place in the chain so all bits of the block are estimated together, participating in operations that try to obscure the possible relationships they had with the original message [22]. This can be seen in Fig. 1.

Symmetric algorithms encrypted text block, but this length change per each algorithm.

In the block cipher four basic operations are performed:

- *Electonic CodeBlock (ECB)*: within it blocks are ciphered separately.
- *Cipher BlockChainning (CBC)*: the cryptogram blocks to relate between them through OR excusive funtions.
- *Cipher feedback (CFB)*: it is performed XOR between character or isolated bits of the text along with the output to the algorithm.
- *Output feedback (OFB)*: equal than CFB, it is performed XOR between character or isolated bits of the text and the algorithms outputs, but this uses the feedback between the output with the inputs, therefore it does not depend on text, it is a generator of the random numbers.

The designer have to exactly know the algorithm to applied the concepts of programming to do each of the blocks and this will be done on PLD device and embedded systems.

## 2.2 Basic Operations of the Cipher Blocks

The basic operations of the cipher block algorithm are shown in the Fig. 2. They could change depending on the proposed algorithm.

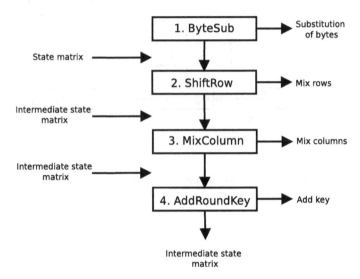

**Fig. 2** Round basic operation of a typical block cipher

For all of the cipher block is generated the combination of operations *ByteSub, ShiftRow, MixColum y AddRoundKey*, or equivalent, when this subroutines execute in one time is named round. the complex and security of the algorithm depending on previous combining operations, the among of rounds made, with the key length used [3, 4].

Each round is composed for three operations based on a uniform and invertible transformations are named *layer*, which have been designed to resist the lineal and differential cryptanalysis, they are shown next:

- *No linear layer*: It makes pass the data through the S box in parallel with non-linearity optimal properties.
- *Linear mix layer*: It guaranties the high level of diffusion along the multiples layers.
- *Key additions layer*: It is applied for XOR operation between the intermediate state and each round key.

After explaining each one, the functional blocks of any cipher. The next section, the functional blocks of the PRESENT algorithm will be shown and analyzed.

## 2.3   PRESENT: Algorithm Structure

In the Fig. 3 the basic structure of the *PRESENT* algorithm is shown, it can see their blocks and each one of their 31 rounds [8].

**Fig. 3** *PRESENT* algorithm round

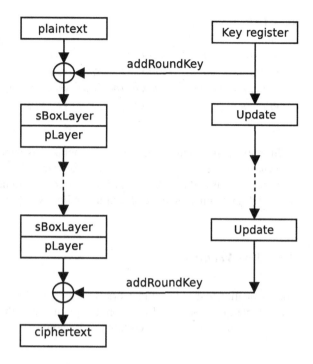

**Fig. 4** *PRESENT* algorithm structure in *Pseudo-code*

*generate RoundKeys()*
*for i=1 to 31 do*
*addRoundKEy(STATE,Ki)*
*SBoxLayer(STATE)*
*pLayer(STATE)*
*end for*
*addRoundKey(STATE,K32)*

## 2.4 Algorithm: Pseudocode

The solution of the algorithm could be summarized in a simple pseudocode compatible with any embedded hardware or software platform [23]. This code is shown next in the Fig. 4:

For each pseudo-code line, a function or method to perform the basic operations of the cipher block algorithm must be created. The student must choose the options between functions or tables for the substitutions operations for nibbles of the *Sbox* layer and the operations of the *Player* [5, 17].

In this case, the use of the dynamic memory increases the execution time to perform the necessary calculus. Otherwise, when the tables are used, the processing time will be reduced, but increasing the among of memory used to storage the substitution tables.

**Table 1** Test vector for *PRESENT*

| Plaintext | Key | Ciphertext |
|---|---|---|
| *00000000 00000000* | *00000000 00000000 0000* | *5579C138 7B228445* |
| *00000000 00000000* | *FFFFFFFF FFFFFFFF FFFF* | *E72C46C0 F5945049* |
| *FFFFFFFF FFFFFFFF* | *00000000 00000000 0000* | *A112FFC7 2F68417B* |
| *FFFFFFFF FFFFFFFF* | *FFFFFFFF FFFFFFFF FFFF* | *3333DCD3 213210D2* |

In embedded systems, the among of memory is short, for this reason the students have to check and analyse what the best way to describe the algorithm is. After that, simulations and resources reports are necessary to determine which is the best methodology to guarantee the algorithm be lightweight type [2, 3, 7].

## 2.5 Test Vectors

For verification to the algorithm works in a good way, the makers of the algorithm [8] give a table in Hexadecimal notation, the plain text with their respective encrypted known text after mixing with certain key is shown in the Table 1:

## 3 Desing Methodology for Embebdded Systems

In the last decade the revolution of the digital programmable devices field has been an significant increase, which has allowed the possibilities to make designs with a high performance in a small devices and low price [11]. For this advantage, it is possible to make designs of the lightweight ciphers algorithms in a embedded systems type hardware [9] and software [5]. For this reason as the universe of possible devices is infinitive, the study case requires a check the most relevance characteristics of the logic programmable devices and microcontroller in our local market [5].

The design meets with the *Top-Down* methodology and requires a certain number of phases, in dependence of the tool, but the general list of this steps are shown next [12]:

- Description a system level.
- Description a behaviour level.
- Validation (verification).
- Co-simulation.
- Estimation performance.

When the designer wants to make a whole system with different sub-systems, He must follow the next steps to get a correct verification of the algorithm. This procedure is called map and mapping and partitioning [11, 12]:

- Functional specification input.
- List of the inputs/outputs of the architectures HW/SW.
- Interconnection mechanisms.
- Optimization functions cost, time, area, communication.
- Hardware synthesis in the different abstraction levels.
- Software synthesis to make simple and concurrent algorithms.
- Interface synthesis a buses level,communications mechanisms.

When the student has the methodology of design cleared, He must generate the standard code, scalable and reusable. This task is achieved when he has followed the steps shown above.

To complete the implementation, the designer requires to make a technical review of the embedded devices and platforms, in this case, the student should make a search in the local market of the microcontrollers, microprocessors and PLD's suitable to bear the *PRESENT* algorithm. The technical characteristics to compare the devices and platforms are shown next: standard architecture, price, available compilers (free or with license), programming languages (standard code), memory, speed, available in the local market [5].

For the PLD's devices, the student will choose a popular hardware description language, the use of the standard tool allows the change of the device in any moment, but the designer has prohibited the use of the IP (intellectual property) cores and another tool with restrictions to change the code to another device manufacturer. After analysing the problem of lightweight block cypher using hardware description languages, the Universidad Distrital advices to use *VHDL* with ISE Xilinx software, since the university has that tool with a student license and a complete laboratory using medium and low cost devices of this brand.

On the other hand, the university has microcontollers of Microchip TM with processor bus width 8 and 16 bits, each one with the compiler license and developer boards, this software has a standard C compiler with the respective support for debugging and simulation. Other advantage is the compatibility of this microcontrollers in theirs different families, it is possible to change de device and it keep the useful of the all pins. As families of 32 bits the microcontroller ARM Cortex-M0 with development board KL25Z have been chosen, this due has a free online compiler available to work in C and C++, the capabilities to manage pointers, dynamic memory addressing, control of versions and formatting code with mercurial, low cost, easy to programming, debugging and load programs, big amount of embedded modules and other advantage is the on-line community which create a libraries every single day, by last, this architecture bear the complete versions of different RTOS (Real Time Operation Systems).

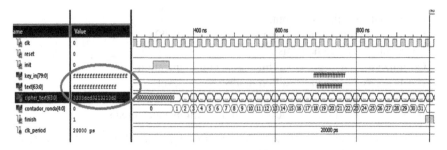

**Fig. 5**  Hardware implementation simulation

# 4   Checking Algorithm: Simulations and Tests

For the result measuring of the *PRESENT* algorithm implementation over microcontrollers and PLD's devices, it will be used the ciphering and desciphering functions with 80 bits key length and 64 bits data blocks.

For each microcontrollers platform is necessary to define certain metrics: program memory (FLASH), data memory (RAM) and throughput would be consider. Depending on the previous metrics, they must be taken into account: the processor bus length, pointers hardware, functions or manage of tables and the speed of the CPU system measured in MIPS (Million Instruction Per Second).

As to PLD's, the metrics are totally different, the throughput importance increases due to the parallelism of the devices, it is searched the reduction of the resources measured in Slices (CLB-GE).

The student must probed the algorithm through simulations using the test vectors, one of this tests is shown in the Fig. 5:

To probe the firmware in the microcontrollers, initially the students should write a standard code in C language, this procedure with the help of the intermediate steps of the hardware simulation shows above en the Fig. 5. The simulation or implementation would change on depending of the device chosen. With the reports thrown from the compiler, it is possible to compare de metrics of the different implementations. Real tests with a measuring of one pin of the device could be made.

# 5   Results

This hardware and software tools were used in a test group of 10 undergraduate students of the seed research annexed to DIGITI (Smart Digital Systems) and ARMOS (Modern Architectures to Power Supply Systems). The research groups had worked with the students in the second semester of the 2015, a quick bonding was achieved with their research projects. Because the students have got the theorical concepts and they have acquired skills designing and working on embedded systems.

A collateral effect was seen with the increase in the academic interest of the students in their different activities and major compromise with the researching group. DIGITI researching group have got big amount of thesis, in the field of the cryptology and cryptography, so far it has two undergraduate thesis, one of them in lightweight block cipher algorithm HIGH implementation in microcontroller, the second one in the HSM implementation in a 32 bits microcontroller. The research group just ended the new work master thesis, this task is going to increases the performance of the CLEFIA algorithm in a 32 bits microcontroller.

# 6 Conclusions

With this educational exercise the research group achieves two undergraduate and one master thesis, without delays on the mathematical foundation; this methodology guarantee the success of the learning process on the seed research. Students in the firsts semesters could work in a real project which to prepare the degree work. The mathematical foundation in the cryptology has been complicated to the undergraduate students, our methodology had decreased the development time of the mathematical knowledge application using the embedded systems over a real problem.

On the other hand it was detected that the design and implementation of these ciphered algorithms on programmable logic devices clinched all previous knowledgement of group students. They improves their skills on combinational and sequential digital circuits, finite state machines and of course, devices description using VHDL and Verilog. That is why, ARMOS and DIGITI researching groups will propose to implement similar issues as practical projects for digital circuits courses for electrical and electronical technology students. It will work as a final course project and it is possible that it improves the learning process in that area.

**Acknowledgments** This work was supported by the District University Francisco Jos de Caldas, in part through CIDC, and partly by the Technological Faculty. The views expressed in this paper are not necessarily endorsed by District University. The authors thank the research groups DIGITI and ARMOS for the evaluation carried out on prototypes of ideas and strategies.

# References

1. Abdellatif, K.M., Chotin-Avot, R., Mehrez, H.: Lightweight and compact solutions for secure reconfiguration of FPGAs. In: 2013 International Conference on Reconfigurable Computing and FPGAs (ReConFig), pp. 1–4 (2013)
2. Alkhzaimi, H.A., Lauridsen, M.M.: Cryptanalysis of the simon family of block ciphers. Technical University of Denmark 1(1), 1–26 (2013)
3. Attridge, J.: An overview of hardware security modules. SANS Institute, InfoSec Reading Room 1(1), 1–10 (2002)

4. Aysu, A., Gulcan, E., Schaumont, P.: SIMOn says: break area records of block ciphers on FPGAs. IEEE Embed. Syst. Lett. **6**(2), 37–40 (2014)
5. Azuero, R., Jacinto, E., Castano, J.: A low-memory implementation of 128 aes for 32 bits architectures. En Congreso Argentino de Sistemas Embebidos CASE, pp. 67–73 (2012)
6. Beaulieu, R., Shors, D., Smith, J.: The simon and speck block ciphers on avr 8-bit microcontrollers. In: LightSec 2014 Proceedings, vol. 1, no. 1, pp. 1–18 (2014)
7. Beaulieu, R., Shors, D., Smith, J., Treatman-Clark, S., Weeks, B., Wingers, L.: The simon and speck families of lightweight block ciphers. Cryptology ePrint Archive, Report 2013/404 (4) (2013). http://eprint.iacr.org/
8. Bogdanov, A., Knudsen, L., Leander, G., Paar1, C.: PRESENT: An Ultra-Lightweight Block Cipher, chap. 5, pp. 450–466. Springer-Verlag Berlin Heidelberg (2007)
9. Camargo, C.: Teching/learning methods for embedded systems using copyleft hardware. IEEE (Rev. IEEE Am. Lat.) Lat. Am. Trans. **9**(4), 503–509 (2011)
10. Chih-Peng, F., Jun-Kui, H.: Implementations of high throughput sequential and fully pipelined AES processors on FPGA. In: International Symposium on Intelligent Signal Processing and Communication Systems ISPACS 2007, pp. 353–356 (2007)
11. Delgadillo, A., Pena, N., Guerrero, M.: Diseno de un criptosistema para redes de sensores inalambricos WSN basado en MPSOC. Master's thesis, Universidad de los Andes (2008)
12. Densmore, D., Passerone, R.: A platform-based taxonomy for esl design. IEEE Des. Test Comput. **23**(5), 359–374 (2006)
13. Engels, S., Kavun, E., Paar, C., Yalcin, T., Mihajloska, H.: A non-linear/linear instruction set extension for lightweight ciphers. In: 2013 21st IEEE Symposium on Computer Arithmetic (ARITH), pp. 67–75 (2013)
14. Fan, C.P., Hwang, J.K.: Implementations of high throughput sequential and fully pipelined AES processors on FPGA. In: International Symposium on Intelligent Signal Processing and Communication Systems, 2007. ISPACS 2007, pp. 353–356 (2007)
15. Feizi, S., Ahmadi, A., Nemati, A.: A hardware implementation of simon cryptography algorithm. In: 2014 4th International eConference on Computer and Knowledge Engineering (ICCKE), pp. 245–250 (2014)
16. Hanley, N., O'Neill, M.: Hardware comparison of the ISO/IEC 29192-2 Block ciphers. In: IEEE Computer Society Annual Symposium on VLSI (ISVLSI), 2012, pp. 57–62 (2012)
17. Kavun, E., Leander, G., Yalcind, T.: A reconfigurable architecture for searching optimal software code to implement block cipher permutation matrices. In: 2013 International Conference on Reconfigurable Computing and FPGAs (ReConFig), pp. 1–8 (2013)
18. Kavun, E., Yalcin, T.: RAM-based ultra-lightweight FPGA implementation of present. In: 2011 International Conference on Reconfigurable Computing and FPGAs (ReConFig), pp. 280–285 (2011)
19. Klinc, D., Hazay, C., Jagmohan, A., Krawczyk, H., Rabin, T.: On compression of data encrypted with block ciphers **58**(11), 6989–7001 (2012)
20. Kumar, M., Singhal, A.: Efficient implementation of advanced encryption standard (AES) for arm based platforms. In: 2012 1st International Conference on Recent Advances in Information Technology (RAIT), pp. 23–27 (2012)
21. Mane, S., Taha, M., Schaumont, P.: Efficient and side-channel-secure block cipher implementation with custom instructions on FPGA. In: 2012 22nd International Conference on Field Programmable Logic and Applications (FPL), pp. 20–25 (2012)
22. Pineda, N., Velasquez, N.: Diseño e implementación de un prototipo criptoprocesador AES-Rijndael en FPGA. Master's thesis, Universidad de Los Llanos (2007)
23. Pospiil, J., Novotny, M.: Evaluating cryptanalytical strength of lightweight cipher present on reconfigurable hardware. In: 2012 15th Euromicro Conference on Digital System Design (DSD), pp. 560–567 (2012)
24. Qatan, F., Damaj, I.: High-speed katan ciphers on-a-chip. In: 2012 International Conference on Computer Systems and Industrial Informatics (ICCSII), pp. 1–6 (2012)
25. Shuangqing Wei, L., Wang, J., Yin, R., Yuan, J.: Trade-off between security and performance in block ciphered systems with erroneous ciphertexts **8**(4), 636–645 (2013)

26. Tay, J.J., Wong, M.M., Hijazin, I.: Compact and low power AES block cipher using lightweight key expansion mechanism and optimal number of s-boxes. In: 2014 International Symposium on Intelligent Signal Processing and Communication Systems (ISPACS), pp. 108–114 (2014)
27. Yalla, P., Kaps, J.: Lightweight cryptography for FPGAs. In: International Conference on Reconfigurable Computing and FPGAs, 2009. ReConFig'09, pp. 225–230 (2009)

# Introducing Smart Technologies for Teaching and Learning of Fundamental Disciplines

**Svetlana Rozhkova, Valentina Rozhkova and Mariya Chervach**

**Abstract** The authors discuss methods and techniques of introducing gaming technologies to the process of education, as well as a merge of intellectual (smart) technologies and serious games (gamification) in the field of education. The main aim and objective of the gamification is seen as the process of motivation allowing disclosing students' creative abilities and stimulating their learning activity. The article presents best practices of implementation of innovative educational approaches, including gamification of educational process, based on the example of a system of elite technical education executed in National Research Tomsk Polytechnic University (TPU).

**Keywords** Gamification · Smart technologies · Mathematics · Innovative learning and teaching methods

## 1 Introduction

Modern society is on the edge of changing the technological paradigm. Information technologies that have set the essence of the XX century give the floor to smart technologies that open the path for development of smart economy, smart education, and smart society.

It is the smart education, which is capable of providing top level of the quality of education correspondent to needs and opportunities of today's world that will allow youth to adapt to the fast-changing environment, and will assure the transition from book content to active teaching and learning.

Nowadays the issues of innovative engineering education are strongly emphasized. The need for improvement of engineering education is induced by rapid

S. Rozhkova · V. Rozhkova · M. Chervach (✉)
National Research Tomsk Polytechnic University, Tomsk, Russia
e-mail: chervachm@tpu.ru

S. Rozhkova
e-mail: rozhkova@tpu.ru

© Springer International Publishing Switzerland 2016
V.L. Uskov et al. (eds.), *Smart Education and e-Learning 2016*,
Smart Innovation, Systems and Technologies 59,
DOI 10.1007/978-3-319-39690-3_45

development of production industry, fundamental and applied sciences, technological, economic and social progress, globalization of world economy and inter-
nationalization of education.

In 2011 Vladimir Putin, a then-Prime Minister of the Russian Federation, noted
that a lack of engineering professionals is an even more important issue, than
corruption and administrative barriers.

One of the elements of innovative education is a system of elite engineering
education for students. In 2004 Tomsk Polytechnic University has developed and
implemented a system of Elite Technical Education (ETE) [1].

The main aim of the elite education is to train highly competitive specialists,
who are able to act innovatively in different spheres of economics and industry. The
system of Elite Technical Education of students includes 3 stages: selection of
students, fundamental education, and professionally-oriented education.

Fundamental education includes thorough learning of basic disciplines of the
curriculum: Mathematics, Physics, Economics, and Foreign Languages for Academic Mobility.

Fundamental education is the basis of higher education; by receiving it students
foster understanding of laws of nature and society progression, form ability to think
logically, analyse and classify facts, take responsible decisions and exploit scientific
approaches to learning phenomena, events and processes. Mathematical education
represents a big part of fundamental higher education.

From the times of Peter the Great, Russian education has always been noted for
its strong emphasis on mathematical background. The body of mathematics is a
universal language that can be used for describing various natural processes and for
incorporation of different factors.

Nowadays, there is an urgent need for the implementation of innovative
approaches in teaching and learning of mathematics. The long-used traditional
teaching and learning methods are not up-to-date for the modern society; they gave
place for new unconventional teaching approaches, i.e. active teaching and learning
methods. Historically, student played a role of an object of study process, receiving
the knowledge through lectures. In the modern times, a graduate is expected to be
mobile, to be able to take quick decisions in the context of uncertainty and
non-standard situations. With respect to these requirements, the traditional teaching
approach (the lecture-seminar form of teaching) does not respond to the identified
challenge. Therefore, there is a rising need for introduction of innovative technologies and teaching methods that would allow involving students in the educational process. Student involvement should provide students with an opportunity to
have a two-way communication not only between each other, but also with professors through active participation in learning process and conduction of various
problem-based tasks [2].

Besides, the disciplines associated with hard sciences are more difficult for
understanding than the human sciences. For the purpose of simplifying understanding and memory acquisition of the study material on these disciplines, a full
variety of smart technologies is being amplified, for instance, distant learning,
personification, interactive study materials, education through videogames, etc.

## 2	Gamification of Educational Process

One of the most efficient methods for teaching fundamental sciences is introducing gaming teaching and learning methods, i.e. the gamification of educational process [3].

Games as a method for adult education allow to [4]:

- foster motivation for education (and, therefore, can be efficient at an early stage of education),
- evaluate students' level of education (can be used both at an early stage of education for entry control, and at a final stage of education as final control of education efficiency);
- assess competency level and transform it from passive form—knowledge, to active form—skills (can be used as an effective form for skills' development right after a theoretical part of a lecture);
- get personal experience of educational game activity, practice skills of design and organization of educational games;
- activate students' self-learning;
- foster pluralistic opinions and actions, multivariance of intellectual operations, interest towards a more efficient professional development;
- develop individual professional thinking, ability to analyze and forecast.

The structure of an educational game should include the following components: aim, subject, scenario, and rules.

When determining the aim of an educational game it is necessary to take into account the reason for the introduction of an educational game, people, who will be involved in the game, what and how will be taught through the game, and which results should be achieved.

The subject of a game is an aspect of the gaming activity that replicates an aspect of the real professional activity.

The scenario of a game is a thorough description of a game, and the game rules are the replication of specific professional processes and events.

The learning outcomes of a game should represent the formation of students' specific professional competences.

It should be noted that when speaking about the formation of professional competences, there should be a distinction made between "competence" and "competency".

"Competency" is a development of student's identity that has been formed within a systematic training process and represents student's professional knowledge, skills, abilities and professional attitudes. "Competence" is an ability to implement the knowledge, skills and abilities on practice. Therefore, educational games are valuable for educational process, since they:

- give an opportunity to get a holistic experience of future professional activity;
- structure existing skills and abilities of a student into a holistic system [5].

It is an educational game that can serve as means for the development of professional competences as a factor of qualified specialists' training.

Participants put into practice the acquired theoretical knowledge and gain experience of solving real problems through playing roles in the game, and through learning and solving game problem situations. The knowledge, skills and abilities acquired through this method have higher degree of accessibility than the traditional teaching and learning methods.

Developers of educational games should take into account SMART requirements towards educational objectives, according to which the objectives should be [6]:

- Specific—distinct and clear description of what should be achieved;
- Measurable—should state how and by what means the success will be rated;
- Agreed—with those, who will work to achieve the set objective and with those, who can influence its achievement;
- Realistic—achievable (taking into account the limitations imposed by the situation and the need to comply with other objectives);
- Timed—expected dates are set for the achievement of the objective.

## 3 Example of an Educational Game "Ruler"

This educational game is intended for students of the Elite Technical Education department of the National Research Tomsk Polytechnic University. The game "Ruler" is aimed at retention of learned materials on "Non-linear operations with vectors" and "Curves and surfaces of the second order" of the course "Linear Algebra and Analytical Geometry" that is executed in the I semester.

Aim of the game:

- To learn key concepts of Vector Algebra and Analytical Geometry;
- To foster ability to use mathematics apparatus for solving engineering problems;
- To foster skills of self-studying necessary for utilization of the acquired knowledge for learning professional disciplines and conducting professional activity;
- To foster mathematical intuition and mathematical culture.

Objectives of the game:

- Educational: to improve and solidify an ability to utilize knowledge of Higher Mathematics' section "Vector Algebra and Analytical Geometry";
- Ampliative: to develop mathematical skills of students and their cognitive independence; to develop creativity; to form the conceptual framework;
- Pedagogic: to foster interest in mathematics.

As a result of the educational game a student should:

1. Know:

   (a) Key concepts of Vector Algebra (non-localized vectors, linear and non-linear operations on vectors, scalars, vectors and triple scalar vectors).
   (b) Key concepts of Analytical Geometry (curves on flat surface and in n-dimensional space, surfaces);
   (c) Curves of the second order: circle, ellipsis, hyperbolic and parabolic curves (their geometrical characteristics, equations and curve plotting);
   (d) General concept of a curve of the second order; curves of the second order in a polar coordinate system; optical characteristics of a curves of the second order; bringing a general equation of a curve of the second order to a canonical form;
   (e) Surfaces of the second order.

2. Be able to:

   (a) Implement concepts and methods of Vector Algebra when solving applied problems;
   (b) Find curve and surfaces equations, use them for solving problems of Mathematics and Physics.

3. Manage:

   (a) Ability to work with non-localized vectors;
   (b) Ability to find curves and surfaces, use them for solving problems of Mathematics and Physics;
   (c) Methods for construction of mathematical models for professional issues and methods for conceptual interpretation of the obtained results.
   During educational process of a study course students should foster the following soft skills and professional competences:

4. General competences (soft skills):

   (a) ability to generalize, analyze, perceive information, ability to set aims and choose paths for their achievement;
   (b) ability to construct oral and written communication logically and reasonably;
   (c) ability to formalize and present results of the work;
   (d) ambition for self-improvement, proficiency enhancement and mastery;
   (e) ability to apply key laws of science in professional activity, apply methods of mathematical analysis and modeling, theoretical and experimental research.

5. Professional competences:

   (a) readiness for independent learning and independent work;
   (b) ability and readiness to solve problems, take responsibility for decisions;

(c) knowledge of key concepts, laws and methods of science; ability to educe scientific essence of problems that occur during professional activity, readiness to apply relevant scientific tooling to solve them;

(d) readiness to apply mathematical apparatus for solving set problems, ability to apply a relevant mathematical model for any process and ability to assess its adequacy.

## 4 The TPU Survey

In 2015, the Department of Higher Mathematics of the National Research Tomsk Polytechnic University has conducted a sociological research that aimed at identifying students' opinion on implementation of modern educational methods for teaching and learning of Mathematics.

75 students of the Elite Technical Education department took part in this survey.

Overall, students showed positive attitude towards introduction of educational games to the teaching of Mathematics. The majority of students—85 % of respondents—showed positive attitude towards implementation of educational games to Mathematics, 8 % of the respondents believed that this discipline did not need active e-courses, and 7 % of the respondents neither agreed, nor disagreed. (Fig. 1)

At the same time this survey allowed receiving students' evaluation of the productivity of certain aspects of educational games. (Fig. 2)

As can be seen from the diagram, according to the students' opinion the most productive aspects were: the formation of a holistic image of professional situation (97 %), the knowledge is acquired not only for future implementation, but within a real process (96 %). Slightly less productive aspects were: the accumulation of social experience (communication, decision-taking, etc.) (84 %), and the ability to apply various solution strategies for set problems (73 %).

**Fig. 1** Students' opinion on the necessity of introduction of educational games to the course "Mathematics"

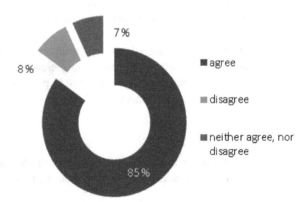

**Fig. 2** Students' evaluation of productivity level of certain aspects of educational games

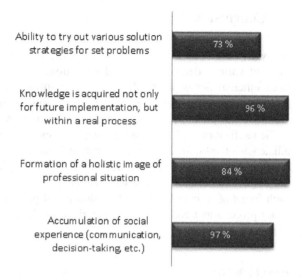

The chart shown in Fig. 3 presents difficulties that students have experienced while participating in educational games.

The chart indicates that all of the problems, mentioned in the questionnaire, have, to some extent, been experienced by students. The majority of students have faced difficulties with understanding rules of the game (65 %) and with lack of time (61 %). Students have also experienced problems with incomprehensible/controversial questions and tasks (54 %) and lack of theoretical materials provided for the solution of a task (27 %).

**Fig. 3** Difficulties that students have experienced while participating in educational games

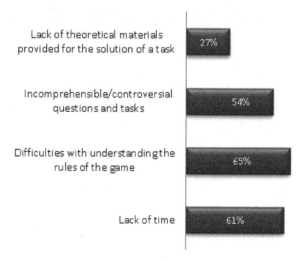

# 5   Conclusion

Innovative methods for teaching and learning of Mathematics allow the development of various demonstrative and educational technologies, games and models. Such efficient developments foster positive attitude towards educational process, provide unintrusive ways for helping and giving an opportunity to choose individual educational speed, encourage successful training for final examination.

Generalization and systematization of the research results shows a high level of efficiency of introducing to the educational process the indicated approaches for teaching Mathematics. This has been confirmed through students' interest raise and the ability for teachers to evaluate students' residual knowledge. Introduction of such forms of active learning to the educational process provides an opportunity to foster professional competencies of engineering HEIs' students efficiently.

# References

1. Chubik, P.S., Chuchalin, A.I., Soloviev, M.A., Zamyatina, O.M.: Training of elite engineering and technology experts. J. Educational Studies. **2**, 188–208 (2013). (in Russian)
2. Pokholkov, Y.P., Rozhkova, S.V., Tolkacheva, K.K.: Practice-oriented educational technologies for training engineers. In: International Conference on Interactive Collaborative Learning, ICL, pp. 619–620 (2013)
3. Zamyatina, O.M., Abdyikerov, Z.S.: Development and assessment of students' competences through gamification of education. J. Koncept. http://e-koncept.ru/teleconf/95147.html. (in Russian)
4. Panina, T.S., Vavilova, L.N.: Modern Approaches for Activation of Education. Academia, Moscow (2006). (in Russian)
5. Verbitskyi, A.A.: Contexts of Educational Contents. Alpha, Moscow (2003). (in Russian)
6. Ishkov A.D.: Diagnostics and technology for improving the adaptation level of HEI faculty. J. Internet-Vestnik VolgGASU, vol. 3. (2012). (in Russian)

# Redefining Knowledge Management Education with the Support of Personal Knowledge Management Devices

Ulrich Schmitt

**Abstract** Knowledge Management (KM) Education is supposed to provide the key opportunities for the 21st century knowledge societies, but Bedford's recent survey results highlight the immaturity in this endeavor. In order to rectify the situation current top-down centralized approaches ought to be substituted by bottom-up concepts focusing on the self-sufficiency and mobility of today's knowledge workers. These efforts need to be supported by smart personal knowledge management devices able to connect with each other as well as to integrate with organizational systems. What is being taught and technologically supplied, thus, creates sustainable impacts on individuals' academic and professional growth and their roles as contributors and beneficiaries of institutional and societal performance. The article introduces a novel Personal KM (PKM) prototype system and educational concept to fill these needs in anticipation of Levy's imagined KM revolution that gives more power and autonomy to individuals and self-organized groups.

**Keywords** Knowledge management (KM) · Personal knowledge management (PKM) · Educational technology · KM education · Memes · Knowcations

## 1 Towards a Personal Discipline for Creative Connection

Wiig defines the root objective of Personal Knowledge Management (PKM) as the desire to make citizens highly knowledgeable. They should function competently and effectively in their daily lives, as part of the workforce, and as public citizens. Accordingly, the quality and extent of their competences and the structural Intellectual Capital (IC) assets available to them matter and also determine the realized performance of enterprises and societies [1]. By the same token, Bedford expects the education in Knowledge Management (KM) to provide the key

U. Schmitt (✉)
University of Stellenbosch Business School, PO Box 610,
Bellville 7535, South Africa
e-mail: schmitt@knowcations.org

© Springer International Publishing Switzerland 2016        515
V.L. Uskov et al. (eds.), *Smart Education and e-Learning 2016*,
Smart Innovation, Systems and Technologies 59,
DOI 10.1007/978-3-319-39690-3_46

opportunities for growing a 21st century knowledge society, just as business, engineering and science education still do for the industrial economy [2].

Levy even envisages a potential KM revolution that gives more power and autonomy to individuals and self-organized groups and puts forward the idea of a personal discipline for collection, filtering and creative connection (among data, among people, and between people and data flows). Consequently, he also regards the sustainable growth of autonomous personal KM capacities as the most important function of future education [3].

Although these reflections focus clearly on the individual, supporting hardware and software solutions are essential to create the smart spaces for mediating the co-evolutions between people, knowledge, and technologies. Accordingly, these "smart spaces need to be designed and, in the process, must empower the users of those spaces" to "confer capability" [4]. What is being taught and technologically supplied, is expected to generate sustainable impacts on individuals' academic and professional growth and their roles as contributors and beneficiaries of institutional and societal performance.

Smart Education is primarily emphasized as a function of intelligent technologies able to adapt, sense, infer, learn, anticipate, and self-organize/create/sustain [5, 6]. But, the same capabilities also need to be conferred to users for individual 'smartification'.

Just as Extelligence (externally stored information) and Intelligence (understanding minds) are driving each other in a complicit process of interactive co-evolution [7], empowered smarter people are able to produce better structured extelligence for intelligent technologies to thrive while decentralized autonomous networked devices (nourished by creative conversations of many individuals' PKM) are able to constitute "the elementary process that makes possible the emergence of the distributed processes of collective intelligence, which in turn feed it." [3].

## 2 A Quest to Identify and Address the Essential Requirements

Such a solution should encompass the following educational and technological aspects:

1. In a society with broad *Personal Competences*, "decision-making everywhere will maximize personal goals, provide effective public agencies and governance, make commerce and industry competitive, and ensure that personal and family decisions and actions will improve societal functions and 'Quality of Life'". Within these related workflows, "small personal 'nano actions' combine with larger departmental actions that combine to create consolidated enterprise actions" which determine the overall performances and viabilities of organizations and societies [1]. Developing and maintaining the personal competencies

of an individual, hence, has to take priority and notice of private (action and reflection), academic (applied competences laid out in qualification frameworks), professional (standards and employability attributes), and societal requirements (developmental goals).

2. Work has suffered from a process of fragmentation which will continue to accelerate accompanied by a slipping control over constant interruptions, the loss of time for real concentration, and less learning by observation and reflection [8]. With specializations and domain-specific knowledge on the rise, one's ability to manage *Complexity and Diversity* (disciplines, mindsets, cultures) based on establishing a shared common ground of methodologies, understandings, professional tools and practices forms a second key issue.

3. An uneven diffusion of digital technologies and their unequal effects have caused detrimental opportunity divides within and across societies worldwide [9]. Hence, a focus on grass roots, bottom-up, affordable *Personal Devices* is a given necessity in order to (1) grow the global community of 'Digerati' independent of space (e.g. development countries) or time (e.g. phase of formal education, lifelong-learning, or career support) and to (2) promote the notion that knowledge and skills are portable and mobile, so that individuals—moving from one project or responsibility to another—are able to keep, maintain, and advance their personal knowledge on their own personal autonomous devices.

4. As the backbone of a society that values individual freedoms, people should be able to *Develop and Maintain Associations* with others to share ideas and augment their creativity [10]. A networking of the personal devices would, thus, strengthen the individual's role as contributor and beneficiary of communal, organizational, and societal performance by removing the barriers that currently prevent potential knowledge suppliers from engaging in a wider sharing and faster diffusion of their ideas for the benefit of more rapid iterative improvement [11].

## 3   A Personal Knowledge Management Concept and System

The novel PKM approach to be introduced addresses these four aspects (a1–a4). At its core, it features a *prototype-software-system* which (after migration to a cloud-based development platform and no-SQL database) allows for developing one's personal knowledge repository (a3) and the voluntary sharing of one's knowledge with acquaintances or the PKM system (PKMS) user community (a4) [12].

The quest for a common-ground system concept and design (a2) has pondered on many methodologies advocated by scholars and practitioners. Fortuitously, what might have appeared initially as difficult to reconcile or at odds (e.g. KM's objectives, philosophies, and methods), has resulted in the integration of over one hundred KM tools and ideas which establishes the baseline for a transparent and coherent educational KM concept, including the rationale of how and why some of the original methods had to be adjusted, extended, re-purposed, or merged (a1).

The educational emphasis of the PKMS concept is further evidenced by, so far, thirty articles and papers including extensive visualizations published in parallel to the ongoing software development. Pitched at chapters in a planned book and envisaged lecture units in a face-to-face or e-learning course design, they already form part of the PKMS repository with its unique meme-based representation ready to be repurposed as a test data set, as an online tutorial, as e-learning content as well as the initial stock of memes to be reconfigured by future PKMS users. In terms of the conference theme, *Smart Digital Futures*, this novel solution to be further introduced provides for a:

- Cloud-based application of networked autonomous personal devices based on development, hosting, and no-SQL platforms across multiple technological environments.
- Conceptual framework for smart education facilitating smart curriculum and courseware design and development in face-to-face, self- or e-learning settings.
- Digital assistant for one's Intellectual, Social, and Emotional Capital, Creative Authorship, and Collaborations throughout one's Academic and Professional Careers.

## 4 The PKMS and Its Educational Concept and Course Design

The development of the PKMS concept and prototype has been validated against accepted general design science research guidelines [13] resulting in an article dedicated to the research methods applied and their relevance, utility, rigor, and publishability in Information Systems research outlets [14]. Earlier, an article employed the systems thinking techniques of the trans-discipline of Informing Science (IS) to align and validate the central models and methodologies incorporated in the PKMS concept [15].

The educational concept and content is closely aligned to this design and applicability context. It can best be pictured (Fig. 1) within the realm of Popper's Three Worlds [16, 17] which differentiate reality into three distinct spheres. *World:1* comprises the concrete objects and their relationships and effects in the real physical world. *World:2* refers to the results of the mental human thought processes in the form of subjective personal knowledge objects. *World:3* represents the thought content made explicit in the form of abstract objective knowledge objects which express the products of *world:2* mental processes. However, to elicit impact on *world:1* physical objects and/or other *world:2* minds, the abstract *world:3*

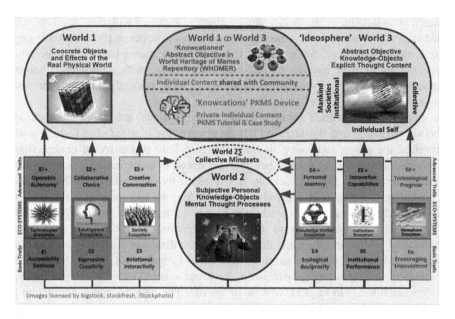

**Fig. 1** PKMS linking popper's three worlds via six digital ecosystems (based on [14])

objects have to be resourcefully combined and physically embodied or realized in concrete *world:1* objects (e.g. books or products).

A *world:2* PKMS user interacts on six ecosystem levels (E1-6 to be further detailed in papers currently under review) with *world:1* physical objects (E1: Technologies), *world:1* encoded/encapsulated knowledge (E2: Extelligence), *world:1 + 2* bodies and collective mindsets (E3: Society), and *world:2 + 3* mindsets and collective knowledge (E4: Knowledge Workers, E5: Institutions, and E6: Ideosphere).

The requirements for operating successfully in any of the ecosystems can be linked to two criteria (Ex and Ex+) which together form a Personal Knowledge Management for Development (PKM4D) framework closely aligned to Maslow's extended hierarchy of needs [18]. The PKMS concept and system supports all twelve PKM4D criteria by applying decentralized autonomous PKMS systems and repositories (Knowcations) supported by a 'World Heritage of Memes Repository (WHOMER)' where voluntarily shared content (memes) are accessible by the individual PKMS community members.

The PKMS concept does not employ the traditional document-centric storage where unique knowledge items are captured in a multitude of diverse containers (e.g. books, reports, files) causing problems of redundancies, fragmentations, inconsistencies, untraceabilities, corruptions, and decay. Instead, its repositories are based on atomic information-structures or memes[1] captured with their relationships, for example, to authors, citations, figures, or topics (a paper or presentation is kept as a meme sequence). Since memes captured can be viewed as the unique concretizations of their explicit objective abstract counterparts representing human thoughts in *world:3*, 'Knowcations' and 'WHOMER' basically introduce the new hybrid *world:1* ∞ *3* (Fig. 1).

Popper's three worlds are highly interactive; "*World:2* acts as an intermediary between *World:3* and *World:1*. But it is the grasp of the *World:3* object which gives *World:2* the power to change *World:1*" [17]. In light of the recent change from information scarcity to a never before experienced ever-increasing *world:1* information abundance, this interactivity is threatened by our finite *world:2* cognitive capabilities. As Simon [21] already noted over four decades ago, "it is not enough to know how much it costs to produce and transmit information; we must also know how much it costs, in terms of scarce attention, to receive it. [...] In a knowledge-rich world, progress does not lie in the direction of reading information faster, writing it faster, and storing more of it. Progress lies in the direction of extracting and exploiting the patterns of the world—its redundancy—so that far less information needs to be read, written, or stored". The PKMS's novel *world:1* ∞ *3* follows this advice and fills a crucial gap.

Figure 2 attempts to provide an overview how the wide-ranging features and functionalities of the PKMS concept and systems translate into the educational context by positioning the articles, papers, and posters published or in-progress accordingly (the individual publications are identifiable by their year-letter-identifier (Schmitt 2016d), fully referenced with mini-abstracts and URLs in a recent paper [22], and accessible via http://www.researchgate.net/profile/Ulrich_Schmitt2/contributions).

The icons on the left hand side of Fig. 2 indicate general issues of knowledge management covering (from bottom to top): matters of *Integration*, an introduction to managing *Knowledge*, knowledge acquisition/preservation *Nescience*, capacity development for *Occupations*, collaboration via *Workflows*, complexity reduction through *Concepts*. Each of these topic has an PKMS application-related counterpart focusing on *Autonomy*, *Traceability*, *Integrality*, *Outreach*, *New Horizons*, and *Systemics* (icons shown in the same row on the right). The six rows connecting the pairs of related themes have subsections corresponding to the six digital ecosystems and their PKM4D criteria already alluded to (bottom row: technologies,

---

[1]Memes were originally described as units of cultural transmission or imitation [19] that evolve over time through a Darwinian process of variation, selection and transmission. In order to survive, memes have to be able to endure in the medium they occupy and the medium itself has to persevere. They can either be encoded in durable *world:1* vectors spreading almost unchanged for millennia, or they succeed in competing for a host's *world:2* limited attention span to be memorized (internalization*) until they are forgotten, codified (externalization*) in further *world:1* objects or spread by the spoken word to other hosts' *world:2* brains (socialization*) with the potential to mutate into new variants or form symbiotic relationships (combination*) with other memes (memeplexes) to mutually support each other's fitness and to replicate together (*-markings refer to comparable SECI Model stages [20].

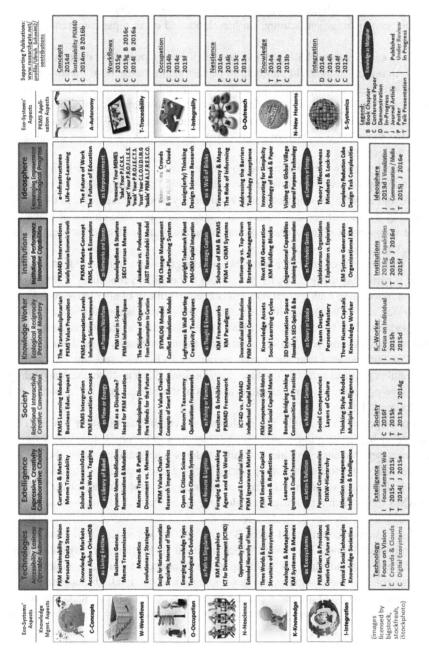

**Fig. 2** The framework for the PKMS educational concept and course design

extelligence, society, knowledge worker, institutions, and ideosphere) with distinctive options for larger scale interventions (scaping, sight setting, socializing, striving, systemizing, and scaling, further elaborated on in a paper currently under review).

While each of these left/right/bottom topics are supported by distinct publications (referenced in the right and bottom boxes), their intersections create a 6 × 6 matrix with cells corresponding to distinct lecturing/learning units and exemplifying content to be covered (including general KM as well as application-related issues). Neighboring cells of the matrix are further joined by ellipses, representing eighteen 'Knowledge as *Metaphor*' themes to be used for intensifying discussions or for assignments.

The design allows for alternative course deliveries column-wise left-right from bottom-to-top or row-wise upwards from left-to-right with an option for covering general and application-oriented issues sequentially or simultaneously. In the future e-learning manifestation, this feature translates into a possible choice of paths for course completion. It also eases partial integration in readers or academic text books, e.g. [23].

As with reports, or any other complex knowledge asset, a tutorial or learning unit is just a particular meme-tracking path through the PKMS repository's "extensive mesh of associative multidisciplinary trails already built-in of alternative pathways" [24] which can be left to handily track any of the other relationships not explicitly incorporated in the tutorial. As the PKM System is meant to support academic and professional learning and performance, any meme encountered or lesson learnt could be incorporated into one's own repository to become subsequently part of one's authorship and properly credited contributions. Such a scope offers appealing and viable opportunities for many stakeholders in the educational, professional, and developmental context.

A prior paper, for example, has explored the hypothetical question: "What if an institute of higher learning would not only succeed in accomplishing the learning outcomes set, but also provide its staff, students and graduates with the means [intellectually and technologically] whose backing and support would encompass the full life span of an individual's academic and professional career?" [25]. A PKMS offers such a potential by following up on Pollard's bottom-up approach [26] and Bush's Memex [24]. As a result, the concept considerably impacts on the six observations/recommendations of Bedford's '2012 km education and training in academic institutions 2012 survey' which included (1) keeping pace with the needs of business and industry, (2) fulfilling the educational, theoretical, guiding and research roles of the discipline, (3) developing common standards supported by a formal association, (4) improving KM education programs, (5) focusing on KM's transdisciplinary nature and (6) on the depth of the broad range of topics covered [2].

## 5  The PKMS as a Technology to Confer 'Smart' Capabilities

The considerable relevance and impact of the PKM concept and devices in the context of developing applied competencies as stipulated in qualification frameworks have been assessed already [25, 28, 29]. The abilities to confer 'Smart Capabilities' [6] to initiate socio-technological co-evolutions as alluded to are not explicitly referred to in these frameworks and, hence, will be looked at in this closing section of the paper by aligning them to the six pairs of PKM4D criteria (Fig. 1) (* points out Johri's and Pal's ICT4D criteria [10] which have been integrated in the PKM4D framework).

1. Adapting (modifying one's behavior to better fit the environment) offers users effective low-cost PKMS devices (Accessibility Easiness*) supporting the notion of portable and mobile knowledge and skills (Operable Autonomy) which, in turn, allows the technology to promote six vital provisions [15] including standardized, consistent, transparent, flexible, and secure meme-based storage formats.
2. Sensing (identifying, recognizing, understanding and/or becoming aware of relevant issues) allows combining one's captured background, know-how, and experiences with own or others' ideas (Expressive Creativity*) together with developing one's emotional capital as a source of self-understanding, self-reflection, and self-determination (Collaborative Choice) which, in turn, suggests the integration of further educational support functionalities (e.g. templates, project management, advice and feedback).
3. Inferring (drawing correct conclusions on the basis of one's knowledge, experience, and observations or suggestions and evidence presented) allows for the better advancement and nurturing of one's social capital as the sum and quality of one's relationships (Relational Interactivity*) which, in turn, allows technology to foster fruitful collaborations within the PKMS community (Creative Conversations).
4. Learning (modifying existing or acquiring new knowledge, experience, skills, etc.) and subsequent authorship provides the opportunity for employing and furthering one's competencies to add productively to the World's Extelligence (Ecological Reciprocity*) providing the means for maintaining and developing one's intellectual capital for career advancement and self-actualization (Personal Mastery) which, in turn, allows for the quantitative and qualitative growth of shared repositories together with their effective curation.
5. Anticipating (thinking or reasoning to predict what is going to happen or what to do next) helps to convert individual into team or organizational performances (Institutional Performance) and eases the wider sharing and faster diffusion of ideas, sources, data, work-in-progress, etc. for the benefit of more rapid iterative improvements or technological substitutions (Innovative Capabilities) which, in turn, further allows WHOMER to evolve in regard to its curation, search, traceability, metrics, and educational services.

6. Self-organizing, self-creating, and self-sustaining (based on reflecting and subsequently consciously and purposefully changing one's situation autonomously) establishes the foundation for assisting and mentoring others in the context of the PKM4D criteria and their self-actualization (Encouraging Empowerment) and for creating enabling spaces for advancing personal and institutional objectives (Technological Progress) which, in turn, allows the PKMS technology to offer further dedicated services for leaders, educators, trainers, and their respective organizations.

After completing the test phase of the prototype, its transformation into a viable PKMS device application and a cloud-based WHOMER server is estimated to take 12 months. Further publications are also in progress or planned covering an update of the PKMS meta-concept and poster visualizations, further demonstrations [27] and tutorials/workshops, a comparison of how the PKMS trail-network compares to traditional hyperlink configurations based on the set of PKMS publications, and also how the PKMS concept compares to, can make use of and add to semantic web technologies.

# References

1. Wiig, K.M.: The importance of personal knowledge management in the knowledge society. In: Pauleen, D.J., Gorman, G.E. (eds.) Personal Knowledge Management, pp. 229–262. Gower (2011)
2. Bedford, D.A.D.: Knowledge management education and training in academic institutions in 2012. J. Inf. Knowl. Manag. **12**, 4 (2013)
3. Levy, P.: The Semantic Sphere 1. Wiley (2011)
4. Frey, J.G., Bird, C., Willoughby, C.: Diverse perceptions of smart spaces. University of Southampton (2012). http://eprints.soton.ac.uk/341019/1/Diverse_Perceptions_of_Smart_Spaces.pdf
5. Derzko, W.: Smart Technologies (2006). http://archives.ocediscovery.com/2007/presentations/Session3WalterDrezkoFINAL.pdf
6. Uskov, V., Lyamin, A., Lisitsyna, L., Sekar, B.: Smart e-learning as a student-centered biotechnical system. In: E-Learning, e-Education, and Online Training, pp. 167–175. Springer (2014)
7. Stewart, I., Cohen, J.: Figments of Reality—The Evolution of the Curious Mind. Cambridge University Press (1999)
8. Gratton, L.: The Shift—The Future of Work is Already Here. HarperCollins, UK (2011)
9. Drori, G.S.: Globalization and technology divides: Bifurcation of policy between the 'digital divide' and the 'innovation divide'. Soc. Inquiry **80**(1), 63–91 (2010)
10. Johri, A., Pal, J.: Capable and convivial design (CCD): a framework for designing ICT for human development. Inf. Technol. Dev. **18**(1), 61–75 (2012)
11. Nielsen, M.: Reinventing—The New Era of Networked Science. Princeton University Press (2011)
12. Schmitt, U.: How this paper has been created by leveraging a personal knowledge management system. In: 8th International Conference on Higher Education Program and Proceedings (ICHE). Tel Aviv, Israel, 16–18 Mar 2014, pp. 22–40 (2014d). http://dx.doi.org/10.13140/2.1.4379.1049. Accessed 5 Oct 2015

13. Hevner, A., March, S., Park, J., Ram, S.: Design science research in information systems. MIS Q. **28**(1), 75–105 (2004)
14. Schmitt, U.: Design science research championing personal knowledge management system development. Accepted Paper at the Informing Science + IT Education Conferences (InSITE2016), Vilnius, Lithuania, 27 Jun—1 Jul 2016 (2016e)
15. Schmitt, U.: Putting personal knowledge management under the macroscope of informing science. Int. J. Emerg. Transdiscipline (InformingSciJ) **18**, 145–175 (2015d). http://www.inform.nu/Articles/Vol18/ISJv18p145-175Schmitt1634.pdf. Accessed 5 Oct 2015
16. Popper, K.: Objective Knowledge—An Evolutionary Approach. Oxford University Press (1972)
17. Popper, K.: Three worlds—The Tanner lecture on human values delivered. The University of Michigan, April 7, 1978 (1978)
18. Schmitt, U.: Making sense of e-skills at the dawn of a new personal knowledge management paradigm. In: Proceedings of the 2014 e-Skills for Knowledge Production and Innovation Conference, Cape Town, South Africa, 17–21 Nov 2014 (2014k). http://proceedings.e-skillsconference.org/2014/e-skills417-447Schmitt815.pdf. Accessed 5 Oct 2015
19. Dawkins, R.: The Selfish Gene. Paw Prints (1976)
20. Nonaka, I., Takeuchi, H.: The Knowledge-Creating Company. Oxford University Press (1995)
21. Simon, H.A.: Designing organizations for an information-rich world. In: Greenberger, M. (ed.) Computers, Communication, and the Public Interest. Johns Hopkins Press, Baltimore (1971)
22. Schmitt, U.: Tools for exploration and exploitation capability: towards a co-evolution of organizational and personal knowledge management systems. Int. J. Knowl. Cult. Change Organ. Ann. Rev. (2016d). http://www.researchgate.net/publication/282852429
23. Research Reading Excerpt: Putting personal knowledge management under the macroscope of informing science. In: Gill, T.G. (ed.): Informing Science Vol. 1: Concepts and Systems (Textbook), pp. 113–118. Informing Science Institute (2015) http://informingsciencepress.com/index.php?route=product/product&product_id=137
24. Bush, V.: As we may think. Atlantic Mon. Issue **176**(1), 101–108 (1945)
25. Schmitt, U., Butchart B.A.H.: Making personal knowledge management part and parcel of higher education programme and services portfolios. J. World Univ. Forum **6**(4), 87–103 (2014g). http://www.researchgate.net/publication/268686737. Accessed 5 Oct 2015
26. Pollard, D.: PKM: A bottom-up approach to knowledge management. In: Srikantaiah, T., Koenig, M. (eds) Knowledge Management in Practice: Connections and Context, pp. 95–109. Information Today (2008)
27. Schmitt, U.: Concept and prototype of a 'next generation' personal knowledge management system. In: Prototype Demonstration at the 6th International Joint Conference on Knowledge Discovery, Knowledge Engineering and Knowledge Management (IC3K), Rome, Italy, 21–24 Oct 2014 (2014i)
28. Schmitt, U.: Innovating personal knowledge creation and exploitation. In: 2nd Global Innovation and Knowledge Academy (GIKA), Valencia, Spain, 9–11 Jul 2013 (2013c). http://dx.doi.org/10.13140/2.1.3717.6003. Accessed 5 Oct 2015
29. Schmitt, U.: The promise of autonomous personal knowledge management devices to be a key educational technology for growing a 21st century knowledge society. Paper Presented at the Emerging Technologies & Authentic Learning within Higher Vocational Education Conference 2015, Cape Town, South Africa, 31 Aug–2 Sep 2015 (2015k). http://www.researchgate.net/publication/277776326. Accessed 5 Oct 2015

# Interactive Educational System Based on Generative Approach, and the Problem of Answer Checking

Vladimir G. Danilov and Ilya S. Turuntaev

**Abstract** In this paper we consider the problem of training tasks organization, delivery and support for courses that involve mathematics. Nowadays, the level of development of IT allows to offer qualitatively new approach to solving this problem, but most of the e-Learning systems do not cover this area well. Several important steps had been made in this direction, but still some problems remain unsolved. In this paper, we introduce our interactive educational system, in which we address those problem. Among others, there is a problem of automatic answer checking. This problem, in fact, refers to the problem of establishing identities in formal mathematical system. And hence it is formally unsolvable. In order to improve the situation, we suggest to reinforce function comparison algorithm with an additional pointwise checking procedure. This, of course, leads to a possible error. In this article, we provide an analysis of the probability of such an error, which appears to be quite low in most cases. In a word, this means that the suggested pointwise checking procedure can be quite successfully used in a case where there is no way to establish equality/inequality for sure.

**Keywords** e-Learning · Training tasks · Constant problem · Computer algebra

## 1 Introduction

Practical classes and trainings on special tasks play essential role in educational process in various mathematical disciplines. The practical part of education usually includes classroom and home practice, and some assignments. This, on one hand, allows students to improve their skills in appropriate knowledge field, and to strengthen understanding of the topic. On the other hand, some reports on how

V.G. Danilov · I.S. Turuntaev (✉)
National Research University Higher School of Economics, Moscow, Russia
e-mail: isturunt@gmail.com

V.G. Danilov
e-mail: vgdanilov@mail.ru

© Springer International Publishing Switzerland 2016
V.L. Uskov et al. (eds.), *Smart Education and e-Learning 2016*,
Smart Innovation, Systems and Technologies 59,
DOI 10.1007/978-3-319-39690-3_47

527

students solve suggested problems give their teacher a picture of how the appropriate material is adopted. Such a feedback may be used in order to decide what the next step should be (whether a class can move on, or some additional job has to be done to work the theme out). Such an approach has been successfully used in mathematics education for a long time, and has proved out well. However, the process of training tasks composition and organization of assessments often appears to be a time-consuming mechanical job. Traditionally, when working on some topic in a classroom, an instructor usually provides a number of training tasks as a homework, often some kind of classroom test follows. This approach has several disadvantages. Consider the following situation. Each student in a group receives a number of training tasks (as a homework, for instance) and works on them. Some students fail in some tasks and succeed in others. After the work is done, some general review is held in a classroom where the most common mistakes are examined and students can get some additional help by asking questions. However, it is impossible provide each student with personal consultation and develop personalized plans in such an approach. At the same time, it is clear that for each student the quality of learning depends on a number of individual parameters. In the traditional approach students may suffer from a lack of helpful information and teachers get few information on the level of understanding in a group.

However, the situation can be improved. It is easy to see that all training tasks on one specific theme have the same general structure. For example, consider training tasks on differentiating a sum of functions. Each task for this rule may look like ≪Calculate the derivative of $f(g(x))$≫, where $f(x)$, $g(x)$ are some concrete functions. So given a theme one may specify a general structure (or maybe several structures) of training task, such that it defines all the concrete tasks on this theme. This idea underlies our interactive educational system for training tasks organization, delivery, and support, which is presented in this paper.

The interactive educational system is a web application which allows teachers to organize training tasks for classes involving mathematics. A teacher can run a classroom and enroll his students into it. Inside a classroom a teacher can create training tasks and assign them to students. The process of training tasks declarations exploits the suggested idea of similar structure of different tasks: the tasks in the system are declared in such a way that a number of different concrete task on the appropriate theme can be obtained from this declaration. In other words, a teacher defines a general structure—training task template—for the tasks he wants to assign to his students. Each time a student requests some task the appropriate task template is compiled into some of the appropriate concrete tasks, and the result is returned to the student. A student then may solve the requested task and return the obtained answer back to the system, which then performs automatic answer checking, and can provide the student with some additional help if the answer appears to be incorrect. More precise overview of training task templates is given in the next section. The system can be used in universities or for self-training.

Such a system structure implies a number of development problems, most of which are technically solvable and are of low interest. However, there is a problem which appears to be formally unsolvable: the problem of automatic answer checking.

Here and further we will consider training tasks with answers being a mathematical expression given in mathematical notation. Automatic answer checking in this case implies the problem of function comparison, which is known to be formally unsolvable. We perform answer checking by comparing a user-obtained answer given by a mathematical expression with the one obtained by the system using a computer algebra system that supports analytic calculations. It is a well known fact (known as the Gödel's incompleteness theorem) that for any consistent, effectively generated theory there exists a valid formula such that neither this formula nor its negation can be inferred in this theory. It is interesting that this idea is being reflected in the problem of comparing mathematical expressions. This is where the problem of identity checking may appear. As a result we end up with a possible inappropriate behavior of the system, that can treat correct answer as incorrect. However, it is possible to improve the situation. In this paper we suggest to reinforce the computer algebra system native algorithm for expressions comparison with an additional pointwise checking procedure when necessary. An error is still possible after such a reinforcement. But in this case the probability of the error can be estimated in a certain sense. We provide a detailed investigation of the probability of such an error.

As will be shown later in the article, the value of the error probability is often lower than any reasonable value. It is always possible to select the additional checking procedure attributes in such a way that this value is, for instance, of order of the probability of obtaining a non-unique result when shuffling a card deck (it happens to be about $0.27 \times 10^{-41}$ for deck of 36 cards). More precisely this is true, when we exclude some exotic cases of frequently oscillating functions.

We end up with the following scheme for checking the answer:

- compare two answers using computer algebra system's standard tools;
- if the result is indefinite, run the pointwise checking procedure in order to make the final decision.

## 2 Training Tasks in Interactive Educational Systems

As it was mentioned in the previous section, in our system training tasks are declared by means of training task template, i.e., the general structure of each task in the appropriate set. A task templates, in a first approximation, consists of three main parts: problem formulation, solution explanation, and answer. Both problem and solution consist of some explanatory text and a set of dynamic attributes (which vary from one concrete task to another). An answer is a mathematical expression (it may also vary from one concrete task to another). A special language (the Small-Task) is used to define dynamic attributes of a training task.[1] Each of such attributes is compiled into its end form (usually a mathematical expression) during the task

---

[1]Here and further, if it is clear from the context, we will use the word "task" to denote tasks by means of the interactive educational system, i.e., tasks templates. We will use term "concrete task" when talking about some concrete implementation of tasks template.

compilation. The SmallTask language was designed specifically for the needs of training task declaration. Its expressions are always kept in the original form so it is always possible to represent them with no evaluation. This property is essential for the system since it allows to use these expressions in problem-like formulations where we always need non-preevaluated results. Another significant demand is that all expressions have to be easily transformed in printing form (LaTeX in our case). The SmallTask language is designed to be as easy and user-friendly as possible so anyone could use it in order to create his own training tasks with the minimal requirements of exterior knowledge.

A teacher declares a task with a special task designer, which allows to define task's dynamic attributes in terms of a random element of some set, rather then some concrete value. All of the SmallTask expressions in some task declaration are evaluated line by line staring from the problem, then continuing with the solution, and ending with the answer. So the attributes, used in problem formulation can be further used in solution and answer. The latter has a reasonable restriction: an answer must be a SmallTask expression which evaluates in a valid mathematical expression.

During task compilation each variable, whose value was declared to be random, is assigned a concrete value picked randomly from the appropriate set. Thus, when a student requests a task he is given a unique task formulation. The appropriate concrete task solution and answer (which is a correct mathematical expression at the moment) are stored on the server in order to perform answer checking and return the correct solution if needed.

For each student a statistics of achievements is stored: for each solving attempt the result (whether a student succeeded or not) is saved to the database with appropriate task id, student id, and the time stamp. An attempt is considered failed if the answer obtained by the student and the one calculated by the system do not match. When failed a student is proposed to try once again (but this time he will get another concrete task, since task compilation includes randomness). So given a student and some task the system provides a full history of the appropriate solving attempts (which can be viewed by a teacher). This allows to observe the progress of each student personally, which forms a clear image of how a theme has been adopted among a group. Compare it to the traditional home assignment, which consists of some fixed number of training tasks. In this case a teacher can retrieve at most a number of yes/no information for each student and each task. Particularly, such an approach gives no information on how many attempts did it take student S to find the correct result for task T. At the same time there is a clear difference between those students who solve T (and similar tasks) in first try, and those who need several attempts to achieve the correct answer. It is most likely that the latter need to do some additional work to achieve an acceptable level of understanding the appropriate material. This implies a new measure of the level of adoption for students: the number of successful attempts to the number of all attempts ratio. This value illustrates the level of understanding well for some reasonable total number of attempts. In the suggested system this allows to formulate home tasks in terms of success percentage: for instance, the traditional ≪Solve 5 tasks on chain rule≫ may become ≪Solve at least 5 tasks on chain rule to achieve 90 % of successes≫.

In e-Learning systems the problem of training tasks organization is seldom considered precisely. However there exist some projects that have relative aims. The most common is the Maple T.A. project by Maplesoft. Maple T.A. is a web-based testing suite for mathematical courses, built on top of the Maple computer algebra system (CAS). This system allows users to organize their own virtual classes and to populate them with different kinds of training tasks (home assignments, quizzes, etc.). Training task development is done with the special question editor, which provides a user-friendly interface for the declaration of various task attributes. The Maple CAS language is used to define complex algorithmic parts of tasks. The latter means that it is possible to create almost any kind of tasks, with some complex calculations, and introduce a desired randomness. However, this also means that one needs to be well up with the Maple language in order to create something more complex than what the interactive editor allows one to do. That may be a serious problem, since in fact many instructors may refuse to spend time on learning a language that they will not probably use elsewhere. Another thing about the Maple language is that it was originally created for another purpose, rather than for training tasks declaration. This implies that many constructions relative to such tasks development will look too complex and sometimes may be really hard to describe using this language. There also are some security issues that follow from the origins of that language.

Among others, there is a problem of automatic answer checking which was mentioned in the introduction. The Maple T.A. developers claim that the system performs correct answer checking and respects mathematical equivalence. This of course is a mandatory requirement for such systems, but such a feature is questionable, since there are some theoretical restrictions, that influence it: the problem is known to be formally unsolvable. In our project a certain CAS is used to perform analytic calculations. Particularly, it is used to provide automatic answer checking by comparing the two given answers. Albeit there is no way to make this checking absolutely correct, there is a way to improve checking result and increase the probability of success. This problem is described in detail in the next section.

## 3 The Problem of Automatic Answer Checking

As it was mentioned earlier, the problem of automatic answer checking when an answer is given by a mathematical expression is formally unsolvable. The reason is that this problem comes to the problem of determining whether two mathematical expressions are semantically equal. We say that two expressions $f_1$ and $f_2$ are semantically equal (we will use the $f_1 \equiv e_2$ notation further in the article), if both $f_1$ and $f_2$ represent the same mathematical object. In contrary, we call $f_1$ and $f_2$ syntactically equal if they have exactly the same representation in mathematical notation. For example, $2^x$ and $e^{x \cdot \log(2)}$ are syntactically different, however they are semantically equal since they both represent the same mathematical function. Here and further we will consider functions of one variable.

Obviously, the problem of establishing semantic equality $f_1 \overset{?}{\equiv} f_2$ is absolutely equal to the problem of determining whether a particular expression equals zero $(f \overset{\Delta}{=} f_1 - f_2 \overset{?}{\equiv} 0)$. The latter is known as the Constant Problem (or the identity problem). It is known that there may exist no algorithm for solving it in a class of functions that include $\log 2$, $\pi$, $e^x$, $\sin(x)$ and $|x|$. (see [1], also [2]) The appropriate result is known as the Richardson's theorem.

The constant problem undecidability has a serious impact on the interactive educational system development. In fact, it means that there is no way to guarantee answer checking correctness. Though the problem may seem a bit far-fetched, it still needs to be examined, since it can appear unexpectedly and it seems impossible to predict such a situation. Here's an example. The CAS we use in our interactive educational system appeared to fail on checking whether $\frac{sec(acsc(x))}{x^2 \cdot \sqrt{1-1/x^2}} \cdot \tan(acsc(x))$ equals $\frac{-1}{(1-\frac{1}{x^2})^{\frac{3}{2}} \cdot x^3}$, while both of them represent the derivative of $\frac{1}{\sqrt{1-\frac{1}{x^2}}}$ with respect to $x$. As a result we received a global failure since the system was unable to find out whether the answer was correct or not.

For most CAS, facing such a problem results in the *None* answer (or something alike) instead of *True* or *False*. Such an answer should be treated as the "I don't know"-answer, which in our case means that we are unable to say for sure whether the answer is correct or not. But still we have to make a decision. At this stage it is too late to just reload the task and hope that the problem will not appear again for a while: the user had already spent his time on solving the task and it would be at least indecently to ignore the work he had done. And, on other hand, there is absolutely no guarantee that the problem case was accidental. It may appear that it is caused by the task template definition itself. So we have to determine our system's behavior on case the CAS is unable to provide accurate result. There are two trivial ways to do this: always consider such an ambiguous answer to be correct or, on the contrary, always return negative comparison result in such situations. While the first alternative seems quite careless, the second one may also lead to critical misunderstanding. A student is unfairly punished in the latter case, and that will definitely lead to bad ratings and rejection of the system among students. Against this background the first alternative is preferable. But still there is a better solution.

In order to cope with the influence of the constant problem on answer checking in an interactive education system we suggest an additional pointwise checking procedure. Consider task $T$ with an answer that is a mathematical function of a single variable. Let $f_{real}(x)$ be the real answer to $T$ problem obtained by the system, and let $f_{user}(x)$ be a student-calculated answer. The formal definition of the additional pointwise checking procedure is given in [3], here we will just take a look at its notion. The algorithm accepts $f(x) = f_{real}(x) - f_{user}(x)$ and tries to determine whether the equality $f(x) \equiv 0$ is correct. In order to do this it checks the given function $f$ value in a number of random points from the given segment ($[A, B]$); and if it encounters a non-zero value then the *False* answer is returned, otherwise it continues until the maximum number of check-points ($m$) is reached. We take these check-points from $\mathcal{U}(A, B)$.[2] The *True* value is returned (which means ≪yes, the function equals 0≫) if and only if all of the check-points ($x_1, \ldots, x_m$) are such that $f(x_i) = 0$ (for $i = 1, \ldots, m$). We also require all of the checking points to be different (we store a set of "used" points for that purpose) since there is no need to check function value at some point twice.

The suggested algorithm is rather simple. But still it has several advantages. Firstly, it is "student-friendly", meaning that it won't fail if the answer is correct. The only case the Algorithm[3] fails is when it occasionally picks all $m$ points from the zero set of the function $f$ being tested, while in fact $f \not\equiv 0$. In this case it will return positive result after checking all $m$ points, which is wrong since in fact student's answer does not match the right one. It is obvious that for $m \to \infty$ the probability of the Algorithm failure converges to 0. The good news are that, as it will be shown further in the article, it is sufficient to take a reasonable number of check-points $m$ to achieve acceptable value of the probability of the algorithm error.

## 3.1 The Probability of the Pointwise Checking Procedure Error

In this section we consider the probability of the suggested algorithm failure. It will be show that this probability appears to be rather low in most cases, and so the algorithm can be successfully used in practice.

Consider $f = f_{real} - f_{user}$, a real-valued elementary function. And let $A, B \in \mathbb{R}$ be such that $A < B$ and $f$ is analytic on $[A, B]$. Assuming that CAS native algorithm for equality checking is unable to determine whether $f \equiv 0$, we use the suggested algorithm to make the final decision.[4] Let $k \in \mathbb{N}$ be an upper estimate for the number of zeros of function $f$ on $[A, B]$ if actually $f \not\equiv 0$. And let $m$ denote the maximum number of check points of the algorithm. Then we have the following statement:

It is easy to see that the Algorithm fails if and only if the given function is in fact non-zero, but it may have more zeros than the maximum number of check points (i.e., $k \geq m$). In order to estimate the suggested additional pointwise checking procedure correctness, we analyze the probability of its failure.

We consider our functions to be analytic functions, and thus a non-zero function is not allowed to have an infinite number of zeros inside a closed segment.

---

[2] $\mathcal{U}(A, B)$ denotes the uniform distribution on $[A, B]$ segment.

[3] Here and further we will denote the suggested pointwise checking procedure by "the Algorithm".

[4] Here and further we assume that all machine calculations of $f$ values are accurate and the $f(d) == 0$ condition works in a pure mathematical way.

Now if $[A, B]$ is a continuous closed segment and the Algorithm picks its check points from a continuous uniform distribution on $[A, B]$, then the failure probability of the Algorithm equals 0, since it is given by the probability of getting into a discrete subset of a continuous set.

But that is not the case. Talking about the computer implementation of the suggested algorithm, we have to take in account the machine arithmetic. So in this case the $[A, B]$ segment becomes a discrete sequence of points defined by the floating-point numerical system we use, a grid of floating-point numbers with spacings determined by the appropriate rounding procedure. So let $AB$ be this set of floating-point numbers inside $[A, B]$, and let $M = |AB|$. And let $Z$ denote the zero set of a function $f$ being examined (i.e., a set of points, such that $f$ equals 0 in each of these points). At this point of view the probability of the algorithm error is given by the probability of picking $m$ points one by one from a $k$-subset of the set of $M$ points.

The probability of choosing one point from set $AB$ of $M$ elements such that it belongs to the $k$-subset of $AB$ is given by (1).

$$\mathbb{P}(d \in Z) = \frac{k}{M} \tag{1}$$

Once the first point $d_0$ had been picked (such that $d_0 \in Z$), there are only $(k - 1)$ points left unknown in the zero set of $f$, and $(M - 1)$ points in $AB$ left unchecked. So the conditional probability of picking the second point from $Z$ becomes $\frac{k-1}{M-1}$. Continuing the same reasoning we end up with the formula (2) which gives the probability of picking $m$ points in a row from $AB$, in such a way that they all belong to $Z$.

$$p_{M,m,k} = \mathbb{P}\left(d_{m-1} \in Z, d_{m-1} \in Z, \ldots, d_0 \in Z\right) = \prod_{i=0}^{m-1} \frac{k-i}{M-i} = \frac{k!(M-m)!}{M!(k-m)!} \tag{2}$$

Formula (2) holds for $m \leq k \leq M$. As it was shown before, the algorithm always returns correct result when $k < m$, and thus the probability of error in this case equals 0. In the case $k > M$ we consider the error probability equal to 1. So finally the full probability of the Algorithm error is given by expression (3).

$$P_{err} = \begin{cases} 0, & k < m, \\ p_{M,m,k}, & m \leq k \leq M, \\ 1, & k > M. \end{cases} \tag{3}$$

$P_{err}$, given by (3), depends on three parameters: the number of floating-point numbers inside $[A, B]$ ($M$), the maximum number of points to be checked ($m$), and the maximum number of zeros of a function on $[A, B]$ segment $(k)$. The exact value of $M$ is defined by the floating-point numerical system used, namely its rounding procedure. Consider a floating-point numerical system with base $\beta$ and precision $p$. Let real number $\tilde{x}$ be represented by $x = d_0, d_1 \ldots d_{p-1} \cdot \beta^e$. Then the error of such a representation can be as large as $\left(\frac{\beta}{2}\beta^{-p}\right)\beta^e = \varepsilon \cdot \beta^e$, where $\varepsilon$ denotes the machine epsilon of the numerical system (see [4] for complete explanation). This implies in turn that the floating-point grid spacing is not decreasing with respect to the increasing magnitude of a real number. So, in order to introduce a lower estimate for the number of points inside a segment we can approximate all spacings inside this segment with the spacing near its right bound. So let $\tilde{M}$ denote the exact amount of floating-point numbers inside $[A, B]$, and let $ulp(x) > 0$ be[5] the distance between the two closest floating-point numbers around $x$, i.e. $ulp(x) = b - a$ for $a \leq x \leq b$, and $a \neq b$. Then the following inequality (4) is correct.

$$M = \left\lceil \frac{B - A}{ulp(B)} \right\rceil \leq \tilde{M} \tag{4}$$

The suggested estimate may seem a little rough but it is sufficient for our needs. The point is that this value is always much bigger then any imaginable value of $k$. For instance, the value of $M$ for the segment $[10, 15]$ and the **NumPy.float64** data type happens to be of order $3 \times 10^{15}$. It seems almost impossible for a student to provide an incorrect answer $f_{user}$ such that it coincides with the correct answer in such an amount of points that is nearly as large as that. So if $k \ll M$ then $\frac{k}{M} \ll 1$ which makes the error probability quite small. And the whole expression given by (2) is getting smaller as we increase the value of $m$.

A more precise analysis of the probability of the Algorithm error is shown in [3]. The result is that the values of $p_{M,m,k}$ are quite low for even huge values of $k$ and relatively small amount of check points. For example, when checking a function with $10^6$ zeros on $[A, B] = [10, 20]$ it is sufficient to check up to 10 points to ensure the probability of error to be lower than $e^{-200}$ (for **NumPy.float64** floating point system). It should also be noted that given some fixed values of $M$ and $k$ the increase in $m$ makes the probability to decrease faster than $\exp(-m)$. In order to illustrate the probability of error we some values of $\log(p_{M,m,k})$ on $[10, 20]$ segment are provided in Table 1. We take logarithm of $p_{M,m,k}$ here since it is much easier to calculate. For example, the first row of the table shows that picking up to 10 points randomly from $[10, 20]$ segment the algorithm fails with the probability of order $e^{-222}$ if the given function may have up to 1,000,000 zeros on this segment.

---

[5]Here we use term *ulp* which refers to the **unit in the last place** as it was defined in [5].

**Table 1** Some values of $\log(p_{M,m,k})$

| $m$ | $k$ | $\log(p_{M,m,k})$ |
|---|---|---|
| 10 | 1000000.0 | −222.281473312 |
| 10 | 2000000.0 | −215.349979006 |
| 10 | 3000000.0 | −211.295320425 |
| 20 | 1000000.0 | −444.563046624 |
| 20 | 2000000.0 | −430.700008012 |
| 20 | 3000000.0 | −422.590674183 |
| 30 | 1000000.0 | −666.844719939 |
| 30 | 2000000.0 | −646.050087019 |
| 30 | 3000000.0 | −633.886061275 |
| 40 | 1000000.0 | −889.126493256 |
| 40 | 2000000.0 | −861.400216026 |
| 40 | 3000000.0 | −845.181481701 |
| 50 | 1000000.0 | −1111.40836658 |
| 50 | 2000000.0 | −1076.75039503 |
| 50 | 3000000.0 | −1056.47693546 |

## 4 Conclusion

When teaching disciplines involving mathematics, it is important that students adopt the key concepts and methods well, especially on early stages. Otherwise they would face serious problems in further education. Training on specially designed problems is a well known way of working these concepts and methods off. In the suggested system we aim to bring this process to the next level by reducing some disadvantages of the traditional approach. It can provide a very large number of similar training tasks given one simple description, which allows to state that each student in a classroom is assigned a personalized set of almost unique tasks. And gathering personal statistics in the system gives teachers a clear image of the level of understanding among students. It is also important to note that traditionally a student does not get specific information when facing problems in solving some task: usually, he can just check whether his answer is right or wrong (comparing it to the correct one). However, such an information often appears to be insufficient for finding the mistake and its origins. In our system each task is assigned an appropriate solution which can be of great help for those who face problems. The system can be used both by teachers (in order to organize training for their students) and by any other people interested in self-education.

Automatic answer checking is the key idea and an essential part of the suggested project and of all similar projects. It is important to understand that in case of answers given in mathematical notation it is impossible to make this check absolutely accurate. This problem is never mentioned and the appropriate examples (where the problem really appears) are likely to be quite exotic. However, the appearance of such examples is unpredictable and this may result into an inappropriate behavior of the systems which provide the functionality of automatic answer checking. The algorithm suggested in this article helps to decrease the risk of mistakes caused by constant problem. We show that it is easy to achieve a very low probability of error of such algorithm, and thus it can be used in practice in e-Learning systems that deal with mathematical expressions.

# References

1. Richardson, D.: Some undecidable problems involving elementary functions of a real variable. J. Symbolic Logic **33**(04), 514–520 (1969)
2. Laczkovich, M.: The removal of $\pi$ from some undecidable problems involving elementary functions. Proc. Am. Math. Soc. **131**(7), 2235–2240 (2003)
3. Danilov, V.G., Turuntaev, I.S.: Reliability of checking an answer given by a mathematical expression in interactive learning systems. J. Interact. Learn. Res. (2016) (in press). arXiv:1602.00243
4. Goldberg, D.: What every computer scientist should know about floating-point arithmetic. ACM Comput. Surv. (CSUR) **23**(1), 5–48 (1991)
5. Harrison, J.: A machine-checked theory of floating point arithmetic. In: Theorem Proving in Higher Order Logics, pp. 113–130. Springer (1999)

# Part V
# Smart Technology as a Resource to Improve Education and Professional Training

# Smart Technologies in Foreign Language Students' Autonomous Learning

Natalya Gerova, Marina Lapenok and Irina Sheina

**Abstract** The article deals with the issues of FL students' autonomous learning organized with the help of Smart technologies. The authors suggest methodological approaches to implementation of computer-based technologies in FL higher education and pre-requisites for integration of computer and communicative competences. Interaction is viewed as a key factor of enhancing motivation, intensification of the teaching and learning process and improving the results of autonomous work. As an example of Moodle-based instruction the authors give a chain of tasks aimed at developing the skill of using the Sequence of Tenses, which is a specific feature of the English language. The article also contains the results of a pedagogical experiment aimed at evaluating the efficiency of Smart technologies for grammar competence development.

**Keywords** Autonomous learning · Competence · ICT · Information activity · Proficiency level · Smart technologies

## 1 Introduction

The structure of a teacher's professional activities is now in the focus of research and the widely accepted approach is that they go beyond the scope of teaching and learning activities within the subject area. They necessarily include acquirement of

N. Gerova (✉) · I. Sheina
Ryazan State University named after S.A. Esenin,
Ryazan, Russian Federation
e-mail: nat.gerova@gmail.com

I. Sheina
e-mail: i.sheina@rsu.edu.ru

M. Lapenok
Ural State Pedagogical University, Yekaterinburg,
Russian Federation
e-mail: lapyonok@uspu.ru

© Springer International Publishing Switzerland 2016
V.L. Uskov et al. (eds.), *Smart Education and e-Learning 2016*,
Smart Innovation, Systems and Technologies 59,
DOI 10.1007/978-3-319-39690-3_48

theoretical knowledge and practical skills in the field of computer technologies. Experience in this field acquired by students as a part of their educational activities in a higher education institution lay foundation for developing the skills of applying computer technologies in their work at school [6, 8, 9]. The Federal on Education emphasizes the necessity to implement modern technologies in the processes of teaching and learning [3].

## 2 Components of Students' Informational Activity

At the first stage of research we identify the structure of foreign language students' informational activity and the main components of their computer literacy necessary for their work as school teachers.

To organize any kind of computer-based professional activity we need to ensure students' computer literacy throughout their educational program. During this period students develop computer literacy and become competent professionals prepared to use computer technologies in all kinds of their activities. Computer literacy is an indicator of professional competence of a would-be teacher.

Nowadays university educational environment includes classroom, laboratory and administrative areas encompassing a number of interconnected divisions whose activities are backed up by computer technologies.

Students master the major educational programs, attend classes of different types, which result is their assessment. They accumulate their achievements that show their efficiency in learning, research, sports, arts, extracurricular activities.

Bachelor students majoring in foreign language education should be prepared to solve a number of professional tasks: to identify their students' abilities and interests, to envisage the results of their learning and personal development, to work out effective computer-based teaching strategies to help their students master the foreign language. Outside the classroom they should be prepared to organize the school's cultural environment for which computer technologies can be very helpful.

Master students are dealing with a broader scope of tasks and master programs presuppose greater autonomy. Students become involved in other types of activities: research, management, project work, development of instructional technologies. Masters should be more competent at designing the route for their self-education and self-development.

In their autonomous research work they analyze and systematize data, do experiments, make generalizations on the basis of a number of research methods, assess the results of their analysis, interact with their colleagues in their home country and abroad.

In the field of educational management master students study the present-day state and the potential of the system of education, learn to organize co-operation with colleagues and social partners, learn how to organize the work and development of an educational institution and to ensure quality management.

Master students' project work includes designing programs and individual trajectories, content for new disciplines and elective courses, methods and forms of assessment, including computer-based methods.

Developing instructional technologies master students analyze teachers' needs and design instructional materials, develop innovative teaching strategies including computer-based tasks and distant forms of teaching and learning.

Designing extracurricular activities master students work out strategies for outreach and awareness-raising programs employing computer technologies to enhance the efficiency of such programs.

## 3 Methodological Approaches to the Development of Students' Computer Competence

Modern educational programs in universities are based on integration of several approaches, such as: the systemic approach, the hands-on approach, the integration approach, and the project-based approach. The structure and the content of the programs ensure integration of methods, strategies and means on different levels of the educational process. Curriculum practical training and teaching practicum as well as autonomous learning include computer-based activities and employment of SMART technologies.

The model of foreign language (hereinafter—FL) students' computer competence includes the components and the content of their computer-based professional activities, the list of computer skills, the types of tasks that require the use of computer technologies, the algorithms of dealing with these tasks, the characteristics of a role model teacher who implements innovative technologies in the teaching process.

Computer competence acquisition is directly connected with use of computer technologies in the process of learning. Students learn how to use imputer technologies at the stage of explanation, training and assessment and it helps to shape their professional experience.

The goals of higher education include development of creative thinking and the ability to gain knowledge autonomously and structure it with the help of computer technologies [7]. This goal can be achieved through combination of individual and group work, discipline-oriented and interdisciplinary activities. Individual work consists in setting up personal objectives of learning, building individual strategies to gain more profound knowledge of the subject, development of personal abilities to face challenges. Group work develops abilities to co-operate in taking decisions, leadership qualities, tactfulness and team spirit. Computer-based technologies help to cultivate the abilities to present the results of autonomous work and participate in discussions, develop the skills of logical argumentation and ensure personal awareness in evaluation of the outcomes. Interdisciplinary tasks help to integrate knowledge obtained from different sources and fields of research and view issues from different perspectives. The interdisciplinary approach also makes computer skills a part of students' professional competence. Within one discipline computer-based technologies can be employed in simulation games, interactive computer games, case studies, debates, etc. [5].

Autonomous work can be done under the teacher's guidance or without it.

A teacher's professionalism is to a large degree determined by his/her ability to interpret information which is significant from the professional point of view and necessary for self-development. A teacher's creative personality is shaped with the help of innovative technologies and is manifested in his/her abilities to apply practical skills at using these technologies in everyday work. At each stage of learning students discover the opportunities that are offered by computer technologies for their professional activities.

FL students' computer competence is developed by adapting methods, strategies and teaching materials to the needs of foreign language acquisition and also be implementing computer technologies in the educational process and in the program management. Students use electronic learning resources, they are instructed how to use the Internet for teaching and learning purposes and for professional networking and how to use computer based forms of assessment.

## 4   Students' Computer-Based Autonomous Learning

Students' computer-based autonomous work requires a number of intellectual skills, such as to analyze, to compare, to find commonalities and difference, to make generalization, to project etc., and also the ability to draw information from all kinds of resources.

Students' autonomous work at a university can is viewed as scheduled (envisaged by the curriculum) independent learning and research activities guided by teachers. The main goal of autonomous work is to facilitate students' intellectual efforts when their communicative and computer competences are being developed and they are acquiring and accumulating experience in computer-based teaching and learning. Autonomous work requires an individual approach, it must be continuous and systemic [6].

Students' autonomous learning management includes the following steps: specifying goals and objectives of autonomous work; identifying students' abilities and motivation for self-teaching, selection of methods, means and ways of achieving the goals; step-by-step planning of students' work; means of formative assessment.

The pre-requisites for students' autonomous learning are the following: effective correlation of classroom part of the curriculum and autonomous work, availability of well-equipped laboratories and modern software, access to the local and global networks, electronic learning resources. Teachers monitor students' autonomous work, check up on it, encourage students' sense of accomplishment and make adjustments if necessary.

Tasks for autonomous learning should be practice-oriented and aimed and developing professional competences. They should be based on previous knowledge and skills and ensure integration of different disciplines including computer science. Regular monitoring maintains feedback which enables the teacher to understand what topics should be given more time. Students become aware of their achievements and professional growth and anticipate their grade at the end of the semester.

Selecting SMART technologies for autonomous learning we should take into account their advantages. SMART technologies are flexible in use, they ensure operating in all the types of communicative activities (speaking, listening, reading, writing), they provide feedback and the opportunity to correct mistakes while doing the task, they open a wide access to authentic materials in the foreign language. They can be used to organize a wide range of professional activities in education: enhance motivation for learning, ensure personal learning styles, construct make-belief situations, integrate knowledge obtained from different subjects, assess students' achievements, make adjustments to the teaching strategies.

But their use should be well-grounded, goal-oriented, functional and systemic; they should widen the scope of teaching strategies. Then only will SMART technologies become an effective tool for developing students' creative approach to their profession and a valuable component of teacher-training programs.

In the recent years SMART technologies have changed their role in education due to rapid development of distant forms of teaching and learning. They are instrumental in meeting the needs of various groups of students in the Information Age when the scope of available information is growing rapidly.

In FL teacher-training programs autonomous learning is focused on foreign language acquisition. Tasks that develop students' communicative competence can be classified as follows:

- Exercises to improve phonetic skills that presuppose listening to model pronunciation of sounds and words, contrastive pairs of words, phrases which are difficult in term of articulation, etc. and trying to imitate the model.
- Exercises to obtain skills at appropriate use of vocabulary items that are intended to teach students to differentiate between meanings of one word and between meanings of words from the same semantic domain, to see the difference between the contextual meaning and the dictionary meaning, to use words according to their lexical and grammatical combinability.
- Exercises to obtain and develop grammatical skills that are aimed at differentiation of grammatical phenomena, learning their forms, identification of their functions in speech, learning how to use them in different contexts.
- Exercises developing reading skills, such as search for the necessary information, identification of the main idea of the text and its logical structure, the author's point of view or attitude, the author' arguments or rhetorical predicates, anticipate information etc.

- Exercises developing skills at listening, such as ability to differentiate sound form of different words, hold fragments of oral speech in operative memory and process them unit by unit, anticipate the incoming information, etc.
- Exercises to develop speaking skills which include abilities to identify the communicative intention of the utterance, to select the necessary lexical and grammatical means to fulfill the intention, to identify the main concept that is to be specified, to structure the utterance logically, ask for information and give it, explain, prove, persuade etc.
- Exercises to develop writing skills which include selecting the topic of the text, the necessary lexical and grammatical units, the necessary rhetorical predicates, identifying the text composition (introduction, the main part, conclusion), etc. which result in writing essays, descriptions, narrations, expository and argumentative texts.

The Internet offers a wide range of opportunities to develop these skills, among them are Foreign Language Education Sites, Foreign Language Lesson Plans and Resources, Pronunciator, sites of the BBC, CNN, ABCNews, Britannica, "Studying phonetics on the Internet", the ERIC Clearinghouse on Languages and Linguistics, electronic libraries and many others.

New interactive methods of instruction require a special electronic environment which can be created with the help of a learning platform, such as Moodle. Moodle is specifically designed to support the delivery of teaching materials and activities which are collected in one online location offering users anytime anywhere access [1]. Moodle provides a number of interactive activities including forums, wikis, quizzes, surveys, chat and peer-to-peer activities. Moodle provides tools that empower both teaching and learning. Using Moodle teachers can:

- Put his/her teaching materials online.
- Add materials from other sources (e.g., electronic textbooks) and links to Internet resources, video courses and video lessons.
- Administer tests which can be time-limited with a limited number of tries.
- Monitor students' autonomous work and control their use of the suggested material (time of the beginning and time of the end).
- Facilitate students' interaction during project work, simulation games, discussions and mini-conferences.
- Tailor tasks to individual learning styles.
- Have operative feedback with students.

When students use Moodle they:

- Study the suggested material online or save them on electronic data storage devices.
- Submit their assignments to the teacher to be checked and corrected.
- Consult with the teacher if they have questions.
- Discuss question in forums with other users.
- Do tests and get their grades immediately after the test is finished.

- Use additional materials and internet resources.
- Use glossaries to expand their vocabulary.
- Put their projects online to get feedback of the peers.
- Participate in online discussions and mini-conferences.
- Have more opportunities of self-checking and reflection.

Using Moodle teachers allow all the participants of the learning situation to share their ideas and their experience and elicit more knowledge from others. They make the teaching and learning environment more flexible in time and space. It also gives teachers the opportunity to readjust the assignments and the schedule, add new activities to help everyone. New ideas and new activities may come up during discussions.

Moodle makes students' autonomous learning more meaningful as they can supplement and update databases themselves, they can share documents, photos and other materials with others, they can even control the process of teaching through their questions and suggestions.

## 5 Results of the Experiment

Our pedagogical experiment was aimed at assessing students' grammar proficiency and evaluating some components of our teaching methodology on the basis of Moodle. Grammar skills acquisition can be time-consuming and requires a lot of drill, which can and should be done outside the classroom.

To assess students' proficiency we identified the main components of grammar competence. They are:

- Ability to recognize grammar phenomena and identify their meaning and function in a sentences.
- Ability to identify the syntactic constructions that are used to compose different types of sentences (simple, complex, compound) and to identify the members of a sentence, their structure and semantic relations between them.
- Knowledge of parts of speech, their categorical meanings and functions in a sentence.
- Knowledge of the rules governing combination of different parts of speech and ability to apply them.
- Ability to compose sentences of different types according to the rules of the foreign language.
- Ability to select and combine the necessary grammar phenomena to achieve the communicative goal of an utterance.

As an example of Moodle-based instruction we can give a chain of tasks aimed at developing the skill of using the Sequence of Tenses, which is a specific feature of the English language.

**Task 1**. If the verb in the principal clause is in the past (*said, asked, wondered, inquired, exclaimed,* etc.), the verb in the subordinate clause will also have one of the past-time forms. This is **the rule of the Sequence of Tenses**. Analyze how the forms are changed and highlight the verb forms in both groups of sentences.

| | |
|---|---|
| I wonder if she really means it | I wondered if she really meant it |
| I will take a holiday in June | He said he would take a holiday in June |
| I've never promised such a thing! | She exclaimed she had never promised such a thing |
| I gave two lectures last week | He said he had had two lectures the week before (the previous week) |
| We'll go by plane if the weather is good | They said they would go by plane if the weather was good |

**Task 2**. Compose meaningful sentences using clauses from the left and right column.

| | |
|---|---|
| I was trying hard to remember | ...what he was thinking about |
| The man asked me | ...if there would be a film on TV |
| I couldn't help wondering | ...if I could tell him how to get there |
| I didn't know exactly | ...where and when we had met |
| He didn't quite understand | ...if I was joking or speaking seriously |
| It wasn't tactful to ask me | ...how much I had paid for it |
| He wasn't quite sure | ...why my (his, her, their) plan had failed |

**Task 3**. Write the sentences in the indirect speech.

"Can you take me home in your car?"—She wanted (hoped)...

"Did Atlantis ever exist or is it no more than a myth?"—He said he had no idea ...

"Oh Steve, thank you ever so much for helping me!"—She...

"I took part in a car race last year."—He boasted that ...

"Does the driver know the way to the stadium?"—I wondered ...

"I saw him leaving the bank at 6 yesterday."—She told the police ...

"Will you phone me from Chicago?" she asked me.—She asked ...

"What speed were you driving at?" the policeman asked. But the man had no idea ...

**Task 4.** Make mini-dialogues: explain why you didn't do something the other person expected you to do.

| Why didn't you ... (yesterday)? | I thought | (to have) enough |
|---|---|---|
| *come in time* | | you (to be busy) |
| *wake me up* | | somebody else (to do something) |
| *phone me* | I hoped | you (to expect) me to come at ... (9, 10) |
| *do the shopping* | | |
| *buy some bread* | | |
| *wait for me,* | I was sure | you (not to work) on Saturday. |
| *finish the translation* | | you (to go) to bed |
| *ask me for help; (turn to me for help)* | I was afraid | you (to leave) home |
| | | you (not to be back) until Sunday |
| *meet me at the station* | I didn't know | I (can or to be able to) ... do it alone |
| *make dinner* | | you already (to buy) some |
| | | you (get) ready for your lecture |
| | | classes (to start) at 12, not at ... |

**Task 5.** Complete the following explanations.

- I'm sorry I didn't wait for you yesterday. I thought.....
- She never spoke to people and she never came to our parties. At first I thought she was too proud and conceited. And then I found out...
- Of course, I refused to go to the country! I knew perfectly well...
- I didn't ask you to help me because I was afraid (that)...
- I am not a busy-body! I never asked her ... I only wondered ...

The tests were drawn up according to the components of the grammar competence. They were designed to assess students' skills at the topics: "The structure of an English sentence" and "Types of predicates in an English sentence".

The experiment was carried out in 2013–2015. The number of participants at all the stages made 182 students. The results were processed with the help of the methods presented in "Proficiency Measurement while Realizing Information Activity of Students Majoring in Pedagogical Education" [4]. The students in the experimental group and in the control group at the beginning of the experiment had the same proficiency in English. At the final stage both groups were tested to measure their proficiency in the afore-mentioned topics. The test consisted of 36 tasks. Each correct choice was given 1 point, each wrong choice was zero.

To find the level of the students' proficiency we used methods suggested by V. P. Bespalko and divided the levels of proficiency into reproductive, adaptive, heuristic and creative [2].

At the ascertaining stage of the experiment the students' English language proficiency was equal. Table 1 shows the results of their proficiency comparison.

At the final stage of the experiment both groups were given tests measuring their grammar skills at the topics: "The structure of an English sentence" and "Types of

**Table 1** Proficiency level at the ascertaining stage

| Proficiency level | Control group | Experimental group |
|---|---|---|
| Reproductive | 38 | 41 |
| Adaptive | 50 | 48 |
| Heuristic | 2 | 3 |
| Creative | 0 | 0 |

predicates in an English sentence". Table 2 shows the difference in their proficiency: the experimental group has a higher proficiency level.

Comparison of the two groups can be shown in the form of a bar graph which is presented in Fig. 1.

The majority of the students in the experimental group achieved higher degrees of the heuristic and creative levels of proficiency.

The results of the pedagogical experiment show that FL students' Smart-based autonomous learning facilitates grammar competence development and is instrumental to interactive and effective instruction.

**Table 2** Proficiency level at the final stage

| Proficiency level | Control group | Experimental group |
|---|---|---|
| Reproductive | 12 | 0 |
| Adaptive | 56 | 24 |
| Heuristic | 20 | 49 |
| Creative | 2 | 19 |

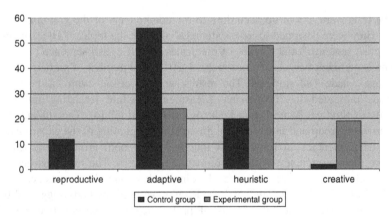

**Fig. 1** Proficiency level comparison between the groups

# 6 Conclusion and Future Work

Development and implementation of Smart technologies in higher education determines the development of foreign language methodology and the strategies of computer-based teaching and learning. It ensures intensification of students' intellectual development, enhances motivation and makes students' autonomous work more meaningful and interesting. Use of Moodle, which presupposes teaching material design, putting them online, students' autonomous work with them, acquisition of skills at work with information systems, helps to maintain interdisciplinary links and shapes students' computer competence as a part of their professional competence.

# References

1. Baichurina A.S.: Using computer-based teaching and learning environment on the basis of "Moodle" to organize students' autonomous learning. In: Organizing Students' Autonomous Work: Thesis of the 2nd All-Russian Internet-Conference. Saratov: "Noviy Proyekt", pp. 29–34, 6–9 Dec 2013)
2. Bespalko, V.P.: On the criteria of the quality of professional education. Vestnik Vysshey Shkoly (1), 3–8 (1988)
3. Federal Law: On Education in the Russian Federation—FL, № 273, 29 Dec 2012
4. Gerova, N.: Smart education and smart e-Learning. In: ICT Proficiency Measurement while Realizing Information Activity of Students Majoring in Pedagogical Education, pp. 309–319. Springer, Italy, Sorrento (2015)
5. Gerova, N.: Methodical system of future teachers' information training in higher education. In: SGEM Conference on Psychology and Psychiatry, Sociology and Healthcare, Education, Bulgaria, vol. 3, pp. 541–547 (2014)
6. Kriuchkov, V.A., Sheina, I.M.: Interculturally Competent. LOGOS Publishers, Moscow (2003)
7. Meshcheryakova, E.A., Strunina, N.V.: Using computer-based technologies to develop autonomous learning skills. In: Organizing Students' Autonomous Work: Thesis of the 2nd All-Russian Internet-Conference. Saratov: "Noviy Proyekt", pp. 49–55, 6–9 Dec 2013
8. Nikitina, G.A.: Development of would-be teachers' professional competence within the framework of the foreign language methodology course. In: Organizing Students' Autonomous Work: Thesis of the 2nd All-Russian Internet-Conference. Saratov: "Noviy Proyekt", pp. 55–59, 6–9 Dec 2013
9. Polat, E.S.: Modern Educational and Computer Technologies in the System of Education: Textbook. M.: Academia, 368 p (2008)

# Formation of the Individual Learning Path in the Information and Educational School Environment

Marina V. Lapenok, Olga M. Patrusheva, Natalya G. Tagiltseva,
Lada V. Matveyeva, Natalya V. Gerova and Valentina V. Makeeva

**Abstract** The article shows that if an individual path in the process of teaching high school students an academic subject, firstly, will be formed from the teaching modules focused on an individual content on a subject matter in accordance with the student-significant learning objectives, Second, teaching modules will be presented as a set of recommended electronic educational resources (EER), the mastering of which is carried out after hours on the basis of Smart-technologies, and thirdly, the basis for the development of content for the EER will be based on academic discipline: the principle of correspondence between the content of student-significant learning objectives, correction of dynamic random content and teaching methods principle, the principle of adaptability and customization, it provides in-depth knowledge achievements of students on the subject, high level of motivation to succeed, high level of readiness to use information and educational environment for self-learning.

**Keywords** Electronic educational resources · Education information environment · Individual learning path

M.V. Lapenok (✉) · N.G. Tagiltseva · L.V. Matveyeva
Ural State Pedagogical University, Yekaterinburg, Russia
e-mail: lapyonok@uspu.ru

N.G. Tagiltseva
e-mail: musis52nt@mail.ru

L.V. Matveyeva
e-mail: lada-matveeva@yandex.ru

O.M. Patrusheva
Ural Federal University, Yekaterinburg, Russia
e-mail: podsnejnik1993@gmail.com

N.V. Gerova
Ryazan State University named for S. Yesenin, Ryazan, Russia
e-mail: nat.gerova@gmail.com

V.V. Makeeva
School Number 20, Yekaterinburg, Russia
e-mail: valvladmak@mail.ru

© Springer International Publishing Switzerland 2016                      553
V.L. Uskov et al. (eds.), *Smart Education and e-Learning 2016*,
Smart Innovation, Systems and Technologies 59,
DOI 10.1007/978-3-319-39690-3_49

# 1   Introduction

Modern educational process is based on the humanistic nature of education, which involves the use and development of individual personal interests and abilities of students in training. SMART-technology creates conditions for the improvement of educational content and teaching methods on the basis of electronic educational resources. Especially important is the task of improving the educational process on the basis of an individual path of training in information and educational environment for the implementation of student-significant learning objectives and achieve substantive results of in-depth level.

The informational educational environment (IEE) in this study refers to a set of tools which ensure the implementation of interactive communication between teachers, students and the EER, as well as the implementation of educational, organizational and administrative activities of the participants of the educational process on the basis of Smart-technologies, including:

- software of training and organizational purposes, including, instrumental packages, electronic educational resources (EER), learning management system (LMS), distance learning system (DLS), etc.,
- hardware for teaching and experimental activities, interfaced with the relevant computer software packages; server complex of educational institution for storage and accumulation of information; delivery means and management information, communication channels, the user equipment; audio and video equipment, combined with the software processing and visualization of information; interactive hardware and software of the voting system for the control and testing of knowledge.

The studies of Robert [1] Starichenko [2] Lapenok [3] disclose methodological objectives that are being successfully implemented in the IEE, including the individualization of education by gradual promotion of students to the planned learning outcomes through lines of varying degrees of complexity.

In connection with the foregoing, it is urgent to develop methods of forming individual paths by using EER to study learning subjects in order to: firstly, improve their motivation to succeed, and secondly, achieve in-depth results by students training in the chosen subject, and in the third, the formation of students' skills by using IEE for self-study.

# 2   Theoretical Foundations of Information and Educational Environment for the Schooling Individualization

Personality-oriented education model (as opposed to knowledge-centric), focused not only on the formation of subject knowledge and skills of students, but also to develop their general and mental abilities. However, in the modern school training content is constructed as a substantive, the learning process is still manageable: the teacher directs learning and cognitive activity of students—organizes it by encouraging self-learning activity.

Complex application of Smart-technology in regular school is the mean of learning individualization, the development of the student's abilities, as well as a mean of improving the content, organizational and educational sides of the educational process. IEE operation in an educational institution (school, college, university) causes a shift from traditional reproductive methods to actively-teaching methods, which make the students to receive the knowledge by themselves and not from a teacher and used methods, forms and means to stimulate the learning process.

IEE provides teachers with a set of technological tools to provide educational content in a variety of formats (text, static image, video, animation, sound, etc.); for the development of various types of test questions (single and multiple choice, open, establishing correspondences between the elements, concepts and categorization of others.) to import teaching materials of different software tools to protect the contents of e-learning materials to navigate the EER and other components. This allows the teacher to focus: on the selection of educational content of EER, focusing on the projected levels of training on the subject; in the choice of means of pedagogical influence in order to create the students motivation to learning, the implementation of variant forms and methods of teaching of the interaction with the students.

Regular school students in connection with training in the IEE on individual paths open additional possibilities:

- joint selection (the interaction of the learner and the teacher) of student-significant learning objectives, means and methods of training in the implementation of an individual path;
- account of the individual characteristics of the student and adaptation services of learning management system (LMS) to the abilities of the student (by volume and the rate of educational information, on the format of the educational content for the diagnosis of learning, etc.);
- systematic and regulated training interaction of a student, classmates and a teacher, not only during the classroom, but also in the period of extracurricular training in real-time and delayed communications;
- a positive motivation for learning, which is achieved due to psychological comfort during the training owing to the individualization of the educational process, objective assessment, opportunities to improve learning outcomes;
- Implementation of Exercises sufficient for the number of repetitions and the variety of tasks for the development of skills, the ability to analyze the level of skills' formation, issue-based recommendations for next steps;
- correction of individual learning paths, which can be carried out automatically on the basis of IEE development deliverables domain (control points), and on the discussion by the teacher and students of educational achievements;

- the situation of "success", which is created by focusing student on completing assignments successfully, fixing educational achievements at independent students completing an electronic portfolio.

This learning process is considered as a process of learning by students of educational material on the subject (the assimilation of knowledge, formation and development of skills) during their training in the IEE. Individual learning path is constructed from so-called "teaching modules", that is curriculum topics, followed by an indication of the recommended methods and forms of development of educational material. Didactic module is implemented as a set of EER, aimed at the development of the theory of the formation of practical skills for problem solving and laboratory works. The didactic module contains text, static and dynamic multimedia components, job-training simulators, video-fragments of experimental tasks.

We have developed a method of forming an individual learning path based on drawing up the routing of the educational process and didactic card of didactic module. The flow chart describes the stages of formation of the student's individual learning path: production, practice-oriented, control and evaluation stage (Table 1).

Didactic card describes: the theme of the discipline through the expanded name of the module; the logical structure of a didactic module by formulating two or three training purposes; materials for diagnostics achieve the stated objectives; options for trajectory correction; dose assignment. Guided didactic cards, student self-studying the topic, check the level of their knowledge on this subject, and, if required, performs the necessary correction of individual path.

The content of teaching modules are the elements of knowledge presented as a concept, law, hypothesis, knowledge of the methods of scientific knowledge. With the development of the didactic content of the module the student forced to use such methods of theoretical knowledge, as modeling, idealization deductive elimination of consequences. The teaching modules include qualitative and quantitative objectives at various levels of complexity, which make implementation of the decision of activities such as analysis, modeling, application of mathematical and logical operations. Applied material (the polytechnic, environmental, providing connection with life) is studied in connection with the consideration of theoretical issues in their respective subject areas. In this application the material is grouped around the main areas of science and technology directly related to the study theory. Thus, the development of a didactic module provides mastering the domain knowledge and the formation of skills to the level of in-depth study of the discipline. The content of the recommended methods and forms of training described in didactic module, selected in accordance with the specific subject area (that is a school subject).

For example, the specificity of physics as an academic subject is necessarily accompanied by a training demonstration of physical phenomena, experiments, laboratory work. Many educators (Shamalo, Usoltsev [4], and others.) Argue that in the absence of the possibility of a real experiment, or in addition to the computer model, implemented using Smart-technology allows the teacher to demonstrate many of the physical phenomenon, experiments effects. Currently used in the practice of teaching physics school digital lab (hardware-software complexes

**Table 1** Flow chart of the formation of individual learning paths in IEE

| | Production stage | Practice-oriented stage | Test and evaluation stage |
|---|---|---|---|
| Goals | Establishing a planned result of training, the formation of a common understanding of the individual path of training, familiarity with the IEE services | Structuring knowledge; deliberate and arbitrary construction of utterance in oral and written form; selection of the most effective ways to solve problems according to specific conditions; reflection methods and conditions of actions, monitoring and evaluation of the process and the results of independent activity | Controlling the process and the results of independent activity, diagnosing students' expected results at each step of developed topics |
| Tasks | Designing of joint teacher and the students educational process; construction of a graphical representation of an individual educational path; organization of staged self-learning, cognitive and reflective student's activity | Formation of the student skills to build a logical chain of reasoning when playing of the subject, represent a chain of events and features in their causation, skills to analyze the object, the process to isolate the essential features, to establish links between new and previously studied concepts; skills to finish the missing characteristics of the process for a particular purpose | Formation of the student skills to finish building the missing links of the educational process for the task; Appreciation result in achieving (or not achieving) the learning objectives |
| Results | The presence of an individual learning path, taking into account the individual characteristics of the student, availability of individual standard training activities (definition included teaching modules certain period of time to the study of the module), the student's self-monitoring system (electronic portfolio) | Separate implementation of student learning activities; the use of models and schemes for educational tasks; conscious mastery of common methods of solving problems; ability to carry out the synthesis as the compilation of the whole the parts; ability to carry out the comparison and classification of criteria, analysis of the process with the release of the essential characteristics | Ability to use technological possibilities of IEE (services) and educational content (EER); mastering the techniques of control and self-mastering of the studied material |
| Universal learning activities | Personality, cognitive, communicative | Personality, cognitive, communicative, regulatory | Personality, cognitive, communicative, regulatory |

(continued)

**Table 1** (continued)

|  | Production stage | Practice-oriented stage | Test and evaluation stage |
|---|---|---|---|
| The form of organization of educational process | Consultation in real time | Independent work of the student in the School IEE | Independent work of the student in the classroom input diagnostic tests in IEE, independent work of the student in the classroom thematic diagnostics. |
| Training duration | Consultation time (40–80 min.) | No time limit. In general: the time devoted to mastering subject content to students in the IEE and the time necessary consultations | Time of input diagnosis (40 min.), the time spent for passing the test by students in each didactic module ($\approx$20 min. $\times$ number of modules), the diagnostic work (80 min.) |
| Means of education | EER embedded in LMS | EER embedded in the LMS; technological capabilities of IEE (services), implemented with the help of Smart-Technologies | EER embedded in the LMS; technological capabilities of IEE (services), implemented with the help of Smart-Technologies |

"Archimedes", "Science Fun", "L-micro", "PROLog" et al.), Provide both the experiment and the creation of a teacher EER containing video experiment. The use of such EER allows, in turn, form the students' research and experimental skills in physics at training outside normal working hours in the IEE. Interactive forms of interaction with the student's EER (such as the management of an interactive model, changing the parameters of objects and processes, navigation through content elements, copying content items to create your own original compositions, multiple choice of content elements, changing the spatial orientation of objects) are accompanied by the need for analysis at each step, and calculations of physical quantities in a given parameter space.

"Passage" of individual learning path and, accordingly, the development of teaching material students consider (without loss of generality) as an example of discipline "Physics":

- studying of the theoretical part of the training material is accompanied by a drawing of the support scheme in the form of abstract or table;
- studying of the practical part of the training material is accompanied by a description of high-quality solutions and (or) quantitative tasks in accordance with a template that determines: schematic entry conditions of the problem; drawing up the necessary drawings illustrating the physical model; description of the sequence of formulas, the use of which leads to the solution; translation units in the standard system; Record of answer;

- studying of experimental work is accompanied by the report on the implementation of laboratory work, executed in accordance with a template that determines: analysis of the conditions of physical experiments and experimental equipment; prediction of the effects of the experiment, and the formulation of hypotheses; identification of possible connections and relationships between physical quantities; analysis of the results of the experiment, and the formulation of conclusions;
- implementation of control tasks by students accompanied by a presentation of the results in accordance with the shown above patterns;
- if there are problems in performing laboratory work at learning, their elimination is carried out through consultation with the teacher in real-time or delayed communications;
- documentary evidence confirming the results of training, students are recorded in electronic form (files) are accumulated in electronic student portfolio, reflecting a retrospective reflection.
- documentary evidence confirming the results of training, students are recorded in electronic form (files) are accumulated in electronic student portfolio, reflecting a retrospective reflection.

The specific application of individual learning paths in the IEE to study subjects humanitarian orientation by the example of music education. Personalized needs of senior secondary school in the systematic implementation of experiences in the field of the favorite destinations of musical art in the composition of songs and instrumentals, to learn playing a musical instrument, to use the background sound of music in everyday life can be met by the saturation of IEE electronic music and educational resources in accordance with the identified areas of high school musical interests.

Creation of teaching modules provides:

- updating knowledge in the field of academic musical art (information component, supported by virtual concerts and tours);
- development of practice-oriented musical and theoretical knowledge (mastery of musical notation program, music terminology, the basics of harmonization and composition);
- formation of skills playing a musical instrument (a program of learning to play the guitar, piano, saxophone and others. Musical instruments);
- creation of music based on music and computer technologies;
- training in the use of music in the art for therapeutic purposes

allows to design individual path of musical self-based EER, providing gradual mastery of musical knowledge and skills with appropriate controls.

At the same time, the need for systematic use of Smart-technologies in the process of "passing" of individual learning path (for example, service LMS, IEE is a technological basis of the school) are responsible for the formation of readiness to use the IEE for self (as a set of skills for independent use of IEE information from various sources, its storage, processing, transformation and presentation of information in different forms).

## 3 Teaching Experiment

The main purpose of the experiment was to evaluate the teaching effectiveness of the developed method of formation of an individual path in the IEE as an example of teaching physics students in secondary educational institutions. The participants of pedagogical experiment were high school students from MBEI (municipal budget educational institution) of SSGE (secondary school of general education) № 20 and MBEI of SSGE № 200 in the amount of 126 people, of whom were formed control and experimental groups, as well as 4 subject teacher of natural science cycle of these schools in the city of Yekaterinburg.

Experimental study included three phases: ascertain, search and control. The aim of ascertaining stage of the experiment was to identify methods and techniques of teaching physics on an individual path in the upper grades of secondary school, the definition of criteria for evaluating students' achievements in the implementation of individual learning paths in the IEE school. The results of this phase of the study have allowed to identify the problem and to confirm its relevance.

The purpose of the search phase of the experiment was to determine the appropriateness of educational EER and determination of EER and their testing during the experiment. As a result of this phase on the basis of services of LMS Naulearning were established electronic educational resources, reflecting the content of the basic and advanced level in physics for educational institutions, structured teaching modules within the divisions: "Kinematics"—7 teaching modules, "Dynamics"—12, "Relativistic mechanics"—1, "Molecular physics"—9, "Electrostatics"—6 teaching modules, respectively. To control the "passage" of individual learning path in IEE students in the experimental group were drawn maps of didactic teaching modules.

The purpose of the control phase of the experiment was to test the effectiveness of training for an individual path in the IEE. Evaluation of subject learning outcomes was based on partial analysis; Estimation of formation of students' cognitive skills and abilities to carry out their own educational activities in the IEE performed on the basis of operational analysis by the method of Usova [5]. The level of motivation of achievement of success is determined by the method of Orlov [6]. As a result of comparing the academic achievements in the control and experimental groups of students it was concluded that the effectiveness of the developed method of formation of an individual path of training in IEE and the feasibility of its use for teaching students of educational institutions. In particular, it is shown to achieve the majority of students enrolled in the individual learning path in the IEE, in-depth knowledge in physics (70 %), high level of motivation to succeed (91 %), a high level of readiness to use IEE in independent learning activities (66 %), confirmed by calculations of statistical significance.

# 4 Conclusions

(1) A method of forming individual paths of students at training school in IEE, which includes three stages of implementation:

- production, where the joint cooperation of the student and the teacher are defined: objectives, forms and methods of training; didactic modules (based on EER placed in school IEE); constructed didactic maps of each didactic module;
- practice-oriented, where the student learns the IEE training materials, developing skills in problem solving, and practical implementation of the experimental tasks, explores video experiment during the consultation, accompanied by a teacher;
- control and evaluation, which evaluates the results of teacher training and, if necessary, adjusts the path of learning for the students to achieve the level of knowledge appropriate to his purposes.

(2) Designed assessment of students' achievements in the implementation of individual learning paths in the school IEE, integrating criteria: Education; motivation to succeed; student's readiness to use IEE in independent learning activities.

(3) The experimental checking of educational achievements of students in teaching the physics of individual paths in the IEE, the results of which showed the majority of students to achieve in-depth knowledge of physics, high level of motivation to succeed, a high level of readiness for learning to use the IEE. The experimental results led to the conclusion about the effectiveness of the developed method of formation of a student's individual path at training in IEE.

# References

1. Robert, I.V.: The development of didactics in conditions of informatization of education. In: International Scientific and Practical Conference on informatization of Education, Omsk, pp. 3–9 (2012). (in Russian)
2. Starichenko, B., Korotayeva, E., Semenova, I., Slepukhin, A., Mamontova, M., Sardak, L., Egorov, A.: The Metod of Use of Information and Communication Technologies in Educational Process. Manual in 4 Parts. Ural State Pedagogical University, Yekaterinburg (2013). (in Russian)
3. Lapenok, M.V., Lapenok, O.M., Simonova, A.A.: Preparation and evaluation of teachers' readiness for creation and usage of electronic educational resources in school's educational environment. In: Smart Innovation, Systems and Technologies, vol 41, pp. 299–308. Springer (2015)
4. Usoltsev, A.P., Shamalo, T.N.: The concept of innovative thinking. In: Teacher Education in Russia, Yekaterinburg, №1, pp. 94–98 (2014). (in Russian)

5. Usova, A.V.: Methodology of Scientific Research. Lecture Course, 130 p. Chelyabinsk State Pedagogical University, Chelyabinsk (2004). (in Russian)
6. Orlov, Y.M.: Need-motivational Factors of Educational Activity Efficiency of High School Students: Dis. Dr. Psychological Sciences: 19.00.07, Moscow, 410, p. (1984). (in Russian)

# Information Education in Russia

Sergey A. Beshenkov, Etery V. Mindzaeva, Elena V. Beshenkova,
Margarita I. Shutikova and Irina I. Trubina

**Abstract** The main tendency of development of a course of informatics within
thirty last years can be expressed as follows: from computer literacy to a subject of
natural sciences and then "meta-a course". Computer literacy (1985). It is, above
all, the learning of algorithms and programming. However, as the practice revealed,
work with algorithms and modern software largely contributes to the development
stereotype pattern of thinking. Of course, the development of operational skills is
essential for the modern man. However, to reduce the informatics as educational
subject to programming skill so—is to deprive students of the future, since any truly
human activity does not fit the templates. Natural science course. The transition rate
of computer science in a qualitatively new state in the mid-1990s was due to two
reasons: (1) further development of science itself, which was the main vector of its
fundamentalization; (2) the need to implement a system of didactic, according to

S.A. Beshenkov (✉) · E.V. Mindzaeva
The Federal State Budgetary Scientific Institution "The Institute of Management
of Education of the Russian Academy of Education", Moscow, Russia
e-mail: iuorao@mail.ru; sgr57@mail.ru

E.V. Mindzaeva
e-mail: 1vega1@mail.ru

E.V. Beshenkova
The Federal State Budgetary Institution of Science Institute of Russian Language
V.V. Vinogradova of the Russian Academy of Sciences, Moscow, Russia
e-mail: evbeshenkova@gmail.com

M.I. Shutikova
State Budgetary Educational Institution of Higher Professional Education
Moscow Region "Academy of Social Management", Moscow, Russia
e-mail: raisins_7@mail.ru

I.I. Trubina
Federal State Budgetary Scientific Institution "Institute of Education Development
Strategy of the Russian Academy of Education", Moscow, Russia
e-mail: uvshp@mail.ru

© Springer International Publishing Switzerland 2016
V.L. Uskov et al. (eds.), *Smart Education and e-Learning 2016*,
Smart Innovation, Systems and Technologies 59,
DOI 10.1007/978-3-319-39690-3_50

which the content is determined by the aggregate structure of the information reality. For these computer components disclosed as follows: information processes (phenomenon), information model (tool), management, technology, society (the application). Meta-course. The current stage of development of computer science course, determined by the global shift from information society to the knowledge society. The basis of the course is to build a full cycle of transformation of the information: "Data"-"Information"-"Knowledge". These tendencies, on the one hand, followed for the friend the friend (from the point of view of domination), on the other hand, today they develop in parallel.

**Keywords** Informatics · Computer literacy · Information literacy · Meta-level · Information principles of the organization of knowledge · Meta-course · Russia

## 1 Introduction

The general objective tendency of development of a general education course of informatics for quarter of the century can be expressed so: *from computer literacy to a subject of a natural-science cycle and from it to "meta-course"*.

Let's consider consistently components of the called triad.

## 2 Computer Literacy

The course of informatics arose in 1985 as implementation of the program of providing "computer literacy of youth". Academician Ershov (1931–1988) was the main ideologist of the maintenance of this course during this period. His personality and ideas have essential impact on the subsequent development of a course of informatics. It is difficult to estimate a result of this influence unambiguously. On the one hand, the informatics received a powerful impulse for the development, with another—its orientation to algorithmization, programming, development of computers far not completely answered the purposes of the general education.

Gradually computer literacy began to associate with information technologies in which a significant amount of people see an informatics essence.

In a modern course of informatics it is the direction it is connected with ICT development—competences. In Russian education it is accepted to allocate the ICT following components—competences:

- competence of the sphere of informative activity: understanding of essence of information approach at research of objects of the various nature; knowledge of the main stages of the system and information analysis;

- competence of the sphere of communicative activity: understanding of features of use of formal languages; knowledge of modern means of communication and the most important characteristics of communication channels; knowledge of ethical standards of communication and basic provisions of legal informatics;
- technological competence: understanding of essence of technological approach to the activity organization; knowledge of features of the automated technologies of information activities; ability to reveal the main stages and operations in technology of the solution of a task;
- competence of the sphere of social activity: understanding of need of care of preservation and enhancement of public information resources; readiness and ability to bear a personal responsibility for reliability of extended information; respect of the rights of others and ability to defend the rights in questions of information security of the personality.

## 3  Natural Science Course

Transition of a course of informatics to qualitatively new condition in the mid-nineties was caused by two reasons:

- further development of the science of informatics towards its fundamentalization;
- need of realization of the system principle of educational content is defined by cumulative structure of a subject of training and structure of the generalized (invariant) activity of the person.

According to rather settled point of view, the informatics is the fundamental natural-science discipline studying regularities of course of information processes in systems of the various nature, and also methods and an automation equipment of these processes.

The understanding of informatics as natural-science discipline introduces in it the certain logic reflecting the main components of knowledge:

- subject of knowledge (phenomenon);
- tool of knowledge (model);
- a scope (where results of knowledge are used).

The researchers allowed to open the maintenance of the called triad in relation to informatics. It was shown that the main phenomenon reflecting information component of reality, information processes, the main tool of knowledge—information models, and scopes which are expedient for considering within comprehensive school—the sphere of management, technologies, society are Rakitina [1]. These three continuous course topics in informatics are covered at every step of learning but of course in various depths.

We will consider below those topics in more details.

The scope of issues traditionally covered by the topic "Information and informational processes" is significant for the course in informatics. At the introductory step, the choice of tasks and learning materials is aimed at providing full description of the concept of usage of [multimedia] in all fields of human activities and problems of subjective interpretation of information. It lays the grounds for further studying of informational modeling, algorithms etc.

In the pre-professional step the procedure of interpretation of information is again in the foreground, but in a different light—as various ways of interaction with technical and social systems. Generally, the steps of learning of informational process are described in Table 1.

The modern stage of the development of education is characterized by increased attention to the notion of the model and the methodology of modeling.

The course in informatics helps, to the largest extent possible, to classify the knowledge that learners have about models and conscious application of informational modeling in their learning and then their professional activities. The creation of models in the courses of mathematics, physics, chemistry and biology is fortified by learning, in the course in informatics, of issues related to the stages of creation of models, analysis of their qualities, validity checks in respect of the model and the object and the aim of modeling, examining the influence of the choice of modeling language on the quality of obtained information etc.

**Table 1** Main learning steps of the topic "Information and informational processes"

| Learning step | Aspects of the course in informatics | Main studied processes |
|---|---|---|
| Introductory | Information as a message in the form of sequence of symbols | Encryption, interpretation |
| Intermediary | Information as a message in the form of sequence of symbols that are kept, transferred and processed through technical equipment | Keeping, transferring, processing |
| Intermediary and pre-professional | Information as a message transferrable via channels which can be kept and processed according to certain rules | Modeling, characterization, algorithmization |
| Pre-professional | Information as data and methods of its processing | Informational technologies, automatization |
| Pre-professional and higher education | Information as semantical feature of matter | Interpretation, methods of learning and communication |
| Higher education | Information as a recourse, product, tool and instrument of professional activities | Methods of practical activities, generation and making of decisions |

**Table 2** Main learning steps of the topic "Informational modeling"

| Learning step | Aspects of the course in informatics | Main studied processes |
|---|---|---|
| Introductory | Modeling as substitute for a subject in the process of learning, interaction and practical activities | Comparison, collating, examination of modelings |
| Intermediary | Modeling as a simplified version of a real subject. Informational modeling as a scheme, image or description of a studied subject | Formalized representation of text, graphic, numerical and audio information |
| Intermediary and pre-professional | Modeling as a new subject which reflects certain qualities of an original subject which are material for the purposes of modeling | Characterization, creation and interpretation of tables, diagrams, flow charts, schemes, formulas and algorithms |
| Pre-professional | Modeling as a way of knowledge, means of communication, tool of practical activities | Structuring of data and knowledge |
| Pre-professional and higher education | Modeling as physical or informational equivalent of a subject which operates in certain characteristics in a way similar to that of an original subject | Creation of valuation criteria, valuation check of modelings |
| Higher education | Modeling as a new subject (real, informational or imaginary) different from an original subject and having and reflecting qualities material for the purposes of modeling which allows it for those purposes to fully replace the original subject | Systematic analysis, design, impact analysis |

The process of learning modeling should be organized in such a way as to enable a learner to try the roles of creator, observer and user of models because trying different roles is especially important in modeling.

The main learning stages for modeling in different learning steps are described in Table 2.

Informational technologies of task solutions are directly related to management techniques. Main learning steps of this topic are described in Table 3.

This approach was realized in a number of textbooks and manuals (Beshenkov, Shutikova, Mindzayeva, etc.) [1–3].

Information models create a basis for high-quality transition of a general education course of informatics in a "meta-course" rank.

**Table 3** Main learning steps of the topic "Informational basics for management"

| Learning step | Aspects of the course in informatics | Main studied processes |
| --- | --- | --- |
| Introductory | Management as handling activities of somebody or something | Work with operators |
| Intermediary | Management as a governing act transferred by way of instructions | Algorithmization, operating of work of computer, operators by way of instructions |
| Pre-professional | Management as directed informational interaction between a managed object and the system of management | Purpose, mechanisms, methods, results, valuation of quality of management |
| Pre-professional and professional | Management as a mechanism of self-organisation of complex systems | Systematic and functional analysis |
| Professional | Management as a sum of principles, methods, forms and ways of influence to an object of management with the purpose of reaching specified characteristics of its functioning and/or expected results of its activities | Preparation, making, realisation of management decisions |

# 4 Informatics as "Meta-Course"

Understanding of informatics as technological and natural-science discipline far don't settle its educational potential. On the contrary, as show numerous philosophical, sociological and pedagogical researches, the informatics reflects the most essential and important lines of a modern civilization.

It is known that signs and the texts made of them gained in the XX century crucial importance for science, culture and human life as a whole. Today the person is almost completely shipped in the world of signs and texts which are speculative (and not always positive) the designs having very weak communications with reality in broad understanding of this term. With the maximum completeness it was reflected in concept of "virtual reality".

As a result of people often doesn't know and doesn't understand world around, first of all the world of physical reality. Manifestation of it are alienation of the person from this reality, inability always adequately to perceive natural phenomena, the facts of cultural and public life.

This situation is reflected and in an education system. The school student can successfully solve various problems, but he, as a rule, isn't able to interpret competently the results received by it—i.e. to work out of the chosen sign system. For example, in the course of the solution of a task on determination of diameter of the globe the pupil can quite receive 1, 5 km in the answer, without feeling thus need for result verification. Similar examples it is possible to give a set.

Thus, driving forces of development of a general education course of informatics at the present stage (development of its "meta-course" aspect) are:

a. virtualization phenomenon—the card of a modern information civilization. Without judgment of virtualization socialization of pupils in the modern world and in general intelligent life and activity of the person is impossible;
b. the cascade of the crisis phenomena of the modern world having mainly information (sign) nature. Became obvious that their overcoming is impossible without accumulation of the certain intellectual potential, capable to generate essentially new ideas, methods, theories.
c. the internal factor connected with need of development of intersubject communications in system of subjects not only +natural-science, but also humanitarian cycles. Only in this case formation at school students of a complete picture of the world is possible that, undoubtedly, is one of the most important problems of the general education [4–7].

"Meta-level" of a training course of informatics allows to put one of main "bricks" in the base for development of intellectually potential of pupils, in formation of a complete picture of the world. Need of such picture is caused by sharp increase in areas of knowledge and human activities, including professional. On the other hand, the conscious perception and intelligent activity are impossible without that general scientific, world outlook representations became integral part of scientific, training and professional work. This fact was quite conscious in the 1930th. In work of the well-known German philosopher M. Heidegger "Time of a picture of the world" (1938) was emphasized that the main process of New time is world development "as pictures", i.e. creation of some uniform image, system.

The informatics role in this process is double. On the one hand, its conceptual framework allows to establish connection between phenomena very far at first sight. With another—the informatics is the methodological base, allowing to allocate the general principles of structuring information in other disciplines.

Many such principles were the knowledge formulated in various areas, and later began to be perceived as general scientific and common cultural. Differently, they to eat began to carry out a role of information beginning of modern knowledge. It is possible to carry to number of such principles:

- principle of systematic character;
- principle of symmetry and related laws of preservation;
- principle of uncertainty and related principle of a complementarity;
- principle of incompleteness of formal system;
- principle of "nonlinearity" (accounting of intersystem interactions).

The called principles are used now far behind a framework of those phenomena for which explanation they were formulated. Today is, first of all, information principles of the organization of knowledge [3, 8].

One of the most important "meta-level" aspects of a general education course of informatics is system and consecutive training of sign and symbolical activity. The informatics is capable to provide information models as means of work with various forms of submission of information. Therefore at lessons of informatics form "metalevel" abilities of work with various forms of submission of information, information models (from creation of model before its use are formed during the solution of a specific objective). These abilities are capable to project meta-knowledge in the field of sign and symbolical activity on other subjects. Training in modeling process at informatics lessons, as a rule, assumes use of examples from different areas of knowledge and activity (linguistics, physics, chemistry, geography, biology, mathematics, theater, music, psychology, etc.). Training in modeling is conducted with use of the most various sign systems: from natural sign systems to systems of high extent of formalization, including programming languages, algebra of logic, etc.

Not less significant "meta-level" aspects of informatics consists in formation of a clear understanding and structuring information surrounding the person, awareness of the social importance of interaction with world around through sign systems and formalization, delimitation of this component. Only in this case it is possible to expect from the person of intelligent and socially significant actions.

## 5   Conclusion

Formulated above trends of the development of educational course of computer science will continue in the near future. However, the content of each of the areas can be transformed. In the coming years we expect significant progress in the field of technology for the development of a set of cognitive technology and robotics.

## References

1. Rakitina, E.A., Matveeva, N.V., Milokhina, L.V.: Continuous course of informatics, p. 43. Publishing House "BINOM. Laboratory of knowledge", Moscow (2008)
2. Beshenkov, S.A., Mindzayeva, E.V.: Educational standards of the second generation. The approximate program on informatics for the main school within standards of the second generations/Materials of cycles of the All-Russian teleconferences concerning federal state educational standards of the second generation. Natural-Science Disciplines, p. 77. Publishing House "BINOM. Laboratory of Knowledge", Moscow (2009)
3. Beshenkov, S.A., Rakitina, B.A., Shutikova M.I.: Humanitarion informatics: from tehnologies and models to the information principles/Information and education, no. 2, pp. 3–7 (2008)
4. Anjewierden, A., Efimova, L.: Understanding Weblog communities through digital traces: a framework, a tool and an example. Lecture Notes in Computer Science, 2006, vol. 4277, pp. 279–289 (2006)
5. Benkler, Y.: The Wealth of Networks, p. 527. Yale University Press, New Haven (2006)

6. Floridi, L.: The informational nature of personal identity. Mind. Mach. **21**(4), 549–566 (2011)
7. Rodogno, R.: Personal Identity Online. Philosophy & Technology (2011)
8. Shutikova, M.I.: Possibilities of Interdisciplinary Informatics. Bull. Cherepovets State Univ. 4 **3** (35), 202–205 (2011). http://elibrary.ru/item.asp?id=17885502

# Implementation of the Internet for Educational Purposes

Irena Robert, Lora Martirosyan, Natalya Gerova,
Vasilina Kastornova, Iskandar Mukhametzyanov and Alla Dimova

**Abstract** The article discusses the possibility of using the Internet for educational purposes. The features provide access to information resources of the Internet and the feasibility of their use for educational purposes. Considered information interaction of the educational purpose in the conditions of realization of the Internet. We describe the power of the Internet for educational services for the certification of specialists for the organization distributed educational projects. It is shown as distributed learning projects are used to organize and conduct educational institutions (geographically distributed) joint educational activities related to collection, processing and analysis of information with the scientific-practical and educational purposes.

I. Robert · V. Kastornova (✉) · I. Mukhametzyanov · A. Dimova
The Federal State Budgetary Scientific Institution
"Institute of Management of Education of the Russian Academy
of Education", Center of Informatization of Education, Moscow,
Russian Federation
e-mail: kastornova_vasya@mail.ru

I. Robert
e-mail: rena_robert@mail.ru

I. Mukhametzyanov
e-mail: ishm@inbox.ru

A. Dimova
e-mail: aldimova@mail.ru

L. Martirosyan
Moscow State Institute for Tourism Industry n.a. Yu. Senkevich,
Moscow, Russian Federation
e-mail: tonashen@yandex.ru

N. Gerova
Informatisation of Education and Methods of Teaching Department,
Ryazan State University Named After S.A. Esenin, Ryazan, Russian Federation
e-mail: nat.gerova@gmail.com

© Springer International Publishing Switzerland 2016      573
V.L. Uskov et al. (eds.), *Smart Education and e-Learning 2016*,
Smart Innovation, Systems and Technologies 59,
DOI 10.1007/978-3-319-39690-3_51

**Keywords** Informatization of education · Internet · Information and communication technology · Information interaction for educational purposes · Distributed learning projects

# 1  Introduction

Informatization, mass network communication of the third Millennium society and
its globalization determine the need for a general trainingof a student in the following areas: autonomy in acquiring education; responsibility for the selection of
the mode of educational activities and information interaction with the source of
educational information; planned progress in studying; participation in research and
project activities of social focus. The use of information resources and Internet
services for educational purposes allows: firstly, to increase significantlythe motivation of learning by visualization and availability of information resource on a
global scale, including audio-visual information; secondly, to enrich and facilitate
the selection of any educational material in preparation of the course; thirdly, to
ensure information security for using Internet resources.

Let's list possibilities of usingthe Internet for educational purposes.

# 2  Provision of Access to Information Resources
of the Internet

The educational use of theInternet information resources can be used—as a source
of educational information for training and education by students; both in preparation    for    classes    and    also    as    a    source    regulatory,    methodical    and
organizational-methodical information by educators; to create individual methods
of learning and educational complexes, comprising electronic educational resources
by administration and methodologists.

Let us dwell on the description of informational educational resources which are
currently available on the Internet. This is primarily Collection of Digital Educational Resources (DER). The present collection is used for occasional use in teaching
as a compilation of digital objects, or links to external sources of information. We
can spot the following as examples: Unified Collection of Digital Educational
Resources (http://school-collection.edu.ru); Directory of Electronic Educational
Resources (http://fcior.edu.ru); Single Window of Access to Educational Resources
(http://window.edu.ru).

Unfortunately, expert assessment and guidelines for their use in the classroom
are optional and often completely absent.

Besides, we can add that there are other very common resourcessuch as Systems
for Testing which determine the level of knowledge and skills of students. Using

**Fig. 1** Test system "Yandex—Unified State Examination"

these systems, control of knowledge, skills, the level of competence of students is exercised or training of students in preparation for the final certification is carried out in the form of computer testing. There are tools through which quizzes or tests can be created.

These are the examples: Yandex—Unified state exam (http://ege.yandex.ru) (Fig. 1); the System of knowledge assessment "Infotest" (http://infotest.by); StartExam System (formerly OpenTest) (http://www.opentest.ru); Single portal of online testing in the field of education (http://www.i-exam.ru).

Educational materials offered on the educational websites are informational educational resources developed by organizations or individuals for the use in education. Unfortunately, their quality leaves much to be desired since they, as a rule, do not undergo an expert assessment, and the developers take no responsibility for their pedagogical soundness. However, there are enough well-designed educational materials and they areof specific methodological importance. We could include in the list the following: Class physics for the curious (http://class-fizika.narod.ru); School mathematics (http://math-prosto.ru); Mathematical studies (http://www.etudes.ru) (Fig. 2); Biology for students (http://botanO.ru); Tutorials "Internet lesson" (http://interneturok.ru).

Additional educational-methodical and scientific-pedagogical information, usually provided on the websites of educational purpose, which contains a huge amount of various information, including photos and videos from personal collections, the latest news from the world of science and technology, etc. is very

**Fig. 2** Website "Mathematical Etudes"

common in the market of electronic educational resources. The search for the required information can be carried out withsearch engines in accordance with certain keywords. However, the information found in this way may be unverified and it usually needs to be clarified and reconciled with other information sources.

Educational materials on websites of publishing houses are widely presented on the Internet. These materials are mainly designed to promote and advertise printed and electronic products, produced by any publishing house. The present materials can be useful to methodologists, teachers and students, as a source of information about new educational and training-methodical literature, sources of further information and so forth. In addition, a number of publishers provide their materials (or part of them) in open access, free of charge. Such sites include the sites of the following publishing houses: "bustard" (http://www.drofa.ru); "BEAN. Knowledge laboratory" (http://www.lbz.ru); "1C"—educational software (http://obr.lc.ru); Publishing house "First of September" (http://1september.ru); Free popular science magazine for 3 D technologies (http://mir-3d-world.w.pw) and etc.

Encyclopedia and reference resources (online encyclopedias) are of particular importance to support educational process. They are the sites of general purpose designed for a wide range of users (not necessarily relevant to the field of education) and for presenting legitimate information. These materials also include electronic copies of official printed reference books. A special place among them is taken by Wikipedia, the authors of which can be any users. While each article is being scrutinized it may be supplemented or corrected by other users. Moreover, Wikipedia contains the most relevant information, which has not yet had time to be published in official printed editions. The examples are: Wikipedia—free encyclopedia (https://ru.wikipedia.org): Yandex Encyclopedia and dictionaries (http://slovari.yandex.ru); Megabook—Megaencyclopediaof Cyril and Methodius (http://megabook.ru); Encyclopedia Krugosvet (http://www.krugosvet.ru).

Tools of the Internet Broadcasting produce an opportunity of organizing dialogue relations and implementing information interaction of participants of educational process. The participants of the educational process may be provided with a variety of educational and methodical information in the form of radio programs, "channels" broadcasting—collections of videos published on the Internet (e.g. YouTube). Good examples can be the following: the Internet radio "Class"/Novosibirsk open educational network (http://www.edu54.ru/radio class); Cognitive and educational TV channel (http://www.znanietv.ru).

A special place is taken by the Regulatory guidance and organizational management information, providing the access of all participants of educational process to the publications presented in electronic form in a variety of regulatory and organizational documents (the official website of the Ministry of Education and Science of the Russian Federation, the websites of educational portals, etc.).

These sources include the following websites: Ministry of Education and Science of the Russian Federation (http://mes.rf); the Portal "Garant–Education" (http://edu.garant.ru/education/law); Official information portal of unified state examination (http://www.ege.edu.ru) (Fig. 3); the Federal portal "Russian education" (http://www.edu.ru); Russian portal of education Informatization (http://www.rpio.ru); the Portal all-Russian Olympiads for schoolchildren (http://rosolymp.ru).

**Fig. 3** The official portal of the Unified State Exam

## 3   Information Interaction for Educational Purposes

Information interaction for educational purposes [3] in terms of using means of information and communication technology is the activity aimed at collecting, processing, using and transferring information by the subjects of educational process (student, teaching, teaching tool, based on ICT) and providing psycho-pedagogical impact-focused on: developing creative capacity of the individual; forming knowledge system of a particular subject area; complex of skills of educational activity, including studies on the regularities of subject domain.

Information interaction between the organizers of the educational process and school staff is also possible. That is information activities between the organizers of the educational process (heads of regional, regional, district, federal education authorities, directors, organizers methodical and educational work, subject teachers, heads of library, health workers, and school psychologists educational institutions) and other school staff of the educational institution. When information interaction between the organizers of the educational process and the staff of the institution is carried out, the collection, processing, storage, transmission, creation of information and methodological materials of various kinds take place. The results of information interaction can serve as certain conclusions on the development of the educational process on the whole, specific conclusions about the progress in the teaching of the individual student, the decision on the further development of the institution and so on. The operation flow of information is carried out in the course of professional work of educators, and in their communication with students and their parents. Thus, the structure of the information interaction between the organizers of the educational process and school staff refers to the internal form of organization of information interaction, serving as the unity of sustainable interrelations between its elements. Realization of information interaction between users is carried out through a variety of communication services that the Internet provides for communication between all participants in the educational process—administration of educational institutions, methodologists, teachers, students and their parents, medical, social workers, etc. This category of Internet services include in particular management information system of educational process ("electronic diaries"), which focuses all the information about the educational process (timetable, curricula, assessment, correspondence, teachers, pupils and parents, etc.).

Let us call the major ones and describe their capabilities.

Management Information Training Systems is specially developed services for conducting "electronic class-book" or "electronic diary", for publication of "timetable", and also means providing communication of the trainer, trainees and other members of the educational process. In addition, the present services provide users with information and methodological support whichallowsmaking the process of record keeping in an educational institution automated. It also maintains specified comfort of the employee's activity in the sphere of education. An important opportunity, provided by this service, is the use of ICT to process information and methodological support and organizational management, the formation and development of information culture of educational personnel appropriate to the stage of Informatization and mass communication of the modern network society [3]. As examples of management information training systems can be called such as: City school information system (https://schoolinfo.educom.ru); Information-analytical system "Moscow register of education quality" (https://mrko.mos.ru/dnevnik).

Communication means of the Internet are electronic mail (e-mail), instant messaging systems (such as ICQ), Internet telephony (Skype), social networks (Vkontakte, Facebook, MoiMir, etc.). They are actively used by various categories of users, and can be applied by educators for rapid communication with learners and other participants in the educational process, including social, medical staff and others.

Educational online communities provide opportunities for communication of individual and group communitiesbased on information interaction and exchange of audiovisual information. The characteristics of this social interaction allow communicatingboth individuals and groups as well as organizations for educational, social and recreational purposes. In educational network communities there is an opportunity for educatorsand learners to interact psychologically comfortably. In a more advanced implementation there is a possible jointdesigning of educational activities for educational purposes, as well as cooperation in scientific-educational sphere [1, 2]. Examples of educational online communities can serve the following: Portal of Proskau.ru (http://www.proshkolu.ru)—teaching network community that represents a free school portal; Web service COMDI is a system of webinars and Web-conferences (http://www.comdi.com); FriendFeed (http://friendfeed.com)— has existed since 2008 as a Web service.

Means of remote presence in information and educational environment are tools that provide a set of conditions for collecting, processing, transfering, reproducing and using of educational information through interaction of the learner (students) and the educator with interactive information resource.

As an example we can offer robotic systems with two way communication managed over the Internet, which allow the student, for whatever reasons, staying at

**Fig. 4** Example robotic telepresence system—Robot "R-Bot" (http://rbot100.rbot.ru)

home to attend classes "virtually", to see the teacher and to hear his or her explanations, to answer questions, communicate with classmates, etc. (Fig. 4).

## 4 Getting Special Education Services

Getting special education services, including additional education, certification of specialists, gives the trainee an opportunity through the Internet, to acquire distance learning (process of transfer of knowledge, formation of skills through interaction between both educators and learners and between the latters and interactive source of information resource [4]. In this case, the possibility of Multimedia technologies, Hypertext provide availability of educational materials, the ability to test knowledge (both self-assessment and final control), rapid communication with the tutor, regardless of the location of the student and the educational institution, and in accordance with the most convenient for the student learning schedule, including foreign institutions.

Information on obtaining educational services on the websites of educational organizations provide opportunities to obtain reliable "first hand" information about a particular educational institution (school, College, University, etc.), to gain access to their "news feeds" etc. Examples of such sites of educational organizations may be the following: the Moscow pedagogical school №1505 (http://gYml505.rn); Moscow State University (http://www.msu.ru); Moscow state technical University (http://www.bmstu.ru); University of the Russian Academy of education (http://urao.edu).

Distance and blended learning, using online resources and network services, provides the implementation of distance and blended learning both for students and for professional development of teachers. Examples of distance learning sites are: the National Open University "INTUIT" (http://www.intuit.ru); Moscow center for distance education (http://bakalavr-magistr.ru); distance education Center "Eidos" (http://www.eidos.ru); Center for distance education Moscow state University (http://de.msu.ru).

Additional education. A number of educational organizations provide with the help of online resources additional services (extra-curricular) training, including tutoring services for exam preparation for final certification, for admission to universities, etc. Examples of web additional education: training Centre "Specialist" MSTU (http://www.specialist.ru); Network Academy "LANIT" (http://academy.ru/?checked=yes ); Higher computer school "Expert" (http://hcse.academy.ru); Academy of additional professional education (http://www.spbapo.ru).

Certification of specialists. Educational organizations in this category provide services of certification of specialists with the issuance of certificates of official sample. There are examples of sites certification: Online certification Retratech (http://certifications.ru); the testing Centers and Microsoft certification (http://www. proinfosystem.com/certification/Microsoft.html).

# 5 Organization of Distributed Learning Projects

Distributed learning projects are for the organization and conducting joint teaching activities on collecting, processing and analyzing data (information) with scientific, practical and educational purpose in educational institutions (geographically distributed). In this case, students and educators are provided with an opportunity to carry out project activities in the conditions of automation: processes of managing an activity of the project participants through social networks; collecting data using a distributed network of sensors and recorders for physical parameters of the studied or investigated objects, processing of results of experiments (including processing and analysis statistical data).

System of distributed learning projects are based on the principle of social networks, but are aimed at solving problems in organizing and conducting training projects. They provide opportunities of registration of students; organization of gathering distributed background information about the project; processing of the received information; publication of the results of the project; communication between the project participants.

An example of such an information system is the Global school laboratory "GlobalLab" (https://globallab.org/ru) (Fig. 5).

**Fig. 5** Global school laboratory "GlobalLab"

# 6 Prevention of Possible Adverse Effects of Internet Use

Considering the power of the Internet, unique in terms of educational applications should highlight the positive effect which is caused by a direct impact on learning in order to create a positive trajectory of socialization, learning the decision of vital problems of the choice of a "virtual" social environment ("virtual" communities) and etc.

The risks caused by the negative impact of the Internet: a violation of the rules and modes of use with a direct and indirect impact on the physical and mental health of the user (for example, direct effect on eyesight and indirectly—on the formation of a psychological dependence Internet). Indirect influence can also be seen in violation of posture, in sedentary lifestyles, individual behavior in isolation.

In general, the use of the Internet currently generates more problems than bright prospects. These problems relate primarily to the restructuring and psychological component settled centuries information exchange between teaching and tutoring. Under the conditions of use of the Internet for educational purposes, it is not based on the authority of training and, most likely, a partnership in a position to establish the teacher with their wards. In addition, there are even more significant difficulties related to the possible negative effects of psychological and medical treatment. The latter are associated with voluntary or involuntary violation or neglect of the user modes that must be followed in the application of the Internet.

No less dangerous possible negative effects of psycho-pedagogical effects; exerted on the student information succinct and emotionally rich content of various Internet sites. Risk related to: the use of unacceptable volume of educational information provided on the screen; mismatch of the on-screen information (structure, quality) age and (or) the individual the opportunity to study; insecurity positive "psychological climate" of information interaction with the objects on-screen virtual worlds, which are abundant in the Internet.

Particular attention should be paid to the fact that the information overload during long-term use of the Internet and the related emotional arousal, deceptive increase efficiency directly behind a computer screen, are dangerous for mental and physical health of the young man.

Coping with health issues carried out by complying with a number of measures of physiological and hygienic nature described in the specific regulations and methods approved by the relevant government authorities. The need for compliance with them there is no doubt, but on the ground, in practice, they do not always comply. This is the case when it is necessary explanatory work among teachers, the school administration. Either way, you must remove a kind of "psychological barrier" to the need to perform has long been established and is constantly modified the rules and regulations governing the online modes.

A more complex and difficult, according to modern psychologists, teachers and doctors are the possible negative consequences associated with an active intrusion into the natural inner life of modern man unnatural, virtual illusory impressions of virtual scenes and interactions at "entry" to the Internet. It is understandable and explainable psychology of the individual human passion bright and unusual, sometimes ghostly, impressions, different from the real, is fraught with many dangers for the unformed psyche of the student. These include, above all, the growing mutual alienation between the modern young people, due to the possibility of light "replacement partner" in the cybernetic and lightweight "communication without any problems". It causes not less concern danger deliberate manipulation of consciousness of a young man to perform certain actions and participates in various subjects, presented on Internet sites.

Pay special attention to the philosophical aspects of the implementation of information exchange on the Internet; in which the world and the communications generated by modern technology, can become so real that the user can finally break away completely from the real world and go into "cyberspace".

For safe and pedagogically appropriate use of the Internet the basic principle should be the strict implementation of the proposed experts—physicians, psychologists, physiologists and hygienists—requirements for the mode, in a pedagogically expedient use of Internet information resources and implementation of information exchange both between learning and students, and between Internet and interactive content [3].

## 7 Conclusion

Thus, resources and services of the Internet (specially designed for the needs of the education system of Russia and abroad, as well as resources and general-purpose services meeting the requirements for the contents and designof educational resources) can and should be actively implemented in the educational process. Their appropriate use develops creative abilities of students, allows significantly increasing the visibility of learning and teaching to provide educators and educated with a large amount of additional information. It also provides opportunities for rapid communication between participants of the educational process and the convenience ofreceiving remote educational services.

## References

1. Vagramenko, J.A., Yalamov, G.Y.: The concept of network of information exchange of students and schoolchildren. In: Educational Informatics, Moscow, vol. 3, pp. 7–12 (2013)
2. Vagramenko, J.A., Yalamov, G.Y.: The implementation of the principle of interaction in a small group of students in a networked environment. In: Informatization of Education and Science, Moscow, vol. 3, pp. 165–180 (2014)
3. Robert, I.V.: Theory and Methods of Informatization of Education (psycho-pedagogical and technological aspects). BINOM Knowledge Laboratory, Moscow (2014)
4. Robert, I.V., Lavina, T.A.: Dictionary of terms of the conceptual apparatus of informatization of education/compilers. BINOM. Knowledge Laboratory, Moscow (2012)

# Assessment of Levels of Formation of Competence of Students as Users of Information and Communication Technology in the Field of Health Care

Iskandar Mukhametzyanov and Alla Dimova

**Abstract** The integration of the educational environment of the educational institution and modern information and communication technology (ICT) changes the traditional environment of education via use of technical means and technologies with poorly studied or unstudied effects on health of a user. It is necessary to speak not so much about the direct negative impacts of ICT on the health of the user, as about those that only come into effect after years or even decades. ICT removes the pre-existing restrictions on education of persons with special educational needs. This requires adaptation of the learning space to the health indicators of the students (visually impaired, etc.), and to technical training facilities used outside of an educational institution. All of the above inevitably affects the health status of graduates of educational institutions and, therefore, the quality of labor resources of the country.

**Keywords** Information and communication technology · The competence of student users of information and communication technology in the field of health care · Requirements for the results of the formation of competence · Requirements for assessment of levels of formation of competence

## 1 Introduction

The analysis of scientific and pedagogical studies of ways of mitigating possible negative effects associated with the use of ICT on the user's health revealed that in recent years domestic scientists have created various instructional materials in the

I. Mukhametzyanov (✉)
Academy of Social Education, Kazan, Russian Federation
e-mail: ishm@inbox.ru

A. Dimova
Institute of Management of Education of the RAE, Moscow, Russian Federation
e-mail: adimova@mail.ru

© Springer International Publishing Switzerland 2016
V.L. Uskov et al. (eds.), *Smart Education and e-Learning 2016*,
Smart Innovation, Systems and Technologies 59,
DOI 10.1007/978-3-319-39690-3_52

field of preservation of health of the student users of ICT. These materials cover the following themes:

- Negative factors associated with the use of ICT;
- Possible negative effects of psychological and pedagogical as well as medical and social nature of the use of ICT;
- Tools, methods and organizational measures aimed at prevention of and compensation for possible negative consequences of ICT use [1, 2, 3].

Familiarization with these instructional materials by the users of ICT in educational institutions would enhance their competence in the field of health care.

## 2 Literature Review

Currently a number of specialists [4] consider competence to be an integrated result of education, expressed through the mastery of certain set of skills, acquired through reflection on the experience. Thus, in order to protect health in the modern conditions of active use of ICT it is necessary for learner—user of ICT to form competence in the field of health care, which is reflected in the acquisition of knowledge, skills and experience of mitigating possible negative consequences associated with the use of ICT. A number of works [2, 5, 6] allows us to conclude that the instructional material on the theme of health protection of the user of ICT is more appropriate to be taught to students in the context of the discipline "Physical education". This discipline has a fundamental role in protection and strengthening of the health of students, it has considerable potential for the formation of competence of the student user of ICT in the field of health care.

However, in the recommendation of the Federal State Education Standards of Higher Professional Education of Russia concerning cultural competence in physical education, there are no requirements for students considering formation of knowledge, skills and experience in the use of means and methods of prevention of and compensation for possible negative consequences for health of an ICT user. There is also no mention of instructional material on the prevention of and compensation for possible negative psychological and pedagogical as well as medical and social consequences of using ICT in education in the general program of the discipline "Physical education".

In this regard, in our opinion, the most promising researches are on these themes:

- Defining competence of the student user of ICT in the field of health care.
- Justification of requirements to the results of the formation of competence and to the assessment of the levels of its formation, as well as to the justification of the content of the course of physical education, that helps to form the competence of the student user of ICT in the higher education institution including stages of its formation.

## 3   Research of Materials

By the competence of the student user of ICT in the field of health care we mean:

- The set of knowledge and skills in the area of prevention and mitigating possible negative psychological and pedagogical as well as medical and social consequences of using ICT in education;
- Experience of implementation of certain activities aimed at prevention of and compensation for the negative consequences for the health of the user of ICT.

Requirements to the results of formation of competence of the student user of ICT in the field of health care are based on the research of the above-mentioned authors and studies of the effects of complexes of activities, which focus on different health aspects, on the health of the student user of ICT [1], were formulated. They include:

- Acquisition of knowledge about:
  - The effects of negative factors associated with the using of ICT;
  - Possible negative consequences of psychological and pedagogical as well as medical and social nature, which might occur because of the using of ICT;
  - Self-assessment and self-monitoring of health status, physical and mental condition;
  - Testing of physical and psychological state; methods of prevention and mitigation of the possible negative effects of ICT use.

- Acquirement of the skills of:
  - Identifying the ways of working on the prevention of and compensation for possible negative consequences of using ICT;
  - Using the methods of self-monitoring and assessment of indicators of state of health and physical and psychophysiological state;
  - Implementing the complexes of activities, which focus on different aspects of health, in accordance with identified diseases.

- Experience in actions aimed at creating a self-oriented complex of actions aimed at prevention of and compensation for possible negative effects of ICT use:
  - identification of individual negative health effects associated with the use of ICT-based methods:

    Self-assessment and self-monitoring of health status, physical development, physical preparedness and operability, mental and psychophysiological state; Testing of indicators of physical and psychophysiological state, state of health, including the utilization of computerized hardware and software diagnostic complexes and systems (an individual health card is created).

- Selection of tools and activities for prevention of negative effects on health;
- Selection of means and methods of compensation for identified negative health consequences;
- Planning of exercises that require the using of compensatory means;
- Monitoring of health indicators (Filling in the electronic individual health card after a preliminary, current and final tests; creation of graphs of dynamics of tested parameters).

## 4   Research Methods

The experiment that is devoted to assessing the level of formation of competencies of student users of ICT in the field of health care was performed on the basis of Modern Academy for the Humanities (Russia) in the period from 2010 to 2014 in three stages: ascertaining (2010–2011 academic year), formative (2011–2013) and final (2013–2014 academic year).

In the first stage of the experiment, 552 students were surveyed. The results showed the necessity of introducing those students to the course of physical education that forms the competence of the user of ICT in the field of health care. The questionnaire was composed, in which students were asked to answer the questions:

- "How do you assess your level of knowledge, skills and experience in the field of health care?"
- "Do you need an intense, short-term course focusing on the formation of knowledge, skills and experience in the field of health care?"
- "What course would you recommend for students of your faculty?"

Analysis of the results obtained during the survey showed insufficiency in training of students in the field of health care: 346 students (62.7 %)—the majority assessed their knowledge and skills in the field of health care as "unsatisfactory"; 158 students (28.6 %)—as "satisfactory"; 48 students (8.7 %)—as "great". Four hundred students (72.5 %) responded positively on the second question. According to collected answers to the third question, the following results were obtained:

- 82 students (14.9 %) considered it appropriate to conduct several additional thematic lectures;
- 128 students (23.3 %) thought an introduction of a course included in the obligatory course of physical education to be a great idea;
- 342 students (61.8 %), which is the majority, also chose a course of physical education, but suggested that it should be an elective one. Thus, the survey helped to identify the need to prepare student users of ICT in the field of health care.

During the second stage, a selection of students for their further training in the field of health protection within the framework of the elective course of Physical education was conducted. A total of 103 second—year students of faculties

"Jurisprudence", "Linguistics" and "Economics" were selected to study in the first (42), second (34) and third (27 people) experimental groups. Taking into account that the formation of competence of the student user of ICT in the field of health care was not previously carried out, the study was conducted only in the experimental groups, without using control groups.

The homogeneity of the formed groups was estimated according to the results of the test, which included 30 tasks, and checked whether students meet the requirements for compulsory primary minimum knowledge and skills in the field of health care. Afterwards, a statistical hypothesis about homogeneity of the three groups based on initial levels of knowledge and skills of their students was suggested.

Testing of the hypothesis was conducted via comparing the samples obtained from the initial testing of students from three groups. Statistics of Pearson's chi-squared test was equal to 2.7. At the significance level of a = 0.05 with six degrees of freedom chi-squared critical value was equal to 12.59, which allowed to accept the hypothesis as plausible. This allowed asserting that judging by the initial levels of knowledge and skills the students of the three experimental groups represented one population.

The training of students was carried out in the framework of an elective course of physical education in accordance to educational programs, which conformed to theoretical framework and organizational and methodical support for the formation of competence of student users of ICT in the field of health care.

The assessment of level of formation of knowledge and skills in the field of health care among students was carried out during the final third phase with the help of the prepared testing tasks.

## 5   The Results of the Study

The results of the test showed that the majority of students in each group assimilated the knowledge and skills in the field of health care at high and sufficient levels (76.2 % of such students in the first group; 82.4 % in the second; and 77.8 % in the third one). The number of students in the combined samples who have acquired knowledge and skills in the field of health protection at high and sufficient levels was equal to 81, which amounted to 78.6 %, i.e. to the majority.

The determination of the level of experience acquired by a user of ICT while studying the use of means of and activities dedicated to mitigating negative consequences was based on the results of the presentations of student-centered complexes of prevention of and compensation for possible negative consequences of using ICT, which were independently prepared by each student. The results of the presentations showed that during the training most of the students in each group acquired sufficient levels of experience (88.1 % of such students in the first group; 85.3 % in the second; 81.5 % in the third). The number of students in the combined sample that gained sufficient levels of experience was equal to 88, which is 85.4 %, i.e. the majority.

The number of students in the combined sample, that assimilated the knowledge and skills in the field of health care at high and sufficient levels, and have gained experience in implementation of ways of mitigating possible negative effects associated with the use of ICT on the user's health was equal to 75, which accounted for 72.8 %, i.e. the majority. Consequently, the results of the pedagogical experiment allow accepting the hypothesis of the study as plausible.

# 6   Discussion

The research proved that the level of knowledge and skills in the field of mitigating possible negative effects associated with the use of ICT in education (further—knowledge and skills) could be assessed by analyzing the results of the pedagogical testing. It was stated that the pedagogical test should contain at least 30 tasks; the correctness of each of them can be evaluated on a dichotomous scale. An indicator $K\alpha$, which is the ratio of correctly completed test tasks to their total number can serve as the quantitative assessment. The set of $K\alpha$ values forms a scale [0; 1], which (according to V. Bespalko) can be divided into four segments: [0; 0, 7], [0, 7; 0, 8], [0, 8; 0, 9] and [0, 9; 1, 0], each quantitatively matching insufficient, requisite, sufficient and high levels of knowledge and skills correspondingly [7].

Two diagnostic tests were prepared:

- A diagnostic test for determining of the level of initial knowledge and skills of student users of ICT, which they need for further training in the field of health care;
- Final diagnostic test.

These tests were prepared:

- Based on the pedagogical requirements for the organization of control of educational activity [8];
- In accordance with the proposed requirements for the results of formation of competence of the student user of ICT in the field of health care;
- Also taking into account the sequence of formation of competence of the student user of ICT in the field of health care in terms of taking mandatory and elective courses of physical education.

Each of the tests consists of six blocks of tasks. First, second and third blocks of the first diagnostic test consist of tasks, the results of which give an opportunity for evaluation of knowledge of the subjects in these fields:

- Conceptual framework of physical culture;
- The main means of physical education;
- Methods of physical education.

Fourth, fifth and sixth blocks of the first diagnostic test consist of tasks, the results of which give an opportunity for assessment of knowledge and skills of subjects in the field of educational technology used in the process of physical education:

- Practical training in physical education;
- Monitoring physical and mental condition;
- Testing of physical preparedness.

In the final diagnostic test first, second and third blocks consist of tasks, judging by the results of which one can evaluate knowledge of the subjects on:

- Typing negative factors associated with the use of ICT;
- Typing possible negative effects on health resulting from the application of ICT;
- The means of and activities dedicated to mitigating negative consequences of ICT use.

Fourth, fifth and sixth blocks of the final diagnostic test consist of tasks, judging by the results of which one can evaluate knowledge and skills of examinees in application of pedagogical technologies, directed towards formation of competence of a student user of ICT in the field of health care:

- Practical training with the use of recreational facilities;
- Having a healthy self-esteem and being good at self-control of both physical and mental state of the user of ICT;
- Testing indicators of physical and psychophysiological state of the user of ICT.

For every correctly done task a student gets one point, otherwise he gets zero points for that task.

# 7  Conclusion

The problems of ICT negatively impacting the health of the user is certainly relevant at the present stage of development of the IEE, not so much for educational institution, but for the user of ICT, taking into account their personal health characteristics. In order to mitigate this negative impact, we have developed a modular structure of the content of a course of physical education that forms the competence of the student user of ICT in the field of health care. The course includes the following modules:

- The impact of negative factors associated with the use of ICT on the body of the user.
- Possible negative effects of psychological and pedagogical as well as medical and social nature, due to the use of ICT.

- Self-esteem and self-control of health status, physical development, mental state and physical health and preparedness.
- Testing of indicators of physical and psychophysiological state and state of health, including the use of diagnostic devices associated with a computer.
- Implementation of preventive and recreational events mitigating possible negative consequences of using ICT.
- Natural universal means of preventing colds and infectious diseases, bad habits. Means of healing the body and prevention of possible negative consequences of using ICT for the user's health.
- Wellness and fitness center and health offices of higher education institution: carrying out of studies, surveys and tests.

Educational programs of the discipline "Physical education" for higher education institutions with traditional and distant learning were created in accordance with the listed modules. A course of health protection is carried out as an elective course of a discipline "Physical education" in higher education institutions and is based on the knowledge, skills and experience of students, which they got during first and second years of a compulsory discipline "Physical education". During the course, the students master theoretical, practical (methodological and practical training) and control sections and subsections of the course. Moreover, medical monitoring is carried out based on the program of testing indicators of physical and psychophysiological state (PPS) of a student. The total course duration is 72 h. Following the course you can talk about the formation of the student's competence discussed in this study. As a result of mastering this course, students form the competence discussed above.

# References

1. Dimova, A.L.: Medical Recovery of the Users of Information Technology: Organization and Technology (Оздоровление пользователей информационных технологий: Организация и технологии). LAP LAMBERT Academic Publishing, Saarbrucken, Germany (2014)
2. Mukhametzyanov, I.Sh.: Health Protecting Information and Communication Technology-Based Environment of an Educational Establishment: Problems and Prospects of Development (Здоровьесберегающая информационно-коммуникационная среда учебного заведения: проблемы и перспективы развития), p. 208. Idel-Press, Kazan (2010)
3. Robert, I.V.: Theory and Informational Technique of Education: Psychological and Pedagogical and Technological Aspects (Теория и методика информатизации образования: психолого-педагогический и технологический аспекты), p. 274. Institute of Informatization of Education of the Russian Academy of Education, Moscow (2008)
4. Golub, G.B., Kogan, E.Ya., Fishman, I.A.: Estimating Key Professional Skills of the Graduates of Basic Professional Training Programs: Approaches and Procedures (Оценка уровня сформированности ключевых профессиональных компетентностей выпускников УНПО: подходы и процедуры). Issues of education, vol. 2, pp. 161–185 (2008)

5. Gorelov, A.A.: Intellectual Work, Physical Efficiency, Motion Activity and Health of the Student Youth (Интеллектуальная деятельность, физическая работоспособность, двигательная активность и здоровье студенческой молодежи), p. 101. Belgorod, Politera (2011)
6. Kondakov, V.L.: System Mechanisms of Designing of Sports and Improving Technologies in Educational Space of Modern Higher Education Establishment (Системные механизмы конструирования физкультурно-оздоровительных технологий в образовательном пространстве современного вуза). LitKaraVan, Belgorod, Russian Federation (2013)
7. Bespalko, V.P.: Foundations of the Theory of Pedagogical Systems (Основы теории педагогических систем), p. 304. Voronezh State University, Voronezh (1977)
8. Slastenin, V.A., Isaev, J.F., Shiyanov, E.N.: Creative Project as Means of Formation of Professional Competence of Future Specialist (Творческий проект как средство формирования профессиональной компетентности личности будущего специалиста). http://nsportal.ru/npo-spo/obrazovanie-i-pedagogika/library/2015/01/20/tvorcheskiy-proekt-kak-sredstvo-formirovaniya

# Intelligent System of Training and Control of Knowledge, Based on Adaptive Semantic Models

Tamara Shikhnabieva and Sergey Beshenkov

**Abstract** Modern information technology and the rapid expansion of the diversity of the network of educational services has caused a stream of innovations on the reorganization of existing educational systems at all levels of education. There are many approaches of improving the educational systems of the destination, one of which is the use of intelligent methods and models for knowledge representation. The paper describes the system we have developed intellectual training and control of knowledge, based on adaptive semantic models.

**Keywords** System of educational purpose · Structuring knowledge · Intelligent tutoring systems · Knowledge control · Adaptive semantic models

## 1 Introduction

It should be noted that such well-structured area such as mathematics, physics, theoretical mechanics are based on a rich mathematical apparatus to describe its regularities, which allows to integrate computer simulation with traditional algorithmic programming (without isolation of the level of knowledge). "Knowledge is important where the definitions are blurred, concepts are changing, the situation depends on many contexts, with high uncertainty, ambiguity of information" [1]. As an example, in our work we consider the subject area "computer science". As we see, computer science is a rapidly developing, ever-expanding new knowledge of the subject area. Uncertainty and vagueness in the terminology, the need to determine the location of new knowledge in the existing system requires the

T. Shikhnabieva (✉) · S. Beshenkov
The Institute of Education Management, Russian Academy of Education,
Moscow, Russian Federation
e-mail: shetoma@mail.ru

S. Beshenkov
e-mail: srg57@mail.ru

© Springer International Publishing Switzerland 2016
V.L. Uskov et al. (eds.), *Smart Education and e-Learning 2016*,
Smart Innovation, Systems and Technologies 59,
DOI 10.1007/978-3-319-39690-3_53

development of theoretical positions and methodical bases of definition of structuring knowledge in learning systems.

The article is devoted to the theoretical and methodological bases of development of intelligent educational systems on the basis of the creation of adaptive semantic model (ASM) [2] weakly structured areas of knowledge.

## 2   On the Influence of Modern Information Technology and Network Services in the Form of Educational and Teaching Methods

Modern information technology and the rapid expansion of the diversity of the network of educational services sparked a flood of innovation on the reorganization of the existing educational systems at all levels of education—from school to University. As a result, changing character and dynamics of interaction-learner—teacher. This has a significant impact on the choice of methods, forms and technologies of training. Training with use of information resources stored in the Internet and Intranet, is a catalyst in the formation of a new, progressive theories of teaching and learning, focused on the development of personality of a student capable of implementing their own, including, educational projects, and striving for self-improvement throughout life.

New mechanisms of transmission of information had a significant impact on the tools, methods and forms of education. As a consequence, there is an urgent need for analysis of special applications of the regularities of the General theory of learning—didactics, in terms of technology computer training and distributed in time and space processes of network learning. The functions of the teacher and students in the educational information environment undergoing radical changes compared to the traditional learning environment. In educational environments using ICT, the student works at their own pace and without constant direct contact with the teacher; the teacher from the main carrier and transmitter of knowledge becomes the Advisor and the Advisor of the student. The teacher controls the learning process, having at its disposal a powerful tool—the computer with its capabilities of delivery, storage and handling of all types of information to demonstrate educational information, training and self-control. The student, in turn, has a powerful technological tool to support independent intellectual work and access to the information environment that is not limited in space and mode of transmission. Be available to the student knowledge about the content and method of training, which until now were prerogative of the teacher.

Intensive work in the information service activates the cognitive activity of students and enhances the creative components of work of the teacher. Means of information and communication technologies in teaching frees the teacher from many functions that have become routine in his daily activities.

However, to use the new opportunities of the educational information environment is essential theoretical understanding and technological support for solving some practical problems related to the reorganization of the educational process.

In this regard, one of the most urgent educational task becomes the problem of efficient use of computer for management of educational process by the students.

For the implementation of the tasks of modern education, need effective, flexible, modular system based on the most advanced technologies and means of learning.

A distinctive feature of the modern stage—search teachers—researchers of ways to apply formal methods to describe the learning process using the apparatus of the system analysis, cybernetics, synergetics, accounting, development and expansion of the concepts, principles and achievements of didactics.

As an example, a weakly structured field of knowledge let us consider the subject area "computer science", which is inextricably linked with information technologies and with the most dynamic resource of the world community. In the process of learning science is manifested in the constant updating of the versions of the tools to learn information technology, the emergence of new user environments and software development systems that are unknown to the teacher. Therefore, it is possible to determine, from our point of view, one of the most important problems of training of specialists in Informatics and ICT: the training system must provide a level which would allow the specialists in their future professional activity to adapt quickly to innovations in the field of information technology.

Informatics as a scientific discipline is a rapidly evolving area of knowledge, some sections of which are already well established and are generally accepted and some are under development.

In addition, the rapid development of ICT and the Internet in recent years has given rise to a number of problems associated with the rapid growth of poorly structured, duplicative of information to be stored and processed.

These shortcomings limit the ability of the semantic information search and access. Over these issues there are numerous scientists and specialists worldwide, in particular, the W3C, where is implemented the concept of Semantic Web. As the study of electronic educational means used in teaching, many of existing electronic courses are closed systems with rigid models, it is not always possible to adapt to specific level of students' knowledge. A disadvantage of existing e-learning tools is the lack of a holistic perception of the educational information by the students. A disadvantage of existing e-learning tools is the lack of a holistic perception of the educational information by the students. The use of intelligent methods and models in the development of training systems can eliminate these drawbacks.

## 3    Comparative Analysis of Models of Knowledge Representation

For knowledge representation in intelligent tutoring systems (ITS) there are different ways.

The various methods are caused primarily by the desire with the greatest efficiency to represent different types of subject areas. Usually, the method of presentation of educational material in intelligent systems is characterized by a model of knowledge representation. Model of knowledge representation is usually divided into logical (formal), heuristic (formal) and mixed.

The basis of logical models of knowledge representation is the notion of a formal theory. Examples of formal theories can serve as the predicate calculus and any particular production system. In logical models, as a rule, use the predicate calculus of the first order, supplemented by a number of heuristic strategies. These methods are systems of deductive type, i.e. they use the model output from a given system of assumptions by using a fixed system of rules of inference. Further development of predicate systems are systems of the inductive type, in which inference rules are generated by the system based on the processing of a finite number of training samples [3]. In logical models knowledge representation of the relationship existing between individual units of knowledge are expressed just using the poor tools that are used by syntactic rules of the formal theory. Unlike formal models, heuristic models have a diverse set of tools that broadcast specific features of a problem domain.

That is why heuristic models outperform logical as possible to adequately represent the problem environment and the effectiveness of the used rules of inference. Heuristic models used in expert systems that include network, framing, production-extraction and object-oriented models. It should be noted that the production models used for knowledge representation in expert systems differ from formal systems of production that they use a more complex design rules, and contain heuristic information about the characteristics of the problem environment, often expressed in the form of semantic structures. As a rule, systems based on knowledge, used not one but multiple views. Executable assertions are represented either in the form of production rules, or in the form of modules (procedures) that are called by the sample. To represent the domain model are used the object approach or network models (semantic networks and frames).

## 4   Justification of the Choice of Adaptive Multi-level Hierarchical Semantic Model for Knowledge Representation in Intelligent Educational Systems

The use of object oriented approach in systems engineering knowledge highlights another feature, namely the possibility of natural decomposition of the task into subtasks, represent quite independent agents working with knowledge. Today it is the only practical possibility of operation in conditions of exponential growth of complexity, typical systems that use knowledge.

We reviewed the features of the most common models of knowledge representation. It should be noted that the model of knowledge representation in a semantic network is a graph structurally. As you know, "the count is very characteristic mathematical object adaptation" [4].

Based on the results of the comparative analysis of mental models, as the main way of representing loosely structured interdisciplinary fields of knowledge in IOS, we chose adaptive semantic model (ASM).

For the design of intelligent educational systems based on semantic models, we were guided by the theory of semantic networks or other heuristic models of knowledge representation, as well as the main scientific approaches to gaining knowledge (constructive, axiomatic, etc.).

The advantage of adaptive semantic models of knowledge representation and the learning process is a clear description of the subject area, flexibility, adaptability to learner goals. However, the property visibility with the increase in the size and complexity of relations of the knowledge base of the subject area is lost. In addition, there are significant difficulties in handling different kinds of exceptions. In addition, there are significant difficulties in handling different kinds of exceptions. To overcome these problems using the method of hierarchical description of networks (emphasis on local subnets, located at different levels).

At the highest level of the hierarchical model of knowledge are located meta—concepts further (below level) are placed macro—concepts (generalized concepts) and at the lowest level are located elementary concepts.

This approach to knowledge organization in the development of training systems shows the relationship of elements of educational material, significantly reduces training time, reduce the amount of memory occupied by the base of knowledge and data.

Model in a hierarchical semantic network, as the logical structure of the studied subject area also shows the sequence of presentation of educational material.

Moreover, the sequence of presentation of educational material can vary. With ACM you can select a particular sequence of presentation of educational material, at the discretion of the teacher. Moreover, you can choose the shortest way to achieve educational goals, reducing learning time.

It is known that training techniques traditionally used in the higher education system as a means of communication and student learning. During training, students comprehend the meaning of messages stored in computers and "interact" with the educational technology. Such use is limited to the computer thinking trained and supervised learning system. It follows that it is necessary to expand the computer's capabilities in terms of presenting information. When creating ASM teaching material, students use a personal computer as a tool to present their knowledge.

Using the computer as a tool to build knowledge involve students in the process of formation of knowledge that contributes to their understanding and assimilation, not just reproduce the memory that is obtained from the teacher.

We note only that the semantic expressiveness and imagery of their networks is an important advantage that makes it easier to identify and show the logical relationships in the learning material.

If the usual method of training can do more ourselves to global structures, when learning using ICT and sequence of elements of the educational material is the core problem. Using semantic models as a tool for building the structure of knowledge, and not as a learning environment, allowing interaction with the computer to transfer to the jurisdiction of the learners themselves, giving them an opportunity to present and express their knowledge. In the process of creating computer semantic networks, students should analyze their own knowledge structure that helps them incorporate new knowledge into existing knowledge structures. The result is an efficient use of acquired knowledge.

The above information, observations and recommendations allowed us to systematize and to summarize the main methodological principles for the presentation and control knowledge in intelligent learning system and served as the basis for its development [5].

## 5  An Example of the Use of Adaptive Semantic Models for Knowledge Representation

Currently, there are different types of educational tools: textbooks, manuals, handbooks, etc., including electronic educational tools. However, existing electronic textbooks on abstract subjects is not significantly different from manuals in hard copy. To find the relationship between the concepts of the discipline have repeatedly flipping the entire textbook and look for the necessary information. Submission of educational material on abstract subjects in a multi-level adaptive hierarchical semantic model (Fig. 1) allows you to create a structured textbook,

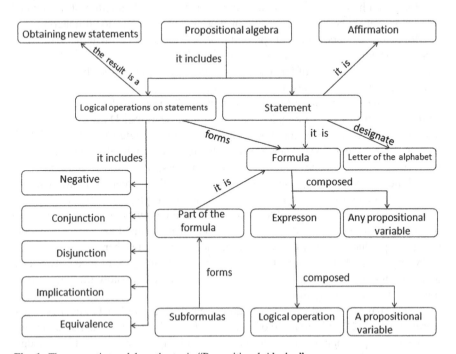

**Fig. 1** The semantic model on the topic "Propositional Algebra"

showing links between the domain concepts, which is important when organizing the training process on the basis of information and communication technologies.

As the experience of developing a semantic model for academic discipline "Mathematical logic", the process of building models promotes efficient knowledge acquisition [6]. Therefore, the training can be run not only on the teacher-developed AFM, but also to give students assignments for their development, that promotes the best mastering of educational material. Shown in Fig. 1 the model of educational material on the topic "Propositional Algebra" presents the basic concepts of this branch of mathematics, and shows causal relationships between them.

## 6 Intelligent Tutoring System Based on ACM

Proposed approaches to the representation and control of knowledge underlying the intelligent tutoring system, which is used in educational process of universities.

Software intelligent tutoring system is implemented in an object—oriented programming environment Delphi. The programming system of Delphi is composed of an instrumental shell with many components. Thanks to the component

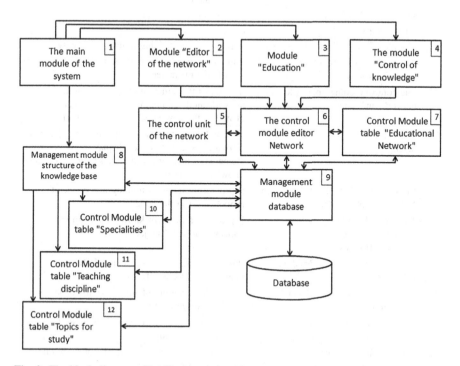

**Fig. 2** The block diagram of intelligent tutoring system

approach to programming, the program can be assembled as a designer, customizing each component to solve a particular problem.

As can be seen from the block diagram (Fig. 2), all database operations are performed through the control module database. The module contains a set of procedures and functions that enable interaction with the database without using SQL language instruction and without direct reference to the database. The control module database is one of the major modules of the system (Table 1). The two other main modules of the system—a control module editor of the network and the control of the network. Other modules provide a convenient interface of interaction with the user.

Network control module editor contains the complete set of functions for network management. Modules "Education", "Edit network" and "Control of knowledge" is used only to its part of the feature set.

**Table 1** Function modules of learning and knowledge control

| № | The name of the module | Appointment |
|---|---|---|
| 1 | 2 | 3 |
| 1 | The main module of the program (uMain.pas) | Provides launch system; Control modules "Edit network", "Education", "Test your knowledge"; Management module structure management knowledge base |
| 2 | Module "Edit network" (uNetEdit.pas) | Implements edit training models |
| 3 | Module "Training" (uNetTeach.pas) | Implements the training function of the system |
| 4 | The module "Control of knowledge" (uNetControl.pas) | It implements a function of the system to verify the knowledge of trainees |
| 5 | The control module objects learning network (uObj.pas) | It contains the structure of objects learning network: node communication, text block, a group of nodes, the new node, the new network |
| 6 | The control module editor Network uNet.pas | The module contains a description of the "Learning Network", which provides display and work with the network in a variety of modes (editing, training, knowledge control) |
| 7 | Control Module table "Educational Network" uNetEditRec.pas | It provides a range of training a network of proposed set in the table |
| 8 | The control module structure Knowledge Base (uThemeHistory.pas) | It provides a range of knowledge from the "Specialization", "subject", "Themes" |
| 9 | Database Management Module uDM.pas | Provides control over the alias ODBC with language instruction Transact—SQL |
| 10 | Control Module table "Specialities" uAreaEditRec.pas | It provides a variety of subject areas for learning and knowledge control |
| 11 | Control Module table "Discipline" uSubjectEditRec.pas | It provides a variety of disciplines to study and control of knowledge |
| 12 | Control Module table "Topics" uThemeEditRec.pas | It provides a choice of the theme of discipline for study and examination |

Control Module editor of the network, as well as the control unit of the network, is based on object-oriented programming. The module describes a single object—learning network. This object is able to manage other objects (of the network) and "draw" their state. The control unit of the network consists of a set of objects, each of which may belong to the parent object—learning network. Control modules database tables are implemented on the basis of standard components for working with databases.

# 7    Conclusion

So, the performance of the educational process in the form of adaptive semantic models allows us to provide: individual pace of learning when implementing feedback; an active approach to the choice of solving the problem based learning situations; the communication of new concepts with existing concepts and ideas, which improves understanding; the implementation of deep processing, which increases the ability to apply knowledge in new situations.

The advantages of our proposed model of knowledge representation are particularly important in the control of knowledge of trainees. We have developed a technique of knowledge control allows to structure the issues and create adaptive tests.

This organization of control of knowledge contributes to the qualitative training, because the trainees analyze the basic structure of the studied notions and ideas, linking them with new concepts.

# References

1. Davydov, V.V.: Types of generalization in learning (logical-psychological problems of the teaching subjects). In: Davydov, V., Longman, M. (eds.), p. 424 (1972)
2. Shikhnabieva, T.Sh.: Methodological basis for the submission and control of knowledge in the field of computer science using an adaptive semantic models. Diss... d-RA PED. Sciences. M., p. 365 (2009)
3. Russell, S., Norvig, P.: Artificial intelligence: a Modern approach. In: Russell, S., Norvig, P. (eds.), p. 1407. Williams, Moscow (2006)
4. Rastrigin, L.A.: Adaptation of complex systems, p. 375. Zinatne, Riga (1981)
5. Shikhnabieva, T.Sh.: On the development of modern educational systems. Information environment education and science. Information environment education and science—2012, No. 10 (2012). http://www.iiorao.ru/iio/pages/izdat/ison/publication/ison_2012/num_10_2012
6. Shikhnabieva, T.Sh.: Training system and knowledge control based on adaptive semantic models. In: International scientific-practical conference "Development of national system of Informatization of education in health terms". The Information environment of education and science—2013, No. 17 (2013). http://www.iiorao.ru/iio/pages/izdat/ison/publication/ison_2013/num_17_2013/

# Use of Electronic Mind Maps for Creation of Flexible Educational Information Environments

Marina Mamontova, Boris Starichenko, Sergey Novoselov
and Margarita Kusova

**Abstract** The article considers the possibility of using smart and e-learning technologies to create flexible educational information environments. A way to combine the modular instruction technology and the technology of building and application of mind maps (mind mapping) is offered. Mind mapping is considered as an instrumental basis for creation of multidimensional information environments. Mind mapping is regarded as instrumental basis of the modular instruction. Mind maps are used for systematic organization of training material and management of the students' activities. The peculiarity of the proposed format is the navigation system that allows, using cross hyperlinks, to direct students to diverse information both within the blocks and modules, and between them, to use sources of information presented on the Internet and in the personal computers of students, to develop common and personal information environments based on the underlying environment created by the teacher. This paper was written as a part of work on Governmental Order 2014/392, Project 1942 of the Ministry of Education and Science of the Russian Federation.

**Keywords** Professional training · Smart training · Student-centered training · Self-control training · Personal information environment · Modular instruction · Distance learning · Electronic educational resource · Mind map · Mind mapping

M. Mamontova (✉) · B. Starichenko · S. Novoselov · M. Kusova
Ural State Pedagogical University, Yekaterinburg, Russia
e-mail: mari-mamontova@yandex.ru

B. Starichenko
e-mail: bes@uspu.ru

S. Novoselov
e-mail: inobr@list.ru

M. Kusova
e-mail: mlkusova@mail.ru

# 1 Introduction

Currently the digital technologies for presentation, transfer and processing of information are actively used in all areas of human activity. Educational system is no exception. The tasks of modernizing the educational process based on modern information and communication technologies are being addressed by many researchers and practitioners both in Russia and abroad [1].

The rate of development of innovative forms of processing, storage and transfer of information is ahead of the rate of modernization of the educational process. In this regard, the process of informatization of education faces a number of challenges. On the one hand, vast experience has been gained in application of innovative educational information formats. On the other hand, the use of interactive, audiovisual, hypertext formats as supplementary tools in teaching methods based on the traditions of book culture does not lead to qualitative transformations in the educational environment. Due to availability of sources of information provided by new technologies the students develop the illusion that acquisition of knowledge is simple and the so-called "mosaic", fragmented thinking is formed which is unacceptable in specialist training. People today need skills of self-study and self-control skills and ability to work with information.

Due to the diversity of information presentation forms the teachers are faced with choices that should be based on specific teaching goals and objectives, on the one hand, and on the peculiarities of students and the environment in which the learning process is implemented, on the other hand. Addressing a most important task—the formation of systemic holistic knowledge of the academic discipline—is to be carried out using methods, tools and technologies that allow, along with systemic specific (subject) knowledge, to form students' self-learning and self-development skills.

To address these problems a comprehensive approach is needed that includes analysis of modern forms of presentation, processing and transfer of information, identification of instructional benefits of innovative formats of the educational resources in the context of traditional and innovative learning and teaching methods.

The project "Methodology of Creation and Application of Innovative Formats of Electronic Educational Resources" is being implemented at the Ural State Pedagogical University. Within the framework of the project the opportunities and limitations for solving various educational tasks pertaining to such electronic educational resources as info graphics, digital storytelling, podcast, learning quest, and mind map are considered. A distinctive feature of the proposed approach is holistic consideration of technological, design and ergonomic, functional, semantic and psychological and pedagogical aspects of the use of innovative learning tools. The study is aimed not at formal mastery of smart technologies, but at identification of relationships, on the one hand, between the quality of information perception and acquisition by students, the learning efficiency, increase of their motivation, development of self-learning skills, and, on the other hand, such features of information presentation and data exchange as semantic capacity, multimedia, possibility to structure information and model processes, the speed of information exchange, and conducting of joint work.

The article considers the possibility of using smart and e-learning technologies to create flexible educational information environments. A way to combine the modular instruction technology and the technology of building and application of mind maps (mind mapping) is offered. Mind mapping is regarded as instrumental basis of the modular training. Mind maps are used for systematic organization of training material and management of the students' activities. The peculiarity of the proposed format is the navigation system that allows, using cross hyperlinks, to direct students to diverse information both within the blocks and modules, and between them, to use sources of information presented on the Internet and in the personal computers of students, to develop common and personal information environments based on the underlying environment created by the teacher. The proposed methodology is used when training the undergraduates of the Institute of Mathematics, Informatics and Information Technologies of the Ural State Pedagogical University within the "Pedagogical Qualimetry" subject area.

## 2  Characteristics of Learners and the Learning Environment

To choose models, methods and tools of learning the peculiarities of the body of learners and the conditions in which the educational process is implemented are to be taken into consideration.

Those features are as follows: Small groups of students (from 6 up to 10 people); Various "starting" level of the undergraduates (they enter masters' programs after receiving education on various educational programs; Various age of the members of the (academic) groups (age from 23 to 45 and higher); Different professional and life experience; Combination of postgraduate study with work; Classes twice a week for 6 h (half of the training time is devoted to autonomous work of students).

To train such groups, it is advisable to choose technologies that provide: Subject-subject relationship between the teacher and the students; Differentiation and individualization of the learning content (substantive flexibility); Dynamic and mobile program structure (structural flexibility); An opportunity to provide training using immersion method (learning) to create conditions for the establishment of systemic holistic knowledge among the students in relatively short time (intervals); Creation of conditions for effective independent and collaborative work; formation of the students' skills of self-learning and self-development.

It is important to note that the technologies are to be accessible and not expensive and to be intended for a small number of users.

The above characteristics, in our view, are met by the modular instruction technologies integrated with such formats of electronic educational resources as mind maps.

# 3   Substantiation of the Choice of Technologies

## 3.1   Characteristics of Modular Instruction Technology

The ideas of modular instruction are presented in the work of Skinner [2]. Modular instruction was theoretically substantiated in the writings of G. Russell, M. and B. Goldschmidt, C. Curch and other foreign researchers. Interest in these systems increased in 70s of the last century. The recommendations of the Paris 1989 UNESCO Conference emphasized the need to establish "open and flexible education and professional structures allowing to adapt to the changing needs of production, and science, as well as to local conditions" [3]. Modular instruction systems meet these requirements since they allow to block the training content, to vary the pace of the discipline study, to integrate various types and forms of learning, and to adapt them to the specific groups of learners.

Modular instruction is high-tech training based on the activity approach and the principle of conscientious training (the student is aware of the training program, its goals, objectives, training content, methods of achievement of the training objectives and his/her own trajectory of learning) that is characterized due to modules by closed management type. The main purpose of the modular instruction is "creation of flexible educational structures in terms of both content and training organization contributing to meeting the educational needs of students, and defining the vector of the new emerging interest" [4]. The key concept of the modular instruction is the use of autonomous units of learning content (teaching units) called modules. J. Russell defines the module as "instructional package dealing with a single conceptual unit of the learning material and actions prescribed to the students" [5]. Yutsyavichene [6] defines the module as "information block which includes logically complete unit of training material, a targeted action program and methodological guide that ensures achievement of teaching objectives."

In relation to a separate academic discipline the module can be considered as "autonomous organizational and methodical structure of discipline including teaching objectives, a logically complete unit of training material, guidance and control system" [7]. This definition of the module was used as a basis for the development of modular instruction program in this project. Modules can be used for training in small groups and in independent work of students. The interest of researchers and practitioners in modular instruction is linked to its capacities to handle a wide range of instructional tasks including integration of various forms and methods of teaching [8], designing flexible problem-modular instruction systems [9], establishment of interdisciplinary links [10] development of individual educational programs [7], forming a holistic system of knowledge of academic discipline [6], and others.

The growing interest in modular instruction emerged in connection with the Bologna process which aims at a paradigm shift in education—"from teaching to learning". In the course of specialists training at higher school the new paradigm is implemented through student-centered and competence-oriented approaches to training. Didactics of higher school acquires new features—"… centering on the learners and their learning process, the changing role of teachers (guidance and educational counseling on environment or situations of learning), orientation of learning towards goals and results, promotion of self-organized and active learning, concentration on motivational, volitional and social aspects of learning" [11]. Modular instruction meets the new needs of the education system.

## 3.2  Mind Map as a Multidimensional Visual Learning Tool

Education develops in the course of improvement of the methods and tools of learning. J.A. Komensky formulated the principles of clarity, consistency and systematicity of training which involves the use of subsidiary tools for presentation of training information. The issue of the development of visual learning tools is discussed in numerous works of Russian and foreign scholars and practitioners (Yu. K. Babansky, D.B. Bogoyavlensky, V.V. Kraevsky, A.N. Leontiev, M.I. Makhmutov, and others). Natural models, symbols and sign & symbol models, diagrams, graphs, drawings, etc. are used as visual learning tools. A special place among the visual learning tools is held by the tools that support cognitive activity of the learners in verbal form through multi-dimensional representation of knowledge in natural language. According to Steinberg [12], "… the problem of multidimensionality is directly linked with one of the narrow, critical points of the majority of learning technologies this is a one-dimensional channel connecting internal and external aspects of training activities due to the inevitable verbalism and portioning information in time". Use of one-dimensional, "conveyor" knowledge transfer mechanisms distorts the original multidimensional material that is adequate to the real world. Multidimensional content, on the one hand, and one-dimensional (linear) form of learning, on the other hand, limits and distorts the knowledge flows between the multidimensional internal and external aspects of cognitive activity. V. E. Steinberg calls the one-dimensionality of the presentation of the training material (use of linear structures) "inherited technological defect of educational processes" [12]. Multidimensional educational tools that allow you to condense and structure information, to present it in the form of multidimensional structurally-logical schemes, provide unity of knowledge content and the method to present this knowledge to students, to process and absorb it. Mind maps whose building and usage for various tasks belongs to Buzan [13] can be counted as such multidimensional educational tool.

The mental map is a means to present visual information, it has a wide range of possibilities for displaying the structure and content of the training material, and facilitates the perception of information by the learners.

Despite the high potential of mental maps, their use in teaching schoolchildren and university students is now limited. However, large volumes of information that you need to be assimilated require more effective means of its presentation and processing. Maintenance of training information (knowledge presentation) in the form of mental maps significantly enhances the level of understanding of information and the depth of its processing.

## 4 Formats of Educational Information Environments

### 4.1 Workbook Format Technology

Modular programs can be implemented with the help of "paper" technologies. The so-called Workbook Format is widely spread. The peculiarities of this format are large fields for comments and notes of students, presence of assignments to be done and recommendations for assessment of tasks' completion. For orientation within a workbook special information retrieval tools are used—a lubricator that as a rule is presented in the list of contents. In the text numerous references to materials both inside the module, and beyond (books, articles, etc.) are used.

### 4.2 Electronic Resources. Distance Learning Systems

Development of information technologies has made it possible to implement the modular instruction model in distance learning systems.

Currently the market offers distance learning systems with various technological and commercial characteristics. The three most common distance learning platforms LMS (Learning Management System) are: Blackboard Learn, Moodle and Sakai.

A common feature of such systems is their ability to create training programs with a block-modular structure of the educational material. Training module can be considered as an instrument for organization of educational material and management of the training activities of the students.

An academic discipline can be represented as a combination of training modules, each including a number of blocks. The blocks can have different structure.

The mandatory elements of the block include: Content and structure of the training material; List of competencies that are to be formed while studying the block; Sources of information to be familiarized with in the course of the block study; Tasks to form necessary skills; Tools for assessment of the students' educational achievements.

As much as any training system, the block-modular instruction using distance learning systems has its advantages and disadvantages.

The undisputed advantages of such systems include: Orientation of the modular instruction towards forming students' holistic system of knowledge of the discipline; each block is considered, on the one hand, as a relatively independent unit of the educational material and, on the other hand, as one of the components of the overall system of knowledge of the discipline; A high degree of structuredness, the "readiness" of the educational material to be perceived by the students; Presentation of the material in concentrated form; a module usually includes the minimum required educational material for assimilation of theoretical material and for doing practical work; The work of the student in his/her spatial and temporal mode.

The limitations (disadvantages) of the distance learning systems include: The need to have continued access to the distance learning system; Lack of technical preparedness of students to the use of the distance learning systems; Impossibility to authenticate the student when assessing his academic achievements; High cost of software (commercial systems require paid subscription) and of the development of distance learning courses; Lack of preparedness of a large part of teachers to master the distance learning technologies (the need for training of teachers working in such systems); The complexity of administration of such systems (in conjunction with the teacher courses are supported in such systems by the organizer; installation and local support are to be conducted by IT specialists), etc.

Knowledge representation in the distance learning systems in the "ready" structured way certainly contributes to a more rapid absorption of knowledge by the students. On the other hand, this may become an obstacle to creation of flexible (developing) personal information environments. The reason for this may be the incompatibility of various formats of information resources.

## 4.3　Electronic Mind Maps as a Format of Flexible Educational Information Environment

An alternative to expensive distance learning systems can be electronic mind maps created using various available and free services.

The use of mind maps is practical in connection with the need to make the process of mastering the course material by the students simple and efficient, and to create the opportunity for self-study. The combination of interrelated mind maps that reflect the content of the program modules and blocks that need to be learned by the students can be called base (underlying) information environment (network). The base environment is created by the teacher. Communication between the modules and blocks is performed using hyperlinks. Modules and blocks can have one link and several links.

The peculiarity of mind maps that is essential for the solution of the task of developing the students' holistic system of knowledge of the (subject) discipline is their ability to be corrected (adjusted), to augment (build up) elements, and to increase the number of levels. This enables each student to create personal infor-mation environment by adding new elements and relationships. Working in such an environment, the students independently "build up" their knowledge with consid-eration of their own educational needs. Thus, a lot of personal information environments are created on the basis of the basic (underlying) structure. It is also possible to exchange files with mind maps and different sources of information among students.

Thus, the use of the proposed format of the modules and blocks with the navigation system using cross-hyperlinks enables (you) to: Guide the learner to diverse information both inside the blocks and modules, and between them; Use sources of information presented in the Internet and in personal computers of students; Make the process of learning the discipline individualized by creating each learner's (personal) information environment; Develop collective and personal information environment based on a basic environment created by the teacher; Distinguish mandatory and elective parts of the training material.

## 5 Selection of an Electronic Resource to Create Flexible Personal Information Environments Based on Mind Maps

Among the many on-line services designed to build mind maps we chose XMind (Russian version) [14]. Xmind allows to create mind maps in logical, tree-type and other views. The program has many options for customizing the appearance of the map and its components—it allows you to select the background color, font settings, add text notes and files, tables, bullets, hyperlinks, annotations, select map templates and themes. The program allows you to focus attention on certain structural elements of mind maps, to create voice memos, check the spelling of the text, use the search of cards components, to export data and publish created maps on the Internet.

The service is provided free of charge. The program can be installed on personal computers of students and used in off-line mode, allowing you to work in a mode independent of the Internet.

Therefore, the presented characteristics of XMind service meet the requirements to teaching conditions for undergraduates (free access, ability to work in off-line mode, ability to exchange information with other students).

## 6 Testing of the Presented Format of Flexible Educational Information Environment

### 6.1 Structure and Content of the Block-Modular Program Based on Mind Maps

The structure and content of the program "Pedagogical Qualimetry" are described in detail in the work of [15]. The program includes five modules: "Introduction to discipline", "Introduction to the problem of the assessment of the quality of education", "Theoretical basis of Pedagogical Qualimentry", "Methods of quality assessment", "Qualimetry methods as instrumental basis of assessment programs". The first module includes blocks presenting the program as a whole: "Learning objectives," "Expected results", "General program map" showing the structure and content of the program with navigation system (links to other cards), "Algorithms for working with the program (training (&) management)", "Calendar-thematic plan", the plan with training modules—with specific dates of module study and control measures, "Self-learning and self-development: methods and tools".

The remaining modules have similar block structure: "Expected results", "Information resources" that are presented in the form of mind maps and reflect the logical structure of interrelated elements of the educational material with links and connections between the elements within the block, between blocks and modules; "Laboratory works" (a map with names of laboratory works and hyperlinks to these works), "List of abstracts" that should be made when working with the module information resources, "Materials for self-control" (a map with names of control and measuring materials and hyperlinks to these materials).

It should be noted that when forming the baseline environment the teacher uses the minimum capacity of the computer program to customize the appearance of mind maps and their elements. When working with maps the students personalize their cards using the broad program features of XMind program (color, shape, font settings, notes, etc., images etc.).

## 6.2 Organization of Learning

The training on the program begins with the "introduction to the discipline". Special attention is given to the conditions of training: XMind hosting on personal computers; Saving of the folder with files representing modules and blocks of the programs on personal computers to work in off-line mode; Creation of an account in the "Cloud" where materials are hosted presenting the base information environment created by a teacher which in the course of training on the program can develop by adding resources created by students.

The first module contains the block "Self-education and self-development: methods and tools". This block presents to the students the mind map technique, its possibilities for structuring training material and self-assessment of the progress in their own knowledge and skills. Guidelines for working with modular program are offered. The program provided in such a way, can be considered as e-workbook with navigation system and learning management.

Each module is designed so that it can be used separately at a single school lesson. The module plays the role of support in the course of independent work of students. Mind maps are becoming a means of deepening and broadening knowledge of students. When working with modules and blocks of the program, the students on the basis of the underlying environments create personal information environment, complementing its with content elements of other disciplines (if it is necessary for understanding and learning the discipline), with additional sources of information and references to them in the maps that represent modules and blocks.

Therefore, on the basis of the baseline environment many personal environments are formed reflecting, on the one hand, the results of student work on the training material, and on the other hand, the individual educational needs of students. Students post completed laboratory works and essays in the "Cloud" in specially created folders to be checked by the teacher.

The materials of interest to the entire group of students could be hosted in the "Cloud" which creates common information environment for an academic discipline.

# 7 Results of Program Approbation (Testing)

The program "Pedagogical Qualimetry" has been implemented since 2011. Two groups of students (14 people, 2011–2012) were trained using traditional training technology. In 2013 and 2014 groups of undergraduates were trained on the presented modular program using mind maps (15 people).

For the evaluation of the results of training final testing was conducted. The final test displays the structure and content of the educational discipline and allows the teacher to estimate the amount of knowledge the students assimilated. Relative test score equal to the ratio of the sum of test scores to the maximum possible score was compared with the criterion score set to 75 %. In criterion-oriented testing exceeding this threshold is considered an indicator of well-structured knowledge base structure [16]. In the first group a prescribed threshold was exceeded by 57 % of the students, in the second—86 %. Reliability of the differences in the two groups was determined using the Fisher criterion [17]. Empirical criterion value is equal to 1.66 Fisher. The differences are statistically significant at $p \leq 0.05$.

The second group of students showed great enthusiasm and interest in learning the course material. Most of them continued using mind-map method in studying other disciplines and in preparation of materials for the master's thesis.

# 8 Conclusions

Therefore, the synthesis of modular instruction technology with the technology for building and using e-multidimensional didactic tools (group of mind maps) allowed each student to create a flexible (capable of adjusting and development) personal information environment for an academic discipline. Working in such an environment contributes to the creation of fairly sufficient system of basic knowledge of discipline and the development of self-learning skills of the students.

# References

1. Uskov, V.L., Bakken, J.P., Pandey, A.: The Ontology of Next Generation Smart Classrooms. In: Proc. of the 2nd int. conf. on Smart Education and e-Learning SEEL-2016, June 17-19, 2015, Sorrento, Italy, Springer, pp. 1–11 (2015)
2. Skinner, B.F.: The Technology of Teaching. Appleton Centery Grofts, New York (1968)
3. The Modular approach in technical education. Paris, Unesco (1989)
4. Vazina, K.Ya.: Human self-development and modular instruction. Nizhny Novgorod (1991) (In Russian)
5. Russell, J.D.: Modular Instruction. Minn., Burgess Publishing Co., Minneapolis (1974)
6. Yutsyavichene, P.: Theory and practice of modular instruction. Kaunas (1989) (In Russian)
7. Borisova, N.V., Kusov, V.B.: From traditional to distance learning through modular instruction. Moscow (1999) (In Russian)
8. Owens, G.: The Module in "Universities Quarterly". Universities Quarterly, Higher Education and Society, vol. 25, № 1
9. Choshanov, M.A.: Flexible Technology of Problem-modular Instruction: Handbook. Moscow (1996) (In Russian)
10. Karpov, V.V., Katkhanov, M.N.: Invariant Model of Intensive Training Technology in Multi-stage Preparation in Higher School. Moscow (1992) (In Russian)
11. Wildt, J.: Vom Lehren zum Lernen. Zum Wandel der Lernkultur in modularisierten Studienstrukturen. In.: Berendt, B, Voss, H.-P., Wildt, J. (eds.) Neues Handbuch Hochschullehre. Berlin (2004)
12. Steinberg, V.E.: Educational multidimensional tools: theory, methods, practice. Narodnoe Obrazovanie (National Education). Moscow (2002) (In Russian)
13. Buzan, B., Buzan, T.: Super-thinking. Potpourri, Moscow (2003)
14. http://soft.mydiv.net/win/download-XMind.html
15. Mamontova, M.Yu.: Development of qualimetric competence of educators in the conditions of reforming of all-Russia system of education quality assessment: content aspect. J. Pedagogical education in Russia. 5, 96–101 (2012) (In Russian)
16. Croker, L., Algina, J.: Introduction to Classical and Modern Test Theory. Cengage Learning (2006)
17. Sidorenko, E.V.: Methods of mathematical processing in psychology. Rech (Speech). Saint Petersburg (2006) (In Russian)

# Exploiting the Characteristics of Lecturers Based on Faculty Performance Evaluation Forms

Thuc-Doan Do, Thuy-Van T. Duong and Ngoc-Phien Nguyen

**Abstract** The faculty evaluation forms can be considered as valuable data source to exploit knowledge which helps to improve the quality of teaching and learning in universities. In this paper, we analyze previous studies on exploiting faculty evaluation forms according to major problems and their solutions. On that basis, we propose and solve the problem of mining useful knowledge about human resource of Ton Duc Thang University. The experimental data are collected from the online faculty evaluation system of our university, with more than 140,000 evaluation forms. We apply the solution to analyze the data set and draw meaningful comments for the exploitation and construction of human resource appropriately and efficiently. The results obtained are compared to the previous study on clustering lecturers based on performance and statistical method.

**Keywords** Student feedback · Faculty performance · Teaching performance evaluation · Faculty evaluation form · Clustering

## 1 Introduction

The quality of education has always been considered as the foundation of the long-term development of all countries. In order to provide people with sufficient knowledge and skills to labor market and enhance their reputation, universities must constantly improve the quality of teaching and learning. Many strategies have been applied to measure the faculty performance, including: student ratings, peer ratings, self-evaluation, videos, student interviews, exit and alumni ratings, employer rat-

T.-D. Do (✉) · T.-V.T. Duong · N.-P. Nguyen
Ton Duc Thang University, Ho Chi Minh City, Vietnam
e-mail: dothucdoan@tdt.edu.vn

T.-V.T. Duong
e-mail: duongthithuyvan@tdt.edu.vn

N.-P. Nguyen
e-mail: nguyenngocphien@tdt.edu.vn

© Springer International Publishing Switzerland 2016
V.L. Uskov et al. (eds.), *Smart Education and e-Learning 2016*,
Smart Innovation, Systems and Technologies 59,
DOI 10.1007/978-3-319-39690-3_55

ings, administrator ratings, teaching scholarships, teaching awards, learning outcome measures and teaching portfolio [1]. Among these strategies, student ratings are considered as the most popular evaluation tool [2].

In this paper, based on the analysis of previous studies on the exploitation of knowledge from faculty evaluation forms to improve the quality of teaching and support stakeholders such as administrators, lecturers and students in making decisions, we propose a new problem that exploits evaluation forms to obtain useful knowledge about human resource of our university and the method to solve that problem. On that basis, administrators can make decisions in salary increase and task assignment; students can choose appropriate lecturers; lecturers realize their strengths and weaknesses. We apply the proposed solution on a real data set including 143,117 forms from the online faculty evaluation system of Ton Duc Thang University. The results obtained are compared to the only study on clustering lecturers based on performance and correlation coefficient analysis method. It provides an overview of human resource in our university, laying the foundation for the exploitation and development of human resource efficiently.

The main contributions of our work are the following:

- Construct a new faculty evaluation form suitable for our university
- Analyze previous studies in terms of main problems solved
- Propose a new problem and the solution to tackle that problem
- Apply the solution to analyze the data set collected from the online faculty evaluation system of Ton Duc Thang University
- Compare the results obtained to the only study on clustering lecturers based on performance and correlation coefficient analysis method

The rest of the paper is organized as follows. Section 2 presents the studies of exploiting faculty evaluation forms in terms of solved problem and proposed method. Section 3 proposes new problem and its solution. Section 4 presents the experiments and results obtained. Section 5 is used for a discussion. Section 6 draws the conclusion.

## 2    Related Works

To the best of our knowledge, there are a few studies on exploiting faculty evaluation forms to improve teaching quality and support stakeholders in making decisions. This section is divided into three parts according to the main problems solved [3].

### 2.1    *Identifying Determining Factors of Faculty Performance*

Regression analysis was applied to find the relationship between one dependent variable, which was the faculty performance in this case, and one or more independent variables such as subject knowledge, communication skills, etc. In [4], the authors analyzed the 4,589 evaluation forms about an online MBA program of a university in 2007 to identify determining factors of faculty performance and course satisfaction. Each form consists of many questions divided into three groups of criteria: personal attributes, learner facilitation and quality of feedback. Two overall evaluation factors are overall performance of the lecturer and overall satisfaction of the course. The result obtained shows that personal attributes are determining factors. In [5], evaluation forms were collected from Management Information System department's courses at Bogazici University between 2004 and 2009 and some other lecturer and course characteristics drawn from the Student Evaluation of Teaching research (SET). Stepwise regression method was used to identify the determining factors of faculty performance. The experimental results show that five factors consisting of the attitudes of the lecturer, the attendance of the student, the ratio of students filled the questionnaire to the class size, the lecturer is a part-time laborer and the workload of the course largely determine the faculty performance.

Statistical tests such as Chi-square test, Info Gain test, Gain Ratio test were used to analyze the impact of each factor on faculty performance. In [6], the empirical data are the faculty evaluation forms from the graduates of a faculty at an engineering university in 3 years. The evaluation factors include: teacher name, speed of delivery, content arrangement, presentation, communication, knowledge, content delivery, explanation power, doubts clearing, discussion of problems, overall completion of course and regularity, students attendance, and result. The result is that content arrangement is the determining factor of faculty performance.

Apriori algorithm was used to find the association rule with the form $A \rightarrow B$ in which A was evaluation factor and B was faculty performance. In [7], the empirical data were collected from a faculty evaluation system in spring semester of 2007–2008. The experimental result shows that the teaching content and teaching attitude have the strongest relationship with the faculty performance. In [8], the authors collected data from a personnel management system and educational evaluation system. Apriori algorithm was used to find the relationship between the personal information of lecturers namely gender, age, certification and overall rating; the relationship between the evaluation factors namely teaching attitude, teaching ability, teaching content, teaching organization, teaching methods and faculty performance. The factors having strong relationship with the faculty performance should be focused to improve the quality of teaching.

Some algorithms were applied to build the model to classify faculty performance based on evaluation factors. In [9], the empirical data were collected from the evaluation forms of an online system based on four groups of factors: subject knowledge, teaching skills and assessment methods, behavior towards students, communication skills. Models for classifying faculty performance using those factors obtained from M5P and REP algorithms were used to identify the determining factors of faculty performance. In particular, the factor at the root of the tree is the determining factor because it helps to split the data into groups with the lowest entropy. The lower level in the tree the factor appears at, the less impact on the faculty performance it has. REP algorithm builds the tree faster and achieves higher accuracy than M5P algorithm in the data set. Subject knowledge is the determining factor of faculty performance in both algorithms. In [5], two CHAID and CART algorithms were used to identify the determining factors of faculty performance. Experimental results generated two different trees. Factors appearing at all levels in the tree are considered as the set of the important factors to faculty performance, in which the attitudes of the lecturer at the root of both trees is the most important factor. In [6], the empirical data were collected from the graduates of a faculty at an engineering university in 3 years. Classification methods consisting of four algorithms: Naive Bayes, ID3, CART, LAD tree were used to build faculty performance classification model based on evaluation factors. These factors include: teacher name, speed of delivery, content arrangement, presentation, communication, knowledge, content delivery, explanation power, doubts clearing, discussion of problems, overall completion of course and regularity, students attendance, and result. The result obtained shows that Naïve Bayes algorithm has the highest accuracy.

## 2.2 Finding the Relationship Among Evaluation Factors

Apriori algorithm was used to find the relationship among the evaluation factors in [10], including: subject knowledge, teaching with new aids, motivating self and students, communication skills, class control, punctuality and regularity, knowledge beyond syllabus, and aggregate.

## 2.3 Adjusting Faculty Performance Based on Clustering Evaluation Forms

Some algorithms were applied to cluster evaluation forms then recalculate the faculty performance based on clusters obtained. In [11], the evaluation factors consist of clear and understandable presentation, methodical and systematic approach, tempo of lecturers, preparedness for a lecture, the accuracy of arrival to the lecture, encouraging students to participate in classes, informing students about their work, considering student comments and answering questions, availability (through individual teacher/student meetings or via e-mail). The authors partitioned students into several clusters based on the similarity on evaluation forms using k-means algorithm then analyzed the faculty performance in each cluster. In [12], the empirical data obtained from the 3,000 student feedbacks about 77 factors to assess 50 Information Technology lecturers of a university. Expectation Maximization algorithm was used to cluster data according to four levels of performance evaluation: very good, good, satisfactory and poor. The number of clusters is 14. The average value of faculty performance was calculated for each aforementioned level based on results obtained from the clusters.

## 3 Problem and Solution

### 3.1 Problem Definition

In terms of the main problem solved as described in the previous section, the studies are divided into three groups: identifying determining factors of faculty performance, finding the relationship among evaluation factors, and adjusting faculty performance based on clustering evaluation forms. In terms of problem-solving methods, the studies on exploiting knowledge from faculty evaluation forms can be divided into three groups: using statistical methods, using machine learning methods, and combining both statistical methods and machine learning methods. While statistical methods are suitable for identifying important factors that influence faculty performance, using machine learning methods in finding relationship among evaluation factors are relevant. However, in general, the exploitation of useful knowledge from evaluation forms is still limited. Therefore we propose the problem of exploiting faculty evaluation forms to obtain characteristics of the human resource in our university.

Let $F_{ijkl} = \langle f^1_{ijkl}, f^2_{ijkl}, \ldots, f^n_{ijkl} \rangle$ be an evaluation form of student $i$ about lecturer $j$, after studying course $k$ in semester $l$, in which $f^m_{ijkl}$ is the $m$th factor of the form and $domain(f^m_{ijkl}) = \{1, 2, 3, 4, 5\}$, equivalent to a Likert scale with intervals of 1–5 (5 = Strongly satisfied, 4 = Satisfied, 3 = Neither, 2 = Dissatisfied, 1 = Strongly Dissatisfied). The form consists of $n$ questions or $n$ evaluation factors, in which first $n - 1$ factors are specific factors while the last factor is the overall rating. A database $D$ contains a set of all evaluation forms.

Let $T_{jl} = \langle t^1_{jl}, t^2_{jl}, \ldots, t^n_{jl} \rangle$ be average rating of lecturer $j$ in semester $l$, in which $t^m_{jl}$ is the average rating of the $m$th factor. This feature vector describes specialized features of each lecturer based on all of the evaluation forms about him/her.

Let $I(j, l)$ be a set of students taught by lecturer $j$ in semester $l$, $K(j, l)$ be the set of courses taught by lecturer $j$ in semester $l$.

### 3.2 Method

In order to solve the problem, firstly we preprocessed data by eliminating inconsistent evaluation forms with the deviation between average rating of specific factors and overall rating being greater than $\delta$ because the reason for the lack of consistence may be that the students did not pay attention to the content of the questions completely and seriously. We then calculated the feature vectors of all lecturers. Thereafter the lecturers were divided into different clusters according to the similarity of those vectors.

After analyzing the advantages and disadvantages of clustering methods including: representative-based clustering, hierarchical clustering, probabilistic model-based clustering and density-based clustering [13], we chose representative-based clustering algorithms because of some reasons. They do not eliminate the outliers which are special lecturers in this case because of low density like density-based clustering. They are less expensive in terms of computational cost than hierarchical clustering. In addition, they assign objects into a single group, which help to partition lecturers into different clusters exclusively. Among representative-based clustering algorithms, we chose X-means algorithm [14] which is extended from k-means and able to estimate the optimal number of clusters and more efficient in terms of computational cost than traditional k-means algorithm.

We analyzed the clusters obtained to outline a picture of the human resource in our university.

The pseudo code of the proposed method is presented in Algorithm 1.

---

**Algorithm 1** Clustering the characteristics of lecturers based on faculty evaluation forms

---

//eliminate all inconsistent evaluation forms
1: **for** $l = 1$ **to** number of semesters **do**
2:     **for** $j = 1$ **to** number of lecturers **do**
3:         **for** $i \in I(j, l)$ **do**
4:             **for** $k \in K(j, l)$ **do**
                //calculate the sum of rating of student $i$ for lecturer $j$ after studying course $k$ in semester $l$
5:                 $sum(i, j, k, l) = 0$
6:                 **for** $m = 1$ **to** $n - 1$ **do**
7:                     $sum(i, j, k, l) = sum(i, j, k, l) + f^m_{ijkl}$
8:                 **end for**
9:                 $avg(i, j, k, l) = sum(i, j, k, l)/(n - 1)$
10:                **if** $|avg(i, j, k, l) - f^n_{ijkl}| >= \delta$ **then**
11:                   exclude $F_{ijkl}$ from $D$
12:                **end if**
13:             **end for**
14:         **end for**
15:     **end for**
16: **end for**
//calculate feature vectors
17: **for** $l = 1$ **to** number of semesters **do**
18:     **for** $j = 1$ **to** number of lecturers **do**
19:         **for** $m = 1$ **to** $n$ **do**
            //calculate the sum of rating for lecturer $j$ in semester $l$ in terms of $m$th factor
20:             $acc\_sum(j, l, m) = 0$
21:             $acc\_count(j, l, m) = 0$
22:             **for** $i \in I(j, l)$ **do**
23:                 **for** $k \in K(j, l)$ **do**
24:                     $acc\_sum(j, l, m) = acc\_sum(j, l, m) + f^m_{ijkl}$
25:                     $acc\_count(j, l, m) = acc\_count(j, l, m) + 1$
26:                 **end for**
27:             **end for**
28:             $t^m_{jl} = acc\_sum(j, l, m)/acc\_count(j, l, m)$
29:         **end for**
30:     **end for**
31: **end for**
32: cluster all feature vectors using X-means

---

## 4 Experiments and Results

We have collected data from the online faculty evaluation system of Ton Duc Thang University for the second semester 2014–2015. The total number of evaluation forms obtained is 143,117. The form consists of 8 groups with 13 closed questions (12 specific questions and a question about overall satisfaction) and two open questions. The form was constructed on the following basis:

- SEEQ evaluation form consists of 10 groups with 33 closed questions and one open question [15] which is widely used in the world
- Evaluation form of the first semester 2014–2015 in our university
- Evaluation form of the second semester 2013–2014 in our university
- Suggestion from departments in our university
- Characteristics of Vietnamese students and our university's students
- Requirements and current situation of our university

For closed questions, we use the Likert scale as mentioned before.

**Table 1** Specific questions or specific factors in the faculty evaluation form

| ID | Question |
| --- | --- |
| Q1 | Are you satisfied with the specialized knowledge/skills of the lecturers |
| Q2 | Lecturers can inspire students |
| Q3 | Are you satisfied with the enthusiasm of the lecturers |
| Q4 | Lecturers often discuss and answer the questions of students |
| Q5 | Lecturers prepare complete and updated course materials |
| Q6 | Lively, clear, easy to understand and take notes lectures |
| Q7 | Lecturers encourage students to give questions, situations, new issues and discuss in class |
| Q8 | Lecturer present and discuss about the development trends and applications of the subject |
| Q9 | Individual assignments and group assignments are given to help students grasp the subject |
| Q10 | Lecturers instruct students the methods of self-study and deeply exploiting the subject |
| Q11 | Lecturers clearly present the forms of examination and assessment to students |
| Q12 | Contents of the lectures are suitable for the tests |

**Table 2** Number of members in each clusters

| Cluster | Number of members |
| --- | --- |
| 1 | 54 |
| 2 | 507 |
| 3 | 28 |
| 4 | 58 |

**Table 3** Values of cluster centroid

| Attribute | Cluster 1 | Cluster 2 | Cluster 3 | Cluster 4 |
|-----------|-----------|-----------|-----------|-----------|
| Q1 | 4.777778 | 4.005906 | 3.392857 | 4 |
| Q2 | 4.351852 | 3.984252 | 3 | 3.052632 |
| Q3 | 4.981481 | 4.021654 | 3.285714 | 3.982456 |
| Q4 | 4.796296 | 4.007874 | 3.392857 | 4 |
| Q5 | 4.333333 | 4.001969 | 3.535714 | 4 |
| Q6 | 4.277778 | 3.988189 | 3 | 3.017544 |
| Q7 | 4.055556 | 3.980315 | 3.071429 | 3.789474 |
| Q8 | 4.037037 | 3.990157 | 3.107143 | 3.824561 |
| Q9 | 4.037037 | 3.994094 | 3.285714 | 3.789474 |
| Q10 | 4.037037 | 3.992126 | 3 | 3.614035 |
| Q11 | 4.407407 | 4.005906 | 3.75 | 4 |
| Q12 | 4.314815 | 4.003937 | 3.571429 | 3.982456 |

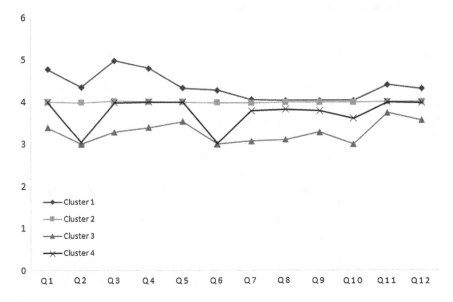

**Fig. 1** Centroid of each cluster

The specific evaluation factors were divided into 6 groups: Specialized performance and study motivation, the enthusiasm of the lecturers, the class organization and teaching performance of the lecturers, the expansion of lectures, assignments/self-study, tests/exams. These groups consist of 12 specific questions as presented in Table 1. Specific questions or specific factors in the faculty evaluation form, corresponding to detailed evaluation factors about the lecturers. Thus, each evaluation form can be considered as a student's perspective on specialized features or the strengths and the weaknesses of a lecturer.

In the preprocessing stage, we eliminated the evaluation forms with the deviation between the average rating of 12 specific factors and overall satisfaction being greater than one ($\delta = 1$). The number of remaining forms after this stage is 139,994 (97.82 %). The value of each faculty evaluation factor is the average of corresponding factor from all relevant forms, rounded to the nearest unit. The results obtained are 647 12-dimensional vectors describing specialized features of 647 lecturers of the whole university.

We applied X-means algorithm for clustering the vectors. The number of clusters obtained is 4. The number of members in each cluster and the values of cluster centroid are presented in Tables 2 and 3, respectively. Figure 1 illustrates the values of cluster centroid on the graph.

# 5 Discussion

## 5.1 Validation

In order to validate the number of clusters obtained by X-means, we used k-means algorithm implemented by Rapid-Miner Studio 6.4 and analyzed results from this tool using the sum of the squared error measure (SSE) [16]:

$$SSE = \sum_{i=1}^{K} \sum_{x \in C_i} dist^2(m_i, x)$$

- $x$ is a vector which belongs to cluster $C_i$
- $m_i$ is a representative vector for cluster $C_i$ (the mean of all vectors in cluster $C_i$)
- $dist$ is Euclidean distance between each vector and representative vector

Figure 2 illustrates the relationship between the number of clusters $k$ and $SSE$. We chose $k$ in the range [2, 17] with 17 being the number of departments in our university. It can be seen from the line graph that the value $k = 4$ creates an elbow, where $SSE$ value starts declining much more slowly. In other words, from the point $k = 4$ the clusters begin to be split into smaller clusters without improving $SSE$ significantly. Therefore the relevant number of clusters is 4. The result we obtained with this value matches the result from X-means algorithm presented.

## 5.2 Discussion

From the results of clustering, we drew the following comments about the human resource in our university:

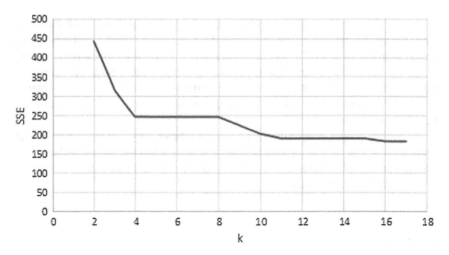

**Fig. 2** The relationship between the number of clusters $k$ and SSE measure

- In general, the lecturers in our university are rated highly. More than 86 % of the lecturers belong to cluster 1 and cluster 2 with the ratings for 12 evaluation factors being greater than or equal to four. The factors getting the highest satisfaction are enthusiasm (Q3, Q4) and knowledge conveyed by the lecturer (Q1). It is quite reasonable for a university that was founded only 18 years ago and the majority of lecturers are young. On the other hand, the ability to inspire students (Q2) and give lively lectures (Q6) is considered as weaknesses of all lecturers. The administrators should pay attention to this problem and try to remedy the situation.
- Most lecturers belong to cluster 2 (78.4 %) and are assessed uniformly for all criteria (4/5 in Likert scale), showing that there is no significant difference in the quality of teaching among the lecturers in the university.
- Cluster 1 consists of the lecturers with the highest rating (8.3 %). There is no remarkable difference between the lecturers of cluster 1 and those of cluster 2 except evaluation factors Q1, Q3 and Q4 in which Q3, Q4 assess the enthusiasm of the lecturers. More than 97 % (38 out of 39) of the lecturers with the average ratings of the overall satisfaction being equal to 5 belong to cluster 1. Therefore it can be seen that the enthusiasm plays an important role in improving the overall satisfaction. The remaining criteria such as the ability to inspire students (Q2) and give lively lectures (Q6), the expansion of lectures (Q7, Q8), and applications and deeply exploiting the subject (Q9, Q10) are not appreciated compared to the aforementioned criteria. It can be explained by the fact that as the lecturers are young, they do not have much practical experience, wisdom and ability to apply academy knowledge in reality.
- Cluster 3 consisting of 4.3 % of the lecturers is assessed almost similar to the lecturers of cluster 2 except two factors: the ability to inspire (Q2) and give lively lectures (Q6).

**Table 4** Correlation coefficient between specific factors and overall rating

| ID | Correlation coefficient |
|----|------------------------|
| Q1 | 0.738 |
| Q2 | 0.745 |
| Q3 | 0.728 |
| Q4 | 0.709 |
| Q5 | 0.676 |
| Q6 | 0.745 |
| Q7 | 0.665 |
| Q8 | 0.675 |
| Q9 | 0.685 |
| Q10 | 0.702 |
| Q11 | 0.677 |
| Q12 | 0.709 |

- Cluster 4 includes the lecturers with the lowest ratings, accounting for 9 %. These lecturers were rated higher in objective factors such as preparing complete and updated course materials (Q5), presenting clearly the forms of examination and assessment (Q11), contents of the lectures are suitable for the tests (Q12). Therefore, they need to pay attention to improve a variety of factors including specialized knowledge and ability to convey knowledge.

To the best of our knowledge, there is one study on clustering lecturers based on performance. In [17], the authors identified 77 factors which influence faculty performance. The empirical data include information about 50 Information Technology lecturers of a university. These lecturers were clustered according to performance, using k-means algorithm. The result shows that there are two clusters: cluster 1 consists of the lecturers assessed distinctively while cluster 2 consists of the lecturers who are similar to the others. It is clear that this study only partitions lecturers according to overall rating, not based on their specific features. Therefore it can not provide valuable knowledge about characteristics of lecturers belonging to each cluster.

In order to examine influence of each evaluation factor on the overall ratings in more details, we conducted the analysis about correlations among them. We calculated Pearson correlation by using SPSS software. The result obtained is presented in Table 4.

It can be seen that all correlation coefficients are greater than 0.6, which are considered as strong correlations. Among 12 factors, Q2 and Q6 are the factors which have the strongest correlation to the overall rating. These are also the weaknesses of lecturers in our university as analyzed before. The next important factors are Q1 and Q3. The interesting thing is that they are also the strengths of our lecturers. Overall, the results obtained by analyzing the correlation coefficient are consistent to comments drawn from clustering characteristics of the lecturers.

## 6  Conclusion and Future Works

In this paper, we analyzed the previous studies on exploiting faculty evaluation forms in terms of the main problems solved and their solutions. In general, these studies only focus on solving a few problems such as identifying factors that have the largest influence on faculty performance or seeking dependencies among evaluation factors. On that basis, we proposed a new problem which clusters evaluation forms according to the similarity in specialized features of the lecturers in order to build an overall picture of human resource in our university. We have applied the solution in analyzing real data collected from the online evaluation system of Ton Duc Thang University. We drew useful comments about the strengths and weaknesses of the lecturers in the university as well as those of the lecturers belonging to each cluster, gave some explanations, and identified evaluation factors which influence the overall satisfaction of the students. These results obtained after comparing to applying correlation coefficient analysis.

In future, we continue to exploit the data source to predict the faculty performance based on personal characteristics of the lecturers such as qualifications, age, gender, etc. In addition, we will also investigate the change of assessment trend over time as well as mining knowledge from open questions in the evaluation forms.

## References

1. Berk, R.A.: Survey of 12 strategies to measure teaching effectiveness. Int. J. Teach. Learn. High. Educ. **17**(1), 48–62 (2005)
2. Kelly, M.: Student Evaluations of Teaching Effectiveness: Considerations for Ontario Universities. COU Academic Colleagues Discussion Paper (2012)
3. Duong, T.-V.T., Do, T.-C., Nguyen, N.-P.: Exploiting faculty evaluation forms to improve teaching quality: an analytical review. SAI **2015**, 457–462 (2015)
4. Wong, A., Fitzsimmons, J.: Student Evaluation of Faculty: An Analysis of Survey Results. U21GlobalWorking Paper Series, no. 003/2008 (2008)
5. Badur, B., Mardikyan, S.: Analyzing teaching performance of instructors using data mining techniques. Inf. Educ. **10**(2), 245–257 (2011)
6. Pal, A.K., Pal, S.: Evaluation of teachers performance: a data mining approach. IJCSMC **2**(12), 359–369 (2013)
7. Qingxian, P., Linjie, Q., and Lanfang, L.: Data mining and application of teaching evaluation based on association rules. In: Proceedings of 4th International Conference on Computer Science and Education, pp. 1404–1407 (2009)
8. Geng, S., Guo, Z.: Application of association rule mining in college teaching evaluation. In: Electrical, Information Engineering and Mechatronics 2011 Lecture Notes in Electrical Engineering, vol. 138, pp. 1609–1615 (2012)
9. Kumar, S.A., Vijayalakshmi, M.N.: A Nave based approach of model pruned trees on learners response. Int. J. Adv. Res. Comput. Sci. Softw. Eng. **4**(9), 52–57 (2012)
10. Singh, C., Gopal, A., and Mishra, S.: Extraction and analysis of faculty performance of management discipline from student feedback using clustering and association rule mining techniques. In: Proceedings of 3rd International Conference on Electronics Computer Technology, ICECT, pp. 94–96 (2011)

11. Kuzmanovic, M., et al.: A new approach to evaluation of university teaching considering heterogeneity of students preferences. High. Educ. **66**(2), 153–171 (2013)
12. Singh, C., Gopal, A., Mishra, S.: Performance assessment of faculties of management discipline from student perspective using statistical and mining methodologies. Int. J. Data Eng. (IJDE) **1**(5), 63–69 (2011)
13. Aggarwal, C.C.: Data Mining: The Textbook. Springer International Publishing Switzerland (2015)
14. Pelleg, D., Moore, A.: X-means: extending K-means with efficient estimation of the number of clusters. In: Proceedings of 17th International Conference on Machine Learning, ICML'00, pp. 727–734 (2000)
15. Marsh, H.W.: SEEQ: a reliable, valid and useful instrument for collecting students' evaluations of university teaching. Br. J. Educ. Psychol. **52**(1), 77–95 (1982)
16. Tan, P.N., Steinbach, M., Kumar, V.: Introduction to Data Mining, 3rd edn. Pearson Education, New Delhi (2009)
17. Singh, C., Gopal, A.: Performance analysis of faculty using data mining techniques. Int. J. Comput. Sci. Appl. 140–144 (2010)

# Teaching Big Data Technology Practices in Cloud Environment

**Alexander Shmid, Boris Pozin and Mikhail Ageykin**

**Abstract** Adoption of the balanced and optimal decisions in business is based on the real time analysis of large volumes of the unstructured or semi-structured information. According to analytical researches, the problem of shortage of the experts capable to deal with big data accrues. The purpose of this work is development of a cloud environment to give students opportunity to get practical access to the Big Data technology using lab works that has been developed earlier [1]. As technologies are new, there are just a few educational programs in the field of Big Data in Russia. The practical use of technology is the focus of the proposed approach. Students improve skills working with practice sets, which are called "lab works". Moreover, access is provided to servers with data for term projects on educational and business problems.

**Keywords** Big Data · Machine learning · Informational system development · Hands-on labs · Learning management system · Cloud systems · Cloud environment · Cloud computing · LDAP

## 1 Introduction

According to IBM strategic forecast, all companies in the next 5 years will be divided into winners and losers depending on quality of making corporate decisions. Research and case studies provide evidence that a well-designed and appropriate

A. Shmid · B. Pozin
MIEM National Research University Higher School of Economics,
Moscow, Russia
e-mail: ashmid@ec-leasing.ru

B. Pozin
e-mail: bpozin@ec-leasing.ru

M. Ageykin (✉)
JSC "EC-Leasing", Moscow, Russia
e-mail: mageykin@ec-leasing.ru

© Springer International Publishing Switzerland 2016
V.L. Uskov et al. (eds.), *Smart Education and e-Learning 2016*,
Smart Innovation, Systems and Technologies 59,
DOI 10.1007/978-3-319-39690-3_56

computerized decision support system can encourage fact-based decisions, improve decision quality, and improve the efficiency and effectiveness of decision processes. IBM Watson, for example. In 2012, IBM opened for the partners the possibility of learning the technologies, on which Watson was created [1].

On a business, social media gives a great opening door to reach much more people for your business needs. Technologies of analytics and forecasting the future situation depending on made decisions (predictive analytics) allow being on some steps ahead of emergence of threats. These technologies are already used. How system for collecting and analyzing the unstructured information should work? [2].

Often one of the main reasons hindering the widespread use of Big Data technology, called the deficiency of appropriate specialists. This is why EC-leasing and the MIEM NRU HSE announced in December'2015 the launch of the pilot operation of the stand for the organization of a practical tools in cloud environment, allowing to develop the Big Data application.

## 2  IBM Big Data Platform

IBM BIG DATA platform is chosen as software products for processing of big data. IBM provides the integrated, high-performance platform that can process practically all types of data—structured and unstructured, stream and stored—and has ample analytical opportunities, including detection, the reporting and the analysis, and prediction.

The new IBM Big Data technologies provide the speed and flexibility for social network data processing and gives the opportunity to develop systems for solving new business challenges using social network data.

IBM InfoSphere BigInsights features Apache Hadoop and its related open source projects as a core component. This is informally known as the IBM Distribution for Hadoop. IBM remains committed to the integrity of these open source projects, and will ensure 100 % compatibility with them. This fidelity to open source provides a number of benefits. For people who have developed code against other 100 % open source—compatible distributions, their applications will also run on BigInsights, and vice versa [3].

IBM InfoSphere Streams is a powerful analytic computing software platform that continuously analyzes and transforms data in memory before it is stored on disk. Instead of gathering large quantities of data, manipulating and storing it on disk, and then analyzing it, as is the case with other analytic approaches, Streams enables you to apply the analytics directly on data in motion [4].

IBM Watson Explorer makes searching across your Big Data assets more *accurate*. The underlying indexes are smaller (compressed), don't need to be maintained as often as other solutions, and you can request more granular index updates instead of having to update everything [3].

IBM Watson Analytics enables control over information with sophisticated natural language processing capabilities to deliver the right information at the right time to the right people. Transforms organizations by uncovering trends, patterns and relationships from enterprise content to drive fact-based decisions [5].

IBM i2—application that enables you to collate and visualize information from many different sources, organize it in a meaningful way, and then analyze it using a variety of techniques [6].

# 3 Architecture of Cloud for IBM Big Data Platform

Developing an architecture for the cloud environment for teaching big data technology based on thesis that IBM big data platform is modular and integrated. For improving students' practical skills four training servers images with various sets of software products mentioned above were created and one image for learning management system with following content:

- instructions,
- practical work manual,
- tests,
- knowledge base,
- forums, chats, etc.

The infrastructure should be scalable in order to add new software products, lab works and practical part of term papers. Widely known and mostly used infrastructure level (IaaS) open source cloud platforms are Eucalyptus, Nimbus, OpenNebula, CloudStack, OpenStack [7]. These platforms use hardware virtualization technologies, available on server. Most popular open source virtualization systems are KVM (Kernel-based Virtual Machine) [8] and hypervisor, based on Xen [9]. In this work, we decided to use KVM, which in our opinion is more user friendly.

As before beginning, we clearly knew the business process how students are going to access practical lab works and practical projects for their term papers. Simplified it shown on Fig. 1.

As we knew business-process, we choose IBM Cloud Orchestrator (ICO) [10] as cloud platform for Its Process Designer tools and opportunity to provide cloud services like IaaS/PaaS/SaaS on different hardware platforms like IBM z Systems, IBM Power Systems and x86 Systems.

Practical work with IBM Big Data Platform involves the execution of asynchronous independent calculations. Such problems are often associated with pre- and post-processing of data, and in many cases performance problems may occur if several researchers, use one Virtual Machine for calculations. As data growth rapidly (Practical Big Data lab works uses raw data from social networks, forums and so on), pre- and post-processing have to be performed on high performance hardware. High Performance tasks well suited for the transfer to the cloud, but require a large capacity and considerable amount of amount of storage. In our case,

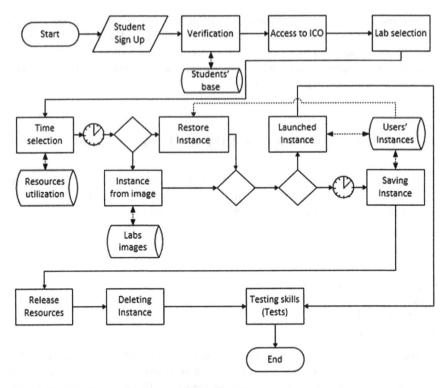

**Fig. 1** Simplified business process of completing practical tasks

we were lucky to have calculations that require a sufficiently high performance virtualization and thus can be effectively performed in a virtual environment.

Network scheme of cloud platform for this purpose is shown on Fig. 2.

## 4 Practical Lab Works in the Cloud

First, 10 practical lab works were launched, focused on data management with IBM Big Data platform in this cloud environment. Target audience for these practical labs are students of the Russian universities, and any interested persons as well, for example, the employees of the IT companies. Here is the list for these lab works:

1. Working with unstructured data in Hadoop Distributed File System.
2. Process control processing of unstructured data using BigInsights.
3. Analysis of structured and unstructured data using BigSheets.
4. Management of large data using Big SQL.
5. The structuring of data using specialized language text processing.

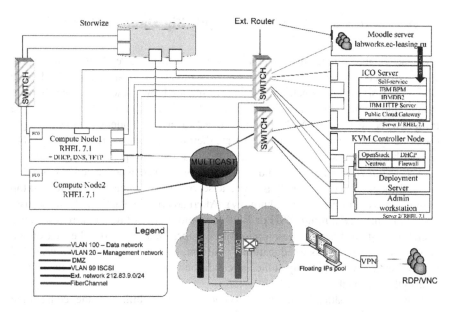

**Fig. 2** Network scheme of cloud platform

6. Analysis of the streaming information using specialized data stream processing languages.
7. Manage the processing of streaming data.
8. Investigation of textual information via Watson Explorer.
9. Analysis of structured and unstructured data using Watson Content Analytics.
10. Identification of hidden relationships based on the analysis of texts using i2.

First students from MIEM NRU HSE, who took part in training Big Data platform with cloud environment, gave the following results, shown in Table 1.

**Table 1** Big Data lab works Results

| Name | Lab work test # | | | | | | | | | | Total |
|------|------|------|------|------|------|------|------|------|------|------|-------|
| | 1 | 2 | 3 | 4 | 5 | 6 | 7 | 8 | 9 | 10 | |
| Nick | 7.1 | 5.0 | 7.9 | 7.5 | 10.0 | 6.0 | 6.7 | 6.7 | 7.0 | 7.8 | 7.2 |
| Elena | 10.0 | 10.0 | 10.0 | 10.0 | 8.0 | 6.0 | 10.0 | 10.0 | 9.6 | 7.8 | 9.1 |
| Mansur | 10.0 | 10.0 | 10.0 | 10.0 | 8.0 | 8.0 | 10.0 | 10.0 | 10.0 | 9.6 | 9.6 |
| Ivan | 10.0 | 5.0 | 10.0 | 10.0 | 10.0 | 8.0 | 10.0 | 10.0 | 8.5 | 10.0 | 9.2 |
| Nastya | 10.0 | 10.0 | 10.0 | 10.0 | 10.0 | 10.0 | 6.7 | 10.0 | 9.6 | 10.0 | 9.6 |
| Nikita | 10.0 | 10.0 | 10.0 | 7.5 | 6.0 | 4.0 | 10.0 | 10.0 | 9.6 | 9.6 | 8.7 |
| Vika | 10.0 | 10.0 | 10.0 | 7.5 | 8.0 | 8.0 | 10.0 | 10.0 | 8.5 | 8.3 | 9.0 |
| Alex | 8.6 | 10.0 | 10.0 | 10.0 | 8.0 | 10.0 | 10.0 | 10.0 | 9.6 | 8.3 | 9.5 |
| Dmitry | 10.0 | 10.0 | 9.3 | 10.0 | 8.0 | 10.0 | 10.0 | 6.7 | 8.5 | 7.9 | 9.0 |

## 5  Access to Educational Cloud with Big Data Platform

The teacher give student there login/password and they are ready to do lab works and/or projects for their term papers. Login and password are stored in MIEM NRU HSE LDAP server and synchronize user information between learning management system and Cloud Orchestrator Self Service Portal. These resources is accessible from HSE network or via VPN from Home or Dormitory. Students, who have login for cloud system have opportunity to create their own instances from images and to reserve hardware resources using a calendar planning and queue management.

Access directly to the IBM big data software products can be carried out as a thin client through a browser (in this case, a server uses a plug-into show desktop through cross-platform Java technology). Alternatively, in PaaS mode when user access his instance through SSH or RDP technology according to current instance operating system. Each of these options has its pros and cons, but the first option allows more users access to educational platform, because in this case browser and an Internet connection is only sufficient condition for training, and users don't need to install additional software.

Figure 3 schematically shows everybody's responsibility in the educational environment. IT administrator creates images of virtual machines with the necessary software for the future work of teachers and students, and add them to the cloud. Students have the opportunity to create their own instances from these images, and access them from the audience/house/dormitory and etc.

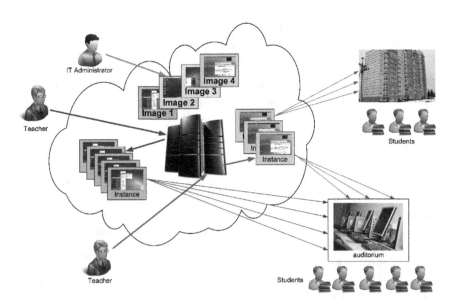

**Fig. 3** Roles and responsibilities in cloud environment

# 6 Tasks, Completed as a Term Papers

In addition to practical lab works, cloud environment allows students to do their projects for term papers. Millions of text messages were processed in these projects. Last term students of MIEM NRU HSE with EC-leasing consulting prepared a list of practical solution of economic tasks.

**Early diagnostics of financial instability of the credit organizations**. More than 4 million messages of bank topic from social networks and popular bank resources are already collected. The model of forecasting the dynamics of outflow of deposits was created, based on predictive analytics. A prototype, which allows to estimate the forecast of outflow of deposits of natural persons, using arriving data, was presented.

**Assessment of degree of satisfaction of clients services**. Log records of Service Desk were analyzed in automated mode. The technology allows revealing regularities of emergence of failures, to trace deviant activity of the personnel, to form recommendations for various incidents based on last experience to receive an assessment of degree of satisfaction of clients.

**Counteraction to insiders and market manipulation**. This technology, applied to identification of the facts of distribution of the classified information on social networks, it is possible to use for further creation of system of continuous monitoring of social networks and blogs for counteraction to the insider activities. More than 14 million messages from social networks were processed for the test use case to test the technology.

**Assessment and forecast of production stability**. Calculation of production stability is made by data for the last 3 years with use of the developed mathematical model based on an average geometrical of three integrated indicators:

- capacity utilization rate
- capital productivity
- profitability.

**Research and analysis of an information field in the Internet**. Monitor information in the network; Research and analysis of an information field allows to collect and analyze the structured and unstructured contents in the documents, e-mail, databases, on the websites and in the other depositaries. Data analyst can use the interface of content analysis for investigate data from different analytical data slices (facets) in the interactive mode and find interrelations between various values of facets and anomalies of values. With the help of search interface enterprise users will be able to run queries to an index, quickly to find and receive the necessary documents in the ranked list of results.

**Search sources of stuffing information**. Performance of a task possibly due to collect and analysis large volumes of information from various sources for finding fake information. For example, if certain information instantly scatters on social networks and even some mass media publish it, as a rule the answer to the following questions is required:

- When and where has it leaked?
- How widespread?
- What interest causes?

Graphical view of data distribution which contains information about news and dates, and also persons who published this information can give the answer to these questions.

**Identification of employees' communications with the use and promotion of drugs**. Prevention of drug addiction in such establishments as: schools, universities, various government institutions—very important and actual problem relating to general antidrug effort. Created collection allowed to reveal communications of these people with bad communities and to analyze their messages using the dictionaries, rules and inquiries with difficult logic, and also visualization tools. With the help of the developed system which takes the obtained data as a basis, we estimated possible participation of each person in the use or promotion of drugs.

## 7 Conclusion

Implementation of the project in MIEM NRU HSE to launch the practical familiarization with Big Data has allowed to facilitate access to new technologies and made it possible either work with pre-planned works (practical lab works) or implement own educational projects. Passage of laboratory work aimed at the familiarization with tools and data handling methods and allows going after passage to the realization of personal educational projects. It should be noted that personal educational projects carried with IBM Big Data platform can get into the production phase earlier, then if they were implemented using open source. As for the cloud, it is possible to use the following tasks:

- Initialization and scaling of cloud resources to obtain images of the desired performance.
- Rapid deployment and scaling of local and remote cloud services for load balancing and minimum hardware downtime.
- Automation of operations performed by IT administrators manually. Creation and deployment of virtual machines, resource reservation, scheduling, manual and automatic removal of instances.
- Reduce the "human factor"—fewer errors.
- Adoption of existing business processes. Existing business processes of the EC-leasing and the MIEM NRU HSE have been moved to the cloud.
- Monitor utilization, capacity and etc.
- Ability to connect a variety of platforms OpenStack (System X), PowerVM, IBM System z and VMware.

# References

1. Shmid, A., Posin, B., Galakhov, I., Ageykin, M., Aleksandrov, D., Kasimov, M., Klemashev, N., Ezhov, G.: New Ways of Working with Big Data: The Winning Management Strategy in Business Analytics. Palmir, Moscow (2015)
2. Galakhov, I., Ageykin, M.: Fields of application social networks data in economic productivity. In: Innovative Information Scientific—Practical Conference. Part 3. HSE, Moscow (2014)
3. Zikopoulos, P., deRoos, D.: Harness the Power of Big Data. McGraw-Hill (2013)
4. Roy, J.: The Power of Now: Real-Time Analytics and IBM InfoSphere Streams. McGraw-Hill Education (2015)
5. Red Book: IBM Watson Analytics. IBM (2015)
6. Quickstart Guide: i2 Analyst Notebook. IBM (2014)
7. Sathi, A.: Big Data Analytics: Distruptive Technologies for Changing the Game. IBM (2012)
8. Endo, P., Gonçalves, G., Kelner, J., Sadok, D.: A survey on open-source cloud computing solutions. In: Brazilian Symposium on Computer Networks and Distributed Systems (2010)
9. Kivity, A., Kamay, Y., Laor, D., Lublin, U., Liguori, A.: KVM: the linux virtual machine monitor. In: Linux Symposium, vol. 1 (2007)
10. Coote, S.: IBM Cloud Orchestrator from A to Z. IBM (2014)

# Author Index

© Springer International Publishing Switzerland 2016
V.L. Uskov et al. (eds.), *Smart Education and e-Learning 2016*,
Smart Innovation, Systems and Technologies 59,
DOI 10.1007/978-3-319-39690-3

Printed in the United States
By Bookmasters